MICROCLIMATE FOR CULTURAL HERITAGE

MICROCLIMATE FOR CULTURAL HERITAGE

Conservation, Restoration, and Maintenance of Indoor and Outdoor Monuments

DARIO CAMUFFO

ELSEVIER

AMSTERDAM • BOSTON • HEIDELBERG • LONDON
NEW YORK • OXFORD • PARIS • SAN DIEGO
SAN FRANCISCO • SINGAPORE • SYDNEY • TOKYO

Elsevier
225 Wyman Street, Waltham, MA 02451, USA
525 B Street, Suite 1900, San Diego, CA 92101-4495, USA

Second edition 2014

Library of Congress Cataloging-in-Publication Data

A catalog record for this book is available from the Library of Congress

Application submitted

British Library Cataloguing in Publication Data

A catalogue record for this book is available from the British Library

For information on all Elsevier publications visit our website at store.elsevier.com

Printed and bound in China.

13 14 15 16 17 10 9 8 7 6 5 4 3 2 1

ISBN: 978-0-444-63296-8

Contents

Preface to the First Edition (1998)

This book has been designed as a useful microphysics handbook for conservators and specialists in chemistry, architecture, engineering, geology and biology who work in the multidisciplinary field of the environment, and, in particular, in the conservation of works of art. It has been especially written following the continuous demand to fill a gap in the literature related to this important application of atmospheric sciences, i.e. to apply the thermodynamic processes of clouds, or the dynamics of the planetary boundary layer, to a monument surface or to a room of a museum. The aim is to furnish them with a background familiarity with the underlying physics behind mathematics and to give a detailed description and interpretation of the main microphysical phenomena that play a fundamental role in practical applications. Correct application of formulae is only possible when all the approximations made in their derivation and the limitations intrinsic to the basic hypotheses are known. In this complex field, an effort is made to substitute scientific demonstrations for common opinions or popular beliefs.

The basis are given for nondestructive diagnostics to evaluate causes of damage and predict outdoor deterioration, determined by meteorological factors, as well as the negative effects in exhibition rooms, due to unsound use of technology and mass tourism. To this aim, suggestions are given on the fundamental principles in designing heating, air-conditioning and lighting and in reducing the deposition of pollutants on works of art. Theory and experience are coupled to describe the complex condensation mechanisms and the fundamental role played by water in stone deterioration and the formation of crusts on monuments. Urban meteorology, air–surface interactions, atmospheric stability, dispersion and deposition of airborne pollutants are also key topics of this book, whose main aim has been to make comprehensible to a wider audience a matter that is only familiar to a few specialists.

This book combines a theoretical background with many years of accurate laboratory research, field surveys and practice. The first part, devoted to applied theory, is a concise treatise on microphysics, which makes a survey on the basic ideas especially on classical, kinetic and statistical thermodynamics, which are necessary for environmental diagnostics and conservation. The second part, *Performing Microclimate Field Surveys*, is devoted to the practical utilization and shows in detail how measurements should be performed, with many suggestions and examples and the indication of some common errors that should be avoided.

Dario Camuffo

Preface to the Second Edition (2014)

When 15 years ago I wrote the first edition of this pioneer book in microclimate, I did my best to share scientific knowledge and my experience to make life easier for students and professionals concerned with the multidisciplinary field of physics, air pollution, atmospheric sciences, meteorology, and especially conservation of cultural heritage. The ambitious goal was to combine sound and comprehensive theoretical bases with many practical issues, i.e. to combine a treatise with a handbook.

Fortunately, the above goals seem to have been reached, at least judging from the funny nickname gained by the book, i.e. 'the Bible', and the number of citations and reprints of the 1998 book, ultimately available in e-book version.

However, we should never be satisfied with (our) previous books; if we lack the driving spring of innovative research and advancement, our knowledge would have stopped at Middle Ages.

Despite its age, this book is still so popular that an updated edition was suggested to keep up to date on cutting-edge research and include the most recent findings in this field.

In addition, this is an excellent opportunity to disseminate some basic information about the novel standards of the European Committee for Standardization (CEN), Technical Committee 346 focussed on cultural heritage conservation (CEN/TC346), to which I have contributed as TC346 member and Convenor of the Working Groups concerning exhibition lighting (WG6) and microclimate control and monitoring (WG7).

This required revising, updating and expanding the text with new sections, graphs and contributions, with over 100 additional pages.

The aim of the second edition remains substantially the same, i.e. to disseminate up-to-date scientific and technological results and field experience but especially to share and construct understanding, to favour critical reviewing and to stimulate further research because constructive criticism and research are the mother and father of novel findings and more advanced knowledge.

Dario Camuffo

Foreword to the First Edition (1998)

Several books have been written on microclimatology and boundary-layer physics, beginning with Rudolph Geiger's *Climate near the Ground* (first German Edition in 1927). Until now, however, nobody has produced a book on the effects of small-scale climate variability around and inside buildings, monuments and other cultural objects. So, Professor Camuffo's well-illustrated account, *Microclimate for Cultural Heritage* will be greatly welcomed by architects, engineers, preservers and restorers of cultural property and the wider community of microclimatologists.

What has 'cultural heritage' to do with microclimatology? About 40 years ago, the response from meteorologists would have been 'very little'. Let me cite a personal example. Back in the 1950s while studying engineering meteorology at the University of Michigan, I wrote a term paper on the weathering of exposed surfaces by atmospheric pollution, which was given an A grade and which was subsequently published (Munn, R.E., 1959: Engineering Meteorology: the Weathering of Exposed Surfaces by Atmospheric Pollution. *Bull. Am. Meteorol. Soc.* **40**, 172–178). But several scientists (including a famous professor of physics at McGill University in Montreal) were alarmed that such a topic should find its way into the meteorological literature and made their views known. About 35 years later, the importance of microclimatological gradients is widely recognized in the preservation and restoration of cultural objects. Credit for this attitude change is largely due to the efforts of a few people like Dario Camuffo, who recognized early on the importance of micrometeorology in studies of the preservation and restoration of cultural heritages.

This monograph contains many practical examples of ways in which micrometeorological knowledge can help in assessments of the deterioration of surfaces that have been exposed to the environment over long periods. This book not only includes interesting outdoor examples from the author's own experience but also includes indoor cases, like horizontal cross-sections of temperature in the Sistine Chapel and of mixing ratio in the Giotto Room in the Uffizi Gallery in Florence.

I know that this book will be widely consulted by specialists in the cultural heritage field and I am pleased to have been involved in a very small way by contributing this foreword.

R.E. Munn
Institute for Environmental Studies
University of Toronto, Toronto, Canada

Reviews to the First Edition (1998)

IL NUOVO CIMENTO VOL. 22, NO. 1, P. 121 (1999)

In today's times when a renewed sensibility and attention towards cultural heritage is developing and spreading over most countries in the world, a book like this undoubtedly offers a basic contribution to a proper formulation and planning of actions devoted to conservation and preservation of cultural objects.

The connection between microclimatology and cultural heritage reflects, in the book of Dario Camuffo, the most relevant involvement that physical sciences can have on the conservation aspect of this field (the other important contribution offered by physics, which is beyond the interest of this review, concerns the restoration).

However, as R. E. Munn recalls in his '*Foreword*' to the volume, the recognition of the importance of microclimatology in the preservation and conservation of works of art had difficulty in asserting itself and being accepted by the community of atmospheric physicists. If this attitude has now changed, credit for it must be given, according to Munn, to the patient and perseverant efforts of scientists like Dario Camuffo who dedicated themselves to serious and rigorous research work and to the diffusion of its principles and results among the scientific community.

One of the merits of the present book is represented (at least in my view) by the careful and accurate description of basic processes and laws of atmospheric physics, which can be relevant to the preservation and conservation of cultural objects. This choice, far from making dull and heavy the reading and belittling the treatment of methods that can assure the best environmental conditions for reducing or avoiding material's deterioration, will help, on the contrary, the conservators not only to develop new ideas and procedures but, most of all, also to increase their awareness in devising better and more effective approaches to their important and arduous task.

The content of the book is divided into two parts: '*Atmospheric Physics applied to microclimate analysis*' and '*Microclimate field surveys*'.

After an introductory chapter, in which a very useful statement is made on the definition of microclimate, the other sections of the first part deal with thermodynamics of air (including saturation processes and conventional temperatures), radiation and light, drop formation, stone weathering, atmospheric stability and pollutant dispersion as well as dry deposition of aerosol (Stefan flows, electrophoresis, photophoresis, inertial impaction, resuspension of particles by turbulence and so on).

Part 2 deals with field measurements, with particular emphasis on temperature, humidity, wind and air motions, rainfall and wind-borne droplets.

An extended and exhaustive reference list, split up chapter by chapter and a clear and detailed Subject Index conclude this highly recommended book.

Arnaldo Longhetto
Professor of Physics, Physics Department University of Turin, Italy
(by kind permission of Società Italiana di Fisica)

2. STUDIES IN CONSERVATION VOL. 45, NO.2, P. 143 (2000)

Even the most casual reader of the conservation literature will have encountered examples of the remarkable body of work by Dario Camuffo wherein he has applied fundamental principles of physics to the study of the microenvironment of such diverse systems as temperature and humidity gradients in the Sistine Chapel, acid deposition and biocorrosion on Trajan's column, and the aeolian erosion of the Great Sphinx.

The present work is no mere compilation or recapitulation of the author's previously published papers but rather a publication of the fundamental principles and theoretical basis for the study of the microenvironments encountered by the most diverse forms of cultural property. It is appropriate that it finds its place in the series, '*Developments in Atmospheric Science*' among such titles as *Physical Principles of Micro-Meteorological Measurements, Man's Impact on Climate* and *Atmospheric Aerosols*. Where rigour is called for, the treatment includes derivations and equations. Yet the author never allows rigor mortis to set in. Always, clear practical examples are provided. Indeed, the theoretical treatment is at the service of the discussions of the many case studies introduced.

It is perhaps useful to consider Camuffo's text in parallel to the widely read classic by Garry Thomson, *The Museum Environment*. In attempting to deal with the different levels of scientific appreciation brought by his prospective readers, Thomson covers the same material twice, offering, in effect, an introductory version and then a more advanced version for each topic. This solution always seemed rather artificial. It is much to the credit of Dr Camuffo that he has written a book that may be the advanced-level version of *The Museum Environment*, while integrating the treatment in such a way that the conservator, the conservation scientist and the atmospheric physicist can all turn to it for reference and even enjoyment.

The book is divided into two parts, '*Atmospheric physics applied to microclimate analysis and conservation*' and '*Performing microclimate field surveys*'. The first part, which forms approximately three-quarters of the text, covers microclimate, air and temperature; humidity, the description of air masses and vertical motions; radiation and light; the physics of drop formation and micropore condensation: atmospheric water and stone weathering; atmospheric stability and pollutant dispersion; and dry deposition of airborne particulate matter. The second part deals with practical aspects of field measurement and survey for parameters associated with temperature, humidity, wind, indoor air motion, rainfall and wind-borne droplets.

A great number of diagrams and photographs with clear explanatory captions support the exposition. The generous inclusion of a few colour plates is a pleasant extravagance.

In summation, Dario Camuffo has provided a definitive treatment of a central aspect of conservation science and a most useful introduction to the application of atmospheric physics to the preservation of cultural property. The practising conservator, especially one who looks askance at the differential equations that not infrequently grace the pages of this journal, would also benefit from an acquaintance with this book. In that regard, it is a pity that the otherwise most thorough subject index does not include the names of the many monuments that form the case studies and practical examples discussed[1].

Norbert S. Baer
Hagop Kevorkian Professor of Conservation,
Institute of Fine Arts, New York
(By kind permission of the International Institute for Conservation of Historic and Artistic Works).

3. AGRICULTURAL AND FOREST METEOROLOGY 111, P. 309 (2002)

This book is volume 23 of the *Developments in Atmospheric Science* series, sponsored by the Environment and Climate Research Program of the European Commission. *Microclimate for Cultural Heritage* is a text of atmospheric physics, applied to conservation and preservation of cultural

[1]added in the second edition.

artefacts. The 13 chapters are divided into two major parts: Part I deals with theoretical and practical treatments of a wide range of physical and chemical phenomena, whereas Part II focuses on measurements of meteorological parameters. Although the primary goal of the book is to provide an information source to those working in the area of cultural heritage, Camuffo has covered almost the whole field of atmospheric physics relevant to the lower troposphere. Hence, the book could be used as a general micrometeorological text.

The uniqueness of this book is the many examples of physical processes operating on monuments, statues and other artwork. Almost all examples are from Europe or North Africa and include both outdoor and indoor (e.g. museums, churches) case studies. These case studies provide nice ideas for someone who may be facing similar problems caused by heat, water, radiation, pollution or air motions. The figures and photographs are clear and instructive. This book is not a descriptive study of micrometeorology for art students! It is full of equations and derivations of basic principles, as well as engineering equations. In fact, the rigorous derivation of many parameters may distract some readers from obtaining the few equations and principles that they require to address a specific problem. For example, the level of detail could be confusing for novices trying to decide whether they should be concerned about specific humidity, mixing ratio, vapour pressure or relative humidity. However, Camuffo describes a wide range of choices and this completeness makes this a good handbook. The book has a good subject and reference index. However, it lacks an appendix that could list fundamental constants, examples of parameters, and the most used equations[2]. Units are in the cgs system.

This book would be a good additional text for a meteorologist interested in European artefacts. However, it may be best considered as a method of teaching micrometeorology and atmospheric physics to European science students. It has the potential to capture their imagination with concrete examples that they can see, much better than many examples from forested or agricultural systems. The section on instrumentation could provide the basis for laboratory studies within buildings and cities close to universities. This application for micrometeorological education could also be appropriate for other parts of the world where urban and cultural examples abound. Hence, the potential for this book could be substantial in many senior undergraduate courses.

Brian D. Amiro
Canadian Forest Service, Northern Forestry Centre Edmonton, Alta., Canada
(By kind permission of Elsevier Sci.)

[2]added in the second edition.

Foreword

About 15 years after the publication of the first edition of *Microclimate for Cultural Heritage,* a second edition is appearing. The addition of the subtitle *Conservation, Restoration and Maintenance* reflects the subtle change of approach to the topic. Conservation is an interdisciplinary field where art historians, architects, conservators and scientists, such as physicists, geologists, chemists and biologists, have to communicate and share their knowledge. It is a challenging field because interdisciplinary communication is not easily achieved. It requires patience and willingness to share information and to learn as much as possible of what we do not know from the other fields. For this reason, books such as this one that bridge the gap between science and conservation are fundamental.

While the core of the book has remained constant, it has been updated and improved by adding sections on new instrumentation available for monitoring critical parameters such as temperature and humidity, results from ongoing research, as well as standards that have been developed in the European Union over the past 10 years, among others. More examples have been included to illustrate points made in the scientific sections. Understandably, most of these examples given come from Europe, but they serve to illustrate how to approach and interpret a given problem to find a solution for it.

At first, the book may appear daunting by the amount and thoroughness of the physics background that deals with temperature, air flow, moisture content, particle deposition, etc. However, the text in between these physicomathematical sections is fascinating. Here, extensive discussions of case studies are presented and it is impossible not to get caught up in them. They make it easy and enjoyable to read. This unique book should really be 'required reading' for students in conservation, both for those focusing on museums and their collections and for those in architectural preservation. That some of the technical material may be beyond their immediate comprehension only serves to remind students that conservation is an interdisciplinary field and we have to learn to accept that we definitely cannot be experts in all fields. But what makes it most valuable is that at no point does the author forget the main aim is the conservation of cultural property. And that is the book's strength.

A real problem that faces the conservation community has been clearly stated by the author: 'a number of various and simultaneous types of risk exist, and the best solution should be to find a compromise between the different needs, and reduce as far as possible the various deterioration mechanisms. This requires a holistic approach and to optimize the synergism between the individual solutions. This always introduces a series of uncertainties to the multidisciplinary decision system.' This problem is illustrated in the following example: 'recommended values of temperature and relative humidity that are in principle suitable for materials, are not always appropriate for individual objects.' And that 'once an object has adapted to a certain RH variability, both in terms of average levels and fluctuations, it requires the respect of the previous conditions as any change, or fluctuations exceeding the historic range, might affect its durability.'

The belief that air-conditioning will solve all problems, as I have had opportunity of witnessing in developing countries in Central and South America, needs to be eradicated. If even in developed countries '*excessive confidence on HVAC (*heating, ventilation, and air conditioning*) and their use has caused several problems'* as the author points out, how can this technology be reliable in places where periodic power outages occur? In developing countries, it is fundamental that preventive conservation measures be taken to ensure long-term preservation of objects considering that many have survived several centuries without air-conditioning where '*local climate was carefully observed and exploited to adapt buildings …to the external ambient and benefit of a natural microclimate*'. There is no doubt that HVAC systems may stabilize indoor climate; however, the installation has to be adapted to the building, the room shape and its contents. As the author points out '*the present-day technological research is focused on making sophisticated devices with improved performances, not in studying and controlling their use and the compatibility with conservation.*'

In summary, this new edition of *Microclimate for Cultural Heritage* reminds us that regular and long-term *monitoring* is requisite for minimizing both the number and intensity of required *maintenance* interventions. These two key words, to which we all agree but seldom properly implement, are the core for the preservation of cultural heritage.

A. Elena Charola Ph.D
Research Scientist,
Museum Conservation Institute,
Smithsonian Institution

Acknowledgments

This book is based on direct experience of a large number of case studies, most of them funded by the European Commission (DG XII: Science Research and Development, Programmes STEP and *Environment*, Contracts FP1-ENV-757-I-SB, FP2-EV4V-0051-I-A, FP3-STEP-CT90-0107-SSMA, *AER* FP4-ENV4-CT95-0088, *Archeo*, FP4-ENV4-CT95-0092, *Friendly-Heating*, FP5-EVK4-2001-00,007, *Sensorgan*, FP6-Contract 022695, *Climate-for-Culture* FP7-ENV-2008-1 (GA 226,973), the COST Action D42 '*Enviart*' and Action IE0601 '*WoodCultHer*', other supported by the National Research Council of Italy (CNR), e.g. Finalized Project '*Beni Culturali*', ENEL, e.g. project *Effects of Air Pollution on Human Health and Cultural Heritage*, the *Consorzio* per *la Torre di Pisa*. Studies also were made on the occasion of special commissions (e.g. European Union, UNESCO, NAPAP, Vatican, Italian Ministry of Scientific Research). This text also utilizes lectures on atmospheric physics, environmental physics and physics for conservation held during the past 35 years at the Physics Department, University of Padova; the Cignaroli Academy, Verona; and the Polytechnic of Milan as well as those on microclimate and physical weathering of monuments at international schools (e.g. European University Centre for Cultural Heritage of the Council of Europe in Ravello, the Community of Mediterranean Universities; UNESCO-ICCROM and COST courses). It also utilizes the experience for standardization concerning cultural heritage in the Italian and European standardization bodies, i.e. UNI-Normal and the European Committee for Standardization (CEN TC346). A number of original contributions that were published in scientific journals, or presented at international symposia, are summarized here.

Special thanks are due to whoever has contributed: all my present and past coworkers, i.e. Dr A. Bernardi, Dr C. Bertolin, Dr A. della Valle, Mr A. Ongaro, Dr E. Pagan, Dr. F. Becherini, Arch. P. Schenal, Dr G. Sturaro and Dr A. Valentino for their cooperation especially in the field surveys and data analysis; the scientific officers of the European Commission, and particularly Dr. A. Sors, Dr. J. Acevedo, Dr. J. Leissner, Dr M. Chapuis, Dr A. Lydon for their interest, kind encouragement and friendly assistance; as well as good friends and colleagues Prof. A. Arnold (Swiss Federal Institute of Technology, Zürich), Prof. N.S. Baer (New York University), Prof. P. Brimblecombe (University of East Anglia, Norwich), Dr L. De Boek (Antwerp University), Prof. M. Del Monte (Bologna University), Prof. R.A. Lefèvre, Université Paris XII, Prof. B. Fitzner

(Technischen Hochschule, Aachen), Dr P. Bacci (ENEL, Milan), Dr C. Price (University College, London), Dr C. Sabbioni (CNR ISAC, Bologna), Prof. C. Saiz-Jimenes (CSIC, Seville), Prof. R. van Grieken (Antwerp University), Dr S. Vincenzi (CNR-ISDGM, Venice), Dr Th. Warscheid (LBW-Microbiology in Conservation, Wiefelstede, Germany), Prof. F. Zezza (Bari Polytechnic), Prof. M. Odlyha (Birkbeck College London), Prof. C.J. Bergstein (GOART Goteborg University), Dr L. Bratasz and Dr R. Kozlowski (Poland Academy of Sciences), Dr H. Schellen (Technical University Eindhoven), Prof. T. Brostrom (Gotland University), Dr M. Strlic (University College, London), Dr C. Cocheo (Fondazione S: Maugeri IRCCS, Padua), Dr A Lenzi, Dr G. Strano and Dr S. Cimmino (Museo Galileo, Institute and Museum of History of Science, Florence) Prof A. Paolucci, Director of the Pontifical Galleries, Museums and Monuments (Vatican), Prof. F. Pedrocco and Dr A. Craievich, Fondazione Musei Veneziani, Venice, Dr V. Fassina, CEN TC346 Chairman, Dr. G. Alessandrini UNI-Normal Chairman and many others for having contributed in different ways.

The climatological analyses of meteorological data for Venice and Rome were based on observations taken by the Meteorological Service of the Italian Air Force.

A grateful thought to the memory of good friends and leading scientists, i.e. Prof. O. Vittori, who was specialist of Atmospheric Physics and Director of CNR-FISBAT; Dr G. Urbani, who was Director of the *Istituto Centrale del Restauro*, Rome, Dr H. Ott, Head of the EC Division for Cultural Heritage, and Prof. C. Petrangeli Director of the Pontifical Galleries, Museums and Monuments (Vatican), for their unforgettable scientific discussions and for their stimulating contributions to apply atmospheric microphysics and environmental field surveys for diagnostic purposes and preventive conservation.

Finally, special thanks and apologies to my family for the taken time.

Credits

Review from *Il Nuovo Cimento* Vol. 22, no. 1, p. 121 (1999) reprinted by kind permission of Società Italiana di Fisica, Via Saragozza, 12, 40123 Bologna, Italy.

Review from *Studies in Conservation* Vol. 45, no.2, p. 143 (2000) reprinted by kind permission of the International Institute for Conservation of Historic and Artistic Works (IIC), 3, Birdcage Walk, London SW1H 9JJ, UK.

Review from *Agricultural and Forest Meteorology* Vol. 111 p. 309 (2002) reprinted by kind permission of Elsevier Science, NL, Sara Burgherharstraat 25, 1055 KV Amsterdam, the Netherlands.

Fig.1.8 reprinted from Camuffo D. and Bernardi, A., (1991): The microclimate of Leonardo's "Last Supper", *European Cultural Heritage Newsletter on Research*, and *Bollettino Geofisico*, joint edition 14, 3, 1–123, by kind permission of European Commission, DG XII, and *Bollettino Geofisico*. Support received by the European Commission, DG Research and Innovation project "AER" ENV4-CT95-0088, 4th Framework Programme, Environment and Climate.

Fig.1.10d is reprinted from Bernardi, A., Camuffo, D., Del Monte, M., and Sabbioni, C., (1985): Microclimate and Weathering of an Historic Building: the Ducal Palace in Urbino *Science of the Total Environment*, 46, 243–260, by kind permission of Elsevier Science - NL, Sara Burgherharstraat 25, 1055 KV Amsterdam, the Netherlands.

Fig.1.16, Fig.1.17 and Fig.1.18: Support received by the European Commission, DG Research and Innovation, project "*Friendly-Heating*", EVK1 2001-00007, 5th Framework Programme, Environment and Climate.

Fig.1.19: Support received by the European Commission, DG Research and Innovation, project "*Sensorgan*", Contract 022695, 6th Framework Programme, Scientific Support to Policies.

Fig.1.21 reprinted from Camuffo, D. and Bernardi, A., (1986): Dinamica del microclima e scambi termoigrometrici tra pareti e atmosfera interna nella Cappella Sistina *Bollettino Monumenti Musei Gallerie Pontificie*, 6, 211–257, by kind permission of Vatican Museums and Galleries.

Fig.1.22 reprinted from Camuffo, D. and Schenal, P. (1982): Microclima all'interno della Cappella degli Scrovegni: scambi termodinamici tra gli affreschi e l'ambiente, *Bollettino d'Arte* special issue "*Giotto a Padova*", pp. 107–209 by kind permission of Ministero dei Beni Culturali ed Ambientali, via di S. Michele 22, Rome and Poligrafico dello Stato, Rome.

Fig.2B.6 reprinted from Camuffo D. and Bernardi, A. (1991): The microclimate of Leonardo's "Last Supper", *European Cultural Heritage Newsletter on Research*, and *Bollettino Geofisico*, joint edition 14, 3, 1–123, by kind permission of European Commission, DG XII, and *Bollettino Geofisico*. Support received by the European Commission, DG Research and Innovation project "AER" ENV4-CT95-0088, 4th Framework Programme, Environment and Climate.

Fig.2B.15 Support received by the European Commission, DG Research and Innovation, project *"Friendly-Heating"*, EVK4-2001-00007, 5th Framework Programme, Environment and Climate.

Fig.2B.17, Fig.2B.19 and Fig.2B.20, by kind permission of Museo Galileo, Institute and Museum of History of Science, P.za dei Giudici, 1, 50122 Florence, Italy.

Fig.2B.21. Unknown author. After 20 years the picture is of public domain according to the Italian laws 22-04-1941 n.633 modified 22-05-2004 n.128 art 92.

Fig.4.8 reprinted from Camuffo D. and Bernardi, A., (1991): The microclimate of Leonardo's "Last Supper", *European Cultural Heritage Newsletter on Research*, and *Bollettino Geofisico*, joint edition 14, 3, 1–123, by kind permission of European Commission, DG XII, and *Bollettino Geofisico*. Support received by the European Commission, DG Research and Innovation project "AER" ENV4-CT95-0088, 4th Framework Programme, Environment and Climate.

Fig.4.10 reprinted from Camuffo, D. and Bernardi, A. (1986): Dinamica del microclima e scambi termoigrometrici tra pareti e atmosfera interna nella Cappella Sistina, *Bollettino Monumenti Musei Gallerie Pontificie*, 6, 211–257, by kind permission of Vatican Museums and Galleries.

Fig.4.15 Redrawn after Wikimedia Commons, Author Chlorophyll_ab_spectra2. PNG: Aushulz. GNU Free Documentation License. https://commons.wikimedia.org/wiki/File:Chlorophyll_ab_spectra-en.svg

Fig.5.8 kindly supplied by emeritus Prof. Marco Del Monte, Department of Geology, Bologna University.

Fig.5.10a,b reprinted from Camuffo, D., (1984): Condensation–Evaporation Cycles in Pore and Capillary Systems According to the Kelvin Model, *Water Soil and Air Pollution* 21: 151–159, by kind permission of Kluwer Academic Publishers, P.O. Box 17, 3300 AA Dordrecht, The Netherlands.

Fig.6.1 is similar to Fig.3 page 44 in Camuffo, D.: Environment and Microclimate; pp. 37-50 in: N. Baer, C. Sabbioni and A. Sors (ed.s): *Science Technology and European Cultural Heritage* (1991) by kind permission of Butterworth Heinemann, Linacre House, Jordan Hill, Oxford OX2 8DP, UK. Support received by the European Commission, DG Research and Innovation, STEP project Contracts ENV-757-I-SB.

Fig.6.2a reprinted from Camuffo, D. (1984): The Influence of Run-Off in Weathering of Monuments. *Atmospheric Environment* 18 (19): 2273–2275, by kind permission of Elsevier Science Ltd., The Boulevard, Langford Lane, Kidlington 0X5 1 GB, UK. Support received by the European Commission, DG Research and Innovation, STEP project Contracts ENV-757-I-SB, EV4V-0051-I-A.

Fig.6.3 kindly supplied by emeritus Prof. Marco Del Monte, Department of Geology, Bologna University.

Fig.6.6 kindly supplied by emeritus Dr Giancarlo Rossi, ENEL, Venezia-Mestre.

Fig.6.14 reprinted with changes from Fig. 3 page 892, Gordon, J. and MacDonald, F. (1953): Anhydrite–Gypsum Equilibrium Relations, *American Journal of Science*, 251, 884–898, by kind permission of *American Journal of Science*, 217 Kline Geology Laboratory, Yale University, New Haven, CT 06,520-8109, USA.

Fig.8.12a,b reprinted from Camuffo, D. (1991): Wall Temperature and Soiling of Murals *Museum Management and Curatorship* 10, 373-383, by kind permission of Elsevier Science Ltd., The Boulevard, Langford Lane, Kidlington 0X5 1 GB, UK.

Fig.8.18 The Sistine Chapel, Rome. Detail of a cleaning test during the restoration in 1986, leaded by Gianluigi Colalucci. Picture taken by the author, by kind permission of Vatican Museums and Galleries.

Fig.8.22, Fig.8.23, Fig.8.24 and Fig. 8.25 Support received by the European Commission, DG Research and Innovation, project "*Friendly-Heating*", EVK4-2001-00007, 5th Framework Programme, Environment and Climate.

Fig.8.29 and Fig.8.30 reprinted from Camuffo, D., (1993): Controlling the Aeolian Erosion of the Great Sphinx, *Studies in Conservation*, 38, 198-205, by kind permission of the International Institute for Conservation of Historic and Artistic Works (IIC), 3, Birdcage Walk, London SW1H 9JJ, UK.

Fig.8.32 is reprinted from www.photographium.com/the-great-sphinxgiza-egypt-1860-1890, by courtesy of Photographium.

Fig.9.2 and Fig.9.4 reprinted from Camuffo, D., (2010): The role of Temperature and Moisture, pp. 9–30 in: Camuffo, D., Fassina, V. and Havermans, J. (Eds): *Basic Environmental Mechanisms Affecting Cultural Heritage - Understanding deterioration mechanisms for conservation purposes*. COST Action D42 "Enviart", by kind permission of Nardini Editore - Nardini Press srl, Via Cavour, 15, 50129 Florence, Italy.

Fig.9.3(a) Reprinted with changes from Nemethy, A. & Scheraga, H. A. (1962) Structure of water and hydrophobic bonding in proteins. A model for the thermodynamic proprieties of the liquid water. J. Chem. Phys., 36, 3382–3400 by kind permission of the American Institute of Physics, College Park, Maryland, USA.

Fig.9.3(b) reprinted from Wikipedia (http://en.wikipedia.org/wiki/Water_cluster), Author Danski 14. This file is licensed under the Creative Commons Attribution-Share Alike 3.0 Unported license.

Fig.10.6, Fig.10.7, Fig.10.14b and Fig.10.15 reprinted from Camuffo, D., Fassina, V. and Havermans, J. (Eds), (2010): *Basic Environmental Mechanisms Affecting Cultural Heritage - U. deterioration mechanisms for conservation purposes.* COST Action D42 "Enviart", by kind permission of Nardini Editore - Nardini Press srl, Via Cavour, 15, 50129 Florence, Italy. Support received by the European Commission, DG Research and Innovation project "*Friendly-Heating*", EVK4-2001-00007, 5th Framework Programme, Environment and Climate.

Fig.11.9 and Fig.11.10 reprinted from Camuffo, D., della Valle, A., Bertolin, C., Leorato, C., Bristot, A., (2011), Umidità e Diagnostica Ambientale in Palazzo Grimani, Venezia. Proceedings of the Symposium: D. Del Curto (Ed.) "*Edifici storici e destinazione museale: conservazione degli edifici e delle opere d'arte progetti* per *il restauro e integrazione di impianti esistenti*". Milan, by kind permission of Nardini Editore - Nardini Press srl, Via Cavour, 15, 50129 Florence, Italy.

Fig.12.4 reprinted with changes from Trotec, *T2000 Practice Handbook* (2004), by kind permission of Trotec Gmbh & Co. KG Grebbener Straβe 7, D-52525 Heinsberg, Germany.

Fig.12.5 Support received by the European Commission, DG Research and Innovation, project "*Climate for Culture*" (FP7-ENV-2008-1, Grant 226973) 7th Framework Programme, Environment and Climate.

Fig.12.6 reprinted from Camuffo, D. and Valcher, S., (1986), A Dew Point Signaller for Conservation of Works of Art, *Environmental Monitoring and Assessment* 6, 165-170. by kind permission of Kluwer Academic Publishers, P.O. Box 17, 3300 AA Dordrecht, The Netherlands.

Fig.13.1 reprinted from Camuffo, D. and Schenal, P. (1982): Microclima all'interno della Cappella degli Scrovegni: scambi termodinamici tra gli affreschi e l'ambiente, *Bollettino d'Arte* special issue "*Giotto a Padova*", pp. 107-209 by kind permission of Ministero dei Beni Culturali ed Ambientali, via di S. Michele 22, Rome and Poligrafico dello Stato, Rome.

Fig.13.2 Support received by the European Commission, DG Research and Innovation, project "*Friendly-Heating*", EVK4-2001-00007, 5th Framework Programme, Environment and Climate.

The Author

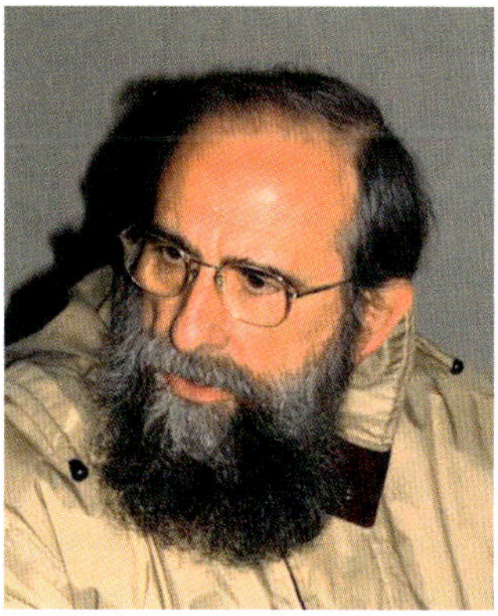

DARIO CAMUFFO. Obtained his degree of Doctor in Physics at the Padua University in 1966, he is currently working in the field of environmental physics. In 1969 he started working at the National Research Council of Italy (CNR), Institute of Atmospheric Sciences and Climate, where his last position was Research Director. Retired in 2008, he is presently continuing his research and teaching as emeritus associate.

Since 1979 he has been lecturer of Environmental Physics and Physics for Conservation at the University of Padua, the Cignaroli Academy, Verona, the Polytechnic of Milan and other institutions. From 1998 to 2008 he was appointed Co-Director of the European Doctoral Course '*Sciences and Materials of the Cultural Heritage*', European University Centre for Cultural Heritage, Ravello.

His activity is mainly devoted to atmospheric physics applied to the conservation of the cultural heritage (he studied several of the most outstanding museums and monuments in the UNESCO World Heritage List) and to climate change, including long-term instrumental series and documentary proxies over the last millennium.

He is active in standardization for cultural heritage having served as vice president of UNI-Normal (Italian Standardization Body), Convenor of two Working Groups (WG6: Exhibition lighting and WG 7: Microclimate), and member of WG4: Collections, at the European Committee for Standardization (CEN) Technical Committee TC346 for cultural heritage.

He has been member of various national and international scientific committees (e.g. European Commission, UNESCO, U.S. NAPAP, Italian Ministries, CNR) on the conservation of works of art, environment and climate and Member of the Directorate of the Italian Geophysical Society (AGI) and the Italian Society for Quaternary Studies (AIQUA).

He also serves on the editorial board of some scientific journals and wrote over 300 scientific peer reviewed papers and some books.

He has led several research projects, some 15 of them funded by the European Commission and/or the European Science Foundation (COST).

PART I

ATMOSPHERIC PHYSICS APPLIED TO MICROCLIMATE ANALYSIS AND CONSERVATION

CHAPTER

1

Microclimate, Air and Temperature

OUTLINE

1.1. MICROCLIMATE

First of all, it may be useful to define the word 'microclimate' to which we refer, as in the everyday practice some terms such as 'global climate', 'macroclimate', 'mesoclimate', 'climate', 'microclimate', 'nanoclimate' and 'picoclimate' are used by people with different meaning, and the same is also true for specialists. It is clear that all these terms have been utilized to define the climate of a specific area, and the prefix is chosen to

Microclimate for Cultural Heritage, *Second Edition*
http://dx.doi.org/10.1016/B978-0-444-63296-8.00001-9

indicate the size of the area involved. However, we cannot proceed in the most apparently straightforward way, that is establishing a scale with the appropriate size unit, for example 1 km in diameter, and then apply the well-known definitions, that is milli $= 10^{-3}$, micro $= 10^{-6}$, nano $= 10^{-9}$, pico $= 10^{-12}$, as in this case the 'microclimate' would apply to a site sized only 10^{-6} km $=$ 1 mm, and this would be ridiculous. Somebody uses the term 'microclimate' for an urban area, 'nanoclimate' for a monument and 'picoclimate' for a very small portion of a monument, but this definition has not gained popularity. In principle, the prefix varies with the actual area size, as determined by geographic, topographic or other local factors, for example the requirement of reaching a basic homogeneity in some key parameters, but it may also vary with reference to the actual interest, in view of a peculiar application, and the list of the subjective elements that intervene in the choice of the appropriate word may continue.

In climate research, meteorology and physical geography, the distinction is rather clear and is determined by the field of interest, that is, the 'global climate' refers to our planet, the 'regional climate' to a geographical homogeneous area and the 'local climate' to a small limited area like a mountain, a valley or a city. Similarly, for conservation, it is useful to use clear terms, derived from the above-mentioned sciences, for example 'regional climate' for the main characteristics of the geographic area where the monument is found, 'urban', 'rural', 'mountain', 'valley', 'coastal' and so on for the next dimensional step, and 'microclimate' relating to the small location, for example a corner of the street, a square, a room, a corner of the room near a painting, where a monument or an object is located. This definition does not imply a precise size, but focuses the attention on a specific artefact (e.g. a historic building, a statue, a small exhibit) and its surrounding, so that the same term can also be applied when studying the interactions between a portion of a monument and the air nearby. In practice, it refers to *the whole ambience which is necessary to study in order to know the factors which have a direct influence on the physical state of the monument and the interactions with the air and the surrounding objects.*

Now that the prefix 'micro' has been explained, it might be useful to clarify the word 'climate'. The following definitions can be found: 'climate is the synthesis of the day-to-day weather conditions in a given area', 'climate is the statistical description of weather and atmospheric conditions as exhibited by the patterns of such conditions, in a given region, over a specified period of time long enough to be representative (usually a number of decades) ', 'climate is the fluctuating aggregate of atmospheric conditions characterised by the states and developments of weather in a given area' (Maunder, 1994). It is evident that in our case the word 'weather' is inappropriate; only in exceptional cases, observations exist over a number of decades, and in general, new observations should be taken in the short term, for example 1 year or 2, before undertaking restorations; the same definition should be applied to either indoor or outdoor environments.

By adapting the above definitions to our aims, the following interpretation can be given: 'microclimate is the synthesis of the ambient physical conditions (e.g. time and space distributions, fluctuating values and trends, average and extreme values, space gradients and frequency of oscillations) due to either atmospheric variables (e.g. temperature, humidity, sunshine, airspeed) or exchanges with other bodies (e.g. infrared emission, heating, lighting, ventilating) over a period of time representative of all the conditions determined by the natural and artificial forcing factors'. When a survey cannot continue for a time interval that is statistically representative of all the conditions, it should at least document one, or a few examples, of key different conditions, just to understand the type of problems we are dealing with.

Another key question is whether meteorological data, taken from a standard weather station located a few kilometres (or less) far from a monument, can be used for the estimation or interpretation of microclimatic situations, or is it always necessary to carry out specific and expensive field tests. It is clear that the recovery of existing data is helpful for a more complete interpretation of the phenomena, but we will see in the following chapter (Chapter 10) that standard weather stations operate by measuring parameters with some specific criteria and methodologies; in conclusion, some measurements will be useful to our aims, others of scarce relevance, and others useless.

In addition, several parameters needed for the science of conservation are not considered in weather stations. For this reason, special field surveys are needed. Only a few parameters, measured by weather stations in a compatible way, might be duplicates and could be omitted.

Generally speaking, it is convenient to record in the same data acquisition system and with the same criteria the whole set of data useful for a specific study and then add or compare further observations, if any. In the field of conservation or in other environmental purposes, observations are made to study certain individual problems, so that the instrumental apparatus as well as the operative methodologies should be specifically tailored to fit the actual problem. On the other hand, it is evident that weather stations are planned and standardized for meteorological measurements as defined by international protocols, so that they are in principle interchangeable.

Not only are the weather instruments and methods generally inappropriate for conservation purposes but also the free position where the standard weather station is located is not representative of the specific site of the monument, which may be perturbed by nearby buildings, trees or other obstacles. For instance, it is sufficient to consider that the wind forms wakes downstream to buildings and that in the wake the wind direction is opposed to the unperturbed flow measured by the weather station. One of the aims of a field test for cultural heritage is just to describe the complex and 'perturbed' situation originated by the presence of all obstacles existing near the monument, whereas the weather station is aimed to monitor the 'unperturbed' situation, that is no obstacles at all.

It is absolutely restrictive to consider the individual parameters separately (e.g. temperature, humidity, sunshine, wind, precipitation) omitting interactions and feedbacks between them and the surfaces. The microclimate is determined by the complex interaction of several factors. The microclimate is not only an accurate description of what is happening but also a good base to forecast what might happen in the future. The microclimate for conservation purposes is not just a mere collection of atmospheric variables (i.e. we should not preserve air) but a description of the complex interactions between air, building and objects, an unicum that includes air and collections and all their mutual relationships.

Air and precious surfaces to be studied and preserved can be found everywhere, either indoors or outdoors. Although the indoor and the outdoor environments are traditionally considered very distinct, in practice, they present similar problems: both undergo daily cycles of temperature and humidity, forced by either the solar cycle or by heating, ventilating and air-conditioning (HVAC) systems; both are exposed to intense shortwave radiation, which may be the direct solar irradiation on the open sky or through window or artificial light; both are affected by advective air movements, that is wind or air currents or air leakage through cracks and openings or turbulence generated by sources of momentum, for example people movements, heat sources, surface roughness in the presence of advective movements. Rainfall and dew are considered typical of outdoor environments, but often rainfall penetrates inside through disconnections, or condensation forms on the window panes, on the surface of cold objects or inside pores. External pollutants can penetrate through windows and doors and are transported from room to room and deposit via the same mechanisms, either outside or inside. The same problems can be found outside or inside, although the scale as well as the level of complexity may change. The most important distinction is that the indoor microclimate can be controlled, at least in principle, and it is very important to know how to do it.

Although the importance of the indoor microclimate has been stressed for a long time (Benoist, 1960; Camuffo, 1983; De Guichen, 1984; Thomson, 1986; Michalski, 1993, 2004, 2009; Padfield, 1994; Camuffo and Bernardi, 1995a; Camuffo et al., 2007), sometimes, inappropriate standards of thermal comfort are used, which are based on human wellbeing and not on the science of material conservation. Are museums oriented towards the wellbeing of visitors or exhibits? Is their aim limited to show objects for cultural and educational purposes or to keep artefacts in the most convenient conditions for conservation, to which visitors should adapt? Of course, it is necessary to find a compromise between the above two conflicting needs paying attention that environmental injuries on works of art are cumulative and irreversible. In the case of conflict, conservation needs should have priority as recommended by the EU standard EN15759-1: 2011 *Conservation of Cultural Property – Guidelines for the Climate Control of Churches, Chapels and other Worship Places.*

In Europe, the situation was better some 50 years ago, when Benoist (1960) wrote as follows: *In winter, museums should be heated not only for visitors and guardians, but also for works of art. In Europe, visitors are content with 15 °C, and in America with 21 °C.* In Europe, the above thermal level was convenient to safely obtain appropriate relative humidity levels without blowing huge amounts of moisture to mitigate dryness as it is necessary to do nowadays to reach compatibility with the thermal comfort level.

In the past, local climate was carefully observed and exploited to adapt buildings and activities to the external ambient and benefit of a *natural microclimate*. For example, Hippocrates (*De locis*), Pliny (*Epistulae*, II, 17, 7–19), Vitruvius (*De Architectura*, VI, 4, 1–2) and Palladium (*Quattro libri dell'architettura*, Book II, Chap. XII) show how a building was built with respect to exposure to sun, wind and precipitation. Rooms were exploited according to the temperature and type of light that could enter through the windows. Nowadays, the modern technology often induces to believe that the climate outside can be ignored and that a new independent *artificial microclimate* can be created inside a building, controlling humidity and temperature with advanced sensors and microprocessors.

By maintaining air intake higher than air exhaust, commercial buildings and museums are kept with indoor pressure higher than the outdoor level. This practice reduces infiltrations of external air and pollutants, but creates an internal atmosphere, with its artificial microclimate, which is usually not in equilibrium with walls, floors, ceilings, and exhibits, and needs many frequent heat and moisture transfers to balance the influence of people, the air leakage and the exchanges between air and surfaces. It has been calculated that about 30% of the moisture supplied to a room is absorbed by the room surface (Rosenhow et al., 1985) with the consequence that the benefit in mitigating air dryness is negatively compensated by moistening surfaces.

The excessive confidence on HVAC and their use has caused several problems. The indoor temperature is mainly regulated taking into consideration the people thermal comfort, sometimes irrespective of the regional climate (except for calculating the HVAC power), the outdoor moisture content and the internal vapour sources. The desired temperature level is assumed as a primary need, and the indoor concentration of water vapour is controlled to create an acceptable relative humidity (*RH*) level in this artificial microclimate. Several systems that control the humidity in historic buildings or museums have been analysed with field tests; however, although these systems may be good in theory, and the machines may operate correctly, their long-term impact has been often found to be deleterious, especially if the main control unit goes out of order. Dry or moist air is blown near walls where paintings, wooden statues or furniture are located. Clouds of moist/dry, cool/warm air are generated, and the internal air motions will move them. The consequence is that objects and collections will suffer for repeated humidity and temperature fluctuations.

The technological limit of HVAC is not in designing new, powerful or sophisticated machines but in being unable to distribute treated air to create an unchangeable uniform climate within all rooms. The homogeneous distribution of heat and vapour in a room needs a capillary distribution of blowers and diffusers, located everywhere at short distances from each other. The blown air may cause undesired air movements that will increase inertial deposition of suspended particles. Unfortunately, the present-day technological research is focused on making sophisticated devices with improved performances, not in studying and controlling their use and the compatibility with conservation.

Environmental monitoring and atmospheric thermodynamics are useful tools in environmental diagnostics and are absolutely necessary for the progress of our knowledge on the basic processes affecting cultural heritage, evidencing causes, effects and deterioration mechanisms.

1.2. AIR, WATER VAPOUR, PERFECT AND REAL GASES

Dry air is composed of a mixture of several gases, mainly nitrogen (N_2, 78.084% volume) and oxygen (O_2, 20.946%), with argon (Ar, 0.934%) and some other minor constituents, that is carbon dioxide (CO_2, variable, 360 ppm in 1998, 399 ppm in 2013), neon (Ne, 18.182 ppm), helium (He, 5.24 ppm), methane (CH_4, 1.77 ppm), krypton (Kr, 1.14 ppm), hydrogen (H_2, 0.5 ppm) and xenon (Xe, 0.09 ppm). Many other trace gases and particles, considered pollutants, are dispersed in the atmosphere. Some of them are nonreactive, and most of them are reactive, for example O_3, SO_2, NO_x. Atmospheric chemistry studies the behaviour and effects of the substances that become reactive in association with atmospheric water and possible synergisms with light. For the sake of simplicity, air is often treated as it were an ideal gas, composed of molecules having mass $M = 28.96$, which is the average weight of the molecules of this mixture.

Water vapour is a variable constituent of the atmosphere, whose concentration depends on weather and other local factors, and generally ranges between 0.5% and 4%. This variability is a consequence of the fact that water vapour may change state, becoming liquid or solid, for example it may precipitate or be in different ways transferred from the atmosphere to the earth's surface or vice versa. The water molecule itself is far from being a 'perfect gas'. It is composed of one oxygen and two hydrogen atoms that are 0.95 Å distant from the oxygen nucleus. Atoms are disposed forming an H—O—H angle equal to 105°. This asymmetrical configuration generates a spatial imbalance between positive and negative charges, so that the water molecule is an electric dipole that can be oriented in an electric field (exerting a strong dielectric action), or may interact with other molecules or bodies exerting van der Waals and electric forces.

However, in a crude approximation, when the water vapour does not undergo changes of state, for several purposes, it may be treated taking advantage of our knowledge and formulas for perfect gases, although some departures may occur and should be considered, as we will see later. Major problems arise when the water vapour approaches saturation, or when water vapour molecules impact on a surface whose temperature is below the dew point, or which is contaminated with hydrophilic salts. Under these conditions, instead of exerting elastic impacts, the vapour molecule will stick on the cold surface, or on the salts (the same holds for condensation nuclei), and the effective number of 'free' gaseous molecules will decrease.

To simplify things, the state of a gas is statistically represented by some key parameters that characterize the average properties of a molecular population. A 'perfect gas' is an ideal reference gas, where intermolecular forces are neglected, all impacts are elastic and the physical state is perfectly described by a simple so-called state equation

$$pV = n\,\Re T \tag{1.1a}$$

where p is the pressure, V the volume, n the number of moles ($n = m/M$, where m is the actual mass of the gas and M its molar mass), $\Re$ the universal gas constant, that is $\Re = 8.3169 \times 10^7$ erg mol^{-1} K^{-1} = 1.986 cal mol^{-1} K^{-1}; and T the *absolute temperature* (expressed in Kelvin, K).

For a particular gas X, the gas constant is defined as $\Re_X = \Re/M_X$, so that for dry air $M_a = 28.965$ g mol^{-1} and $\Re_a = 0.2870 \times 10^7$ erg g^{-1} K^{-1} = 0.06857 cal g^{-1} K^{-1}; for water vapour $M_v = 18$ g mol^{-1} and $\Re_v = 0.4615 \times 10^7$ erg g^{-1} K^{-1} = 0.1102 cal g^{-1} K^{-1}. For the gas X, the state equation becomes

$$p_X V = m_X \Re_X T \tag{1.1b}$$

where p_X and m_X are the partial pressure and the actual mass, respectively. Although a real gas may depart from the simplified perfect gas model, this is, however, the basic equation that will be useful in the following and, although within limits, to the water vapour too, but when it is far from the saturation point.

When more precise calculations are needed, or saturation is approached, we should consider the equation of state for real gases:

$$pV = n\,\Re T + n\,B(T) \tag{1.1c}$$

where $B(T) > 0$ at relatively high temperatures and $B(T) < 0$ at low temperatures. $B(T) < 0$ means $\partial T/\partial p > 0$, that is when the gas is expanded (i.e. the pressure decreases), it will become cooler as in the well-known Joule–Thomson experiment where a gas compressed in a vessel expands in another volume decreasing its temperature for the expansion. If it were a perfect gas, no change in temperature would occur. For an ideal gas, the molecular population would distribute within the allowable volume without work. Therefore, an expansion without work would have occurred without any change in the temperature level. The same can be expressed by saying that in a perfect gas, the internal energy of the system is independent from the space distribution of the molecules. However, this is contrary to the actual experience with real gases.

For a perfect gas, the equation of the *kinetic energy* E_c for the whole molecular population is

$$E_c = \sum_1^n \frac{m_i\,v_i^2}{2} \tag{1.2a}$$

where the sum $\sum$ of the individual kinetic energies is extended to the whole population of n molecules. A perfect gas is like a geometrical point, that is no size, no rotation and no vibration. For a real gas, kinetic translation energy (and temperature as well, see later) is affected by molecular interactions, that is:

$$E_c = \sum_1^n \frac{m_i\,v_i^2}{2} + U \tag{1.2b}$$

where U represents the *energy of interaction* between molecules or between molecules and surfaces. To the translational energy, one should add the kinetic energy due to *rotation* and *vibration* that will increase with the number of degrees of molecular freedom related to the number of atoms of which the molecule is composed.

The graph of U is represented in Fig. 1.1. $U > 0$ means repulsion, $U = 0$ no interaction and $U < 0$ attraction. It is evident that two molecules cannot approach each other at a shorter distance than r_o, where $U > 0$. When $r > r_o$, $U < 0$ and the molecules attract each other, with a more stable position at $r = r_A$.

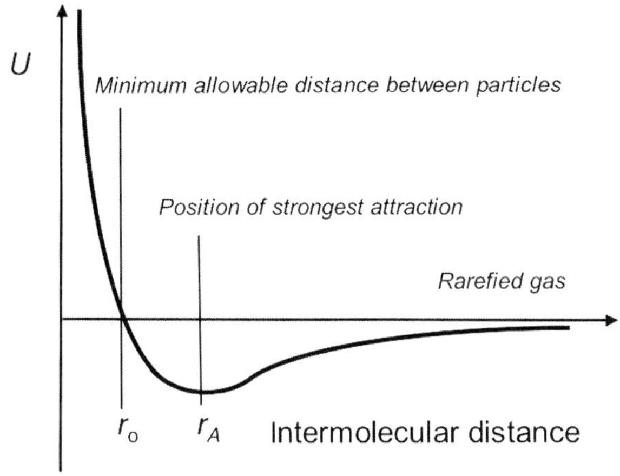

FIGURE 1.1 Energy of interaction U between molecules versus intermolecular distance.

Only when molecules are far away (i.e. rarefied gas), intermolecular forces become weaker and weaker and the perfect gas approximation becomes acceptable. U explains why molecules redistribute energy between each other and nearly assume the temperature of the wall where they impact.

1.3. TEMPERATURE

Temperature is the condition that determines the direction of the net flow of heat between two bodies, that is from the warmer to the colder one. For this property, a thermometer can be put into equilibrium with a body, to read the temperature of the body on the thermometer, if the thermometer does not perturb the original temperature of the body and is not influenced by other factors.

From the thermodynamic point of view, in a gas, the *temperature* T represents the *average translational kinetic energy* E_c of molecules, according to the principle of equipartition of the energy. The latter states that the average kinetic energy of a molecule is equally distributed among its degrees of freedom.

The *average translational kinetic energy* E_c of a molecule equals the kinetic energy of an ideal monoatomic gas having the same mass as the given molecule and is related to T by the equation

$$E_c = \frac{3}{2} k\, T \tag{1.3}$$

where $k = 1.38 \times 10^{-16}$ erg K^{-1} is the *Boltzmann constant*, given by the ratio $\Re/N$ where $\Re$ is the gas constant and $N = 6.023 \times 10^{23}$ is the *Avogadro number*, that is the number of atoms or molecules that form a *mole*. In other terms, N is the amount of substance whose weight (expressed in g) equals the atomic or molecular weight of the substance. As the air is predominantly composed of diatomic molecules, namely N_2 and O_2, characterized by 5 degrees of freedom, the total kinetic energy E_t is

$$E_t = \frac{5}{2} k T. \tag{1.4}$$

For these relationships, T is also called *molecular temperature*. In meteorology, it is also called the *dry bulb temperature* as opposed to the *wet bulb temperature* that will be seen later.

In the following, the *absolute thermodynamic temperature* (K) will be indicated with the capital letter T and the temperature in degree Celsius (°C) with the lower case t. As $t = T - 273.15$, then $\Delta T = \Delta t$. The value 273.15 is the thermodynamic temperature of the triple point of water at one atmosphere pressure and is usually approximated 273.

The concept of temperature can be easily extended from a gas to a liquid or a solid (e.g. thermal vibrations).

Another empirical definition is popular: *the temperature is the variable measured by a thermometer*. As it will be discussed in the following, this definition is obscure and is affected by all instrumental and observational errors. However, even in the absence of errors, the thermometer measures only the temperature of its bulb, which is not necessarily the same of the air or the object under investigation. In fact, the thermodynamic equilibrium between thermometer bulb and air involves a balance between conductivity, convection and radiant heat exchange, which are different for air and for each body. In particular, radiant heat is exchanged with other external bodies either nearby or far away, and this contribution is not included in the definition of temperature. The temperature is a consequence of the present and past energy balance that also includes advective contributions due to the transport of air and only in rare cases is uniformly distributed in a body or in a room.

1.4. MECHANISMS OF TEMPERATURE DEGRADATION

Temperature is an important factor in the conservation of works of art for a number of mechanical, chemical, mineralogical or biological mechanisms. Changes in temperature induce differential expansions in materials and tensile strengths between the surface and the subsurface layers.

Thermal expansion may be relevant for not only objects but also structural stability of monuments or buildings. The simplest example of thermal expansion of a Spanish *Azulejo* is shown in Fig. 1.2 where the

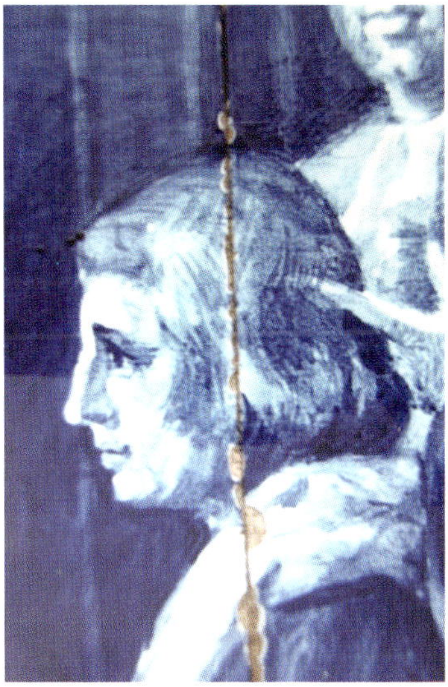

FIGURE 1.2 Tile of a Spanish *Azulejo* damaged for thermal expansion after exposure to solar radiation and overheating.

solar radiation overheats tiles while the underlying masonry remains colder. Tiles, expanding in size, will enter into contact with one another, and the exerted pressure will cause cracks and break borders (flaking).

The Leaning Tower of Pisa (Fig. 1.3) is composed of a cylindrical body contoured by six orders of loggias, having thin columns. In the sunny

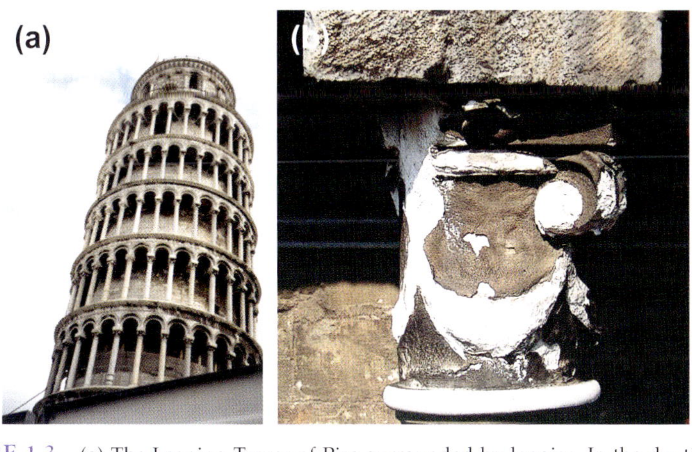

FIGURE 1.3 (a) The Leaning Tower of Pisa surrounded by loggias. In the daytime, the solar heating causes the expansion (and compression) of the columns and an additional temporary bending of the tower that follows the apparent course of the sun, like a sunflower. (b) A severely damaged capital for the compression between the thermally expanding column and the fix lintel.

days of the warm season, the stone temperature of the thick central body of the tower remains nearly unchanged, but the external parts, and in particular the columns exposed to direct solar radiation, undergo daily cycles of some 20 °C, which cause the expansion of the heated stone (Camuffo et al., 1999).

Considering that the expansion coefficient for limestone is 8×10^{-6} °C and that the tower extends for some 50 m in height, the above daily cycles cause 8 mm vertical expansion in the warm side. This overheating causes a temporary bending of the Tower towards the colder side, which follows with some delay the solar motion, like a sunflower. The daily movements of the top of the sunflower tower have been measured with a pendulum, and the maximum excursion is some 4 s of arc from East to West (Jamiolkowski, 1995), which corresponds to about 1 mm horizontal displacement. The vertical expansion of columns and the unchangeable distance between basement and lintel produces a compression of the thin columns and their capitals: most of them are severely damaged or broken or substituted in the past.

Temperature cycles: The most popular examples of dangerous daily cycles are the freezing–thawing cycles or the temperature cycles of objects directly exposed to sunshine. Temperature cycles induce internal stress and accelerate fatigue failure in vulnerable materials. Materials act as a low-pass filter that attenuates the penetration of fast fluctuations: the shorter the duration of the fluctuation, the thinner the layer affected by it (Camuffo et al., 1984). However, it should be remembered that the key part of the artistic value of monuments lies in the surface layer. For these reasons, daily (or shorter) temperature cycles might be more important than the seasonal ones.

Thermal cycles may cause weathering and mechanical damage to the outer layers of stones, beginning at the discontinuities included into the rock and the interfaces between the different minerals of which the stone is composed. The thermal anisotropy of the crystalline lattice, the size of the granules and their spatial association determine a system of internal tensions that result in surface damage. Sedimentary rocks are characterized by a more regular structure and composition, but the nature of the binding cement between granules acts as a discontinuity factor (Veniale, 1995).

In sunny days, monuments are overheated by the solar radiation and dramatic temperature changes (thermal shocks) occur when the sun appears or disappears; in addition, marked short-term (3–15 min) temperature fluctuations are generated by variations in wind speed and cloud cover (Camuffo, 1981; Jenkins and Smith, 1990).

Granular disaggregation is frequently found on stones with granular or crystalline texture, for example granite or marble, where stresses generated between grains or large crystals with differently oriented crystallographic axes, or having different expansion coefficients, produce fatigue failure along grain or crystal interfaces. For example, calcite crystals expand along the principal axis and contract along the secondary one in the case of a temperature rise. In the long run, heating–cooling cycle will slightly displace crystals from their original position forming a less regular weaker structure, which will lead to the disaggregation and loosing of granules called *sugaring*. The damage is irreversible and cannot be restored (Fig. 1.4).

Air movements and surface soiling: Temperature variations over space affect the temperature (and density) of the air coming into contact with the colder or warmer surfaces. The air looses or gains buoyancy and convective cells develop. The air movements will be responsible for the transport and deposition of gaseous pollutants and airborne particulate matter and soiling of paintings and monument surfaces.

Glass–rubber transition: It is well known that, supplying heat at a certain constant rate, the temperature of a crystal will increase until it will reach the melting point. At that level, the temperature will remain unchanged until the whole crystal turns from the solid into the liquid the liquid phase. After this transition, the liquid temperature will continue to increase. The *crystal stiffness* (represented by the *Young's E-modulus*) will remain almost unchanged until the melting point; afterwards, it will drop to zero level, typical of the liquid phase. This behaviour is typical of all crystals, either metals or crystalline polymers. The situation is different for polymers, either natural or synthetic, for example cellulose, varnishes (Fig. 1.5(a)), rubber, Bakelite, neoprene, nylon, PVC, silicone, wool, silk. In the case of amorphous polymers, or polymers composed of a mixture of crystalline and amorphous domains (e.g. cellulose), the transition from

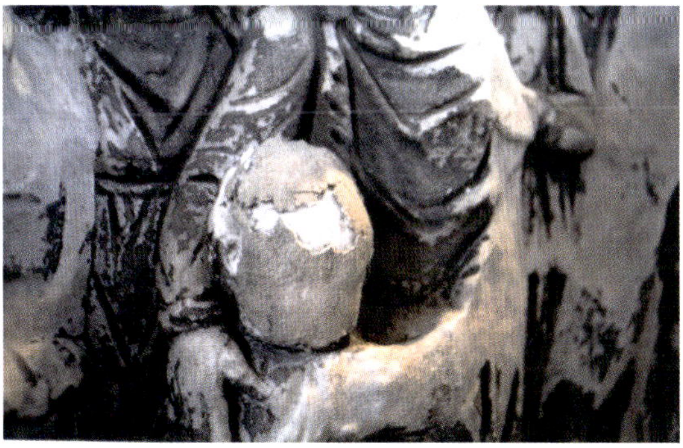

FIGURE 1.4 Granular disaggregation of marble. An inappropriate restoration with a remaking of the external layer, made with not fully compatible materials (e.g. less permeable to vapour, different expansion coefficient), had dramatic consequences. Aurelian Column, Rome.

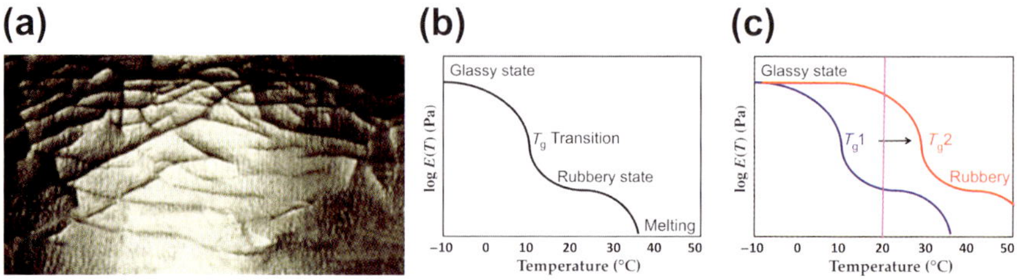

FIGURE 1.5 (a) Aged painting on canvas, with brittle and severely damaged colour coating. (b) Transition state plot of stiffness (E-modulus) versus temperature, showing the progressive transition from the glassy to the rubbery state and melting. T_g is the glass transition point. (c) Ageing implies a shift of the transition plot (from blue to red) towards higher temperatures. An object that was at 20 °C was initially rubbery (flexible) then will become glassy (brittle).

the solid to the liquid phase is more complex with intermediate phases passing through the *glass transition temperature* (T_g), also called the *glass–rubber transition temperature* (i.e. the point in which the acceleration of the stiffness change rate inverts sign), that is kept as a reference (van der Vegt, 2002).

At low temperatures (Fig. 1.5(b)), polymers have properties similar to glass: they are rigid and relatively brittle (i.e. *glassy state*). In the glassy state, the volume is slightly smaller and the molecules have a limited degrees of freedom: they have thermal vibrations, but they cannot rotate or slip over each other. Stiffness is the rigidity of a material, that is the extent to which it resists deformation in response to an applied force. It is described by means of the elastic modulus (E-modulus), also called Young's modulus. At increasing temperatures, the stiffness will progressively diminish at an increasing rate. In the case of polymers, the stiffness varies with continuity representing the bulk of the molecular mobility and motions (i.e. rotation, sliding motions), flexibility, free volume and so on, which are characterized by a continuous transition, differently from metals or minerals. The E-modulus will decrease at increasing temperatures, but not at a constant rate: first, it will accelerate until it reaches the T_g point; afterwards, it will decrease its change rate until the polymer reaches a viscous state called 'rubbery state' in which the polymer is soft and flexible. If the temperature will continue to increase, the polymer will undergo another similar transformation, passing from the rubber to the liquid phase. Aged polymers become hard and brittle because the material undergoes a slow transition from the rubbery to the glassy state.

Ageing factors are time, high temperatures, temperature cycles, humidity cycles, ultraviolet (UV) radiation and so on. In particular, time and temperature appear to be equivalent in their effect on the behaviour.

This means that higher temperatures, or repeated temperature cycles, will have the same impact as a long time, and the life time of the object will be reduced accordingly. The mechanism is illustrated in Fig. 1.5(c): ageing will shift the stiffness diagram from left (lower temperature, blue line) to right (higher temperature, red line). At ambient temperature (e.g. 20 °C) at the beginning, the material diagram was in the rubbery state, and after ageing (time or environmental forcing), the plot has been displaced, falling now on the glassy state that means rigid, brittle and increased vulnerability.

Changes in relative humidity and moisture content: It is to be noted, however, that the pure thermal effect is an academic abstraction, as various water activities are always synergistically linked with temperature, and several deterioration mechanisms are caused by water, but triggered by temperature, for example surface heating, moisture evaporation, internal salt migration and efflorescence formation.

The most popular effect is that when the temperature increases, the relative humidity decreases and vice versa. Relative humidity is a key parameter in conservation and is governed by temperature in addition to the moisture content in the air, as we will see later. Temperature variations cause changes in the degree of saturation of water vapour, as well as moisture content in bodies. Several materials, for example wood, parchment, ivory, plaster, change their dimension with water content, swelling or shrinking, with deformations, micro or macro fissuring and so on. The effects of temperature forcing are in general very complex. For instance, wood is characterized by low thermal conductivity, but the propagation of heat is faster than the propagation of moisture, and this causes a redistribution of moisture absorbed into the grains. As a consequence, delayed differential stresses and shrinkage are induced. Again, temperature changes inside porous stones cause changes in *RH* within pores, which in turn is related to the evaporation of water, the concentration of dissolved salts and their precipitation in crystalline form when the solution becomes supersaturated.

In addition to temperature, wetting–drying cycles too cause swelling–shrinkage cycles in some kinds of stones, especially sandstones including clay. However, although not full consensus exists on the efficiency of solar radiation in stone weathering, the opinion is that most granular disaggregation especially occurs in combination of synergisms with other factors or when stones have been previously weakened by some other cause. Weathering mechanisms act in synergistic combination or are the cumulative results of many cycles and have different efficiency depending on the intrinsic rock properties (e.g. albedo, thermal conductivity and heat capacity, mechanical strength, porosity and specific surface), thermal variations, repeated stressing of the material and role of moisture and dissolved salts (Smith, 1978; Warke and Smith, 1994).

Chemical kinetics: The influence of temperature on oxidation, corrosion, hydrolysis and other chemical reactions is clearly described by the Arrhenius (1889) equation (Chapter 9) concerning the equilibrium coefficient κ_A for chemical reactions:

$$\kappa_A = Ae^{-\frac{E_a}{\Re T}} \tag{1.5}$$

where the pre-exponential factor has units s^{-1} and can be physically interpreted as a frequency factor in some way dependent from the number of molecular collisions per unit time or in terms of the entropy of the system; E_a is the so-called activation energy, that is the energy threshold a molecule should have for a reaction to occur or, which is the same, the energy barrier that the reactants must surmount to react. It should be considered that in gases and liquids, molecules are distributed following the Maxwell–Boltzmann distribution of velocities. This means that the molecular population has many different levels of energies, from the less energetic to the most energetic ones, the mode being at intermediate levels. Arrhenius concluded that a molecule must possess some threshold energy E_0 before it can react, and the fraction of molecules that possess this amount of energy is given by the function $\exp(-E_0/kT)$. Chemical reaction rates increase with increasing temperature for the simple reason that at a higher temperature a larger fraction of reactant molecules possesses energy in excess of the reaction energy barrier.

The Arrhenius equation is illustrated in Fig. 1.6 that reports the logarithm of the equilibrium coefficient ($\ln \kappa$) versus the inverse of the absolute temperature ($1/T$). The choice of the variables is made to represent the equation as a straight line and obtain the equation after two experimentally observed reaction rates κ_1 and κ_2 made at two different temperatures T_1 and T_2, where $T_1 < T_2$.

It should be noted that the Arrhenius equation was originally found for chemical reactions in liquid phase and it is useful to describe several weathering mechanisms where the presence of water is necessary for a chemical reaction to occur on the surface or inside an object. On the surface, rain and dew are the major sources of liquid water. Even in apparently dry conditions, but at elevated humidity levels, the surface of objects is covered with a number of layers of water molecules or microdroplets that will make faster oxidation and corrosion rates. The moisture content in paper or wood may increase cellulose hydrolysis, oxidation and other deterioration mechanisms. Therefore, the Arrhenius equation should always be considered in synergism with the presence of water. If a bronze monument exposed to sunshine overheats, all the surface water evaporates and no chemical reaction with gaseous pollutants may occur in the dry phase.

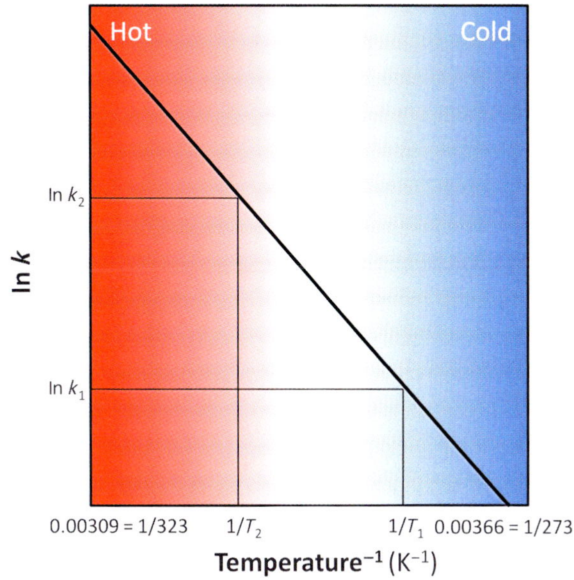

FIGURE 1.6 The Arrhenius equation expressed as ln k versus $1/T$ to obtain the equation represented as a straight line and obtain κ after the observation of two reaction rates made at two temperatures T_1 and T_2. In this inverse scale, cold is on the right side (cyan) and hot on the left side (red). In this example, the range is from 0 to 50 °C, that is from 0.00366 to 0.00393 K^{-1}.

At ambient temperature, the Arrhenius equation can be compared with the rule of thumb established by *van't Hoff* for chemical reactions, which states that the conversion rate is increased two-fold or more for each rise of 10 °C in temperature.

Biodeterioration: In combination with moisture, temperature is a key factor in determining the habitat for biological life and in controlling metabolisms. The Arrhenius equation, and the derived *van't Hoff* rule for chemical reactions, at least in the range above 0–35 °C, might in general describe the response of biology to temperature, but they cannot be simply adapted to all bio-organisms.

At temperature below 0 °C, the ice crystals that were formed inside micro- or macro-organisms may kill them. In the interval above the freezing point up to 20 °C, the metabolic processes are favoured with increasing temperature; the biodegradation due to bacteria can be often prevented with an appropriate choice of low or mild temperature and other environmental variables (e.g. humidity, light, ventilation). However, although the temperature range from 20 to 35 °C particularly favours the microbiological activity, the variable response and adaptability of microorganisms to lower or higher temperatures, as well as to other extreme and stressing environmental conditions (e.g. water activity, pH value, ionic/osmotic strength), should be considered when preventive remedies against microbial or fungal attack should be undertaken (Sedlbauer, 2002; Sedlbauer et al., 2011). Above 40 °C, most microorganisms suffer from the unfavourable hot habitat. As a consequence, biodeterioration becomes less relevant when temperature increases above this threshold. For microorganisms that have colonized stone, masonry or other materials, the temperature of the support where they are living is even more relevant than the air temperature.

Microbial biofilms covering the surface of stones or other materials may have several negative consequences: they may enhance the deposition of airborne particulate matter, and the deposited material, as well as the biofilm forms a composite layer, which changes the albedo of the surface, the porosity and water vapour diffusion inside the material, the thermal conductivity and the water balance, especially in the outer, endangered uppermost layer (Warscheid and Krumbein, 1996). On the other hand, in some cases, biofilms exert a protective function with their polymeric matrix, so that it is difficult to formulate an accurate balance between negative impacts and positive factors. A problem is the variable response of the microflora infesting surfaces (Caneva et al., 2009).

1.5. TEMPERATURE IN A BUILDING, A ROOM

The European project 'Assessment of Environmental Risk Related to Unsound Use of Technologies and Mass Tourism' (AER) was devoted to identify and analyse the complex effects on the condition of works of art, which result from the technologies used in museums and the growing number of visitors they attract. The use of HVAC, exhibition lighting, building use and presence of visitors all contribute to determine the indoor climate in addition to the natural exchanges of heat and moisture across the building envelope. Under this project, a number of European museums were monitored to discover potential sources of risk and suggest mitigation remedies. In a building, the external forcing parameters (e.g. solar radiation, heat conduction across roof and walls, air exchanges through openings) depend upon the architectural features and the materials choice. Thin or conductive walls are sensitive to the apparent daily course of the sun; windows may allow the penetration of solar beams and behave as a greenhouse; in addition, they can regulate exchanges of external air. Different exposures in a building have a different heat balance, and not all the rooms have the same temperature. The inner rooms are more shielded and the external forcing is smoothed out; and this applies especially to the ground floor, where the soil has an enormous heat capacity. The opposite holds for the last floor, being topped by a roof that receives solar radiation during the day and looses infrared (IR) radiation during the night. HVAC or people may completely change the natural equilibrium.

Although rooms are provided with thermostats for the temperature control, the temperature in a room cannot be represented by only one time-dependent variable, but is a four-dimensional variable, that is three to describe the spatial variability at any given point (x,y,z) and one for time. As the air is mobile and has a very small specific heat, the indoor temperature will be determined by exchanges with floor, ceiling, walls, windows, doors and all the other sources or sinks of heat, for example heaters, air-conditioning systems, solar radiation, lamps, people. If there are internal sources (or sinks) of heat and/or moisture, open windows or doors, or forced flows of air at different temperature, the situation becomes complex and various factors should be analysed, and a distinction should be made between natural factors and those derived form the building use and climate control.

In a closed room, warm air rises until it reaches the ceiling. The air density regulates the distribution of the air inside the room, that is warm and lighter air will reach the top, and the cold and denser air the floor. For this reason, a stable atmospheric stratification with temperature increasing with height, that is thermal layering, tends to form. However, if there are some sources of sinks of heat, or all the surfaces are not exactly in thermal equilibrium with the air at the same height, the mass conservation requires that the ascent of warm air is always associated with an equal flow of descending air and vice versa. This may happen in a number of ways, determined by the boundary conditions and room features, as in the following examples.

1. Everything is in equilibrium, except for the presence of a heat source inside the room, for example a heater, a lamp. The heat source will develop an upward warm air column and an indoor air layering. If the heat source is powerful enough, the upward flow will reach the ceiling, and the upward flow will be compensated by a slow downward motion of the whole air volume. If the heat source is modest and the indoor air is stratified, the convective motion will stop and will horizontally diverge at the level where it will loose buoyancy, that is where the density of the local air is the same or less. All upper layers will be left unaffected and a fresco on the ceiling will remain untouched, protected by the upper layering. Similarly, if the heat source is not located on the floor but at a middle height, the descending flow will stop at the level of the heat source because below this level it will find colder and denser air. The lower air layers will remain unaffected and paintings placed below the lamps will remain protected against the convective air movements and related soiling. As the air−mass balance remains unchanged, the ascending air flux equals the descending one. As a consequence, the ratio between the ascending and the descending velocities equals

the inverse ratio of the cross-sections of the upward and downward flows.

2. Everything is in equilibrium, except for the floor which is colder (e.g. ground floor in the summertime) or the ceiling which is warmer than the adjacent air (e.g. metal roof or domes in a sunny day during the warm season). The air–surface exchanges increase the intensity of the atmospheric layering, and the air remains motionless.
3. Everything is in equilibrium, except for the floor which is warmer (e.g. floor heating) or the ceiling which is colder than the adjacent air (e.g. metal or glass roof in the winter). The air in contact with a warm floor will form convective rising cells, associated with other descending air, like the convective movements inside a pot of boiling water. Similarly, the air in contact with the cold ceiling becomes denser and will sink, forming rivulets of descending cold air associated with convective rise, which result in a continuous mixing of the whole atmosphere.
4. Everything is in equilibrium, except for the walls that are warmer than the air. The heat exchanges will form an internal boundary layer (IBL) of warm air rising along walls. This warm air will feed the layer of warm air below the ceiling and will slowly displace downwards the whole mass of stratified layers. If the walls are colder, the IBL flow is directed downwards and cold air will accumulate above the floor and will raise the whole mass of stratified layers.

From the above examples, it appears that, in general, a natural layering is expected in an inside atmosphere and some air motions may derive from the presence of bodies with different temperature. Thick walls of historic buildings have an enormous heat capacity, tend to keep the same equilibrium, and the typical condition is thermal layering, except for the presence of perturbing factors, for example HVAC systems, lamps or people. Thick walls are very effective in reducing the daily temperature cycles and also, to a minor extent, the seasonal cycle, so that the internal microclimate naturally tends to become homogeneous, weakly dependent upon the daily variability and external weather.

The best condition is achieved when the thermal inertia of walls, floor and roof are similar between them; when they are different, the internal stability changes seasonally. For instance, in the case of churches, the ceiling follows the seasonal variations with a short time-lag and the floor a longer one. As a consequence, in the summertime, the relatively warm ceiling and the cool floor will generate internal layering with a vertical thermal gradient. As opposed, in the wintertime, the relatively warm floor and the cool roof tend to destroy the air stability with some internal mixing shown by an isothermal vertical profile.

An example is given for the Basilica of S. Maria Maggiore, Rome (Fig. 1.7). The daily cycle of external temperature is some 10 °C and the

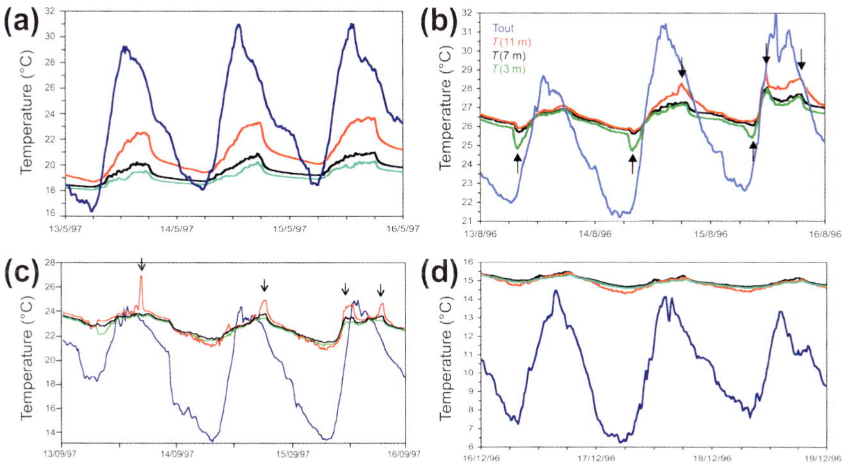

FIGURE 1.7 External and internal temperature in the Basilica of S. Maria Maggiore, Rome. Blue: outdoor temperature; Red, black and green: temperature measured at the levels 3, 7 and 11 m above the floor, respectively. (a) In spring (13–16 May 1997), the thermal inertia of the thick walls keeps the indoor temperature lower than outside. The external air, entering the room, is warmer and tends to stratify with thermal layering. (b) In summer (13–16 August 1996), the indoor temperature is close to the average of the external cycles. When in the morning the doors are opened, cold air enters and is visible as temporary drops at the 3 m level (upward arrows). Celebrations with powerful illumination generate peaks at the 11 m level (downward arrows). (c) In autumn (13–16 September 1997), the heat accumulated during the warm season makes the internal temperature higher than the external average of the daily cycle. Temperature peaks are visible during the liturgical offices on Saturday evening and Sunday (downward arrows). (d) In December, the indoor climate is mild for the heat accumulated during the warm season.

internal is one order of magnitude less, being governed by the limited exchanges with the exterior, the walls, the floor and the ceiling, and the heat accumulated (or lost) by these structures in the previous months. In spring (Fig. 1.7(a)), the indoor temperature is fresh for the thermal inertia of the thick walls. The external air is warmer, and when it enters, it tends to form thermal layering. When the outdoor air is close to saturation, the air entering the church may leave some vapour to condense on the indoor cold surfaces. In late summer (Fig. 1.7(b)), the inside temperature is near the average of the external temperature cycles. The external range has a limited influence on the indoor microclimate, whose changes are mainly governed by the doors and windows openings, the solar radiation and nocturnal cooling of the metal domes. In early morning, the opening of the two front doors is visible, when cold air enters and spreads layering, lowering the temperature of the sensor at 3 m level, but not of the sensors at 7 and 11 m. In autumn (Fig. 1.7(c)), the indoor temperature is higher than the average of the external cycle, being close to the daily maxima, which equal the temperature of the walls, still warm for the heat accumulated during the warm season. Internal condensation is hardly probable in this season. Peaks of temperature were found in all seasons during the liturgical offices celebrated on Saturday evening and Sunday morning and evening for the lighting operated with incandescent lamps and massive churchgoer participation. In winter, the heat accumulated during the warm season makes the indoor climate mild, that is 5–6 °C warmer than the outdoor air.

Also interactions between rooms should be considered. Often contiguous rooms have air flowing through doors and spreading from one room to another (Camuffo and Bernardi, 1991a; Fig. 1.8). The circulation may be forced by HVAC systems or existing pressure differences that external winds will generate blowing through windows or doors.

Similar interactions may occur between rooms at different floors. For example, a staircase connects the ground floor of the San Rocco Oratory (Fig. 1.9), Padua, with an underground cellar. Both floors have no heating,

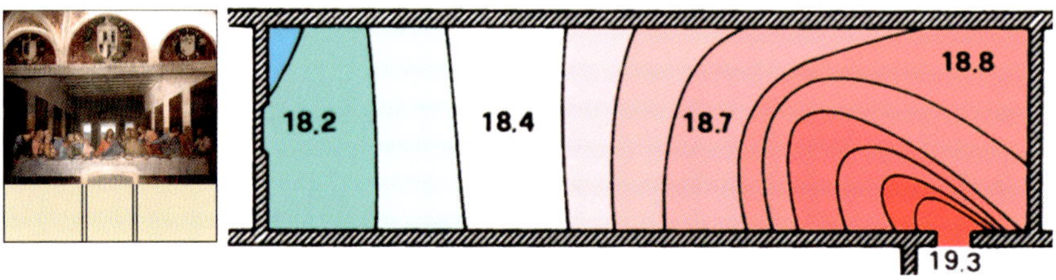

FIGURE 1.8 Warm air flowing through the entrance door and spreading in the refectory of the Leonard's *Last Supper*, Milan (11 May 1982). The mural is on the left side of the room. *Source: By kind permission of European Commission, DG XII, and* Bollettino Geofisico. *See credits.*

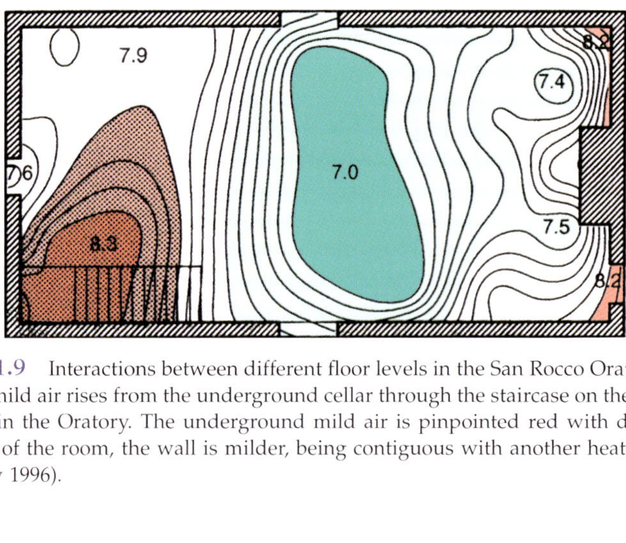

FIGURE 1.9 Interactions between different floor levels in the San Rocco Oratory, Padua. In winter, mild air rises from the underground cellar through the staircase on the left corner, spreading in the Oratory. The underground mild air is pinpointed red with dots. On the right hand of the room, the wall is milder, being contiguous with another heated building (18 January 1996).

but the cellar is less sensitive to the daily temperature cycles and attenuates the seasonal one too. In the summertime, and especially during the daytime, the cellar is much colder than the Oratory, so that the cold air remains entrapped in the cellar without any exchange with the upper floor. In the wintertime, and especially during the night and morning hours, the cellar is milder, and this generates a continuous exchange through the staircase: cold air descends and mild air rises with some entrainment and mixing with ambient air up to the lighter air reaches the ceiling and then spreads horizontally. In addition to the mild area near the staircase, another mild area is found near the opposite wall in the bottom of the room as this wall is in common with another building that is heated.

An example of interactions between an architectural structure and its surroundings was found at the Ducal Palace in Urbino, Italy (Fig. 1.10). This Renaissance building has a courtyard and the staircase of honour

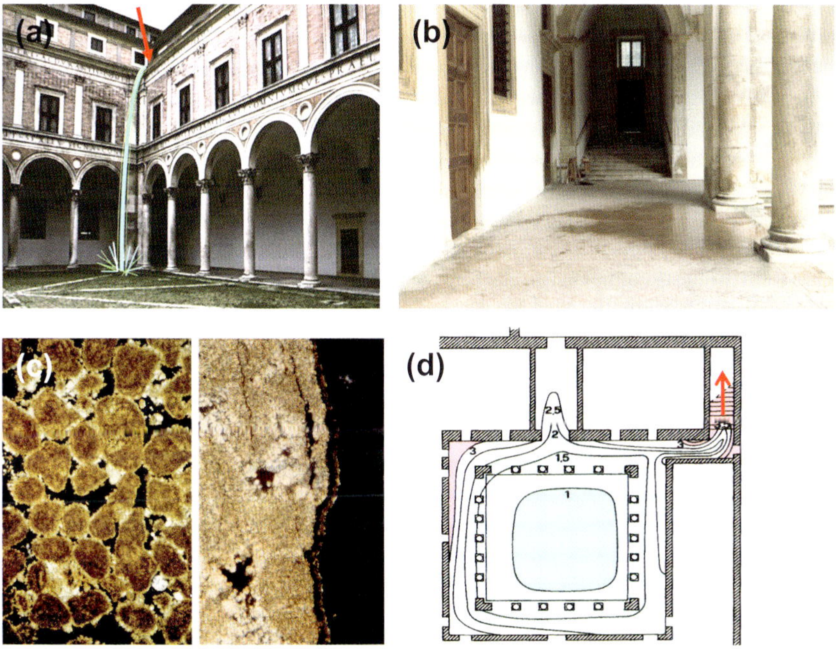

FIGURE 1.10 (a) The courtyard in the Ducal Palace, Urbino, sketched with the rainwater stream freely falling at each corner where it is released in the absence of gutters. (b) Detail of the gate to the honour staircase with deteriorated decorations on both sides. The staircase acts as a gigantic chimney, and sucks in airborne rain droplets, some of them splashing on the stone decorations. (c) Thin section of the oolitic limestone (large magnification) and thin section of the subsurface layer (small magnification). (d) Air temperature inside the courtyard and cold air entering the staircase (19 December 1982). *Source: By kind permission of Elsevier Science. See credits.*

that connects the courtyard with the upper floors. The staircase starts from a corner, below the loggia. The edge of the staircase, with magnificent bas-reliefs in a local oolitic limestone (sedimentary rock composed of spherical grains called *ooids*, i.e. eggs), is particularly damaged. Seasonal field surveys were made to establish the cause of decay. In clear days, the courtyard and the loggia were conditioned by the solar cycle. The middle of the courtyard was generally colder (Fig. 1.10(c)) due to more intense ventilation. A cold area was found near the open gate of the honour staircase. Open gate and staircase behaved as a gigantic chimney and created a violent flow of air that was sucked in from the courtyard at ground level and blown out to the top floor. This stone is not resistant to weather injuries, especially rainwater. The problem was not the transfer of heat, but the transport of droplets. All the decorations in proximity of the staircase gate were damaged for the action of airborne water droplets.

Monitoring excluded condensation, capillary rise and the hypothesis of dampness for a contamination from a back room that in the past was used to store sea salt. The cause was found in rainy days as the buildings lack drainpipes. The water collected by the roof and gutters flows down at each corner of the courtyard. When water freely falls, the water stream explodes into a myriad of droplets for the friction with the air. The airborne droplets were transported by the above flow through the staircase, and many of them splashed against the decorations. Two remedies were considered: to apply a drainpipe or a glass pane to protect decorations and a glass door to stop the airflow. The latter was preferred.

Modern buildings, made of metal and glass, have higher conductivity in comparison with traditional bricks and are more sensitive to the external climate. As the thermal capacity of walls and roof is relatively small, the envelope does not smooth out the seasonal wave. In wintertime (and during the night), the external envelope cools and generally reaches an intermediate temperature between the heated interior and the cold exterior, so that the cold ceiling and the cold walls generate a continuous internal mixing.

For example, the Sainsbury Centre for Visual Arts, Norwich (Fig. 1.11), which is built with metal and glass with some insulating panels, during a winter survey in a foggy week in December 1996, was found to have the internal air temperature $T_i = 19.5$ °C; the ceiling panes made of metal and glass and walls were between 14.2 °C and 15.7 °C; the outside temperature was $T_o = 7.2$ °C. The continuous mixing of the air masses generated by the contact with the cold structures is also artificially increased, and even in a much more enhanced extent, by the many fans distributed all along the walls to blow warm air into the room. To reach the interior of the wide room, and obtain a uniform temperature, the ventilation rate is very high. However, this violent mixing is not sufficient to produce uniform temperature (Fig. 1.11) and exchanges of heat and moisture are favoured,

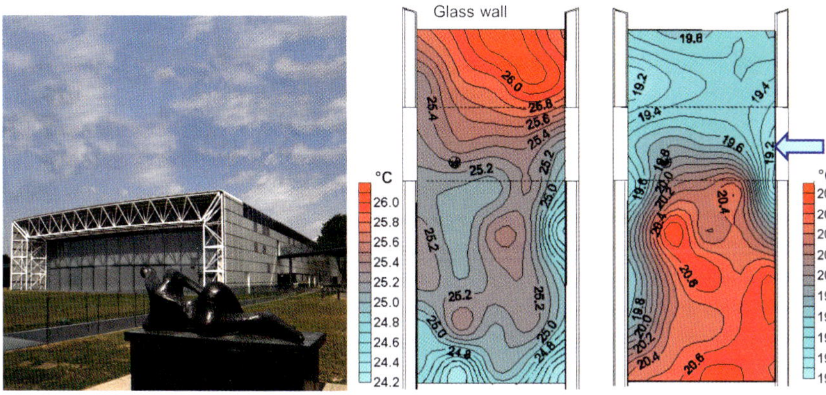

FIGURE 1.11 The Sainsbury Centre for Visual Arts, Norwich (U.K.), and the indoor temperature distribution in a horizontal cross-section in 1997 before the renovation works. Middle: summer, right: winter. The arrow indicates one of the side doors. Modern buildings, made of metal and glass, do not benefit of the inertia of the thick wall, and the microclimate is conditioned by heat exchanges and blown air mixing. However, fans are not sufficient to obtain a very homogeneous temperature distribution.

as well as the deposition of airborne particles. Possibly for this, for safety or other reasons, the main parts of objects were appropriately protected by an individual Plexiglas case. On the other hand, in the sunny days of the warm season, the ceiling became warm forming internal layering, and the glass panes generated some greenhouse effect.

1.6. TEMPERATURE IN A SHOWCASE

A showcase is an example of a box-in-box model, that is of a small controlled volume inside a room. We have seen that building envelopes protect from weather injuries, which inside buildings the walls smooth out and delay the 'primary' external heat wave, and the heat spreads in different ways within the rooms may be outside the desired thresholds. For this reason, a further natural filter is often useful to control, or to smooth out, this 'secondary' temperature change, and showcases accomplish to this goal (Fig. 1.12), in addition to the aim of protecting vulnerable exhibits from dust deposition and accidental shocks.

In this example taken from the Uffizi Gallery, the external daily wave reaches some 20 °C, and both the room and the showcase have waves with some 4 °C amplitude, but the showcase has a temperature that is smoothed out with a 2 h time lag. The exhibits in the case are exposed to a slightly smaller temperature span as the unshielded objects in the room, but are protected against rapid fluctuations or temperature (and

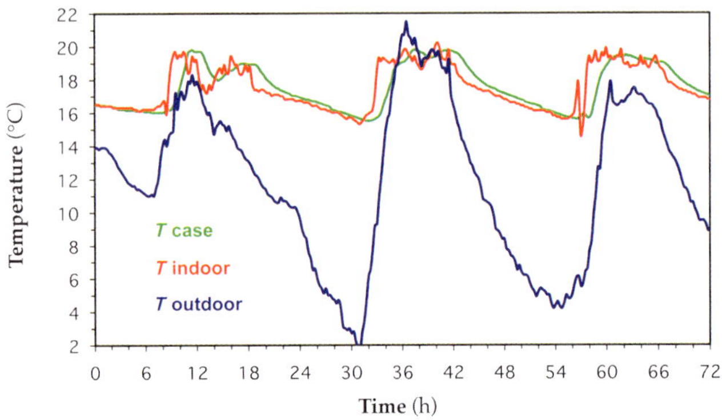

FIGURE 1.12 Comparison between the outdoor temperature (blue), the indoor value (red) and the temperature inside a showcase (green). Pollaiolo Room, Uffizi Gallery, Florence, 11–13 March 1997.

humidity) changes. The reason why the temperature range is almost the same is due to the fact that the single glass pane is conductive.

It might be useful to comment that the outside temperature was measured with a sheltered thermometer, fixed the wall at the second floor, 30 cm external to the wall. When a wall is hit by solar radiation, the surface temperature may increase by some 20–30 °C, depending on cloud cover, wind speed, plaster colour, wall orientation, level above the soil (i.e. different ventilation), hour of the day, season and so on (Fig. 1.13).

On a building, the air flowing in proximity of the warm plaster is heated by some 5–10 °C; the warm air gains buoyancy and forms a rising IBL of warm air slipping along the surface (Fig. 1.14). This air warms windowpanes and enters the room for leakage or through open windows. At the Uffizi Gallery during microclimate monitoring, our thermometer

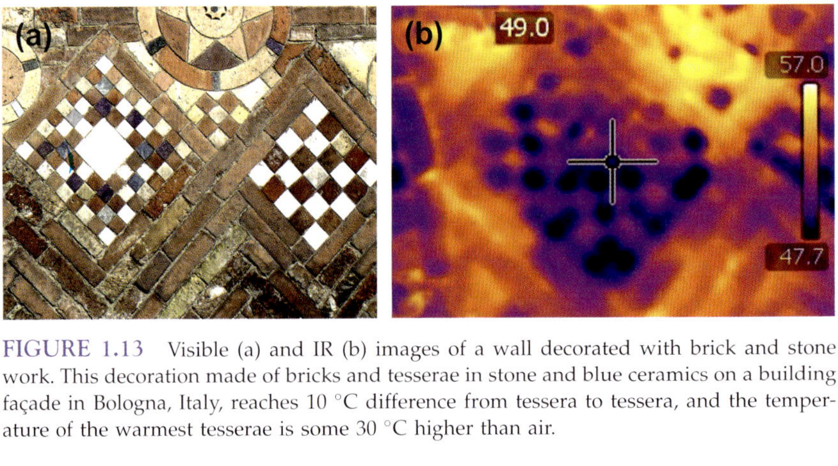

FIGURE 1.13 Visible (a) and IR (b) images of a wall decorated with brick and stone work. This decoration made of bricks and tesserae in stone and blue ceramics on a building façade in Bologna, Italy, reaches 10 °C difference from tessera to tessera, and the temperature of the warmest tesserae is some 30 °C higher than air.

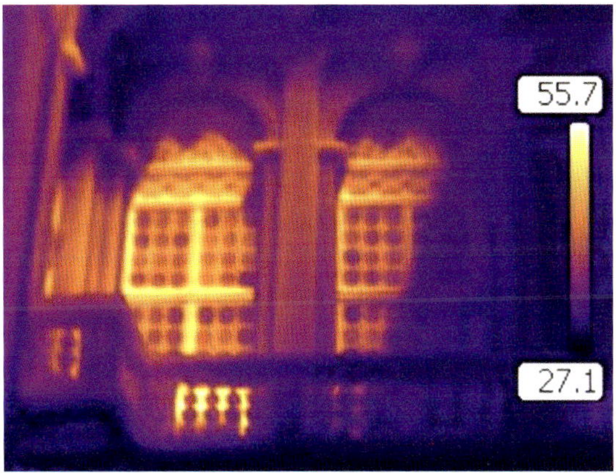

FIGURE 1.14 IR thermography of a balcony facing on a courtyard. Part of the building is in the shade, and part is reached by sunshine. The surface temperature may vary some 20–30 °C from point to point depending on the material type, colour, etc.

was immersed into this IBL and was representative of the air that entered for window leakage. The external temperature of leakage is the temperature of the IBL air, with values different from those of a standard weather station located in the free air, over countryside grass, at 2 m height. Records for standard weather station for airport, agriculture or other purposes are hardly representative of the air interacting with building envelopes.

Lighting should be obtained with lamps placed outside the showcase; on the contrary, a lamp inside will act as powerful heater, and most of the heat dissipated will remain entrapped inside the case. However, although the light source is external, when a showcase is made of a material that is not fully transparent to the IR, it becomes a greenhouse. Air and exhibits are overheated, and the relative humidity may be lowered at dangerous levels.

It is popularly claimed that glass is relatively opaque to the IR radiation and that more appropriate are polymethylmethacrylate (i.e. Plexiglas), polycarbonate, polyethylene and polypropylene. However, for all these materials, the transmittance in the IR band is neither one nor homogeneous but is generally good, except for some narrow absorption bands (Touloukian and DeWitt, 1972; Saint-Gobain, 1977; Michalski et al., 1991), whose relevance changes with the intensity of the spectral band of the IR radiation having the same specific wavelength. For this reason, referring only to absorption spectra could be ambiguous. A laboratory test is much more useful to decide the most suitable material. Some identical showcases have been built, sized 20 × 10 × 10 cm and with panes 5 mm thick, made one of glass, one of Plexiglas and one of polycarbonate, as well as others with panes having twice this thickness. In the bottom of these cases, a black sheet of paper has been placed to transform absorbed light into IR radiation and a thermometer. All these cases were lit from the outside, with a tungsten incandescent lamp, supplying 500 lux at the top of the boxes to obtain a clear effect. The panes, being partially transparent to the IR radiation, absorb part of the IR radiation incoming from the incandescent lamp as well as part of the outgoing radiation, and the heat accumulated in the panes is redistributed part inside and part outside. The result of this balance is shown in Fig. 1.15.

Plexiglas and polycarbonate practically behave in the same way, and their overheating (i.e. 2.9 °C) is only slightly less than that of glass (i.e. 3.5 °C). Making twice the pane thickness, a delay is introduced because of the initial advantage of a greater portion of the IR that is stopped (and accumulated) in the top pane and cannot penetrate inside; after some 2 h, this advantage is lost and the thicker thickness generates a more efficient greenhouse effect, as expected. After some 6 h, the overheating is 3.1 °C, that is only 0.2 °C more than the box having panes with half thickness. This experiment clearly shows that, in the case of external lighting with incandescent lamps, Plexiglas is only slightly better than glass. A further advantage for short-term lighting can be obtained increasing the thickness of the pane. This would reduce and make more gradual the internal overheating and, consequently, the drop in relative humidity. However, after a certain time interval, the situation will worsen.

It is thus necessary to avoid the IR radiation, as far as possible, and fluorescent lamps or LED might be a relatively good approach. The same experiment discussed above but with 500 lux generated with a fluorescent lamp led to no detectable overheating. However, these lamps should be controlled for harmful UV emission, but the main drawback is that their irregular spectrum confers unpleasant tune to colours. The best method

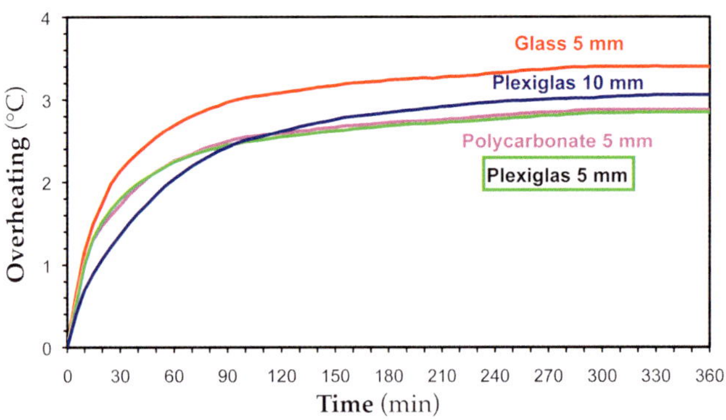

FIGURE 1.15 Overheating, as a consequence of the greenhouse effect, of showcases made of glass, Plexiglas and polycarbonate with 0.5 and 1 cm thickness, as indicated. The showcases are lighted 500 lux, with a tungsten incandescent lamp.

consists of using fibre optic lighting, which can filter noxious UV and IR radiation, as it will be discussed in Chapter 4.

Showcases, wall display, display tables and so on are useful only if they are suitably built, using materials that are inert and no off-gassing and are appropriately managed. Cases with forced ventilation do not meet the aim of smoothing out room temperature fluctuations. To be kept free of dust, the airtight cases should be built with materials that do not release and accumulate noxious substances or favour biological infection.

1.7. IS IT POSSIBLE TO COMBINE PEOPLE COMFORT, CONSERVATION NEEDS AND SUSTAINABILITY?

When the natural microclimate of a building is not comfortable for people (less frequently the problem is posed whether it is suitable for conservation too), HVAC systems are installed to obtain the desired comfort. Traditional systems are used, for example hot water radiators, fan coil convectors, radiant panels, humidifiers, following the everyday practice of keeping a temperature fluctuating around the selected level or switching on/off the system according to the activity periods with sudden peaks or drops in temperature (and, consequently, in relative humidity). All these systems are characterized by intermittent use and are located in spot areas, so that they continually generate microclimate perturbations. Fans tend to destroy thermal layering but generally worsen the situation, forcing air currents in the rooms.

The worst situation is reached in winter in buildings used only at times, as in the case of churches attended weekly for the Saturday and Sunday liturgy. The first need for conservation is a constant climate, people need a mild climate, and a constant mild microclimate seems the most obvious conclusion, but it is expensive and it is very difficult to keep *RH* unchanged. As it is not easy to combine man comfort, conservation needs and low costs, various solutions have been attempted, often disregarding conservation.

The best compromise is the so-called local heating (Fig. 1.16), that is to localize the heating zone and reduce heating at minimum levels, warming only pews with electric wires or heating foils at low temperature when people are in. This method is aimed at giving a comfortable contact with pews when churchgoers are sitting or standing, and is acceptable for heavy dressed people and for a relatively short time. The air should remain unaffected or so. The methodology has been especially studied with the EU Friendly-Heating Project (Camuffo et al., 2007, 2010). This method is not common in Italy for the elevated cost of the electric power but is becoming more popular in Northern and Central Europe.

With local heating, central heating or whatever else type of heating, when the indoor air temperature is higher than the ceiling or the wall

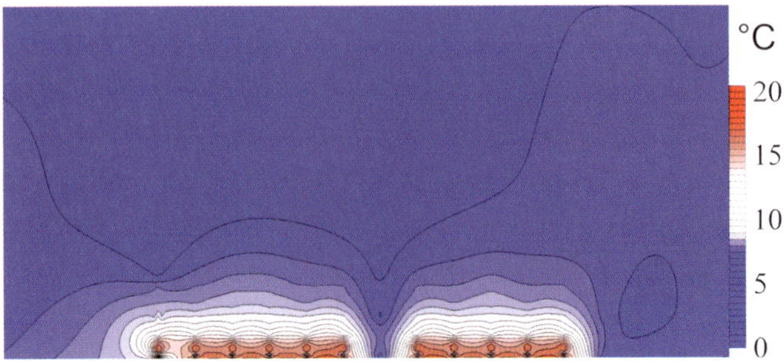

FIGURE 1.16 Local heating made with the EU Friendly-Heating Project. Church Santa Maria Maddalena in Rocca Pietore, Italian Alps.

temperature, convective motions are generated, associated with downward currents of cold air (draughts) flowing along cold surfaces, especially windows and walls. Draughts are unpleasant to people, and these internal air motions lead to increased deposition rate of candle smoke and other suspended particles.

Another popular compromise solution is to heat the floor in the pew area, just to mitigate the temperature where people stay. The floor heating system uses pipes that carry hot water, placed over an insulating layer and embedded in a conductive layer, which constitutes the floor. Heat is transferred from the pipes to the floor, and the room is heated by low-temperature IR emission from the floor. An advantage of the floor heating is that a large mild-temperature heating surface produces comfort at lower air temperature (about 2 °C), therefore reducing heating requirement (Porges, 1995). However, when the warm surface is reduced to the pew area, the comfort is diminished, and in addition, pews intercept the IR radiation, leaving people in a relatively cold environment. In any case, floor heating is characterized by low risk for damage to frescoes, paintings, statues and other church decorations, but by high risk to historic pews, which undergo enhanced temperature and humidity changes.

When such a system is designed for daily use in business buildings, the highest efficiency is obtained with the combined effect of IR radiation and air heating via floor conductivity that generates ceaseless air convective motions. However, in the case of weekly use in churches, when walls and ceiling are much colder than the air temperature, the air that comes into contact with the cold surfaces sinks, forming down droughts of cold air, which are harmful to both people and conservation. To avoid dangerous convective motions, the floor should ideally radiate without transferring heat via conductivity to the air or limited to a minimum extent. For this reason, materials with high IR emissivity and low surface conductivity should be used. Several materials are good emitters, for example dolomite that reaches 96% emissivity; granite 93%; brick, 93%; oak 90%, and have a different conductivity, for example the above materials have, respectively,

1.5, 2.9, 1.4, 0.16 W m^{-1} K^{-1}, so that granite is much more conductive than wood, which is quite an insulator. In most cases, inappropriate materials are used for the floor, having a too low conductivity, which reduces (or vanishes) the system efficiency, or having too high conductivity and generating enhanced convective motions or being poor emitters. For example, the medieval church of Colle S. Lucia in the Italian Alps, 1400 m above sea level, has footboard heating. The top of the footboard is wooden, which makes difficult heat transmission. During a field survey in January, with the feeding water in the pipe at 30 °C, the footboard top was at some 15 °C and the air at 1 m was between 8 and 9 °C. When the water temperature in the boiler was passed from 30 to 60 °C, the footboard temperature rose to 23 °C, but the air temperature was almost unaffected, that is only 1 °C or so (Fig. 1.17).

Systems based on floor heating need several hours to warm up the surface layer and reach the regime efficiency. This requires almost 1 day before the use, determining dangerous environmental cycles and reducing the spare of money. In practice, to abate costs and shorten heating times, these systems are often integrated with other faster systems, for example with inflows of warm air, which have the negative consequence of increasing the amplitude of environmental cycles.

If costs were neglected and the floor heating system were continually kept into operation for the whole season, in theory, the whole room, walls and ceiling would reach equilibrium and the problem would be easily solved, at least for the wellbeing of people, but such a practice is not completely safe for conservation. An example of that was found in the refectory of the Leonard's *Last Supper*, Milan (Camuffo and Bernardi, 1991a), with a terracotta floor heating. The air, heated from below, was continually mixed by convective motions that caused heavy deposition of suspended particles and blackening of the painting. The blackening occurred at such a fast rate that the

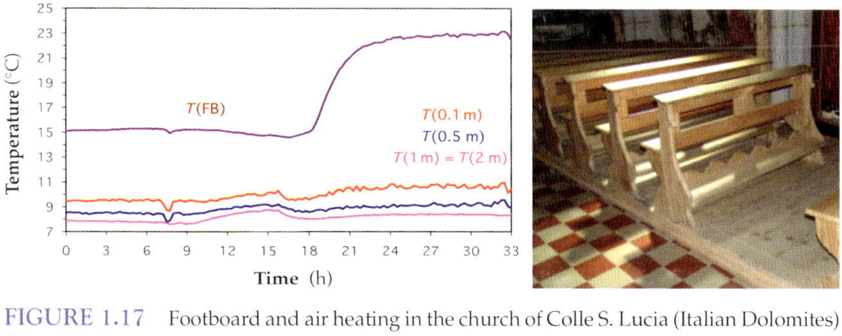

FIGURE 1.17 Footboard and air heating in the church of Colle S. Lucia (Italian Dolomites) from 00.00 h, 16 January 1997, to 09.00 h, the next day. Footboard temperature T(FB) and air temperature at 0.1, 0.5 and 1 m above the floor level; the measurement at 2 m shows the same temperature as found at 1 m. The sharp rise in footboard temperature at 18:00 h is due to a change in the boiler temperature from 30 to 60 °C.

restoration works were interrupted for a certain time waiting for an improvement of the ambient conditions. A number of mitigation measures were undertaken, for example to lower the floor temperature and insulate the ceiling to reduce the sources of air mixing and to forbid the local traffic, which was an important source of soot.

Returning to the heating system, although the floor temperature was kept as low and homogeneous as possible, the temperature distribution inside the room was often characterized by strong gradients determined by the penetration of warm air from the nearby room, where the heating was planned at comfort level for custodians and visitors (as already seen in Fig. 1.8). In general, the floor heating, generates air mixing with the consequence of transporting and depositing suspended particulate matter on indoor surfaces. For this reason, it is not recommended in the case of historic buildings with murals or frescoes, or should be kept operating at a very low temperature levels. To avoid soiling, exhibits may appropriately protected in showcases. However, although it is not the best choice, in certain cases, it might be accepted, with a very soft and carefully controlled use.

A popular system, which is preferred for its low cost, is the warm air heating. Warm air is blown inside before the people entrance, and the operation is stopped when people go off. The blown air partially mixes with the indoor air, generating a wide range of air temperatures: the buoyancy of warm air favours upward flows and thermal layering below the ceiling; on the other hand, the denser cold air originally present in the room remains confined in the lower part above the floor; finally, turbulence and mixing characterize the intermediate levels. Especially in the upper part of the room, where air temperature is higher, and where air movements are faster, warm air systems may cause severe soiling, even worse than floor heating. For instance, in a small mediaeval church in the Alps at 1140 m above sea level, temperature was 20 °C at the height of the altarpiece, paintings and statues (Fig. 1.18(a)) and only a few degrees at the lower congregation level. When warm air heating is operated, temperature is characterized by sharp heating peaks with strong atmospheric layering, followed by relaxation cooling when the system is off. Churchgoers have little advantage (some 4 °C heating) while the upper part of the church is overheated. On the other hand, the *RH* shows a complex behaviour (Fig. 1.18(b)). When heating is on, the warm air buoyant aloft causes a sharp drop in *RH*. The cold ceiling and the upper part of the walls start to evaporate, and the moisture is soon dispersed and mixed for the droughts and the high diffusivity of moisture in air. In the lower part of the church where heating is very little, the additional moisture causes an increase in *RH*. At the middle level, at the beginning, the *RH* drops following the increase in temperature, but soon, it changes trends and increases when the effect of the increasing moisture content

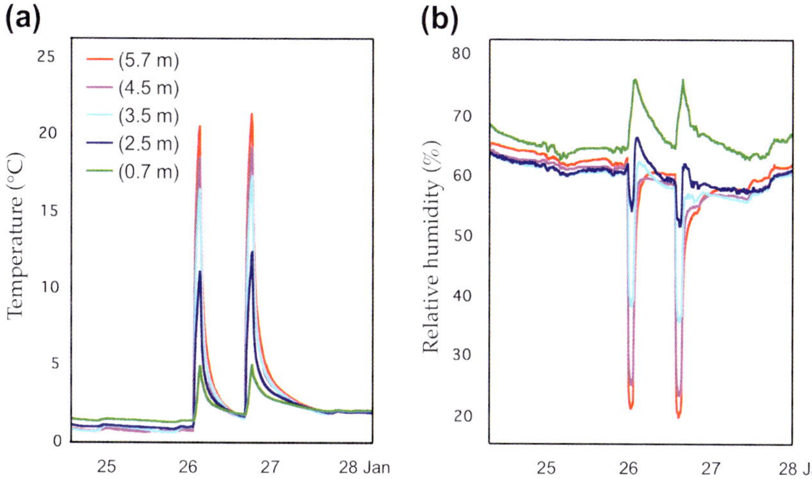

FIGURE 1.18 Two episodes of occasional warm air heating in the church of Rocca Pietore, Italian Dolomites. The indoor air temperature and *RH* were measured at different levels, see legend for the colour codes. (a) Sharp peaks in heating temperature showing indoor thermal layering, followed by thermal relaxation when heating is switched off. (b) Relative humidity perturbation caused by the above temperature peaks. When heating, the *RH* drops in the upper part of the church that is above 3 m; it has a mixed mode characterized by a starting drop and final increase at 2.5 m; it increases at the low level.

becomes dominant. This complex behaviour is amplified when churchgoers are inside, releasing moisture for breathing and transpiration. However, this behaviour has been observed even when the heating system is operated in the absence of people (Camuffo et al., 2010).

The sharp peaks in temperature and drops in relative humidity cause internal stress to canvas paintings and wooden statues, which will undergo noxious thermal expansion and moisture shrinkage. The problem becomes particularly complex because objects aloft suffer for hot and dryness and objects near the floor for dampness not compensated by the poor heating. The severity of the problem increases with the height above the floor, but air humidification is not an acceptable remedy to dryness aloft because it would cause condensation in the bottom and on all cold surfaces.

In the case of objects composed by a number of different parts and materials, changes in temperature may cause unpredictable effects for the different response of the various parts. An example is shown for a pipe organ (Fig. 1.19) where the metal pipes, the wooden structure and the walls are characterized by different heating rates. Taking into account that the moisture content too is variable, the result is that some of these materials will approach to and others will depart even more from the Dew Point (see Chapter 2A) with the consequence that the former will absorb moisture from the air and the latter will loose some

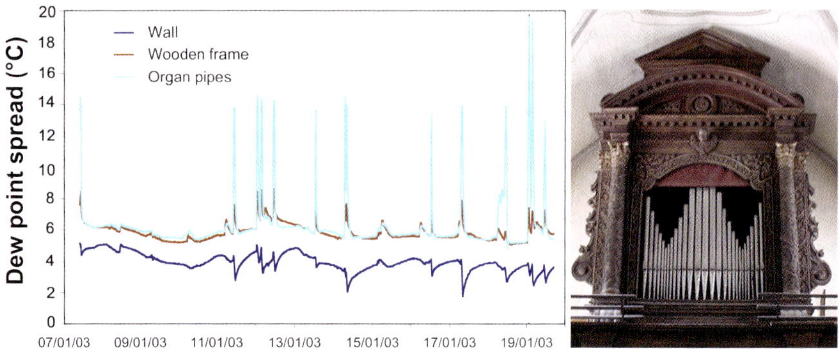

FIGURE 1.19 Effects of warm air heating on a pipe organ: difference between the observed temperature of the wall surface (blue), the wooden frame of the organ (brown), the metal pipes (cyan) and the dew point. This difference is called dew point spread. When the spread vanishes, or becomes negative, condensation occurs.

moisture content. In this case, the wall and especially dissolved salts will absorb moisture; the organ pipes will loose some monolayers of water absorbed on the surface and the wooden frame will loose some bound water and shrink.

In warm air heating, blown warm air generates turbulence with the result of increasing the deposition rate of airborne particles and candle smoke that will blacken paintings and frescoes on cold walls. This method is not the best for conservation purposes.

A mitigation measure could be to mix vertically air with fans to counteract the thermal layering. This would reduce the dramatic temperature rise aloft with the associate humidity drop, but would increase the deposition rate of particles on the surfaces. To reduce the room turbulence, the fan can be inserted into a vertical tube, with the lower opening near the floor and the upper near the ceiling. This mounting will generate turbulence only near the two ends of the tube, that is in the zones of convergence and divergence of the forced flow.

A particularly risky system, which was especially widespread years ago, is based on mantle stoves, which burn methane (CH_4), liquefied petroleum gas (LPG) or other liquid fuel. The chemical reaction of the liquid gas with the atmospheric oxygen generates carbon dioxide and water vapour, for example $CH_4 + 2O_2 = CO_2 + 2H_2O$. The most negative aspects of this system are the risk of fire and explosion, the abundant production of moisture and other pollutants due to bad combustion, for example CO, various hydrocarbon chains or catalytic reaction between O_2 and N_2 in air to generate NO_X. Only a small fraction of this vapour is visible when it condenses on the cold windows forming rivulets of water; the main part is absorbed by walls and decorations, oxidizes metals, weakens stuccoes, swells wood, fades tapestry and favours microbial colonization.

From the above examples, it is evident that it is not easy to combine human comfort, conservation needs and low costs and that a compromise is required, where the conservation needs should dominate in proportion with the cultural heritage value and the building use. For instance, it could be said that the choice of the microclimate in a museum, which should be aimed at conservation, and where many important items are concentrated, should be more rigorous than in a church that is more oriented to people use.

It is also clear that every system presents a number of negative aspects, some of which may have a major impact in certain circumstances or minor one in others. Therefore, it is important to carry out a careful analysis of pros and cons and choose, time by time, the system that causes the minimum damage. Sometimes, a combination of different systems might be considered to avoid the excessive impact of specific adverse factors; for example in particularly severe climates, the combination of soft radiation from the floor or other external emitters with an electrical pew heating might reach an acceptable comfort with a modest ambient perturbation and might be better than the more common solution of a radiant floor heating associated with warm air inflows.

As a general comment, it might be preferable to reduce interventions to a minimum level, to reduce negative impact to a minimum sustainable level. It can be argued that many objects have survived till today just because the modern heating was not yet invented and now it constitutes a new challenge. This is true in several cases; nevertheless, it is also possible to use modern technology to improve natural negative situations, as for example when a room temperature is below the dew point and condensation forms everywhere. The conclusion is that heating systems should be installed by experienced engineers, but under the strict control of experts in conservation science. The use and operation of these thermotechnical devices should be managed under the strict control of experts in conservation. The real problem is that conservation needs are many, complex and often disregarded.

Note: Readers interested to this topic will find much more in the subsequent developments of this research and should consult:

Camuffo et al., 2010: An advanced church heating system favourable to artworks: a contribution to European standardisation. *Journal of Cultural Heritage* 11, 205–219.

Camuffo et al., 2007: *Church heating and preservation of the cultural heritage: a practical guide to the pros and cons of various heating systems* Electa, Milan.

EN 15757: 2010, Conservation of Cultural Property – Specifications for temperature and relative humidity to limit climate-induced mechanical damage in organic hygroscopic materials. European Committee for Standardization, CEN TC346, Brussels.

EN 15759-1: 2011. Conservation of Cultural Property – Guidelines for the Climate Control of Churches, Chapels and other Worship Places. Part 1: Heating. European Committee for Standardization, CEN TC346, Brussels.

1.8. MONITORING AIR TEMPERATURE TO STUDY AIR–SURFACE INTERACTIONS AND FOR ENVIRONMENTAL DIAGNOSTICS

The use of a thermo-hygrograph sited in a corner of a room is very common, but this instrument furnishes data representative only of the point where the sensor is located, not of the whole room, which is generally characterized by spatial gradients continually variable, with fast or slow rates. When the heating or cooling systems are turned on/off, they provoke an abrupt change of temperature, which in general reach few degrees Celsius (Fig. 1.20).

Frequent and dangerous sharp changes, when they are relatively modest, pass nearly unobserved in the thermograph strip chart records, whose resolution is generally ±0.5 or ±1 °C, in that they are smoothed out

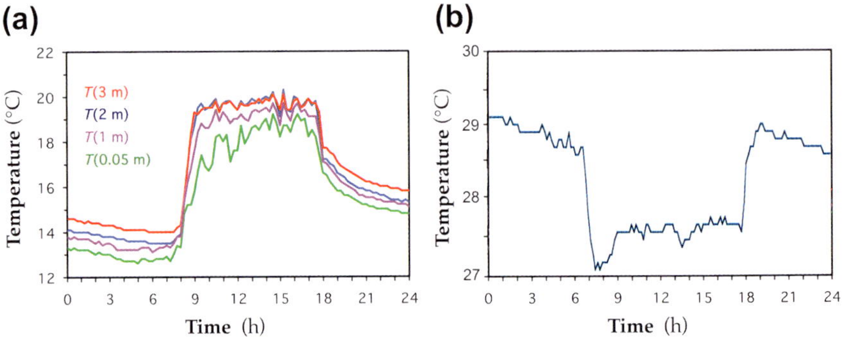

FIGURE 1.20 (a) Sharp increase and drop in temperature in winter (18 February 1997) when the heating system was turned on in the morning and off in the evening. Measurements, respectively, taken at 0.05 m (green), 1 m (pink), 2 m (blue) and 3 m (red) above the floor. During the daytime, the air above 2 m is well mixed and the temperature is uniformly distributed between 2 and 3 m. (b) In summer (11 August 1992), similar sharp changes are generated by the air-conditioning system, but the air is well mixed and the curves at the different heights are practically indistinguishable. For this reason, only one has been reported here. Pollaiolo and Giotto rooms, Uffizi Gallery, Florence. After the measurements, the HVAC was better tuned.

by the friction of the pen on the strip chart or the loose mechanical coupling. Electronic records generally are accurate and not affected by mechanical friction and are convenient to diagnostic and control aims. Mechanical strip chart recorders belong to the past, although some thermohygrometers still survive for a long tradition in some museums. Strip paper and ink pen recorders are dated, with low sensitivity and can only monitor important changes, as for example those induced by warm air heaters, which in the extreme case of churches heated once a week, generate impressive temperature rises, for example 20 °C in 1 or 2 h.

Measurements of the air temperature and humidity distribution, taken in a series of points in proximity of surfaces or in horizontal cross-sections of rooms, can be used as a diagnostic tool to individuate air–wall interactions, causes of microclimate perturbation, space gradients, moisture and heat exchanges, the path of the air masses and some deposition mechanisms.

Taking readings made on a regular grid on horizontal or vertical cross-sections of a room and mapping the results, one obtains an instantaneous view of how heat or moisture is distributed inside. If such monitoring is repeated at regular time intervals, for example every 2 h, it is also possible to follow the temporal evolution of the above variables. It is impossible to disseminate a large number of sensors in a room, each having a wire connected with the data logger for two reasons. The first is that connecting wires will form an entangled spider's web in which people cannot move, but the problem is solved with wireless sensors connected via radio to a receiver. The second problem is that not all the sensors will exactly have the same response or calibration, and the instrumental departures will be misleadingly interpreted as microclimate anomalies.

It is preferable to sample temperature and *RH* moving the same fast response instrument along a chosen regular grid with points close to the walls as well as inside the room. Repeating these runs, also the time evolution of the temperature distribution is obtained. The method is very effective and has been originally devised for the evaluation of indoor environments (Camuffo, 1983) to find risk areas, causes and effects and has been applied in many circumstances either indoors or outdoors (Camuffo and Schenal, 1982; Bernardi et al., 1985, Bernardi and Camuffo, 1995a,b; Camuffo, 1986, 1991, 1994; Camuffo and Bernardi, 1988, 1991a,b; 1993, 1995a,b, 1996, 1997).

It is essential to avoid intercalibration errors, and for that reason, it is preferable to use the same fast response instrument, that is a precision electronic psychrometer (Chapter 12). The spacing between measuring points and hence the number of observations are conditioned by the size of the room and the rate of ambient change. It is essential that in the time employed for the observations, the ambient conditions do not change too much. A species of *uncertainty principle* holds: the greater the spatial resolution, the larger the number of observation, the longer the time elapsed and the less representative the whole monitoring. In several cases, some 40 observations have represented a reasonable compromise solution. The series of measurements can be repeated every 2 or 3 h during the daytime. During the night, the whole ambient system relaxes without violent perturbing factors, and all the gradients tend to flatten or disappear. The intervals between consecutive measurements can be therefore increased. Ideally, the first observation should be made near sunrise to monitor the less perturbed situation; then, the next observation should be made during the cleaning time when the windows are open; then, 2 or 3 h later and so on, to arrive at the end of the visiting time; after the museum closure, a few observations are needed. In practice, security reasons limit the possibility of doing measurement outside the museum business time; fortunately, the most important situations are found during the business time.

All the cross-sections must be monitored exactly following the same order for two reasons: (1) first of all because if there is a time-lag effect in the monitoring, it repeats identically in all the distributions, and for each point, the interval between two successive readings is exactly the same; and (2) then because it is less easy to do errors, and the data must be filed with the same order in the computer.

It is also important to repeat for some days the measurements for two reasons: the former is to see whether the results are repetitive and really representative of the seasonal situation; the latter is that small variations can be found, and their repetition helps to distinguish regular trends from casual fluctuations, that is small repetitive variations are not casual.

The grid of the measuring points in the horizontal plane is composed of two interconnected series of data: a first series of points at the same height, measured at regular intervals close to the walls, that is at 2 mm and at 20 cm from the walls, to detect the existing gradients and the exchanges of heat and vapour; the second series (always at the same height) is aimed to show the temperature and humidity distribution inside the room and is composed of other points, distributed following a regular grid, the points being the vertex of rectangles with the same dimension. All readings were made exactly at the same level, although the choice of the level is not so important. The height of 1 or 1.5 m from the floor is generally preferred for being comfortable for the operator and being representative for the most part of the exhibits or for the thermal comfort of visitors.

After the observed values of temperature (or humidity) are reported in the cross-section, it is possible to use computer graphics or manual interpolation to show the space distribution of the parameter, the anomalies, the intensity and the shape of gradients. Close isolines show gradients. Closed isolines spreading like water waves will enclose a source of heat if the temperature is radially decreasing; they will enclose a sink of heat if temperature is increasing. Alternatively, they can show the zone of divergence (or convergence) of air when there is a three-dimensional convective motion. In a room, isolines shaped like a tongue that begins from a door or a window show the penetration of external air masses and their path inside. Isolines parallel to a wall surface show that the wall is adsorbing heat if the temperature decreases approaching the wall or vice versa. In the same way, a gradient of temperature is an index of transport of heat, and a gradient of moisture content is an index of transport of vapour and condensation (if the minimum is at the air–surface interface) or evaporation from the surface (the maximum at the interface). Gradients of air temperature are also gradients of air density. A thermal minimum along a wall means that the air in contact with it is denser, looses buoyancy and will sink. A thermal maximum means that air is lighter, gains buoyancy and will rise. Therefore, two-dimensional horizontal gradients of temperature may also help to interpret the three-dimensional dynamic state of the air inside a room, the transport and (partially) the deposition of pollutants.

An example is shown in Fig. 1.21 reports two horizontal temperature distributions in the Sistine Chapel, Rome (Camuffo and Bernardi, 1986, 1995a). The former is in the early morning when the Chapel is open for

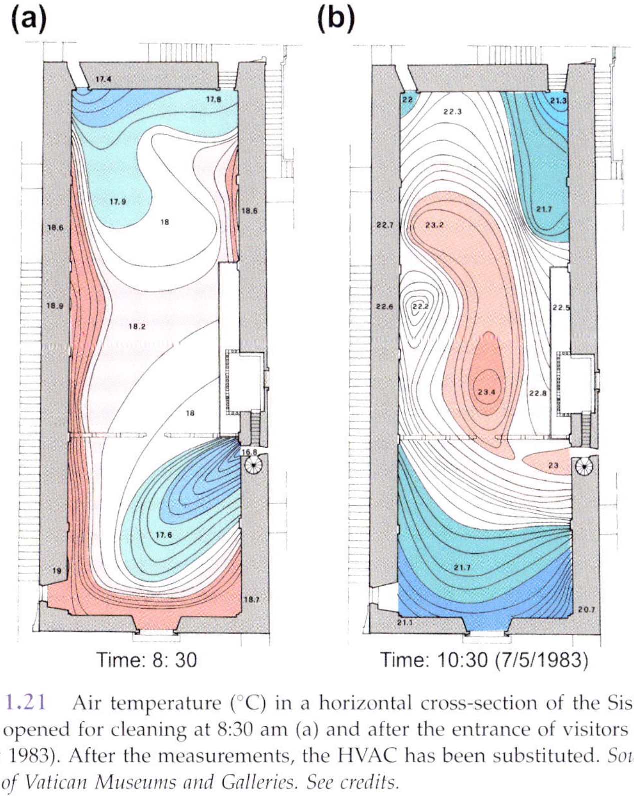

FIGURE 1.21 Air temperature (°C) in a horizontal cross-section of the Sistine Chapel when it is opened for cleaning at 8:30 am (a) and after the entrance of visitors at 10:30 am (b) (7 May 1983). After the measurements, the HVAC has been substituted. *Source: By kind permission of Vatican Museums and Galleries. See credits.*

cleaning. The latter is a few hours after the entrance of visitors. In the first map, the inside atmosphere was originally in equilibrium with the thick walls, but the opening of three doors (two on the top and one on the right) causes the entrance of cold air: chiefly from the door on the right. The air in the middle of the room, far from the doors, is warmer. However, the air close to the walls is even warmer because of the heat released from the walls. This air gains buoyancy and forms an upward current along the walls. This airflow tends to increase the deposition rate of airborne pollutants via inertial impaction (see Chapter 8), although the positive gradient of temperature tends to partially counteract it. In the second map, the situation has changed because of the external atmospheric warming, the presence of visitors and the heat released by lamps: the air is now warmer than walls, and a heat island is formed in the central part of the room, where visitors stay longer. Temperature gradients near walls are negative, and heat is now flowing from air to walls. The air in proximity of walls becomes denser and sinks forming downward currents. The deposition rate is now increased as the negative temperature gradient generates a thermophoretic transport that increases blackening in synergism with the inertial impaction.

It is clear that the two-dimensional temperature plots should be considered under a three-dimensional point of view, including the air movements generated by the temperature differences: where air is warmer, it gains buoyancy and rises; where it is colder, it has greater density and falls. The result is that inside the room, a three-dimensional circulation of air will be established, and the motor is the uneven temperature pattern.

The same method can be applied to investigate the vertical distribution of the air temperature. An example with three different situations found in summer in the Giotto Chapel, Padua, is shown in Fig. 1.22 (Camuffo and Schenal, 1982; Camuffo, 1983).

In the early morning, when the door is open for cleaning, cold external air enters through the door and fills the lower part of the room up to the height of the door. Above this cold layer, the warm (and less dense) air that was before inside the building remains entrapped there, opposing that new air enters and fills the volume above the height of the door. The isolines are not horizontal, being tilted by the dynamic effect of the external air mass that is entering the room transporting momentum and inducing oscillations in the cold layer. At mid-morning, external air enters through the door and goes out through some windows in the apse. This air has temperature (and density) that is midway between the cold air tongue that extends from the northern apse to the middle nave and the warm air layer that is entrapped below the ceiling. After noon, the solar radiation through the three-mullioned window on the façade hits the floor in the front part of the nave, and also heat is released from the warm

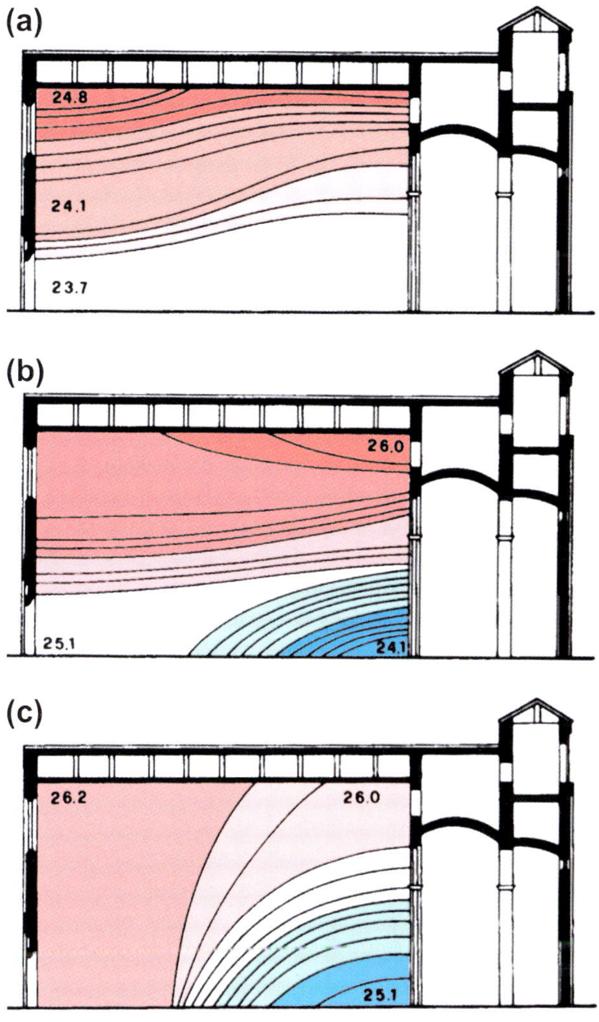

FIGURE 1.22 Vertical cross-sections of the Giotto Chapel in the summertime. (a) In the early morning (8 am, 9 July 1977), cold air penetrates through the door, and above the cold layer, the warm air remains entrapped. (b) At mid-morning (11.30 am, 8 July 1977), external air penetrates through the door and has a density intermediate between the cold tongue near the floor on the apse and the warm air entrapped aloft. (c) In the early afternoon (14.25 pm, 9 July 1977), the warm floor generates mixing and isotherm in the front part of the nave; after a transition zone in the middle of the Chapel, the air in the apse remains unaffected and thermally stratified. *By kind permission of Ministero dei Beni Culturali ed Ambientali. See credits.*

door. The front part of the nave has the same temperature along the vertical, showing a continuous mixing of the air in the homogeneous region. However, the part of the Chapel near the apse remains unaffected by this perturbation, and there, the air is thermally stratified (horizontal isotherms), showing a transition zone (curved isotherms) in the middle of the Chapel. The more vertical the isolines, the more intense the mixing.

Sometimes, the architectonic features of the rooms are more complicated or the situation is apparently more complex. An interesting example (Fig. 1.23) is found in the *Cour Marly*, Louvre Museum, Paris, which was obtained by closing with a glass roof a courtyard and the floor is divided into three levels connected with stairs. The top level (on the left) is C shaped and embraces part of the middle level floor; this is Y shaped and surrounds the rectangular bottom level (on the right). Summer air-conditioning is accomplished with violent flows of air blown from long linear slits sited at the edges of the floor, and suction is made with other horizontal slits on walls. The high speed of the airflow facilitates the mixing with the ambient air up to a distance of a few metres from the slits. In addition to the cold air blown from the floor slits, cold air conditioned from some exhibition rooms that face the top level enters the *Cour* through open windows and doors. The cold air tends to descend to the lower levels of the *Cour* flowing along the staircases and jumping over the parapets of the upper floors and is sucked out by the intake slits.

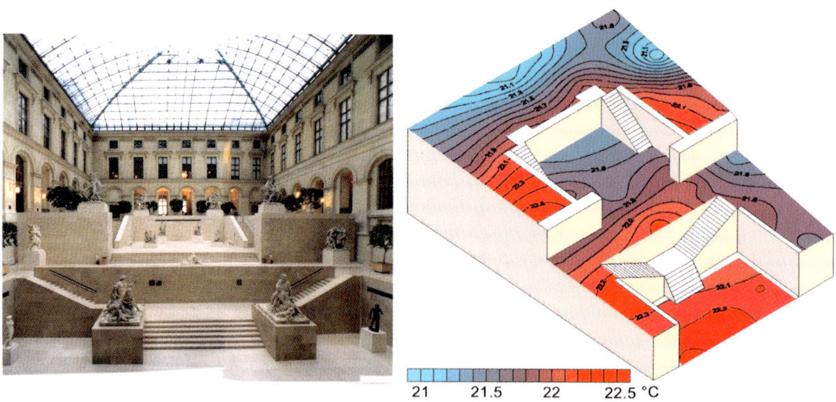

FIGURE 1.23 Air temperature (°C) measured above each of the three levels of the *Cour Marly*, Louvre Museum, 11 August 1993. Warm air enters the bottom floor level through the four doors that connect to the entrance hall of the Museum. The warm air rises immediately forming a convective motion and descent of colder air on the opposite side. Cold air penetrates also through windows and doors of the exhibition rooms faced to the top floor level in the opposite side.

At the bottom level of the *Cour*, four doors allow free air exchanges between the *Cour* and the corridor to the entrance hall. When the corridor is colder, cold air enters the *Cour Marly* and forms a cold lake of stable air in the lowest part of the *Cour*, which is in front of the four doors. When the corridor is warmer (as in this example), warm air enters, gains buoyancy and rises immediately, forming an upward flow. This warm current ascends until it is stopped at a certain height by warmer air layer stratified below the warm glass roof or by the roof itself. Whatever is the level aloft where the vertical motion stops, at that level, the airflow changes direction and diverges without loosing momentum: it becomes horizontal, crosses the *Cour* and other colder air sinks, closing the cell.

The presence of warmer and buoyant air, entering the bottom layer from the corridors, and colder air, that is denser and heavier, on the top for the cold ceiling, determines an unstable situation. For that reason, the air is continually moving to mix and reach an equilibrium that is impossible to attain until the forcing factors (i.e. warm inflow through the doors and the cold ceiling) persist.

In addition, in summer, when the sun is high over the horizon and solar beams reach the edge of the three floors, warm spots were found, which generate secondary local upward cells.

References

Theory and General Applications

Arrhenius, S., 1889. Über die reaktionsgeschwindigkeit bei der inversion von rohrzucker durch säuren. (on the reaction velocity of the inversion of cane sugar by acids.) für physikalische chemie. Stöchiometrie und Verwandtschaftslehre 4, 226–248.

Maunder, W.J., 1994. Dictionary of Global Climate Change. UCL Press, London, 257 pp.

Michalski, L., Eckersdorf, K., McGee, J., 1991. Temperature Measurement. Wiley, New York, 514 pp.

Porges, F., 1995. HVAC Engineer's Handbook. Butterworth Heinemann, Oxford, 278 pp.

Rosenhow, W.M., Hartnett, J.P., Ganic', E.N., 1985. Handbook of Heat Transfer Applications. Mc Graw-Hill, New York.

Saint-Gobain, 1977. Manuale Tecnico del Vetro. Fabbrica Pisana, Milano, 331 pp.

Touloukian, Y.S., DeWitt, D.P., 1972. Thermal Radiative Properties of Nonmetallic Solids. In: Thermophysical Properties of Matter, vol. 8. IFI/Plenum, New York.

van der Vegt, A.K., 2002. From Polymers to Plastics. DUP Blue Print. Delft University Press, Delft, The Netherland, 237 pp.

Applications to Conservation

Benoist, L., 1960. Musées et Muséologie. Presses Universitaires de France, Paris, 128 pp.

Bernardi, A., Camuffo, D., Del Monte, M., Sabbioni, C., 1985. Microclimate and weathering of an historical building: the Ducal Palace in Urbino. Science Total Environment 46, 243–260.

Bernardi, A., Camuffo, D., 1995a. Uffizi gallery in Florence: a comparison between two different air conditioning systems. Science and Technology for Cultural Heritage 4 (2), 11–22.

Bernardi, A., Camuffo, D., 1995b. Microclimate in the Chiericati palace municipal museum, Vicenza. Museum Management and Curatorship 14, 5–18.

Camuffo, D., 1981. Hot-horse anemometry. Atmospheric Environment 15, 1767.

Camuffo, D., 1983. Indoor dynamic climatology: investigations on the interactions between walls and indoor environment. Atmospheric Environment 17, 1803–1809.

Camuffo, D., 1986. Deterioration processes of historical buildings. In: Schneider, T. (Ed.), Acidification and Its Policy Implications. Elsevier, Amsterdam, pp. 189–221.

Camuffo, D., 1991. Environment and microclimate. In: Baer, N., Sabbioni, C., Sors, A. (Eds.), Science Technology and European Cultural Heritage. Butterworth, Oxford, pp. 37–50.

Camuffo, D., 1994. Effects of air pollution on historic buildings and monuments. Scientific basis for conservation: case studies in the deterioration of stone monuments in Italy. European Cultural Heritage Newsletter on Research 8 (1), 7–15.

Camuffo, D., Bernardi, A., 1986. Dinamica del microclima e scambi termoigrometrici tra pareti e atmosfera interna nella Cappella Sistina. Bollettino dei Monumenti, Musei e Gallerie Pontificie 6, 211–257.

Camuffo, D., Bernardi, A., 1988. Microclimate and interactions between atmosphere and the Orvieto Cathedral. Science Total Environment 64, 1–10.

Camuffo, D., Bernardi, A., 1991a. The microclimate of Leonardo's "Last Supper", joint edition European Cultural Heritage Newsletter on Research, and Bollettino Geofisico, special issue, 14 (3) 1–123.

Camuffo, D., Bernardi, A., 1991b. Indoor and outdoor microclimate: the Trajan Column and Sistine Chapel. In: Baer, N., Sabbioni, C., Sors, A. (Eds.), Science Technology and European Cultural Heritage. Butterworth, Oxford, pp. 295–305.

Camuffo, D., Bernardi, A., 1993. Microclimatic factors affecting the Trajan Column. Science Total Environment 128, 227–255.

Camuffo, D., Bernardi, A., 1995a. The microclimate of the Sistine Chapel, joint edition European Cultural Heritage Newsletter on Research, 9, 7–32 and Bollettino Geofisico, 18 (2) 7–32.

Camuffo, D., Bernardi, A., 1995b. Study of the microclimate of the giant hall of the Da Carrara's Royal Palace, Padova. Studies in Conservation 40, 237–249.

Camuffo, D., Bernardi, A., 1996. Deposition of urban pollution on the Ara Pacis, Rome. Science Total Environment 189/190, 235–245.

Camuffo, D., Bernardi, A., 1997. Controlling the microclimate and the particulate matter inside the historic anatomic theatre, Padova. Museum Management and Curatorship 15, 285–298.

Camuffo, D., Schenal, P., 1982. Microclima all'interno della Cappella degli Scrovegni: scambi termodinamici tra gli affreschi e l'ambiente. In: Ministero dei Beni Culturali ed Ambientali: Giotto a Padova, Special Issue of Bollettino d'Arte. Poligrafico dello Stato, Rome, pp. 107–209.

Camuffo, D., Sturaro, G., Valentino, A., 1999. Urban Climatology Applied to the Deterioration of the Pisa Leaning Tower, Italy. Theoretical and Applied Climatology 63, 223–231.

Camuffo, D., Vincenzi, S., Pilan, L., 1984. A first-order analysis of the heat wave in the soil. Water, Air and Soil Pollution 23, 441–454.

Camuffo, D., Pagan, E., Rissanen, S., Bratasz, Ł., Kozłowski, R., Camuffo, M., della Valle, A., 2010. An advanced church heating system favourable to artworks: a contribution to European standardisation. Journal of Cultural Heritage 11, 205–219. http://dx.doi.org/10.1016/j.culturher.2009.02.008.

Camuffo, D., Pagan, E., Schellen, H., Limpens-Neilen, D., Kozlowski, R., Bratasz, L., Rissanen, S., Van Grieken, R., Spolnik, Z., Bencs, L., Zajaczkowska-Kloda, J., Kloda, P., Kozarzewski, M., Mons. Santi, G., Chmielewski, K., Jütte, T., Haugen, A., Olstad, T., Mohanu, D., Skingley, B., Sáiz-Jiménez, C., Bergsten, C.J., della Valle, A., Don Russo, S.,

Bon Valsassina, C., Accardo, G., Cacace, C., Giani, E., Giovagnoli, A., Nugari, M.P., Pandolci, A.M., Rinaldi, R., Acidini, C., Danti, C., Aldrovandi, A., Boddi, R., Fassina, V., Dal Prà, L., Raffaelli, F., Bertoncello, R., Romagnoni, P., Camuffo, M., Troi, A., 2007. Church Heating and Preservation of the Cultural Heritage: A Practical Guide to the Pros and Cons of Various Heating Systems – Il Riscaldamento nelle Chiese e la Conservazione dei Beni Culturali: Guida all'Analisi dei Pro e dei Contro dei Vari Sistemi di Riscaldamento. Electa Mondadori, Milan, 240 pp.

Caneva, G., Nugari, M.P., Salvadori, O., 2009. Plant Biology for Cultural Heritage – Biodeterioration and Conservation. The Getty Conservation Institute, Los Angeles.

De Guichen, G., 1984. Climate in Museums. ICCROM, Rome.

Jamiolkowski, M., 1995. Leaning Tower of Pisa – description of the behaviour. In: Zezza, F. (Ed.), The Conservation Project: Knowledge of the Functional Elements for the Planning of the Interventions and Geotechnical Aspects of the Protection. Community of Mediterranean Universities, Adda, Bari, pp. 203–226.

Jenkins, K.A., Smith, B.J., 1990. Daytime rock surface temperature variability and its implications for mechanical rock weathering: Tenerife, Canary Islands. Catena 17, 449–459.

Michalski, S., 1993. Relative Humidity: A Discussion of Correct/Incorrect Values. ICOM Committee for Conservation, 10th triennial meeting: Washington, DC, 22–27 August 1993: preprints. Bridgland, Janet (Ed.). ICOM Committee for Conservation, pp. 624–629.

Michalski, S., 2004. Care and preservation of collections. In: Running a Museum: A Practical Handbook. International Council of Museums, Paris, pp. 51–90.

Michalski, S., 2009. The Ideal Climate, Risk Management, the ASHRAE Chapter, Proofed Fluctuations, and Toward a Full Risk Analysis Model Contribution to the Experts' Roundtable on Sustainable Climate Management Strategies, Held in April 2007, in Tenerife, Spain. The Paul Getty Institute, Los Angeles.

Padfield, T., 1994. Role of standards and guidelines. In: Krumbein, W.E., Brimblecombe, P., Cosgrove, D.E., Stainforth, S. (Eds.), Durability and Change. Wiley, New York, pp. 191–199, 307 pp.

Smith, B.J., 1978. The origin and geomorphic implications of cliff foot recesses and tafoni on limestone hamadas in the northwest Sahara. Zeitschrift für Geomorphologie 22 (1), 21–43.

Sedlbauer, K., 2002. Prediction of mould growth by hygrothermal calculation. Journal of Building Physics 25, 321–336.

Sedlbauer, K., Hofbauer, W., Krueger, N., Mayer, F., Breuer, K., April 2011. Material specific isopleth-systems as valuable tools for the assessment of the durability of building materials against mould infestation – the "isopleth-traffic light". p. 8. In: Proceedings of the XII DBMC, International Conference on Durability of Building Materials and Components. Porto, Portugal.

Thomson, G., 1986. The Museum Environment, 2nd Edition. (Butterworth-Heinemann Series in Conservation and Museology), Routledge, London.

Veniale, F., 1995. Minerali costituenti le rocce: processi e sequenze di alterazione. In: Lefèvre, R.A. (Ed.), La pietra dei monumenti nel suo ambiente fisico. Istituto Poligrafico e Zecca dello Stato, Rome, pp. 11–32.

Warscheid, T., Krumbein, W.E., 1996. Biodeterioration of inorganic nonmetallic materials – general aspects and selected cases. In: Heitz, E., Sand, W., Flemming, H.C. (Eds.), Microbially Induced Corrosion of Materials. Springer Verlag, New York, pp. 273–295.

Warke, P.A., Smith, B.J., 1994. Short term rock temperature fluctuations under simulated hot desert conditions: some preliminary data. In: Robinson, D.A., Williams, R.B.G. (Eds.), Rock Weathering and Landform Evolution. Wiley, New York, pp. 57–70.

CHAPTER

2A

Theoretical Grounds for Humidity

OUTLINE

2A.1. PARTIAL PRESSURE OF WATER VAPOUR

The popular term *humidity* is often used when speaking of the moisture present in the atmosphere; however, it is an ambiguous, generic word. Several specific terms exist, each of them representing a different parameter useful to describe a peculiar property. The first hygrometric parameter is the *partial pressure of water vapour*, conventionally indicated

Microclimate for Cultural Heritage, Second Edition
http://dx.doi.org/10.1016/B978-0-444-63296-8.00002-0

with the letter *e*. The partial pressure exerted by water molecules may increase until a certain limit is reached; at this threshold, the number of water molecules evaporating from the liquid equals the number of those returning to it from the atmosphere, establishing a dynamic equilibrium between evaporation and condensation, called *saturation*. After this threshold, condensation prevails. The saturation threshold is determined by the temperature *T* but is irrespective of the dry air pressure according to the *Dalton* law of the independence of the *partial pressures*, i.e. the total pressure *P* of a mixture of nonreactive gases equals the sum of the

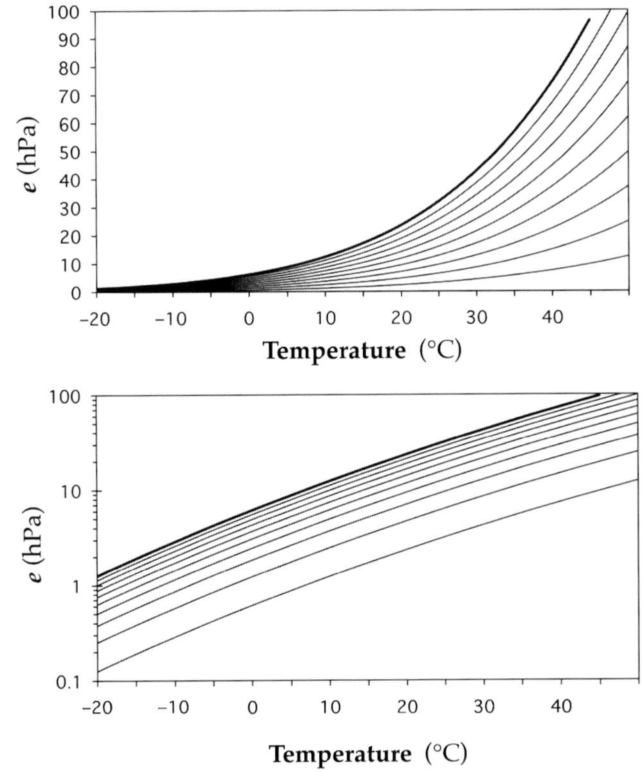

FIGURE 2A.1 Saturation pressure (e_{sat}) of the water vapour (thick line) and partial pressure (e) of the water vapour at different values of relative humidity (RH), i.e. at $RH = 90\%$, 80%, 70%,...10%. Of course, the thick line is for $RH = 100\%$. The first graph is with linear scales in order to be more immediate to nonspecialists and present a better resolution of the ordinary vapour partial pressures. Specialists may prefer the second graph: ordinates in logarithmic scale better represent the physical relationship between vapour partial pressure and air temperature. The lines are not exactly straight and the departure shows how much the Magnus and Tetens equation departs from a purely exponential function, giving an idea of the approximation made using such a simplified formula.

individual pressures p_i that each gas would exert if it were present alone in the same volume V. In other words, in a mixture of gases any gaseous species behaves independently of the presence of other gases, the identity of any gaseous species being unimportant in the P-V-T domain. The state of being *saturated* is a characteristic of all vapours, not of the gases, and of the air in particular. We remind that '*vapour*' is a substance in the gas phase at a temperature lower than its P-V-T critical point (visible in Fig. 2A.2). This means that *vapours* can be condensed and pass to the liquid phase by increasing their pressure without reducing the temperature but *gases* can not.

The *saturation pressure*, also called *vapour tension*, $e_{sat}(t)$, is computed with the empirical formula attributed to *Magnus* (1844) and *Tetens* (1930):

$$e_{sat}(t) = e_{sat}(0) \times 10^{at/(b+t)} \quad (\text{hPa}) \tag{2A.1}$$

where $e_{sat}(0) = 6.11$ hPa (1 mbar = 1 hPa), $a = 7.5$ and $b = 237.3$ °C. The graphical representation of this equation is shown in Fig. 2A.1.

In presence of ice, the saturation tension should be calculated with reference to the solid phase, i.e. $a = 9.5$, $b = 265.5$ °C. As the tension for ice is lower than that for liquid water, if the two phases coexist, water molecules will evaporate from the liquid and sublimate onto the solid phase. This equation gives accurate values at the usual atmospheric temperature but is less accurate near the boiling point, where $e_{sat}(100) = 1013$ hPa.

It can be noted that Eqn (2A.1) is independent of V, so every isothermal compression causes a faster condensation rate until the dynamic equilibrium is established. In the case of isothermal expansion, the evaporation continues until equilibrium is reached; only when all the liquid has evaporated does the water partial pressure decrease, according to Eqn (1.2).

2A.2. DERIVATION OF THE LATENT HEAT

The first law of the thermodynamics states that, in a system that undergoes a change of phase, the increase of the specific energy U equals the mechanical equivalent of the heat absorbed Q plus the work W expended in producing the change.

$$(U_v - U_l) = Q + W \tag{2A.2}$$

Let us consider a liquid in equilibrium with its vapour; V_v is the specific volume V/M of the saturated vapour and V_l is the specific volume of the liquid, $p_v = p_l$ are the pressures of the vapour and the liquid (please note that p_v has been previously indicated as $e_{sat}(t)$, but in this context a new symbol has been introduced for uniformity with the liquid phase and to

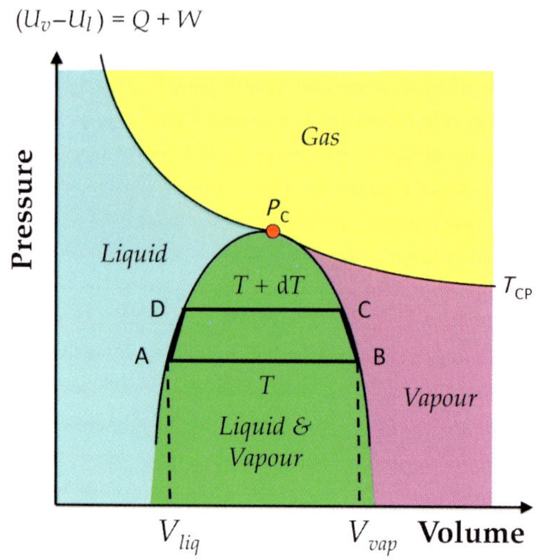

FIGURE 2A.2 Thermodynamic cycle ABCD in the Andrews phase transition diagram to derive the Clausius–Clapeyron equation. P_C is the critical point where the liquid, the vapour and the gas areas meet; T_{CP} is the critical isotherm that separates the gas from the liquid and the vapour areas.

make possible the following mathematical treatment), and U_v and U_l are the two specific energies U/M of the two phases. All these quantities are only functions of the temperature T of the system. In a p versus V diagram (Fig. 2A.2), a reversible evaporation–condensation cycle close to the Carnot cycle can be applied to a sample of water. The cycle is as follows: first the system undergoes an isothermal and isobaric evaporation with increasing V, from A to B, then some adiabatic heating by compression from T to $T + dT$ from B to C, then isothermal and isobaric condensation with V decreasing from C to D, and finally cooling by adiabatic expansion to the original state. During the evaporation, the variation of the specific energy of the system is $(U_v - U_l)$ and the external work, which is negative, is $W = p_v(V_v - V_l)$. The heat Q absorbed in this phase is the *latent heat of vaporization* L_v of the liquid, i.e.

$$Q = L_v = (U_v - U_l) + p_v(V_v - V_l). \tag{2A.3a}$$

After one complete cycle, the net work done is given by the area ABCD between the two isobaric and the two adiabatic transformations and in the first approximation, by the rectangle $dW = (V_v - V_l)dp$. However, the efficiency η of the Carnot cycle is

$$\eta = dT/T = dQ/Q = dW/Q \tag{2A.4}$$

so that $dW = Q\,dT/T$ and, combining this with the above equations

$$dW = (V_v - V_l)dp = Q\frac{dT}{T} \tag{2A.5}$$

$$Q = T(V_v - V_l)\frac{dp}{dT} \tag{2A.3b}$$

By equating the two values of Q in the equations (2A.3a) and (2A.3b) we get

$$(U_v - U_l) + p_v(V_v - V_l) = T(V_v - V_l)\frac{dp_v}{dT}. \tag{2A.6}$$

Remembering that $Q = L_v$, the following equation is obtained:

$$L_v = T(V_v - V_l)\frac{dp_v}{dT} \tag{2A.7}$$

This equation for the latent heat of vaporization was deduced by Clapeyron from Carnot's theory (for this reason, it is also called the *Clapeyron equation*) and was proved by Clausius. This equation is used to calculate L_v at any temperature when we know the specific volumes and the relationship between the increase of saturation pressure and T. For example, for pure water at boiling point $T = 373$ K and standard pressure $p_v = 1013$ hPa, $V_v = 1674$ cm^3, $V_l = 1$ cm^3, $dp_v/dT = 36.15$ hPa K^{-1} = 3.62×10^4 dyne cm^{-2} K^{-1}, one obtains $L_v = 2260 \times 10^7$ erg g^{-1}.

The Clapeyron equation shows that the latent heat of vaporization is due partly to the increase of specific energy and partly to the external work. In order to find a relationship between these two quantities, it is necessary to find the ratio between W and L_v

$$\frac{W}{L_v} = \frac{p_v(V_v - V_l)}{L_v} = \frac{p_v}{T}\frac{dT}{dp_v} \tag{2A.8}$$

and in the above conditions, $W/L_v = 0.075$. This means that the external work forms only a small part of the latent heat of vaporization.

If we consider that the ratio of the volumes of a molar mass of water in the liquid and the gaseous states is 18:24,000 cm^3 cm^{-3} = 0.75×10^{-3}, a simple approximation of the Clapeyron formula is obtained by neglecting V_l in comparison with V_v and assuming that the state equation (Eqn (1.1)) for perfect gases is still valid (Fermi, 1937). Under these reasonable assumptions, the formula becomes

$$L_v = \frac{\Re_v T^2}{p_v}\frac{dp_v}{dT} \tag{2A.9}$$

and in the critical case of boiling water, we obtain $L_v = 547.5$ cal g^{-1}, slightly larger than the observed value $L_v = 538.7$ cal g^{-1}. This difference is due to the fact that the specific volume of the saturated vapour at 100 °C is less than the value calculated using the equation for perfect gases. For usual atmospheric temperatures, this approximation is good.

The above equation can be rewritten as

$$\frac{\mathrm{d}\ln p_v}{\mathrm{d}T} = \frac{L_v}{\Re_v T^2} \tag{2A.10}$$

and, if we assume L_v is constant over a wide temperature interval, this equation can be solved as:

$$p_v = A\exp\left(-\frac{L_v}{\Re_v T}\right) \tag{2A.11}$$

and this gives the exponential relationship between the saturation pressure and the temperature, theoretically derived.

Another approximated formula (Planck, 1926) is obtained by substituting in Eqn (2A.6) the formula for the specific energy, $U_v = c_v T + const_v$ where c_v is the specific heat at constant volume, which holds for perfect gases in the isothermal processes, when $Q + W = 0$ and is given by $c_v = 3\ \Re_v = 0.331$ cal g^{-1}K^{-1}. If for the liquid the specific heat at constant pressure, $c_l = 1$ cal g^{-1} K^{-1}, is assumed to be constant and the external work is neglected, one gets $U_l = c_l T + const_l$ and $(U_v - U_l) = (c_v - c_l)T$. In addition, if the specific volume of the liquid V_l is neglected in comparison with that of the vapour, V_v, with the help of the equation for perfect gases (1.1), the approximations $p_v(V_v - V_l) = \Re_v T$ and $(V_v - V_l) = \Re_v T/p_v$ are found. By substituting these findings into Eqn (2A.6), it follows that

$$(c_v - c_l)T + \Re_v T + const = \frac{\Re_v T^2}{p_v}\frac{\mathrm{d}p_v}{\mathrm{d}T} \tag{2A.12}$$

Multiplying both sides by $\mathrm{d}T/T^2$, this equation can be integrated and one obtains

$$p_v = A' T^{(c_p - c_l)/\Re_v}\exp\left(-\frac{A''}{T}\right) \tag{2A.13}$$

where A' and A'' are positive constants and $c_p = 0.441$ cal g^{-1} K^{-1} is the specific heat of the vapour at constant pressure. The exponent of T has been obtained considering that $\Re_v = (c_p - c_v)$ and, consequently, $1 + \frac{c_v - c_l}{\Re_v} = \frac{c_p - c_v + c_v - c_l}{\Re_v} = \frac{c_p - c_l}{\Re_v}$.

Equation (2A.11) can be rewritten in other forms, of wide use, as follows.

$$e_{sat}(T) = e_{sat}(273) \times \exp\left[\frac{L_v}{\Re_v}\left(\frac{1}{273} - \frac{1}{T}\right)\right] \tag{2A.14}$$

$$e_{sat}(T) \approx e_{sat}(273) \times \exp\left[\frac{L_v}{\Re_v 273^2}(T-273)\right] \tag{2A.15}$$

$$e_{sat}(T) \approx 6.11 \times 10^{0.0318t} \tag{2A.16}$$

where $0.0318 = (L_v/\Re_v 273^2)\ \log e$ (e is the *Neper* number 2.718281 82845904…). All the above formulae are in hPa. From the above formulae, the *latent heat of vaporization* (or *condensation*), L_v, *latent heat of fusion* (or *melting*), L_f, and *latent heat of sublimation*, L_s, can be derived, as follows:

$$L_v = 597.3 - 0.57\,t, \quad L_f = 79.7 + 0.52\,t, \quad L_s = 677 - 0.05\,t \tag{2A.17}$$

with all the units in cal g^{-1}. Naturally, $L_v + L_f = L_s$.

2A.3. MIXING RATIO OF WATER VAPOUR AND DRY AIR

The *mixing ratio* w of moist air (i.e. dry air and water vapour) is *the (dimensionless) ratio of the mass of water vapour* m_v *to the mass of dry air* m_a and this ratio represents *the ponderal mixture of these two gaseous substances*, i.e.

$$w = \frac{m_v}{m_a} = \frac{n_v M_v}{n_a M_a} = \frac{n_v}{n_a}\frac{18.016}{28.966} = 0.622\frac{n_v}{n_a} \quad (\mathrm{g\,g^{-1}}) \tag{2A.18}$$

where $M_v = 18.016$ is the molecular mass of water, $M_a = 28.966$ is the molecular mass of dry air and n_v and n_a are the number of molecules of water vapour and dry air, respectively, present in the mixture. Therefore, the mixing ratio is also representative of the relative mixture of water vapour in dry air in terms of molecular population.

If e is the partial pressure of water vapour and p is the atmospheric pressure, then the partial pressure of dry air is $p_a = p - e$ and substituting this in the equation of state for perfect gases in the form (1.2), the previous equation can be written in terms of pressure:

$$w = \frac{e}{p-e}\frac{\Re_a}{\Re_v} = 0.622\frac{e}{p-e} \approx 0.622\frac{e}{p} \tag{2A.19}$$

where 0.622 is the above ratio between the molar masses of water and dry air and, consequently, also of $\Re_a/\Re_v$.

It is evident that w is independent of the temperature T, volume V and atmospheric pressure of the air parcel and remains constant except when condensation, evaporation or mixing with other air masses occur. As a consequence, w can be considered as a characteristic value, which is useful to recognize an air mass and its hygrometric exchanges with the environment, being invariable to either adiabatic (i.e. without exchange of heat) or diabatic (i.e. with exchange of heat), isobaric or nonisobaric heating or cooling. For example, in the air masses aloft a decrease of w gives an indication of the amount of water that has been removed by the system, having been precipitated as rainfall.

This parameter is dimensionless because the unit is g g^{-1}, i.e. a pure number that represents the fraction of grams of vapour mixed with 1 g of dry air. It might be expressed in percentage too. However, as the numerical value of w is very small, it is common to multiply it by 1000 and

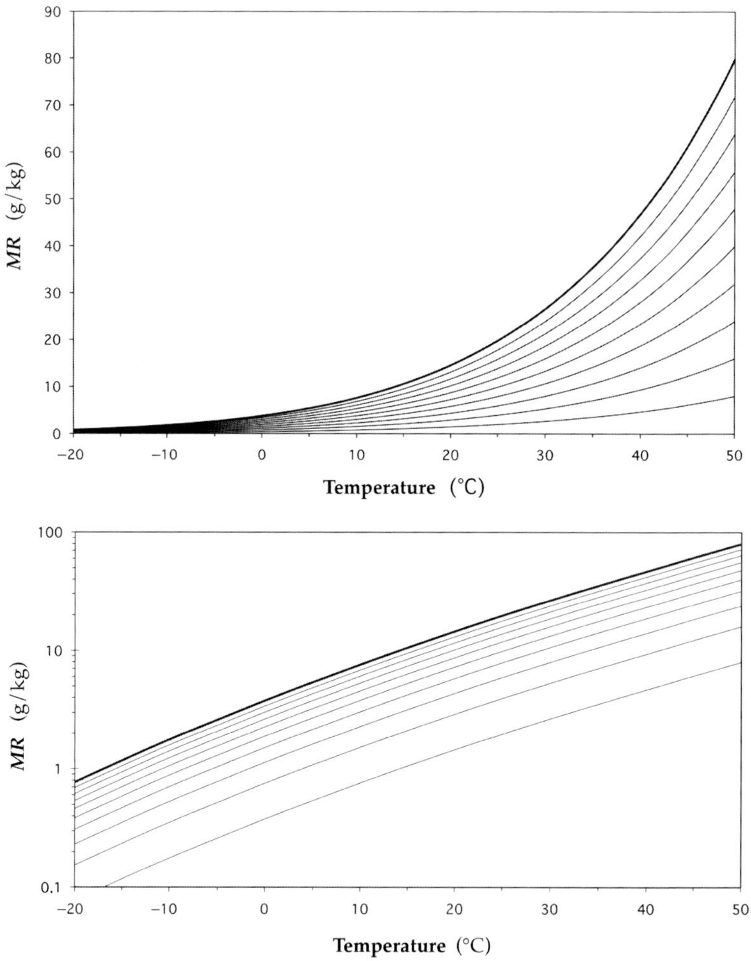

FIGURE 2A.3 Saturation mixing ratio (*MR*, thick line) of the vapour in air and actual value of the *MR* at different values of *RH*, i.e. at *RH* = 90%, 80%, 70%,… 10%. Of course, the thick line is for *RH* = 100%. As in Fig. 2A.1, the first graph is with linear scales and the second has the ordinate with logarithmic scale.

use the practical unit $MR = 1000\,w$, expressed in g kg^{-1}, which represents the number of grams of vapour mixed with 1 kg of dry air. A plot of this parameter is shown in Fig. 2A.3. As the atmospheric pressure p is fairly constant (i.e. ±4%), a common approximation is to write 1000 hPa instead of p, which simplifies calculations.

At every environmental temperature T, the MR increases proportionally to the mass of vapour that is supplied to the atmosphere, until the saturation limit is reached, i.e. when the vapour pressure e equals the top level $e_{sat}(t)$, given by the Magnus and Tetens equation Eqn (2A.1), and relative humidity $RH = 100\%$. At saturation, the MR is indicated as MR_{sat} and is computed by means of Eqn (2A.14), using $e_{sat}(t)$ instead of e (for a quantitative evaluation of the error, see the psychrometric chart). Again, as for every T, e is proportional to RH, by measuring RH it is possible to calculate MR, i.e. when $e_{sat}(t)$ is computed, the saturation value MR_{sat} is obtained using Eqn (2A.14) and the actual MR is $MR = RH \times MR_{sat}$. By dividing the graph of MR_{sat} versus temperature into fractional parts, one obtains graphically the values of MR at different values of RH. The parameter MR_{sat} is an increasing function of T, with a trend similar to that of $e_{sat}(t)$. At usual meteorological conditions, in temperate climates (i.e. with $t \leq 30°\text{C}$), $MR \leq 30$ g kg^{-1}.

When two different air masses mix together, the final value of the MR depends on the two initial MR_i values and the relative proportion of mixing. The mixture is calculated as a weighted mean, i.e. by mixing x grams of air having MR_x with y g of air having MR_y the resulting MR is

$$MR = \frac{x\,MR_x + y\,MR_y}{x + y}. \tag{2A.20}$$

In meteorology, this parameter is used to distinguish different air masses, e.g. to evidence the change between the land (dryer) or sea (moister) breeze or to monitor the evapotranspiration from the soil or the evaporation over large water bodies. A quantitative evaluation can be made by measuring the vertical gradient of this parameter and the exchange coefficient for the vertical transport in the atmosphere. Similarly, in the conservation science, this parameter can be used to recognize a number of important features and interactions.

2A.4. SPECIFIC HUMIDITY

The *specific humidity s* of moist air is *the (dimensionless) ratio of the mass of water vapour m_v to the mass of moist air $m_a + m_v$* and this ratio represents the *ponderal dilution of the vapour in the atmosphere*, i.e.

$$s = \frac{m_v}{m_a + m_v} \quad (\text{g g}^{-1}) \tag{2A.21}$$

It is also called *mass concentration* or *moisture content* of moist air. Substituting Eqn (2A.21) into Eqn (2A.18),

$$s = \frac{w}{1 + w} \tag{2A.22}$$

and, operating similarly to Eqn (2A.19):

$$s = 0.622 \frac{e}{p - 0.378\,e} \approx 0.622 \frac{e}{p} \tag{2A.23}$$

Like the mixing ratio, this parameter is dimensionless, the natural unit being g g^{-1}. It might also be expressed in percentage; however, in order to avoid the use of small decimal numbers, the practical unit $SH = 1000\,s$, expressed in g kg^{-1}, has been introduced. This represents the number of grams of vapour dispersed in 1 kg of moist air. In practice, it can be observed that in the denominator of Eqn (2A.22) $w << 1$ and is therefore negligible; similarly in Eqn (2A.23), $0.378e$ can be neglected in comparison with p. As a consequence, the values of w and MR are very similar to s and SH, respectively, with differences of the order of 1%. The same properties of the MR can be extended to the SH and the saturation specific humidity SH_{sat}; also, the same graphical representation can be used at a first approximation. Both MR and SH are conservative for adiabatic or diabatic changes of temperature, pressure or volume but are not conservative for evaporation or condensation, as changes of m_v are involved.

2A.5. ABSOLUTE HUMIDITY

The equation of state (1.1) for the water vapour can be written in the form

$$e = \frac{m_v}{V} \Re_v\, T = a\, \Re_v\, T \tag{2A.24}$$

where the last term defines the absolute humidity a as *the density of the water vapour,* i.e. *the mass of vapour contained in the unit volume of moist air*

$$a = \frac{m_v}{V} \quad (\mathrm{g\ cm^{-3}}) \tag{2A.25}$$

From the definition, it follows that a varies with m_v, i.e. in the case of condensation, evaporation and mixing with other air masses, as well as with V, i.e. compression or expansion, due to, e.g. changing atmospheric pressure or height; from Eqn (2A.24), it is also evident that a is directly proportional to e and inversely proportional to air temperature. From the above formula, we note that the presence of dry air is not necessary and the definition of a holds even when some moisture is present in a given volume V from where dry air has been removed. However, this is not relevant to our aims.

Of course, a is expressed in g cm^{-3}, but in order to avoid the use of small decimal numbers, the practical unit $AH = 10^6 a$, expressed in g m^{-3}, has been introduced. AH represents the mass of vapour contained in 1 m^3 of moist air. As at standard pressure and temperature, 1 m^3 of atmosphere contains 1.255 kg of air, the values of AH are numerically similar but always greater than those of MR or SH. In fact, MR and SH represent the mass of vapour contained in 1 kg of dry or moist air, which occupies some 80% of 1 m^3.

The AH can be computed by means of the equation of state (2A.24) by considering that the vapour density is the inverse of the specific volume V_v, i.e.

$$a = \frac{1}{V_v} = \frac{1}{\Re_v}\frac{e}{T} = \frac{1}{\Re_v}\frac{e}{273(1+\alpha t)} \quad \text{(2A.26)}$$

where the last term comes from the transformation from Kelvin to degree Celsius, i.e. $T = 273(1+\alpha t)$ where $\alpha = 1/273 = 0.00366$.

Depending on the units in which e is expressed, the formula to compute AH assumes one of the following forms. If e is in hPa or mbar,

$$AH = 220\frac{e}{T} = 0.806\,\frac{e}{1+\alpha t} \approx 0.806\,e \quad (\text{g m}^{-3}) \quad \text{(2A.27)}$$

If e is in dyne cm^{-2}:

$$AH = 0.22\,\frac{e}{T} = 0.806 \times 10^{-3}\,\frac{e}{1+\alpha t} \approx 0.806 \times 10^{-3} e \quad (\text{g m}^{-3}) \quad \text{(2A.28)}$$

or, if e is in mm Hg:

$$AH = 290\frac{e}{T} = 1.062\,\frac{e}{1+\alpha t} \approx e \quad (\text{g m}^{-3}) \quad \text{(2A.29)}$$

Eqn. (2A.29) shows that at $t = 16.4$ °C, the AH is numerically equal to e measured in mm Hg; at usual meteorological values of t, these two parameters (in the above units) are numerically similar.

By using the saturation value $e_{sat}(t)$ instead of e, the saturation absolute humidity AH_{sat} is obtained (Fig. 2A.4). This represents the maximum quantity of vapour that the unit volume of atmosphere can carry under specified temperature conditions. For this reason, it is also called (moisture) *capacity*. In Chapter 9 we will explicitly see that the moisture capacity uniquely depends upon the interaction existing between the water molecules at any given temperature. From the above formulae, it appears that AH_{sat} is a function of air temperature but not of pressure; in the case of adiabatic compression or expansion, the change of AH_{sat} is governed by the change of t (although induced by the change of p), according to the Poisson equation. The value of AH_{sat} allows the computation of the quantity of the liquid water that may evaporate or the quantity of vapour that condenses during an adiabatic transformation.

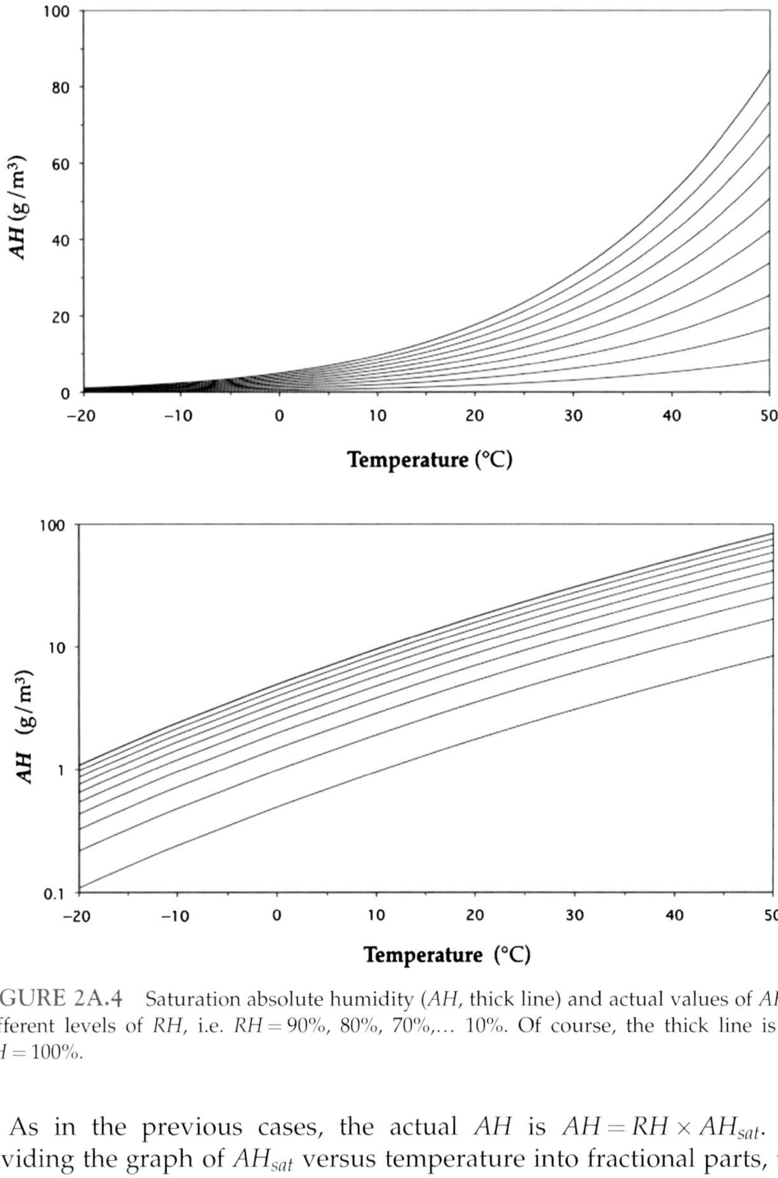

FIGURE 2A.4 Saturation absolute humidity (*AH*, thick line) and actual values of *AH* at different levels of *RH*, i.e. $RH = 90\%$, 80%, 70%,... 10%. Of course, the thick line is for $RH = 100\%$.

As in the previous cases, the actual *AH* is $AH = RH \times AH_{sat}$. By dividing the graph of AH_{sat} versus temperature into fractional parts, the values of *AH* at different *RH* levels are graphically obtained.

2A.6. RELATIVE HUMIDITY

It is useful to state beforehand that in meteorology, for practical purposes, one makes the conventional use of attributing all the properties of the atmosphere to only one fictitious gas, called '*air*', which behaves as the actual mixture of gases and vapours that form the atmosphere. In order to describe particular phenomena, a distinction is made between *dry air* and *moist air*. The adjective *dry* may have two meanings, which are clarified by the context: the former is literal, i.e. *without vapour*, whereas the latter means *without condensation*, i.e. the mixture behaves as an ideal gas, with no changes of phase. Similarly, *moist* indicates the opposite of these two situations. From this point of view, the vapour component is formally ignored and it is improperly expressed in terms of relative humidity *of the air* or *degree of saturation of the air* instead of the correct form: relative humidity of the water vapour and *degree of saturation of the vapour*.

From the physical point of view, it is meaningless to speak in terms of saturation in the case of a mixing of two (or more) gases, as they are absolutely miscible from 0% to 100%, according to the Dalton law. The saturation arises because one of the components is not a gas but a *vapour* and all vapours become saturated when their partial pressure reaches the limiting value determined by the ambient temperature, called *saturation tension*. In the case of water vapour, this is given by the Magnus and Tetens formula (2A.1). It is independent of the presence of other gases and their pressure and the definition of relative humidity is the same in the presence of air or in absence of it (e.g. for some vapour in a vessel deprived of air).

Relative humidity u (popularly termed *humidity*) was originally defined as

$$u = \frac{e}{e_{sat}(t)} \tag{2A.30}$$

i.e. *the (nondimensional) ratio between the actual partial pressure of the vapour and its saturation vapour pressure at the same temperature*. Of course u ranges from 0 to 1. For practical reasons, relative humidity is popularly expressed in percent (%) unit i.e. $RH = 100\ u$.

From formula (2A.30), we note that the presence of dry air is not necessary and the definition of u holds even when some moisture is present in a given volume from where dry air has been removed. However, this is not relevant to our aims.

So far we have defined the common use of RH (%). However, the story is complex and several misleading ideas are related to this fundamental parameter that will be thoroughly discussed in the following section.

Below 0 °C, the saturation value $e_{sat}(t)$ is conventionally computed with reference to supercooled water or to ice if the temperature is far below 0 °C, and this fact introduces ambiguity.

The above definition (2A.30) is not any more rigorous when the saturation is approached. In fact, in this condition, the vapour departs from the ideal behaviour of a perfect gas in that some molecules may aggregate among them forming small clusters. As a consequence, the actual number of free particles n that describe the equation of state decreases, as well as the partial pressure e of the system. Near saturation, the partial pressure of a vapour departs from the ideal condition and another unaffected property should be used instead.

In order to avoid the above problem, in 1947, the International Meteorological Organization (later World Meteorological Organization) reformulated the definition of u in terms of the mass m_v of the vapour contained in the air parcel, which is a conservative property, by substituting $e = (m_v/V)\Re_v T$ and $e_{sat}(t) = (m_{v,sat}/V)\Re_v T$, where $m_{v,sat}$ is the mass of water vapour present under saturation (List, 1966). The relative humidity u was defined as *the (nondimensional) ratio between the mass of vapour m_v actually present in whatever volume of atmosphere to the greatest amount possible at the same temperature $m_{v,sat}$*, i.e.

$$u = \frac{m_v}{m_{v,sat}} \tag{2A.31}$$

and physically represents *the degree of saturation of the vapour*, also ranging from 0 to 1. From the above definitions, it is easy to recognize that

$$u = \frac{m_v}{m_{v,sat}} = \frac{w_v}{w_{v,sat}} = \frac{\frac{m_v}{m_a}}{\frac{m_{v,sat}}{m_a}} = \frac{a_v}{a_{v,sat}} = \frac{\frac{m_v}{V}}{\frac{m_{v,sat}}{V}} \approx \frac{s_v}{s_{v,sat}} = \frac{\frac{m_v}{m_a+m_v}}{\frac{m_{v,sat}}{m_a+m_{v,sat}}} \tag{2A.32}$$

and it depends on both the amount of moisture that is actually present m_v (to which it is proportional) and the temperature t, which acts exponentially in the denominator $m_{v,sat}$. In Chapter 9 we will give a further physical explanation of this fact. The dependence of u on t is very strong and at first sight the diagrams of u and t seem to be nearly symmetric, as the reflected image in a mirror.

In 1966, the World Meteorological Organization returned to the original definition in terms of the molar fraction of water (WMO, 1966, 1983, 1988). The molar fraction of water vapour (x_v) in moist air is the ratio between the number of moles of water vapour ($n_v = m_v/M_v$) and the total number of moles ($n_a + n_v$) present in the sample of moist air.

$$x_v = \frac{n_v}{n_a + n_v} = \frac{\frac{m_v}{M_v}}{\frac{m_a}{M_a} + \frac{m_v}{M_v}} \tag{2A.33}$$

The symbol n_a is the number of moles of dry air ($n_a = m_a/M_a$) in the sample, m_v is the mass of vapour, m_a is the mass of dry air in the sample, $M_v = 18.01$ is the molar mass of water and $M_a = 28.96$ is the average molar mass of dry air. The molar fraction at saturation is indicated x_{vw}.

In terms of molar fraction, the definition of u is practically given by the ratio between the number of moles of water vapour and the highest possible number at saturation, in formula:

$$u = \frac{x_v}{x_{vw}(t)} = \frac{e}{e_{sat}(t)} \tag{2A.34}$$

A careful reader might note a subtle methodological difference between Eqn. (2A.30) and Eqn. (2A.34). Both arrive to the same result but (2A.34) passes through an unnecessary mixture of water vapour with dry air, although dry air is ultimately neglected.

The two definitions of relative humidity, in Eqn. (2A.31) dated 1947 and Eqn. (2A.34) dated 1966, that in the following will be indicated with the date subscript, i.e. u_{1947} and u_{1966}, respectively, or with RH_{1947}, RH_{1966} for the values expressed in %. These are slightly different between them and are related to each other as follows:

$$\frac{u_{1966}}{u_{1947}} = \frac{0.62198 + w_w}{0.62198 + w} = \frac{0.62198 + w_w}{0.62198 + w_w u_{1947}} \tag{2A.35}$$

The departure between the two definitions increases with temperature and dryness (Fig. 2A.5). At low and moderate temperatures and at high humidity levels they are practically indistinguishable. Near the freezing point, the largest departure is smaller than 0.5%; at 20 °C and $RH = 50\%$, is 1%; and at 50 °C and $RH = 50\%$, is about 5%. Practically, this difference is generally negligible and smaller than the instrumental error except for tropical climate under arid conditions.

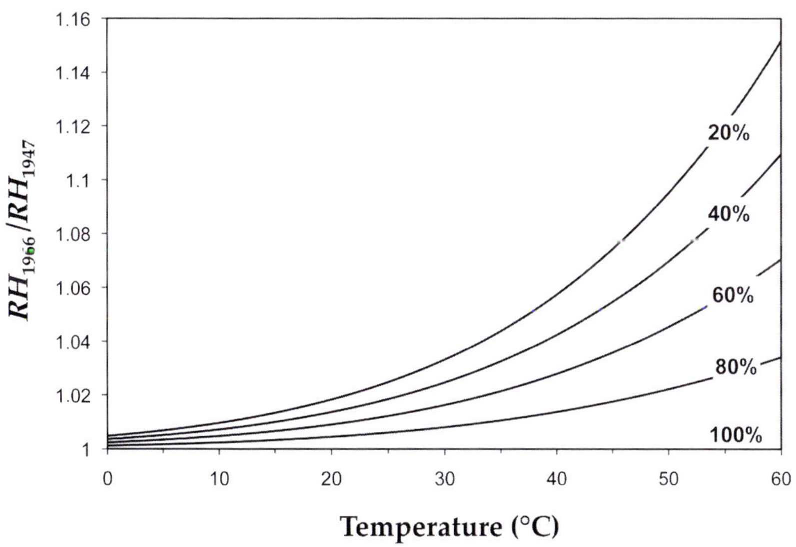

FIGURE 2A.5 Comparison between the two *RH* definitions established in 1947 and 1966, i.e. RH_{1947} and RH_{1966}, respectively. The two definitions are almost identical at low and moderate temperature levels and high humidity levels. They differ in tropical climate especially at very low *RH* levels but their departure is smaller than the observational error at usual room temperature.

With the help of Eqn (2A.19), Eqn (2A.32) and the Magnus and Tetens equation (2A.1), it is possible to obtain practical formulae to compute the *MR* or *SH* from measurements of *t* and *RH*:

$$MR \approx SH \approx \frac{37.95}{p} RH \times 10^{at/(b+t)} \approx 0.03795\ RH \times 10^{at/(b+t)}\ (\text{g kg}^{-1}) \tag{2A.36}$$

$$e = 0.0611\ RH \times 10^{at/(b+t)}\ (\text{hPa}) \tag{2A.37}$$

$$AH = \frac{13.44\ RH \times 10^{at/(b+t)}}{273.3 + t}\ (\text{g m}^{-3}) \tag{2A.38}$$

The *RH* can be measured directly or indirectly. In the following discussion, a very accurate method for the determination of the *RH* will be described, based on the measure of *T* and the wet bulb depression; by substituting *RH* in the above formulae, the values of *MR*, *e* and *AH* can be calculated. It is customary to compute Eqn (2A.36), Eqn (2A.37), and Eqn (2A.38) with the Magnus and Tetens coefficients $a = 7.5$ and $b = 237.3\ °\text{C}$, referred to (supercooled) water also when $t < 0\ °\text{C}$. This convention is partially justified as in mild climates and daytime, ice is covered with a film of liquid water. However, when the wet bulb temperature is below the freezing point and the sensor is covered with ice, the appropriate coefficient for ice should be used.

Climatology requires average values. It might be useful to underline that *RH* is a parameter derived from other elementary physical variables, e.g. *MR* or vapour pressure, which is proportional to it, and air temperature. The equations that link the *RH* to the basic parameters are not linear and for this reason, a mean value obtained as the arithmetic average of a set of *RH* observations is not physically correct. Mixing two air masses with different *RH* values, the *RH* of the final mixture is not the arithmetic average of the two initial *RH* values but is determined by the final temperature and *MR*. The 24-h or the monthly averages should be computed after the averages of *MR* (or vapour pressure) and air temperature. The common use is to sacrifice correctness to convenience and data are handled in terms of averages; however, in this field, the most correct statistical form is to abandon averages and to present data in term of percentiles, which are based on frequency distribution (see Chapter 10).

2A.7. DEW POINT: THE TEMPERATURE OF CONDENSATION

The *dew point temperature*, commonly termed *dew point, DP, is the temperature to which a parcel of moist air must be cooled at constant atmospheric pressure and constant water vapour content in order for saturation to occur.* It can be alternatively defined as *the temperature at which the actual pressure of the vapour contained in an air parcel equals the saturation pressure, under constant atmospheric pressure and MR*. Although it is popularly called *DP of the 'air'*, it is a property of the vapour that might be extended to the *'air parcel'*, i.e. the little mass of mixture of dry air and vapour taken into consideration. From the definition, it is a conservative property of the air parcel with respect to isobaric heating or cooling without addition or subtraction of vapour. It is nonconservative with respect to adiabatic expansion or compression. Of course, in a completely dry atmosphere, there is no any temperature at which water can condense and this parameter does not make sense.

This parameter can be easily computed from the *RH* and air temperature, starting from the consideration that the *DP* is reached with an isobaric process, so that the vapour pressure at the original dry bulb temperature equals the saturation pressure at *DP*, i.e. $e(T) = e_{sat}(DP)$. By substituting this finding in the formula (2A.20), one obtains with the help of the Magnus and Tetens formulae:

$$u = \frac{e(t)}{e_{sat}(t)} = \frac{e_{sat}(DP)}{e_{sat}(t)} = \frac{e_{sat}(0) \times 10^{aDP/(b+DP)}}{e_{sat}(0) \times 10^{at(b+t)}} = 10^{[aDP/(b+DP)]-[at/(b+t)]} \tag{2A.41}$$

and hence

$$\log u = \frac{a\,DP}{b+DP} - \frac{at}{b+t} \tag{2A.42}$$

and

$$DP = \frac{b+DP}{a}\log u + \frac{b+DP}{a}\,\frac{at}{b+t} \approx \frac{b+t}{a}\log u + t \tag{2A.43}$$

where the last approximate finding has been obtained substituting t to *DP* in the right-hand side of the first identity. Of course, the first term is negative as $u < 1$ and $\log u < 0$.

Another formula can be deduced considering what happens to air over an evaporating surface. The air temperature lowers, while the increase of *MR* raises the *DP*. The air temperature t continues to decrease until the temperature of the evaporating surface, called *wet bulb temperature*, t_w, is reached (see section 2A.9). When the evaporated vapour reaches saturation, $t = t_w$. Starting from the Clapeyron equation and the definition of w and always considering the difference $DP - t_w$, after some steps and approximations, the following formula is obtained:

$$DP \approx b\frac{b\log u + t\log u + at}{ab - b\log u - t\log u} \tag{2A.44}$$

where a and b are the Magnus and Tetens coefficients for vapour in equilibrium with the liquid phase. Equation (2A.43) is a better approximation. The above formulae can be used once the RH is known, and obviously $log\ u = \log(RH/100) = \log RH - 2$.

$DP \leq T$ and $DP = T$ only when $RH = 100\%$. The DP is determined once the air temperature T and the RH are both known, or also when only the MR (or SH) is known. In particular, maxima of MR correspond to minima of DP, and vice versa, so that the DP can be used for diagnostic purposes instead of the MR and may be useful to express the moisture content in °C.

The *dew point spread* (also called *spread*), i.e. the difference $\Delta DP = T - DP$ basically depends on both the actual air temperature T and the MR. Following the approximation (2A.43), it can be expressed as a function of air temperature and RH

$$\Delta DP \approx -\frac{b+t}{a}\log u. \tag{2A.45}$$

It physically shows how much the air temperature is close to, or far from, the DP. The zones having the smaller ΔDP are more prone to form condensation and to allow microbiological life and weathering to occur. Useful maps of this parameter can be easily drawn for diagnostic purposes. However, although the RH is a very different but related parameter, the areas with maximum RH are the same as those in which the ΔDP is minimum, and if the critical cooling is not requested, maps of RH are sufficient to give a qualitative description of these microclimatic problems.

Dew has the typical form of droplets and especially forms on leaves during the nocturnal cooling due to the infrared (IR) emission. The formation of dew on leaves is favoured by the local excess of moisture due to stomatal transpiration. The surface tension of water tends to displace the larger droplets to the edges of the leaves and in particular to the points of leaves, especially the lance-shaped ones. The upward IR loss during clear nights is a very effective cooling mechanism. The surfaces on which dew forms are free from any upper shield and in practice are the same that are reached by rainfall. This is the reason that people often believe that dew falls similarly to drizzle. Dew is favoured over vegetated areas, but it occurs on monuments as well, when their surface temperature falls below the DP. We will see examples in Chapter 2B.

2A.8. FROST POINT: THE TEMPERATURE OF FREEZING

The *frost point* (FP) is defined in the same way as the DP, but reference is made to *ice* and in the Eqns (2A.43) and (2A.44), the Magnus and Tetens coefficients are $a = 9.5$ and $b = 265.5$ °C. From these equations, it follows that $FP \le T$ and $FP = T$ only when $RH = 100\%$. In addition, $FP = 0$ °C only in the case that at this temperature $RH = 100\%$. In general, the frost forms at a temperature below 0 °C, which depends on the moisture content in air.

Frost (also called *hoar, hoar frost, crystalline frost or white frost*) is generated by the direct sublimation of the vapour, which forms needle-shaped crystals of ice. It is different from the *frozen dew* (also called *white dew or silver frost*) that has the appearance of ice spherules. Similar to dew, frost also prevalently covers the horizontal surfaces that loose heat by *IR* radiation towards the clear sky.

Soft rime results from the rapid freezing of very small supercooled water droplets in fog or cloud when they impact on (prevalently) vertical surfaces with temperature below 0 °C. In this way, wind-borne droplets stick forming a dendritic accretion and all the trees are white on the side facing the wind, and iceless on the opposite side. The ice is white, porous, and composed of very small granules separated by few or many air inclusions. Similar to soft rime, but formed with drizzle droplets of micrometric size, is *hard rime*.

In the case that the concretion of ice has been generated by ice needles already formed in the atmosphere, this phenomenon is known as *advection hoar frost* (or also *ice fog, frozen fog, frost fog, air hoar, or rime fog*).

If large raindrops splash on a surface at temperature below the freezing point, a homogeneous and translucent ice coating will form on it, known as *glaze* (or also *glazed frost, clear ice* or *ground ice* for the soil). Wind-borne drops and splashing droplets may cover differently oriented surfaces forming layers with different thickness. Trees may break branches due to the weight of the accumulated ice.

2A.9. WET BULB TEMPERATURE: THE TEMPERATURE OF EVAPORATION

The *wet bulb temperature* T_w (*or* t_w) *or isobaric wet bulb temperature, is the temperature an air parcel would have if adiabatically cooled to saturation at constant pressure by evaporation of water into it, all latent heat being supplied by the parcel*. This temperature is directly measured by the wet *bulb of a psychrometer*, or can be obtained indirectly by means of a psychrometric diagram or formula, after the dry bulb temperature and any one hygrometric value (i.e. *e*, *MR*, *SH*, *AH*, *RH* or *DP*) are known. From the thermodynamic point of view, T_w *is the temperature that an air parcel would have when some liquid water is supplied gradually, in very small quantities and at the same temperature as the environmental air, and then this water is evaporated into the air adiabatically* (i.e. *the latent heat being supplied by the air*) *at constant pressure, until saturation is reached*. Saturation is reached for the combined action of two factors due to the evaporation: the increase in MR and the drop in T. Consequently, T_w is the lowest temperature that an air parcel would have by evaporating water, the latent heat being subtracted from the air and utilized for the change of state of water from liquid to vapour, until saturation is reached. T_w is also *the equilibrium temperature of an evaporating surface of water.*

Applying the first law of the thermodynamics to an air parcel formed of 1 g of dry air with a mass of vapour m_v, i.e. with mixing ratio $w = m_v$, and experiencing the above process,

$$\int_{T}^{T_w} c_{pm}(1 + m_v)\mathrm{d}T = \int_{m_v}^{m_w} L_v \mathrm{d}m_v \tag{2A.46}$$

where c_{pm} is the isobaric specific heat of the moist air that can be expressed in terms of isobaric specific heat of dry air c_{pd} ($c_{pd} = 0.240 \text{ cal g}^{-1} \text{ K}^{-1} = 1.003 \text{ J g}^{-1} \text{ K}^{-1}$), i.e. $c_{pm} \approx (1 + 0.8\, m_v)c_{pd}$ and L_v is the latent heat. After integration, by dividing both sides by ($c_{pd} + <w> c_{pv}$), the *wet bulb depression* $\Delta T_w = T - T_w$ is obtained

$$\Delta T_w = \frac{(m_{sat,w} - m_v)L_v}{c_{pd} + <w>\, c_{pv}} \tag{2A.47}$$

where $<w>$ is the average mixing ratio during this process, $m_{sat,w}$ is the saturation mixing ratio at the temperature T_w and c_{pv} is the isobaric specific heat for the water vapour ($c_{pv} = 1.81 \text{ J g}^{-1} \text{ K}^{-1}$). A further approximation of the wet bulb depression is obtained by using the formula (2A.19) for w and considering that $c_{pd} + <w> c_{pv} \approx c_{pd}$ (i.e. $<w> << 1$), so that the wet bulb depression becomes

$$\Delta T_w \approx \frac{0.622\, L_v(e_w - e)}{c_{pd}\, p} \tag{2A.48}$$

where e_w is the vapour pressure at T_w. This equation can be solved with successive approximations for every initial set of T, p and e. Graphic solutions are popularly used.

Saturation occurs at temperature T_w for the dynamic equilibrium that follows the increase of MR for the water that is added to the system. From the definition, T_w is conservative with reference to the evaporation of falling raindrops. In the case of an evaporating porous surface, the evaporation rate is generally modest, so that T tends towards T_w but hardly reaches it; the MR increases at the interface, generating a negative gradient of MR (i.e. MR decreasing for diffusion from the surface) and a positive gradient of T in the air close to the evaporating surface. In the opposite case of evaporation occurring in the internal pores of a wall, the latent heat is supplied by the masonry and the air temperature remains unchanged, but the evaporation is still evidenced by a negative gradient of *MR*.

An analogy exists between *DP* and T_w: both are based on isobaric cooling until saturation is reached but *DP* is reached without changes in *MR*. On the contrary, T_w is reached with the addition of external water, which raises the *MR* of the air parcel, favouring saturation. For this reason, *DP* cannot be reached with evaporation cooling, even in the case a forced ventilation is applied. *DP* is the temperature typical of condensation and T_w is typical of evaporation.

In addition,

$$DP \leq T_w \leq T \quad \text{and} \quad \Delta DP \geq \Delta T_w \geq 0 \tag{2A.49}$$

where the identity holds only for $RH = 100\%$ when $DP = T_w = T$.

In atmosphere, saturation is usually found on foggy days or at night when dew is forming. In general, during fog, the *RH* is $95\% \leq RH \leq 100\%$. During short rainfalls or showers, saturation is only seldom reached. Summer showers may occur with low *RH* and the precipitated water then evaporates in a short time.

T_w can be easily measured by means of a *psychrometer*, which is composed of a couple of ventilated thermometers, i.e. a normal one (i.e. the *dry bulb*) and another having the bulb (i.e. the *wet bulb*) covered with muslin, which supplies water adsorbed for capillarity from a reservoir. The speed of blown air generally lies between 3 and 5 m s^{-1} (see Chapter 12). It is evident that, for a given temperature, the greater the *MR*, the higher the *DP*. The same can be said for T_w that lies between *T* and *DP*. It is also obvious that T_w is related to the degree of saturation of the vapour in air, in that the higher the *RH*, the smaller the amount of water that should be evaporated to reach saturation and, consequently, the cooling ΔT_w caused by the absorption of the latent heat for vaporization. When the wet bulb is ventilated, its temperature will be lowered until the equilibrium is reached.

The latent heat Q_v lost by evaporation from the wet bulb is

$$Q_v = C\, L_v\, S \frac{e_w - e}{p} \tag{2A.50}$$

where *C* is a proportionality coefficient, L_v is the latent heat of vaporization, *S* is the surface area of the evaporating surface, *e* is the actual vapour pressure, $e_w = e_{sat}(t_w)$ is the saturation vapour pressure at the temperature of the evaporating surface and *p* is the atmospheric pressure. This equation is known as the *Dalton law of evaporation*. On the other hand, the sensible heat Q_s transferred from the ambient air (at temperature *T*) to the colder wet bulb (temperature T_w) is

$$Q_s = B\, S(T - T_w) \tag{2A.51}$$

where *B* is another proportionality coefficient.

At equilibrium, the loss of latent heat Q_v equals the gain of sensible heat Q_s. Equating the two above equations, the basic psychrometric formula is obtained

$$e = e_{sat}(T_w) - A\,p(T - T_w). \tag{2A.52}$$

where $A = (B/C\,L_v)$ is the so-called *psychrometer coefficient*, which is not really a constant and depends, inter alia, on the ventilation rate (see Chapter 12).

In the (2A.30) definition of relative humidity, substituting e with the psychrometric formula, e_{sat} with the Magnus and Tetens formula (2A.1), and the coefficient $A = 6.667 \times 10^{-4}\ \mathrm{K}^{-1}$ and then dividing numerator and denominator by $e_{sat}(0) = 6.11$ hPa, a useful equation is derived

$$RH = 100\frac{10^{at_w/(b+t_w)} - 1.09 \times 10^{-4}\,p(t - t_w)}{10^{at/(b+t)}} \tag{2A.53}$$

which allows a precise determination of RH after measurements of t and t_w and, consequently, precise calculations of MR (Eqn (2A.33)), e (Eqn (2A.36)), AH (Eqn (2A.38)) and DP (Eqn (2A.28)). This equation has been reported in the form RH versus ΔT_w in Fig. 2A.6 for selected values of T.

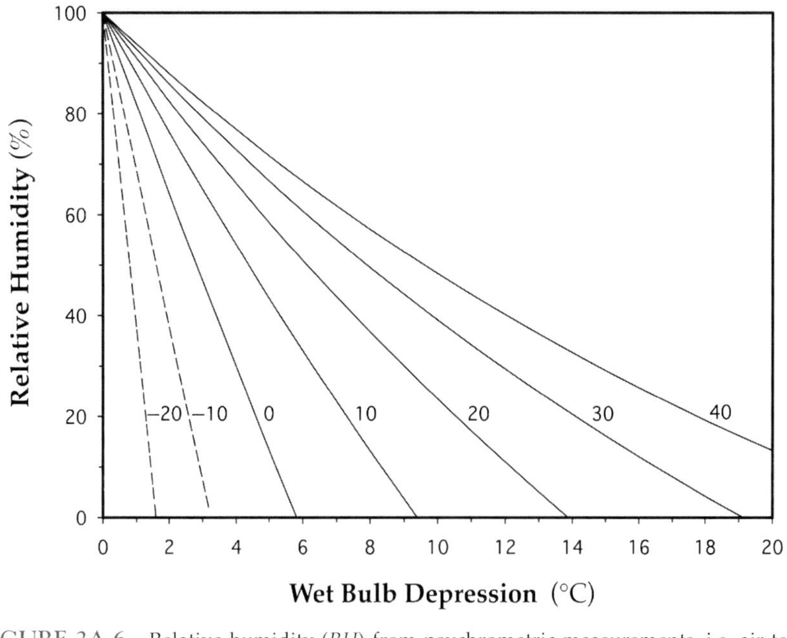

FIGURE 2A.6 Relative humidity (RH) from psychrometric measurements, i.e. air temperature t and wet bulb depression $t - t_w$. Lines refer to different values of air temperature, i.e. from $t = -20$ to $t = 40$ °C. Values below 0 °C (dashed lines) are calculated for vapour in equilibrium with supercooled water.

The ΔT_w depression is also calculated after the readings of T and DP with the following empirical formula:

$$\Delta T_w = \Delta DP(-1.281 \times 10^{-4}\, t^2 + 0.01786\, t + 0.3682) - 0.00889\, \Delta DP^2. \tag{2A.54}$$

where the DP spread is computed using Eqn (2A.45) in the case of known RH instead of DP. Conservators do not commonly use this inverse formula because it gives values that they can directly observe with instruments. However, meteorological services furnish the DP instead of the ΔT_w or the RH and one could use such data to calculate other basic parameters.

2A.10. THE PSYCHROMETRIC CHART

The following thermohygrometric parameters can be graphically computed with the help of the psychrometric chart (Fig. 2A.7).

- The abscissa is the actual air temperature T (i.e. the *dry bulb temperature*) and the ordinate is the MR (or the SH).
- Vertical lines are *isotherms*. Going upwards, the MR increases (e.g. addition of external water, e.g. by evaporation). Going downwards, it

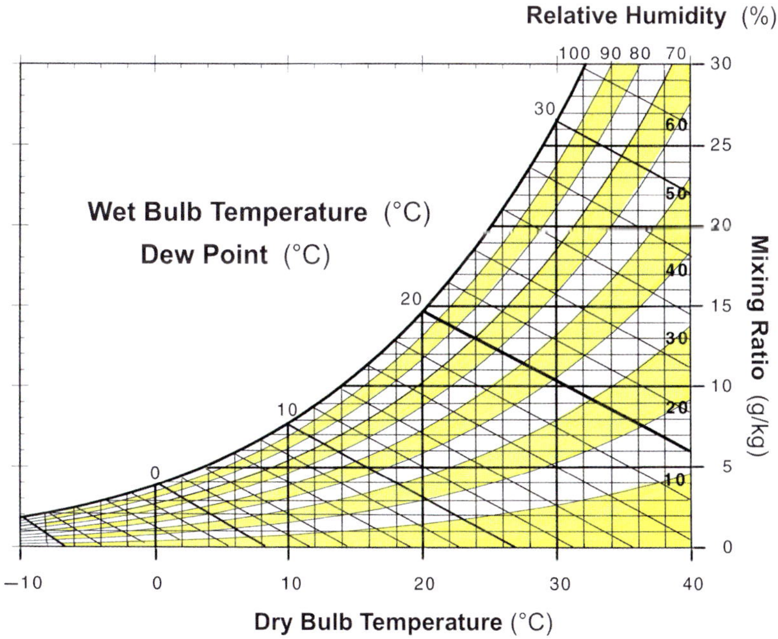

FIGURE 2A.7 The psychrometric chart.

decreases (e.g. condensation, absorption or adsorption of vapour that is subtracted from the atmosphere; the term '*adsorption*' implies the condensation of the vapour on the surface or in internal porosities of solids).

- Horizontal lines are *isohumes* in terms of constant MR. A displacement to the right indicates warming of the system without change in MR; a displacement to the left, cooling, and according to our previous definition, the final point of this "horizontal" cooling is the DP.
- The nearly exponential curve on the left is the *saturation curve* and represents simultaneously the MR_{sat}, the $RH = 100\%$, the DP and the T_w.
- The (vertical) distance between each point of the saturation curve $RH_{sat} = 100\%$ and the abscissa ($RH = 0\%$) is divided into 100 parts. Each of them, by definition (or under the approximation discussed in section 2A.6), shows a given percentage of RH_{sat} or MR_{sat}. Matched values of T and MR used as orthogonal Cartesian coordinates in a psychrometric diagram determine the actual value of the RH of the same air parcel. In the diagram, the nearly exponential curves with $RH = 10\%$, 20%, 30%,... 100% are evidenced. These lines are named *isohumes* in terms of RH.

Tilted lines, where both the T decreases and the MR increases, are characterized by $T_w =$ constant and are nearly *isenthalpic*. In fact, there is a small departure, being drawn for a nonperfect gas. The value of T_w of each thermodynamic point can be read on the saturation curve. Sometimes, the scale for enthalpy of the system is also added.

The *enthalpy* $\mathcal{H}$ of a system in equilibrium at pressure p and temperature T is defined as

$$\mathcal{H}(p, T) = U + pV \quad (2A.55)$$

where U is the internal energy of the system and V is its volume. Therefore,

$$d\mathcal{H} = dQ + p\,dV \quad (2A.56)$$

where dQ is the amount of heat gained or lost. For a process that is *isobaric* and *reversible*

$$d\mathcal{H} = dQ \quad (2A.57)$$

and for a perfect gas (i.e. no change of state)

$$d\mathcal{H} = c_p\,dT \quad (2A.58)$$

the enthalpy is a function of T only, i.e. the change in enthalpy measures the heat imparted to the system. In meteorology, the enthalpy is considered synonymous of *sensible heat* (i.e. heat, in opposition to *latent heat*) as exchanges of enthalpy coincide with exchanges of heat. Indicating by

$\mathcal{H}_{mv}$ the enthalpy per mole of pure water in the vapour phase and by $\mathcal{H}_{mc}$ that in the condensed phase, we get

$$\mathcal{H}_{mv} - \mathcal{H}_{mc} = L_v \quad (2A.59)$$

i.e. the *molar heat of evaporation* denoting the *latent heat of vaporization* for 1 mol of the associated condensed phase. Therefore, changes of enthalpy always represent transfer of heat, either in the sensible or in the latent form.

For an isolated air parcel, all the processes that are simultaneously *adiabatic* and *isobaric* are also isenthalpic. A process that is isenthalpic is also *adiabatic*, as well as *isentropic*, as the *entropy* $\mathbf{S}$ is thermodynamically defined as

$$\mathbf{S} = \int \frac{dQ}{T}. \quad (2A.60)$$

In fact, all these processes are characterized by $dQ = 0$. By definition, a process that is *adiabatic* but not *isobaric* (e.g. the upward rise of an air parcel) is *isentropic* (i.e. $dQ = 0$) but not *isenthalpic*, because of the work done against the external pressure during the variation of the volume of the air parcel. The sum of enthalpies of the phases of a closed system is conserved in an adiabatic isobaric process.

Sometimes, the *isochoric lines*, i.e. the lines with equal specific volume, are also reported, expressed in $m^3\ kg^{-1}$.

Each point in the chart shows directly, or by interpolation, the complete thermodynamic state of an air parcel defined by the following variables:

- *T*, obtained by following downwards the isotherm to the abscissa, and the *MR*, by following the horizontal isohume to the ordinate;
- *DP*, obtained by following the same isohume to the saturation curve;
- T_w, obtained by following the iso wet bulb line (i.e. the isenthalpic) to the saturation curve;
- *RH*, obtained by considering on the same vertical the distance between the thermodynamic point and the abscissa and expressing it as a percentage of the whole distance between the abscissa and the saturation curve, or interpolating with the nearly exponential curves with $RH = \text{const}$ passing close to the point.

A few examples may be useful to clarify the method and to become familiar with the chart.

1. Let us consider the thermodynamic point (Fig. 2A.8(a)) characterized by $t = 25\ °C$ and $RH = 70\%$; the vertical isotherm and the curve of equal saturation 70% meet at a point where the horizontal isohume shows on the right $MR = 14\ g\ kg^{-1}$ and on the left $DP = 19\ °C$. Moisture can be isothermally added to $MR_{sat} = 20\ g\ kg^{-1}$, or isenthalpically to $MR_{sat} = 15.75\ g\ kg^{-1}$, and the corresponding temperature is $T_w = 21\ °C$.

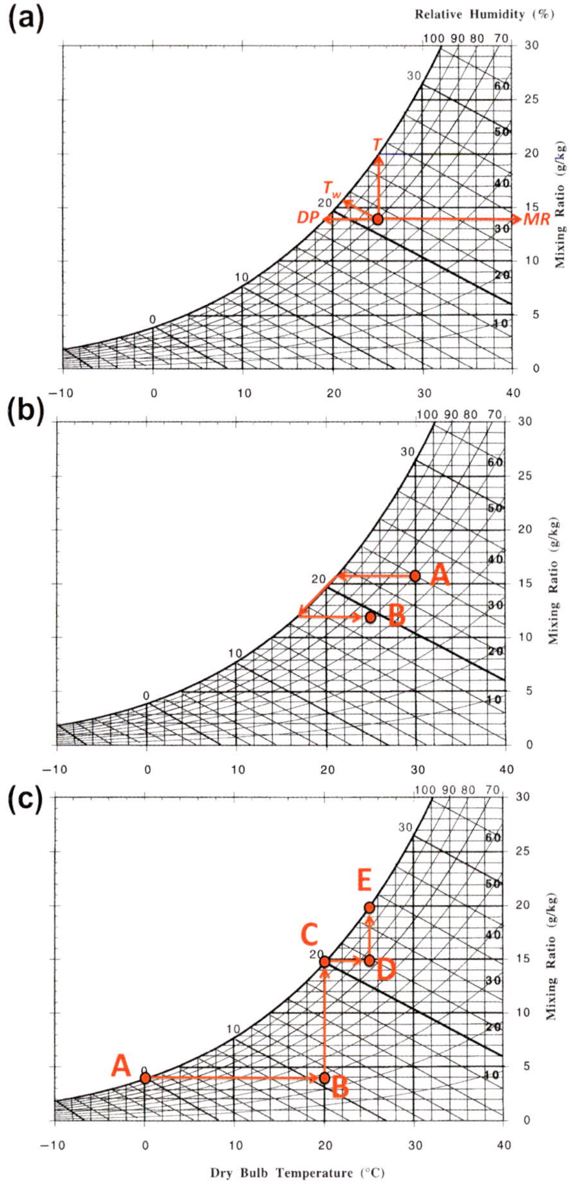

FIGURE 2A.8 (a) Thermodynamic characterization of an air parcel with $t = 25\ °C$ and $RH = 70\%$. (b) Thermodynamic transformation needed to cool an air mass initially at $t = 30\ °C$ and $RH = 60\%$ to $t = 25\ °C$ but with the same RH. (c) Example while drying cloths in a bathroom, testing various drying strategies.

2. In Fig. 2A.8(b), we want to lower the air temperature in a room, initially at $t = 30$ °C and $RH = 60\%$ (Point A), in order to reach 25 °C but with the same *RH* (Point B). We should remove some heat and some moisture by condensation on a cold surface. The air-conditioning system will start cooling the air. Initially, the cooling will occur with unchanged *MR* (i.e. 16.3 g kg^{-1}, with a horizontal displacement along the isohume to left); at around 21 °C the *DP* will be reached. Further cooling will cause the thermodynamic system to follow the saturation curve $RH = 100\%$ with condensation on the refrigerating surface and subtraction of vapour from the air. The system will condense water until the moist air will reach 17 °C and $MR_{sat} = 12$ g kg^{-1}.

   Removing the fraction of liquid water that has been condensed and heating the air parcel to 25 °C with a horizontal displacement, the desired temperature and humidity will be reached.
3. Suppose (Fig. 2A.8(c)) to hang out the damp laundry in a bathroom, when the external conditions are characterized by $t = 0$ °C and fog, i.e. $RH = 100\%$; (Point A) the *MR* is 3.8 g kg^{-1}. The volume of the bathroom is 25 m^3 and contains the total mass 25 m^3 × 1.255 kg m^{-3} = 31.4 kg of air. For simplicity in this example, we will neglect the small density changes. The external air that entered the room was heated to $t = 20$ °C (Point B); as there is no subtraction or addition of moisture, the indoor *RH* is $RH = 25\%$, i.e. very dry. At this indoor temperature, the maximum allowable *MR* is $MR_{sat} = 14.8$ g kg^{-1} (Point C) and the amount of water that can evaporate is given by the number of kilograms of air contained in the bathroom, multiplied by the moisture capacity of each air kilogram, minus the initial moisture content, i.e. 31.4 kg × (14.8 – 3.8) g kg^{-1} = 345 g, which is a very little amount, corresponding, e.g. to one sock. Heating the bathroom to 25 °C (Point D), the MR_{sat} will increase to $MR_{sat} = 20$ g kg^{-1}, which allows a further evaporation (Point E) of 31.4 kg × (20 – 14.8) g kg^{-1} = 163 g, i.e. half a sock. This shows that the critical factor for indoor evaporation is the limited ambient volume and little advantage is obtained by room overheating. Damp laundry needs open windows and continual air exchanges to dry.

References

Fermi, E., 1937. Thermodynamics. Prentice-Hall, Upper Saddle River, New Jersey, 160 pp.

List, R.J., 1966. Smithsonian Meteorological Tables vol. 114. Smithsonian Institution Press, Washington DC.

Magnus, G., 1844. Über die Ausdehnung der Gase durch die Warme (On the expansive force of steam) Poggendorff Annalen der Physik und Chemie. vol. LXI, pp. 225–247, translated from German by W. Francis, pp. 218–239. In: Taylor, R. (Ed.). 1846: Scientific Memoirs Selected from the Transactions of Foreign Academies of Science and Learned Societies and from Foreign Journals vol. IV. R. & J. Taylor, London.

Planck, M., 1926. Treatise on Thermodynamics. Dover, New York, 297 pp.
Tetens, O., 1930. Uber einige meteorologische Begriffe. Zeitschrift Geophysik 6, 297–309.
WMO, 1966. International Meteorological Vocabulary. No 182 TP 91. World Meteorological Organisation, Geneva.
WMO, 1983. Guide to Meteorological Instruments and Methods of Observation No 14. World Meteorological Organisation, Geneva.
WMO, 1988. General Meteorological Standards and Recommended Practices No 49. World Meteorological Organisation, Geneva.

Books consulted for the definitions of meteorological variables in this and the next chapters

Harrison, L.P., 1965. Fundamental concepts and definitions relating to humidity. In: Wexler, A. (Ed.), Humidity and Moisture, Measurement and Control in Science and Industry, Fundamentals and Standards vol. 3. Rehinold, New York, pp. 3–69.
Huschke, R.E., 1959. Glossary of Meteorology. American Meteorological Society, Boston, 638 pp.
List, R.J., 1971. Smithsonian Meteorological Tables. Smithsonian Institution, Washington DC, 527 pp.
Parker, S.P., 1988. Meteorology Source Book. McGraw-Hill, New York, 304 pp.
Saucier, W.J., 1989. Principles of Meteorological Analysis. Dover, New York, 438 pp.
UK Meteorological Office, 1991. Meteorological Glossary. HMSO, London, 335 pp.
WMO, 1966. International Meteorological Vocabulary. WMO No.182, World Meteorological Organization, Geneva, 276 pp.
WMO, 1987. International Cloud Atlas, vol. II. World Meteorological Organization, Geneva, 212 pp.
WMO, 2008. Guide to Meteorological Instruments and Methods of Observation. WMO Technical Publication No 8, World Meteorological Organization, Geneva.
WMO, 2008. Guide to Hydrological Practices. In: Hydrology – from Measurement to Hydrological Information, sixth ed. vol. 1. WMO Technical Publication No 168, World Meteorological Organization, Geneva.

CHAPTER

2B

Humidity and Conservation

OUTLINE

Microclimate for Cultural Heritage, Second Edition
http://dx.doi.org/10.1016/B978-0-444-63296-8.00003-2

2B.1. AIR–SURFACE INTERACTIONS AND ENVIRONMENTAL DIAGNOSTICS

The humidity mixing ratio (*MR*) is very useful for diagnostic purposes, to provide evidence of the action of heating, ventilating and air conditioning (HVAC) systems or air–surface interactions. By measuring this parameter along a horizontal cross-section of a room, it is possible to see the advection of external air penetrating through open (or through fissures of) doors or windows (Fig. 2B.1), the moisture released by visitors, or where the walls are evaporating or adsorbing moisture. The effects of humidifiers are also pointed out, as they do not generate a homogeneous increase of vapour content but a cloud of moist air that might damage objects in proximity or moves into the room borne by the advective currents between adjacent rooms. A very common, but inappropriate, position for a humidifier is near corners or walls where works of art are located.

Humidifiers generate clouds of moisture that grow in sheltered space (Fig. 2B.2). Then these clouds diffuse into the room, are displaced by local air currents, or are broken and transported in several directions by the turbulence generated by the motion of visitors. Works of art are alternatively immersed into, or emerge from, these clouds or portions of them.

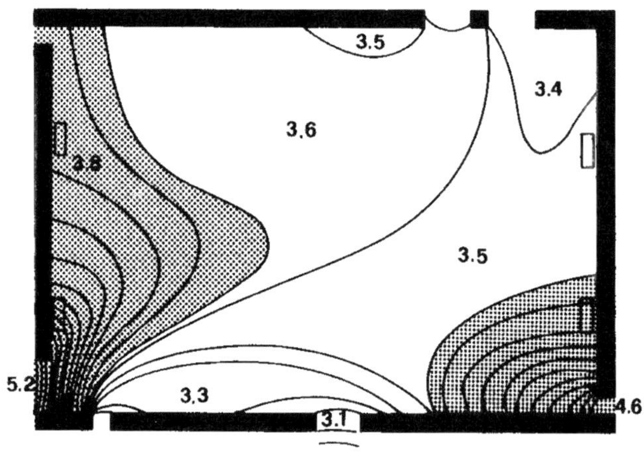

FIGURE 2B.1 Mixing ratio (g kg^{-1}) in a horizontal cross-section of the Giotto Room in the Uffizi Gallery, Florence, showing the penetration of external air masses evidenced with shading (1 February 1993, 4 p.m.). Sometimes the air coming from elsewhere is only a bit different and can be observed as a useful tracer of different air masses. In this case, the emission of moisture in the different rooms is not balanced in comparison with the size and needs of each room. Therefore, the air coming through doors or the vertical slits opened for the passage of paintings has different hygrometric characteristics and may perturb the exhibits causing internal stresses.

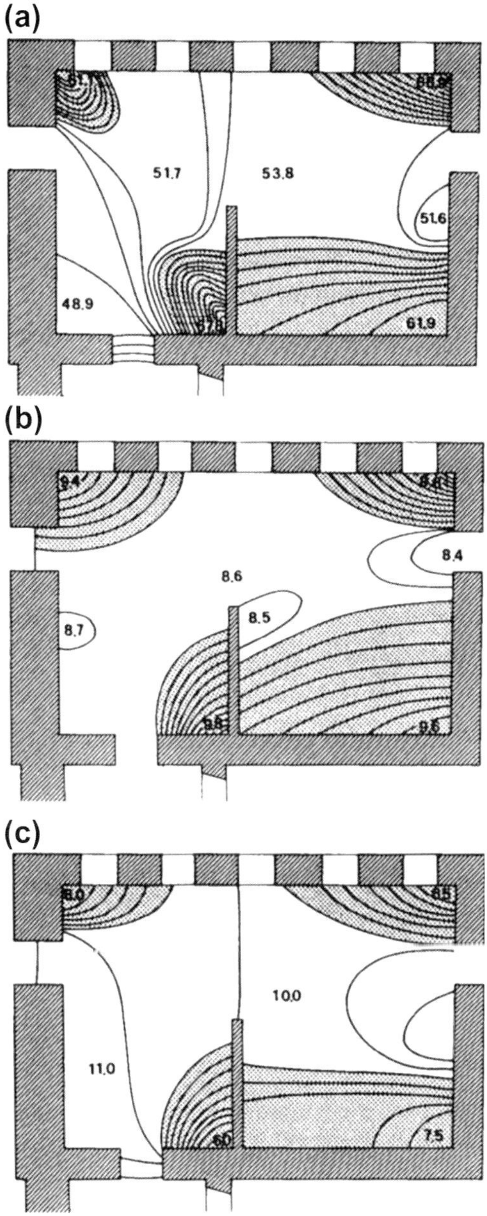

FIGURE 2B.2 Micromapping in a horizontal cross section of the *Salon Carré*, Louvre Museum, Paris, showing clouds of moisture (evidenced with shading) diffusing into the room, which are generated by humidifiers. Four active humidifiers are placed in corners of the room. Others are in stand-by position. This situation is here described by means of three parameters, in order to gain familiarity with different representations, i.e.: (1) relative humidity (*RH*, %), (2) mixing ratio (*MR*, g kg^{-1}), (3) dew point spread (*DPS*, °C). The clouds of moisture appear as local maxima of *RH* or *MR* and minima of *DPS* (21 February 1995).

The result is continuous stress and fatigue of objects that lead to weakening, microfissuring and flaking of their outermost layer of their colour coating. A more appropriate position is far from the exhibits, in the central part of the room, either on the floor (e.g. Kunsthistorisches Museum, Vienna or Uffizi Gallery, Botticelli room) or on the ceiling, although the latter position presents a further problem: the suction outlets placed on the lower part of walls determine a preferential path of the moisture into the room and a nonhomogeneous distribution.

Gradients of the *MR* close to a surface show that the surface is exchanging vapour with the atmosphere. If the surface is evaporating, the air is locally enriched with water molecules and the *MR* locally increases; on the other hand, if the surface is adsorbing water molecules or is condensing, the air in proximity to it is depleted of water molecules and the *MR* locally decreases. On monitoring this parameter near walls directly lighted by solar radiation or artificial lamps, release of moisture can be seen for a short time during heating and it stops when the plaster comes in equilibrium with the hot spots (Camuffo and Bernardi, 1991, Fig. 2B.3).

The opposite occurs when the wall cools again after the input of radiant energy has stopped. The repetition of daily cycles of direct light, always on the same area, may be noxious to vulnerable objects.

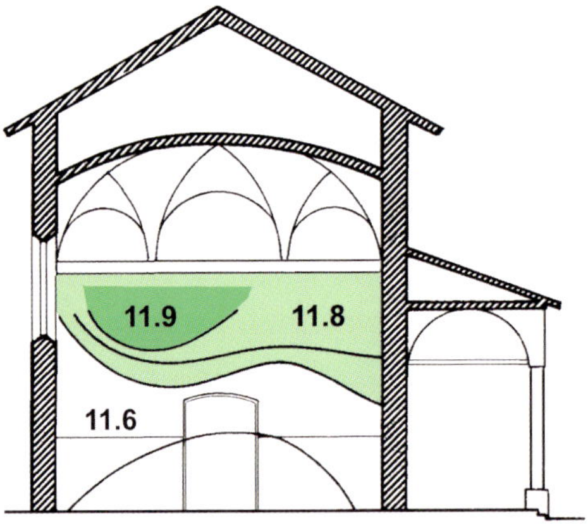

FIGURE 2B.3 Mixing ratio (g kg^{-1}) in a vertical cross-section in front of the Leonard's *Last Supper* (Milan) showing the local release of water vapour from the area hit by direct lighting with incandescent lamps, 29 July 1982. The vapour released from the wall causes a local enrichment of *MR* in the air nearby, i.e. the maximum found in the upper part of the wall, on the left. The radiant energy warms the wall surface and forces some evaporation. When the equilibrium is reached to the micropores, the evaporation stops and the *MR* maximum disappears. *Source: By kind permission of European Commission, DG XII, and* Bollettino Geofisico. *See credits.*

If a cold surface is put in contact with the environmental atmosphere, initially the air in contact with the surface decreases its temperature *T* and the *MR* remains unchanged. This increases the relative humidity (*RH*) level. As soon as the condensation begins, the *MR* starts to decrease and this tends to invert the trend of *RH*. After some time, the air in the proximity of the surface will be characterized by $\Delta MR < 0$ and nothing can be said about the final value of *RH*, which is subjected to two counteracting effects due to *T* and *MR*. In general, the thermal effect prevails and $\Delta RH > 0$. When a wall is hit by direct solar radiation it warms ($\Delta T > 0$), evaporates ($\Delta MR > 0$) and the opposite situation occurs. These obvious facts appear as a paradox when improperly described: *the air is dryer in proximity to a surface where condensation is occurring and moister near an evaporating surface*. Misinterpretation may derive when 'dry' or 'moist' is not specified *in terms of MR* and people generally associate the concept of *dry* and *moist* with *RH* (Chapter 2A). From the definition, the air is dry or humid in terms of *MR* as a consequence of its water vapour content only, independent of the drying rate capacity or the temperature *T* of the system. In contrast, 'dry' or 'moist' in terms of *RH* means how far or close to saturation the vapour is and this depends on two factors: *T* and *MR*. A 'dry' air parcel in terms of *MR* becomes 'humid' in terms of *RH* when it is cold, whereas one 'moist' in terms of *MR* becomes 'dry' in terms of *RH* when it is warm.

Plots of *RH* constitute a very useful tool for environmental diagnostics and risk assessment and the methodology is the same for air temperature and *MR*. In a room, it should be advisable to keep the *RH* level unchanged, both in time and in space. In the example of the historic Anatomy Theatre, Padua, a humid corner is evidence for the presence of a damp wall (Fig. 2B.4).

2B.2. THE EQUILIBRIUM MOISTURE CONTENT AND DIMENSIONAL CHANGES IN WOOD

Several materials are very sensitive to their *moisture content* (*MC*) and this may reach equilibrium with the local microclimate. Changes in *MC* are responsible for material swelling and shrinkage. In materials science, a concept similar to the *MR* is used in defining the *MC*, i.e. the mass of water per unit mass of anhydrous material, and is generally expressed in percent of weight. The *equilibrium moisture content* (*EMC*), is also defined, i.e. the *MC* reached at equilibrium with ambient temperature and humidity. The water molecules may be adsorbed, as in many organic (e.g. wood, paper, parchment, leather, ivory, bone and soap) and some inorganic materials (e.g. plaster, stucco and stones containing abundant clay minerals) or may be transformed into hydration or crystallization

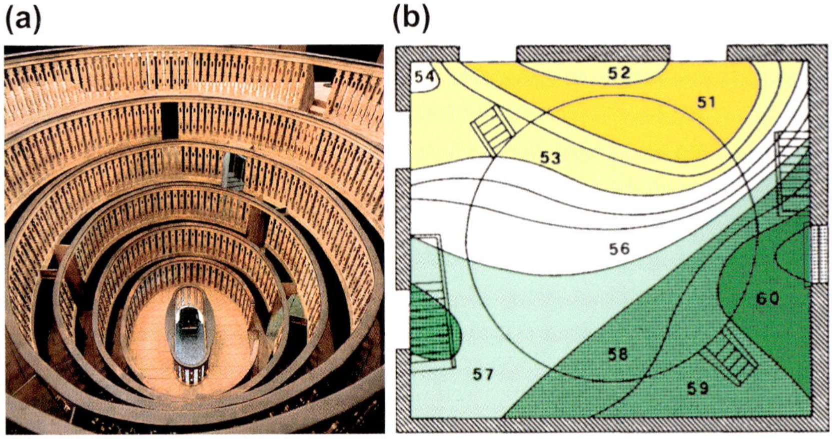

FIGURE 2B.4 (a) The historic Anatomy Theatre, University of Padua. (b) *RH* (%) in a horizontal cross-section of the theatre (3 July 1992, 5 p.m.). The humid corner is green.

water, as in some inorganic salts (e.g. sodium sulphate, copper sulphate and barium chloride). *EMC* varies with ambient temperature and *RH*. It is nearly zero for nonporous, nonhygroscopic materials; for other materials, it vanishes only at null *RH* level. In some inorganic (e.g. clay minerals) and many natural organic materials, microclimate changes induce *EMC* and, consequently, swelling, shrinkage and internal stresses.

Wood is composed of a combination of three chemical polymers: *cellulose, hemicellulose and lignin*. The first two are linear polysaccharides that form chains of glucose subunits and may break by hydrolysis. Each glucose subunit possesses three hydroxyl groups that have affinity for water. *Cellulose* forms long unbranched chains and long twisted microfibrils whose structure is complex and not yet known in detail. *Hemicellulose* forms short-branched unstable chains and absorbs water to any extent. When wet, hemicellulose may break by hydrolysis; when dry, it may break down into carbon dioxide and water. This degradation process is accelerated with higher temperature and in the presence of acids, bacterial attack and visible and ultraviolet radiation. Dimensional changes of wood, when *RH* changes, are mostly due to the amount of water adsorbed by hemicellulose. When water is adsorbed, the material will increase in size and decrease in strength for the attenuation of the hydrogen bonds and the van der Waals intermolecular forces. When a certain threshold is exceeded, irreversible yield and cracks might be generated. *Lignin* is a complex and heterogeneous mixture of polymers that form an amorphous phenolic adhesive that is rather stable because it has a lot of internal H bonds. It encrusts and stiffens the cellulose and

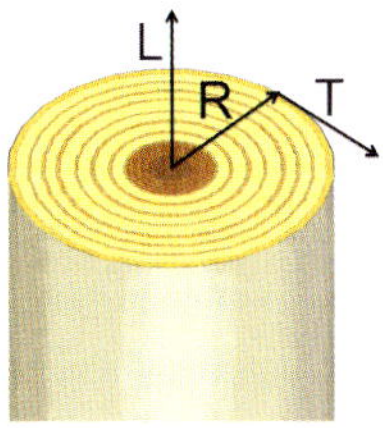

FIGURE 2B.5 The three key directions in wood, i.e. T, *tangent* to the tree growing rings; R, *radial* from the centre of rings to the exterior; and L, *longitudinal*, along the direction of the trunk.

hemicellulose polymers and contributes to the formation of woody cell walls of plants and the cementing material between them (Copedé, 1991; Richardson, 1993; Winandy, 1994; Gelbrich et al., 2008; La Mantia and Morreale, 2008; Vanholme et al., 2010).

For the peculiar anatomic structure of plants, three key directions should be considered in wood: (1) *tangent* to the tree growing rings; (2) *radial*, i.e. from the centre of a cross-section of a trunk to the exterior; and (3) *longitudinal*, or parallel to grains, i.e. along the direction of the trunk (Fig. 2B.5). Anisotropic deformations generated by changes in temperature and humidity result in strain and strain-induced stresses. They force internal tensions, deformation, fatigue and often fractures. Common effects are configurational strain (e.g. warping, cupping and bowing) or fracture (cracking and splitting).

The *MC* is 30% or more for green wood and around 20% for seasoned wood at ordinary temperature and *RH*; in indoor artefacts, the *MC* typically decreases to between 7% and 15%, although with departures at extreme *RH* changes.

When the ambient *RH* decreases, the *EMC* too decreases and the wood shrinks with important wood dimensional changes and the reverse occurs when *RH* increases. The dependence of wood *EMC* on temperature and *RH* is shown in Fig. 2B.6, where it is evident that wood is weakly sensitive to temperature changes but very sensitive to *RH* changes. Changes in temperature should not be disregarded because they might have a small direct impact on wood but a large indirect impact for the change in *RH*.

Wood looses moisture when the ambient temperature increases because the increasing kinetic energy of the water molecules counteracts the H bonds with the result that a number of molecules will return to the gaseous phase. In the temperature range 0 °C−40 °C, the average amount of *EMC* lost per each 1 °C temperature increase is shown in Fig. 2B.7. The loss increases at increasing *RH* levels, i.e. after *RH* = 75%, with a trend showing acceleration at high humidity levels, similar to the plots in Fig. 2B.6.

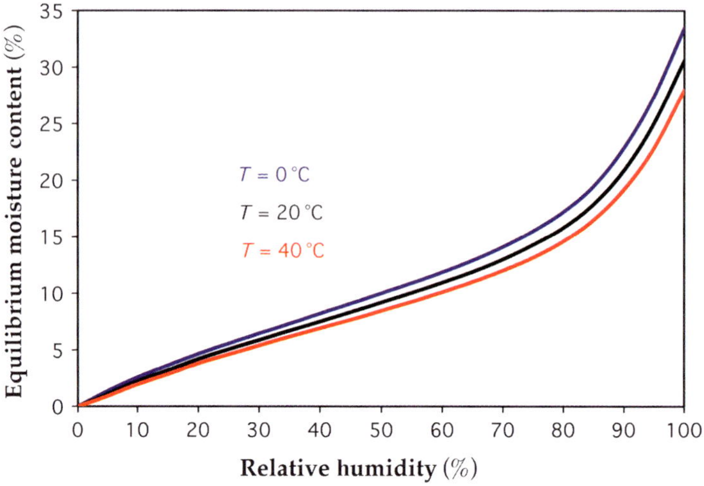

FIGURE 2B.6 Dependence of wood equilibrium moisture content (*EMC*) on ambient temperature ($T = 0\,°C$, blue; $T = 20\,°C$, black; and $T = 40\,°C$, red) and *RH*. Wood is very sensitive to humidity changes, especially in moist environments, e.g. $RH > 75\%$.

Returning to Fig. 2B.6, the *EMC* increases at a higher rate at high *RH* levels (e.g. $RH > 75\%$) and this is crucial because the absorbed water exceeds the critical level to allow mould colonization. In the humidity range $10\% < RH < 70\%$ the plot of *EMC* versus *RH* is almost a straight line or slightly bent to form a more or less horizontal plateau in the *RH*

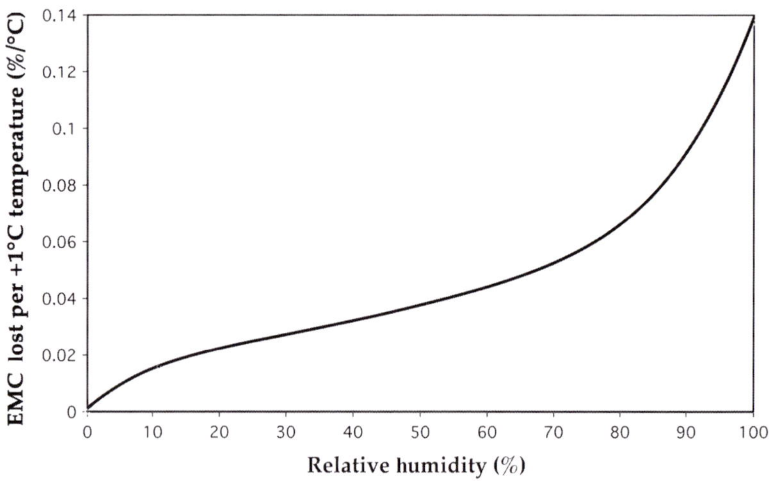

FIGURE 2B.7 Equilibrium moisture content lost in wood per each 1 °C temperature increase versus ambient *RH*.

interval from 40% to 60% for some wooden species (e.g. poplar). In this humidity interval and at 20 °C room temperature, for a number of wooden species, the *EMC* observed by Giordano (1993) in average obeys the equation

$$EMC = 5.9\,RH - 4.7\,(\%) \tag{2B.1}$$

but the coefficients may vary with the wood species, seasoning and temperature.

For practical purposes, in the *EMC* range between 0% and 20%, the relationship may be assumed to be linear (Summit and Slicker, 1980). The shrinkage, expressed as percentage of the reference size at *EMC* = 20%, is shown in Fig. 2B.8 for two wooden species, i.e. oak and pine, and for the two main shrinkage axes: the direction tangent to the tree rings and the radial one.

The wood shrinkage (or swelling) versus the equilibrium *RH* is evidenced combining the findings in Fig. 2B.6 and Fig. 2B.8. Figure 2B.9 shows the response of wood (with initial *EMC* = 20%) to ambient humidity changes: wood swells in damp environments and shrinks when drying. Poplar remains the most stable species.

From the above figures it is evident that the tangential change is twice the radial one and that both are greater than the longitudinal one. The coefficients for tangential, radial and longitudinal dimensional changes follow the respective proportion 1: 1/2: 1/10 or less for longitudinal one. For wooden panels bound together, the most important factor is the

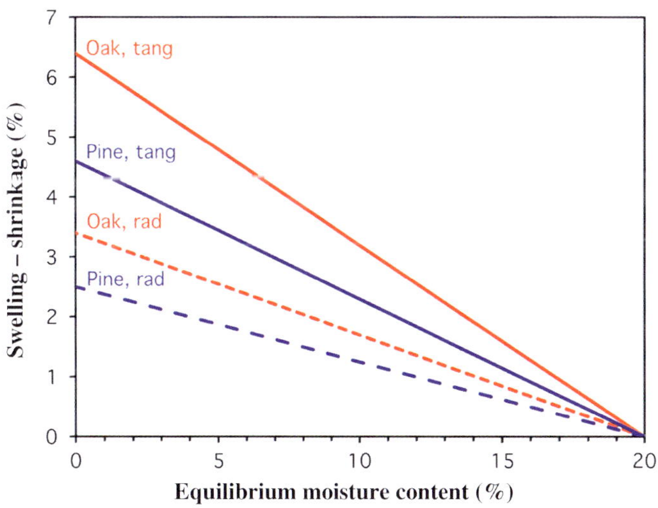

FIGURE 2B.8 Wood dimensional change (shrinkage for decreasing *EMC* and swelling for increasing *EMC*), expressed as percentage of the reference size at *EMC* = 20%, for oak (red line) and pine (blue line) and for the two main axes, i.e. the tangential (continuous line) and the radial (broken line) ones. The tangential change is about twice the radial one.

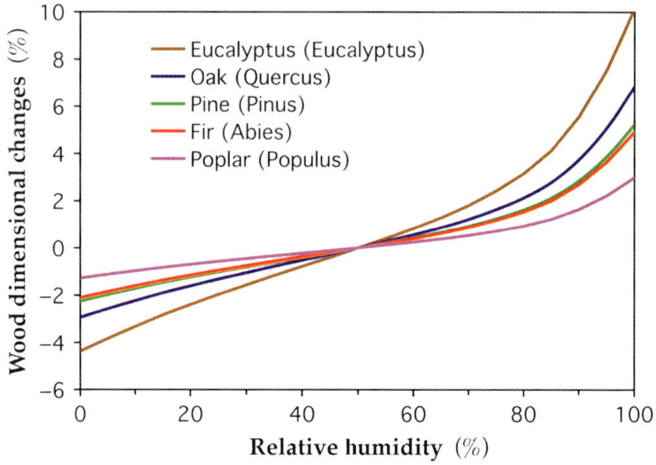

FIGURE 2B.9 Wood dimensional changes (swelling or shrinkage), expressed as percentage versus the equilibrium *RH*. Departures are calculated from the initial condition at *RH* = 50% and *EMC* = 20%. Lines refer to tangential swelling or shrinkage for a number of wooden species as indicated in the legend.

combination of tangential, radial, or mixed cuts. For roof timbers, the most critical change is in length, which fortunately is also the smallest one. However, the considerable length of the timbers may induce large changes in size.

Poplar is the most stable species and this justifies the use in panel paintings, or for stringed instruments, especially the famous lutes cellos and violins from Cremona, Italy, built in the seventeenth century by Nicola Amati, Andrea Guarnieri and his grandson nicknamed '*del Gesù*' (i.e. Jesus Christ) for the splendid voice of his instruments. Antonio Stradivari, also from Cremona, made his instruments with spruce fir, which is the second in stability, but he improved stability against *RH* change and close breathing of the player by applying an impermeable varnish on both the external and the internal sides of the instrument. This practice was followed in Cremona and is one of the reasons of the high quality of these instruments. We should exclude the hypothesis that these instruments were of special quality because Stradivari used red fir grown on the mountain slopes of Val di Fiemme, Italian Alps, in the middle of the Little Ice Age and the cold might have improved wood density, strength and acoustic properties. Regular temperature observations made in Central and Northern Italy for the 1654–1670 period show that the temperature increase was modest in comparison with the 1961–1990 reference period (Camuffo and Bertolin, 2012), so that the temperature belt that characterized the seventeenth century now has been slightly displaced upwards. Therefore, trees grown under the same climatic conditions can still be found on the same mountain slope but displaced towards higher elevations by some 100–200 m. The Stradivarius secret does not consist in the past climate, or in various varnishes that were used, but in the genial skill of the instrument maker.

Temperature and moisture changes will affect the equilibrium and induce gradients with dimensional changes and internal stress to wood. Heat migrates quite rapidly in comparison with moisture, and moisture moves 12–15 times faster along the grain than across it. Any temperature or *RH* unbalance caused by heaters, air-conditioners or humidifiers will cause exchanges of heat and moisture starting from the wooden surface and extending to the interior. This migration and the related shrinkage/swelling will generate internal stress in panels, frames, furniture and so on. This means that temperature peaks for short-term heating or exposition lighting will only affect the surface or subsurface layer, maybe with cracking/swelling of colour coating. Impermeable varnishes or colour coatings might attenuate moisture exchanges.

On the other hand, long-term climate unbalances will affect the deeper layers causing structural stress, possibly forming macrocracks. The different migration rate of moisture along or across grains is not very relevant for wooden panels where the effect is limited to edges but becomes relevant in statues or decorations where the carved surface is characterized by areas with cuts along grains and areas with cuts across, determining areas with different exchange rate, which are potential sources for structural stress. If the microclimate changes exceed certain thresholds that will be discussed later, they will generate differential expansions in wooden structures, which may undergo severe stress and cracks or permanent dimensional changes (Fig. 2B.10).

It should be noted that any warming will cause a modest thermal expansion in wood but a consistent lowering in *RH* and this, in turn, will cause a decrease in the *EMC* followed by wood shrinkage, and vice versa. By the way, the thermal expansion and the *EMC* shrinkage are opposed but the latter is dominant. As the diffusion coefficients of heat and moisture of wood are different, the former being larger (i.e. heat moves faster) and the latter smaller (i.e. moisture moves slower), wood will start reacting to the temperature change and will stop reacting to the moisture change.

It is clear that it is best to keep wooden objects in their historic climate, avoiding any *RH* change or fluctuation. However, this is often impossible or too expensive. A fundamental question is what is the maximum allowable *RH* fluctuation that may occur without causing damage? The answer is not easy. ASHRAE (2007) and Michalsky (2009) introduced the concept of the '*proofed fluctuation*', that is the largest *RH* fluctuation that

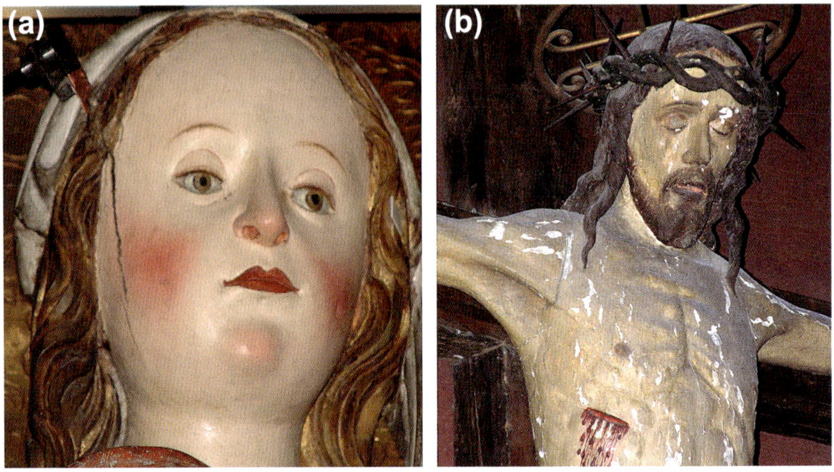

FIGURE 2B.10 (a) Crack formed on the face of an altarpiece sculpture. (b) Wooden shrinkage and blistering of the gesso and the colour coating layer. Both cases had occasional warm-air heating.

an object has accidentally experienced passing apparently untouched. The idea is that this was a severe unplanned test. If the object survived in good conditions, the probability that it will suffer damage if the same fluctuation will be repeated is very low. However, it has been observed that the repetition of *RH* cycles begins to form inadvertent internal microfractures with detectable acoustic emission (Jakiela et al., 2007) and that in the long run these grow internally until macroscopic cracks form, reaching the surface (Bratasz et al., 2007; Bratasz, 2010).

Mecklenburg, Tumosa and Erhard (1998) and Bratasz (2010) made useful laboratory tests to assess the vulnerability of wooden samples at some selected *RH* changes. They found that wood is more resistant in the middle of the *RH* scale, i.e. $40\% < RH < 60\%$, and more vulnerable at the extremes, i.e. in dry or moist environments where even a small fluctuation might be noxious. A synthesis of the above results is represented in Fig. 2B.11. They found that wood has reversible dimensional changes when *RH* fluctuates within the green area indicated in the figure. When fluctuations overpass the yield borderline and enter the yellow area, deformations become irreversible and the deformations are larger when the *RH* fluctuations deeply penetrate within the yellow area. When the environment becomes too dry, and the *RH* overpasses the failure borderline and enters the red area, macroscopic cracks will develop. However, it is useful to underline that the above results apply to small wooden samples used for laboratory tests.

The situation is different for complex artefacts made with wooden boards bound with each other with glue and nails in a different combination of

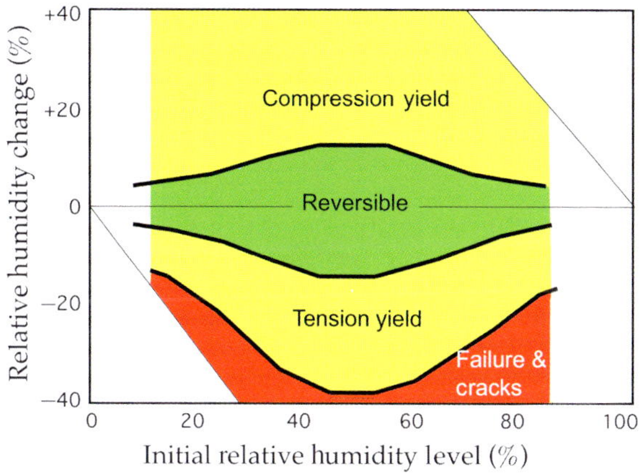

FIGURE 2B.11 Sustainable yet nonsustainable *RH* changes for a commercial wooden sample, laboratory test. Green area, reversible dimensional changes; yellow area, irreversible changes, i.e. compression and tension yield; and red area, failure and cracks.

tangential, radial and longitudinal cuts. The problem is worsened for aged wood that loses strength for rotting or insect tunnelling. In such a case, even small departures from the historic climate may be noxious and failure and cracks may form as expansion joints as an adaptation of the object to respond to the nonsustainable microclimate variability.

Another difference concerns aged wood with extensive insect tunnelling. In principle, tunnelling does not affect dimensional changes (i.e. shrinkage and swelling) but the material loss weakens the wooden structure as well as the internal tensions generated by dimensional changes. These two effects will tend to compensate among them, but not equally, the internal tunnelling distribution being critical.

2B.3. MECHANISMS OF HUMIDITY DEGRADATION IN PAPER AND PARCHMENT

2B.3.1. Mechanical Effects

Paper, cardboard, parchment, leather, rope and other materials are very sensitive to *RH* changes. Dimensional changes are so remarkable that around 1664 Francesco Folli invented a hygrometer based on a paper ribbon and Vincenzo Viviani made some improvements to it (Fig. 2B.12).

In the Folli hygrometer, a paper ribbon serving as a hygroscopic sensor is fixed at both ends that are kept some centimetres above a wooden basement (Fig. 2B.12(a)). The ribbon is weighted in the middle with a small pointer that confers a V shape to it. Dry air causes the paper ribbon to shrink thus raising the pointer, while humid air causes it to stretch thus

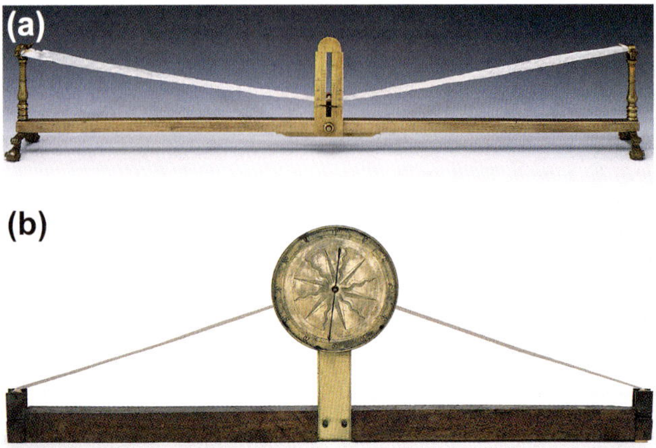

FIGURE 2B.12 Two types of the Folli & Viviani hygrometer, based on the dimensional changes of a paper ribbon (later substituted with parchment), which is in equilibrium with the ambient *RH*. (a) V-shaped type, with a weighted pointer placed in the middle of the ribbon and moving vertically. (b) Rotating pointer type, where the stretching and shrinking ribbon drives the pulley fixed to the pointer. *Source: By kind permission of Museo Galileo, Institute and Museum of History of Science. See credits.*

lowering the pointer. Displacements from the midposition are related to dryness or dampness. In another similar type (Fig. 2B.12(b)), the paper ribbon is fixed at one end of a wooden tablet and is tight with a small weight, or a spring, at the other end (not visible in the figure). Changes in ribbon length drive a pulley with a pointer that rotates over a circular graduated scale fixed on the middle of the tablet, where humidity levels were marked in degrees of dry and wet. Both instruments were improved by substituting paper with parchment, which is much more resistant to traction and use.

The Folli hygrometer, based on paper or parchment, makes clear that humidity changes are particularly dangerous to books that are composed of a number of parts and materials bound to each other but with different properties, e.g. paper leaves, hard cover in leather, cardboard or cloth, sewing tapes or cords and glue. All these constituents will react differently to moisture changes with dramatic mechanical effects.

An example is given in Fig. 2B.13 where a book, kept for centuries in a room at high humidity levels (above 70% *RH*), was moved to another location, with lower humidity level (around 40% *RH*). Different humidity levels can be found suggested in the literature; the most popular level is 45 and the rule of thumb is '*if RH is halved the life expectancy is doubled*' (Michalski, 2000; Canadian Council of Archives, 2003). Apparently, the novel location had better humidity levels to resist acid-catalysed hydrolysis, free radical-mediated oxidation, mould colonization and pest insect attack, in line with the above-mentioned rule of thumb that is valid for an

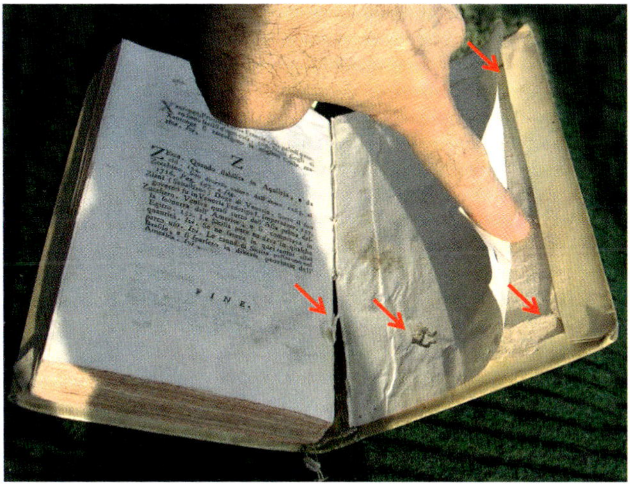

FIGURE 2B.13 Damage caused to a book on moving it from its historic climate. Arrows point out various types of mechanical damage on the binding leather cover, finishing paper leaf and binding cord.

individual loose leaf. However, in the complex case of a book, the tensions generated by the different shrinkage of the various components tore and damaged the binding leather cover and finishing leaf and broke the binding cord. In this case, the change from the historic climate to a novel condition may have been better for the cellulose degradation but was nefarious to the book integrity.

Not only dampness but also dryness may affect the molecular structure. Paper has a structure that is weaker compared to wood and is more sensitive to humidity changes. Moderate cold and dryness is better for storage and conservation but makes paper brittle. Moderate humidity causes the absorption of a number of water molecules, which ensures mobility to the cellulose chains, thus increasing flexibility, and this environmental condition is more convenient for use. Therefore, when a book that is kept in a controlled repository of an archive or a library is moved to be consulted, it should pass from a cold dry storage, to an intermediate transition climate, and then to the reading room, which should be moister to make paper more elastic and warmer to be more comfortable for the reader. For this reason, the choice of the *RH* may change according to the use.

2B.3.2. Chemical Damage to Paper

Using temperature (Arrhenius equation) and *RH* as inputs, Sebera (1994) calculated a set of curves of equal durability, called *'isoperms'* making reference to the reaction rate at 20 °C and 50% *RH*. Durability is the reciprocal of the constant reaction rate i.e. $1/\kappa$ where κ is the reaction rate. From the reciprocal of isoperms one gets similar lines but characterized by the same degradation rate. Such lines are called *'isoburns'* and give the same rate of decay for hydrolysis. The above lines are useful in evaluating risks, as well as the costs and the most convenient strategy, i.e. to reduce temperature, *RH*, or both, in order to increase the lifetime of materials sensitive to hydrolysis.

Another consequence of dampness is oxidation, especially for modern paper (i.e. after mid-nineteenth century) that is composed of wood pulp and contains several free radicals and other acidic elements. Factors such as the presence of inks, temperature, pH, humidity and the application of cleaning techniques may be implicated in accelerating cellulose degradation. Damage is particularly severe in the presence of atmospheric pollutants and especially ozone. In addition, when paper is stored in a humid environment, it releases volatile organic compounds (VOC) and sulphuric acid that accelerates the deterioration (Lattuati-Derieux et al., 2004; Bulska and Wagner, 2005; Strlič et al., 2007; Ramalho et al., 2009; Havermasns, 2010). The typical smell in repositories is due to paper degradation.

Similar to paper, the collagen in parchment is arranged with polymeric chains with problems similar to those discussed for cellulose in paper and dampness, or extreme dryness, may irreversibly alter the structural composition.

From about the fifth century to the nineteenth century and later, the most popular writing and drawing ink in Europe was the so-called *iron gall ink*. This is a purple-black or brown-black ink made from iron salts and tannic acids from vegetable sources, especially oak galls. In humid environments, the water absorbed inside the paper may act as a solvent for the ink with two major consequences. The first is that solute ink will migrate spreading on and crossing the paper leaf, forming stains on the verso (Fig. 2B.14). The second is that water activates the tannic acids that in the presence of catalysts, especially metals, will corrode the paper, especially in

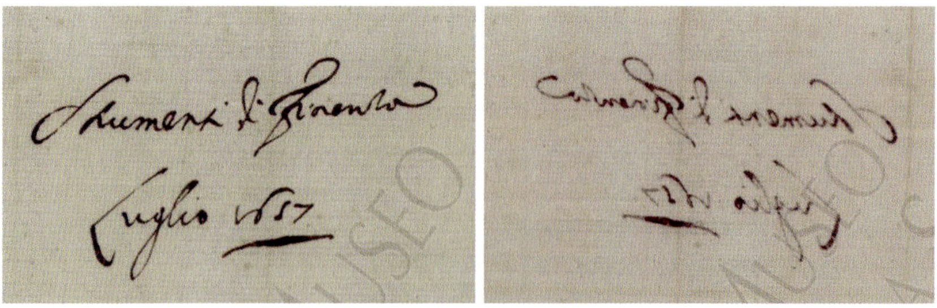

FIGURE 2B.14 Recto and verso of a leaf with the writing 'Strumenti di Fiorenza – Luglio 1657' (Instruments of Florence, July 1657). The iron gall ink has crossed the leaf and is visible on the verso. *Source: By kind permission of Museo Galileo, Institute and Museum of History of Science. See credits.*

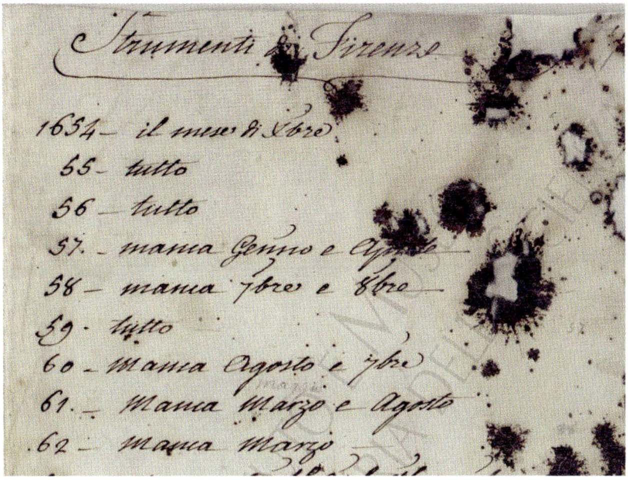

FIGURE 2B.15 Corrosion of paper caused by iron gall ink stains. *Source: By kind permission of Museo Galileo, Institute and Museum of History of Science. See credits.*

correspondence to thick lines or stains where the concentration of acidic radicals is higher (Fig. 2B.15). Some EU projects were focused on clarifying this complex mechanism (Kolar et al., 2004, 2005, 2007; Malešič et al., 2005; Strlič et al., 2009).

2B.3.3. Extreme Events

Flood water, water used to stop fire, rain or pipe breakage may invade libraries and archives. The key problem is that moulds will infest damp books and documents in 1 or 2 days, especially when water is mixed with mud, sludge, and other contaminants including bacteria. Rapid reaction helps to minimize the loss of soaked books. In the past, the only possibility was to remove all leaves, rinse and dry them in a warm, ventilated room, hanging leaves from a rope. In the case of the dramatic flood that invaded Florence in 1966, Prof. Umberto Baldini, who was in charge of the emergency recovery and subsequent restoration, requisitioned lemon greenhouses to store soaked furniture, wooden objects, statues and polychromy on panels for slow drying and controlled shrinkage; in contrast, he used pottery factories and tobacco leaf drying rooms for quick drying of the huge stuff of rinsed documents and book leaves, as they were in a ventilated, gentle drying oven (Baldini, private communication; Fig. 2B.16). Only the National Central Library in Florence had over 1,300,000 books and documents affected and the works to recover and restore books continued for decades.

Nowadays the technology has changed and the so-called *'vacuum freeze drying'* has become popular, although some problems remain unsolved

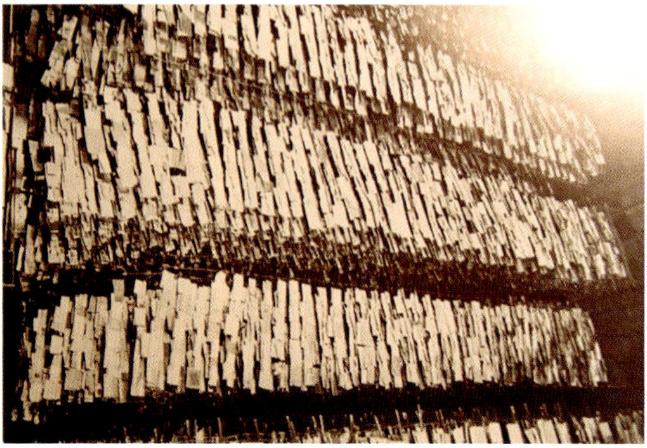

FIGURE 2B.16 Leaves removed from books, rinsed and hung to dry on ropes after the flood of the Arno River in Florence, 3–4 November 1966. Leaves are hung with clothes pegs, visible in the lowest row. Unknown author.

(Mc Clearly, 1987; Day and Stacey, 2007). Leather and vellum items may be damaged and the polymers forming paper may also be degraded and the effects of this chemical decay will appear in the long term.

The first step of vacuum freeze drying consists in rapid freezing of books to avoid swelling, mould formation, hydrolysis and oxidation. Frost will stop the progress of ongoing deterioration mechanisms, allowing time for recovery and restoration. Frost is usually made at freezing temperature from −10 to −30 °C and the rate at which books freeze is critical for further mechanical or chemical deterioration mechanisms. Micro ice crystals are less dangerous than macrocrystals but they require very low temperatures. However, fast cooling rates generate freezing fronts inside books that alter the penetration of cold. The progressive advancement of the freezing front causes a hardly controllable situation characterized by a sharp temperature discontinuity at the front and moving with it.

The second step is drying under vacuum conditions, i.e. lyophilization, and conversion of solid, frozen water to vapours while bypassing the liquid phase. Water vaporizes without passing through the liquid phase, avoiding additional wetting or swelling. Ink will not smudge and the book has better chances of survival. The block of frozen leaves will dry without sticking together. In some cases, no rebinding is necessary and mud, dirt and soot are brought to the surface, allowing a faster cleanup. However, once again, temperature, vacuum and lyophilization rate are crucial. The progressive recession of the sublimation front alters the sample temperature for the sublimation cooling and increases the resistance to vapour migration. Some vapour can be reabsorbed into the dried material. The changes in temperature and moisture in proximity of the front might cause structural softening or collapse of the material structure (Adams and Irons, 1992; Day and Stacey, 2007).

It is clear that vacuum freeze drying is particularly useful for a private library, where the expectancy of life of books is related to that of the owner. Although some damage is caused by this methodology, this is considered negligible compared to the potential damage caused by water. However, things are different for precious books and documents that should be preserved for centuries and this practice needs further research to be refined and to establish precise freezing and drying protocols. A number of articles have been published about how different libraries, or specialized service companies, dry books after a disaster; however, very little research has been carried out on this subject.

2B.4. BIOLOGICAL HABITAT

For all materials, the microclimate – and especially dampness – should be continually controlled in order to reduce or eliminate the risk of biological attack of fungi, bacteria or other parasites. The association of dampness with mild to moderate temperatures and dust deposits creates a favourable habitat on all materials, so that cleaning is a highly recommended preventive conservation strategy (Sterflinger and Pinzari, 2012). For example, algae and cyanobacteria need high *RH* and fungi generally prefer $RH > 65\%$, but in already contaminated materials, it should be considered that, due to the humidity-conserving biofilm, microorganisms might survive or even act at $RH > 50\%$. High *RH* levels, especially when associated with $T > 20\ ^\circ C$, favour biological rotting especially in the case of organic materials whose composition is appreciated by parasites or may offer a good substratum for the development of microbiological life.

Laboratory and field studies have determined the risky combination of matched levels of temperature and *RH* that will form a favourable habitat for mould growth (Sedlbauer, 2002; Sedlbauer et al., 2011). In the case of a biological substrate (e.g. wood, paper), at ambient temperature below the ice melting point, i.e. $T < 0\ ^\circ C$, the formation of ice crystal inhibits the biological life and *RH* may reach 100%. However, when the temperature increases, *RH* should decrease to reduce the *MC* available for microorganisms, e.g. at $T = 10\ ^\circ C$, $RH = 82\%$; at $20\ ^\circ C$, $RH = 77\%$; and at $30\ ^\circ C$, the critical *RH* level is 75% (Fig 2B.17). Uncontaminated masonry and stones allow slightly higher *RH* levels (e.g. $RH = 80\%$ at $30\ ^\circ C$); good biological substrates are more easily attacked (e.g. $RH = 70\%$ at $30\ ^\circ C$). In practice, in cold regions, cultural heritage objects may resist to relatively elevated humidity levels; in temperate regions, the *RH* level should prudentially remain below 70%.

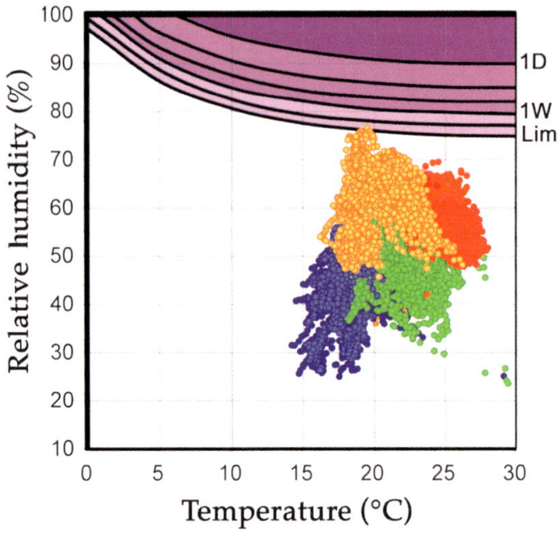

FIGURE 2B.17 Risk analysis to predict mould infestation in a room. The pink area on the top indicates the dangerous combination of temperature and relative humidity that may determine a favorable habitat for mould colonization. 1D (dark pink): possible colonization after one day; 1W (medium pink): risk after one week persistence in these conditions; Lim (light pink): threshold limit. White area indicates no risk. Coloured dots indicate indoor microclimate records of the case study. Temperature and *RH* readings in winter (blue dots), spring (green dots), summer (red dots), and autumn (orange dots). The plot shows that in this case study winter and spring are absolutely safe against moulds, summer is still safe but a bit less, and autumn rarely exceeds the attention threshold.

Generally speaking, the attack of microorganisms is basically controlled by a favourable habitat (e.g. surface temperature, *RH* and *MC* in the material), the structure and the chemistry of the substrate (e.g. porosity, inner surface and biosusceptibility of the material) and is secondly determined by air pollution levels (e.g. salts and organic materials) as well as further environmental conditions (e.g. ventilation, light, pH and redox potential) (Warscheid and Braams, 2000; Mandrioli et al., 2003; Caneva et al., 2009; Sterflinger, 2010). Special biological niches are formed near corners and edges, which are hardly reached by ventilation and remain colder and moister, thus accumulating *MC* in the subsurface layer. Another common problem occurs with objects wrapped into plastic foils not allowing an exchange of vapour: the latter may reach saturation when the ambient temperature decreases (Sterflinger and Pinzari, 2012).

Microflora live at the interface between a solid substrate and the air: these two media may have a different temperature and *MC*, e.g. with water being adsorbed or trapped into pores. In certain cases, the dynamic situation generated by daily cycles, which are transmitted with a phase delay in the two media, may lead to the fortunate circumstance that microorganisms may benefit alternatively from one or the other medium, so that considering only the physical characteristics of the air is analysing only a part of the problem, generally less relevant than the material substrate.

Wood is conducive to decay, mould rotting and insect damage in moist, warm conditions. Strength is rapidly reduced in even the early stages of decay, e.g. 5% weight loss from insect tunnelling or rotting can result in strength losses as high as 50% (Winandy, 1994). Lignin is highly resistant to biodegradation and only higher-order fungi are capable of degrading the polymer via an oxidative process called peroxidase (Schoemaker and Piontek, 1996).

As far as the atmospheric part is concerned, the appropriate environmental conditions should be carefully chosen as a whole, including not only *RH* and *T* but also ventilation and turbulence (which regulate the deposition and removal of spores), light, pH as well as air pollution levels and the relevance of each of these factors may change in the case of indoor or outdoor environments.

For instance, in the past, when the air pollution was modest, lichens colonized many monuments especially in the Mediterranean belt, where solar radiation and rainfall were most appropriate, and many lichens can still be found in rather unpolluted zones of Portugal, Spain, Italy, Greece and so on. Later, when the SO_2 and sulphates reached elevated levels, the atmosphere became toxic for them and they disappeared from most cities, leaving on stone monuments typical signs of their past presence, e.g. pitting and oxalates. However, the microflora of stones represents a complex and highly adaptable ecosystem, metabolizing inorganic and organic substrates from natural or anthropogenic sources. Because of their metabolic flexibility and remarkable tolerance against osmotic stress (e.g. salts) or toxic compounds, air pollution might cause considerable shifts in the composition of the stone-colonizing microflora but will hardly control or stop biodeterioration.

2B.5. METALS, PIPE ORGANS AND OTHER MATERIALS

2B.5.1. The Formation of a Surface Water Layer on Metals, Glass, Stones

It is well known that high humidity levels increase the deterioration rates in several ways; in any case, however, this is a consequence of the moisture transferred from air to the object. A list of some of the most important phenomena recurrent in conservation has been reported by Thomson (1986) or can be found in special studies. In order to clarify the action of moisture, it is useful to summarize some key mechanisms.

If we keep a material in a ventilated oven, at above boiling temperature, e.g. 103 °C, it will release its *MC* until it will become completely dry.

If starting from the initial condition $RH = 0\%$ and $EMC = 0\%$ we increase the ambient *RH*, some water will be adsorbed by the material and the adsorbed amount will increase with the *RH*. If the material is hydrophilic and porous, the moisture will distribute from the surface to the interior; if it is a metal, or glass, some monomolecular water layers will form on the surface and will remain on it. As the *RH* will increase, other H_2O molecules will adhere to the surface building up an increasing number of successive monomolecular layers covering the material surface. This number will increase with the ambient *RH* level and will reach the maximum at saturation (Fig. 2B.18). On metals, some 10 layers may form at saturation (Leygraf and Graedel, 2000) and the number of molecular layers (*NML*) increases with increasing *RH* following the equation

$$NML = 0.7153 \exp(0.0263\, RH) \tag{2B.2}$$

The water molecules that form the first layer on a metal surface are bound to the surface with strong bonding forces and may be considered to be in the solid state. The successive layers will be bound to the surface with weaker and weaker forces, progressively passing from the nonreactive solid state of the first layer to the liquid state in the most external layer that will form in damp environments. For this reason, at low *RH* levels chemical reactions do not occur and biological life is impossible. In opposition, at elevated *RH* levels, the water layer on the surface may favour corrosion, may act as a solvent, forming aggressive solutions when reached by aerosols, or may feed microbiological life.

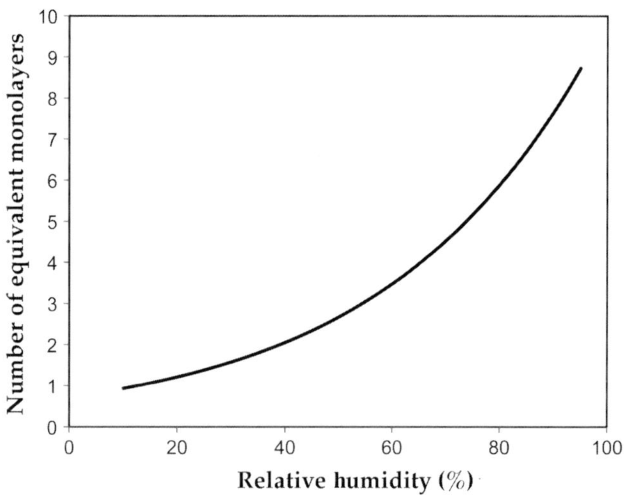

FIGURE 2B.18 Number of equivalent monolayers composed of H_2O molecules that cover a metal surface in equilibrium with the ambient *RH*. *Source: Equation (2B.2) has been obtained after findings by Leygraf and Graedel (2000).*

2B.5.2. Chemical Forms of Decay

The presence of liquid water on the surface, or in the interior, favours chemical reactions that varies from material to material.

Metal corrosion (Jones, 1996; Leygraf and Graedel, 2000) is the destructive result of a chemical reaction between a metal and its environment and therefore also includes oxidation that is the first form of metal decay. Metallic corrosion involves transfer of electronic charges in aqueous solutions. For this reason, in the presence of water, metals tend to combine with other chemical compounds and return in a form similar to the natural minerals from which they are extracted, making free the energy required for their extraction.

The transformation of marble or limestone into gypsum is probably the most popular example (for an extensive discussion of this mechanism, see Laurenzi Tabasso and Marabelli, 1992). Gypsum is mainly formed in heterogeneous phase on the monument surface when, in the presence of a film of liquid water, the absorbed sulphur dioxide (SO_2) as a primary pollutant is first converted into sulphurous acid (H_2SO_3) and/or sulphuric acid (H_2SO_4). The oxidation of S(4) into S(5) is favoured by the presence of O_3, NO_x and catalysts such as metals or soot; the reaction with the calcium carbonate (i.e. calcite, $CaCO_3$) forms directly, or as a final product, the hydrated crystal of calcium sulphate, i.e. gypsum ($CaSO_4 \cdot 2H_2O$). The chemical reaction is easier when atmospheric SO_2 is dissolved and oxidized to sulphuric acid into rain droplets (acid rain) or sulphate aerosols deposit on the marble or limestone surface and in the presence of water, they are transformed into sulphuric acid and gypsum.

Stones, bricks, pottery, plasters, frescoes and other materials having water-soluble salts in the pores are sensitive to microclimate changes. Environmental conditions leading to cycles of *RH* may generate crystallization and dissolution of salts by means of two key mechanisms: (1) the water evaporation, which is controlled by ambient *RH*, causes a concentration of the solute, supersaturation and precipitation of salt crystals; (2) hygroscopic salts, which are in crystalline form only in a dry environment, become deliquescent or precipitate following the ambient *RH* cycles. These mechanisms generate cumulative mechanical damage, efflorescence, subflorescence, delamination, spalling, flaking, etc.

For glass (Newton and Davison, 1989; Schreiner, 2005), the situation is complex, as there are several types of glass and deterioration mechanisms. Water originated by condensation or that arrives with rainfall, or simply by vapour adsorption, is necessary for the replacement by protons of the diffusing alkali ions and subsequent hydration of the silica network. Deterioration begins with leaching, i.e. an ion exchange between penetrating ions from the moisture (i.e. H^+ and/or H_3O^+) and an alkali metal ion (e.g. Na^+ or K^+ that are typical constituents of ancient glass), which is removed from the glass. Modern glass made with Ca^{2+} is more resistant to leaching. Water-originated radicals may diffuse inside the glass causing ion exchange and alkali extraction. Corrosion may occur when the silica constituent is affected. The state of the outer layer and the penetration of water radicals are crucial (Cailleteau et al., 2008). Chemical solutions (and the pH of the aggressive solution) may provoke chemical changes on the surface, which may then spread to the whole of the glass. The presence of pollutants, or adverse weather condition, may severely affect glass leaching and corrosion (Munier et al., 2002). In dry condition, or under the action of solar radiation, the loss of the absorbed water causes dehydration. For this reason, humidity cycles are dangerous, even in the absence of air pollution.

2B.5.3. Complex Structures: The Pipe Organ

An example of complex interactions is given by the church microclimate and the preservation of pipe organs. This has been the object of the EU-funded project Sensor System to Detect Harmful Environments for Pipe Organs (SENSORGAN) (Bergsten et al., 2010). The Project objectives were to develop three types of sensors for real-time monitoring and detection of harmful environments for organs. The early detection system was based on three key sensors placed inside the organ without disturbing the use, even in the case of concert.

The first sensor was an electronic dosimeter to detect organic acids that are corrosive to metal organ pipes (Odlyha et al., 2010a). A coating of lead (very sensitive to organic acids) was applied on a piezoelectric quartz crystal for the first time. When the coating reacts with the organic acids and starts to corrode, the mass of the coating increases and a change in frequency of the oscillation mode of the crystal can be detected. The dosimeter was placed in a holder under the pipe without affecting the sound from the pipe. It was then used in the EU-funded project Improved Protection of Paintings during Exhibition, Storage and Transit (PROPAINT) to monitor emissions in paint frames but its first use and calibration was performed in the SENSORGAN Project.

The second sensor was a detector of acoustic emissions emitted from the wood under stress to indicate risk of damage (i.e. cracks) to wooden parts of organs (Jakiela et al., 2007, 2008). Acoustic emission detected microfracturing of wooden parts of the organ thereby constituting an early warning system for emergence of cracks in the wood. When a microcrack develops inside a wooden structure, a sound pulse is emitted. This sound is not in the audible range but is detected by the sensor and recorded.

The third sensor detected dew or frost formation inside organ pipes (Baldini et al., 2008). The principle of the dewing–frosting detector is based on the change of the fraction of light reflected at the end surface of an optical fibre, when droplets, or ice crystals, form on its end surface. When this occurs, the fraction of reflected light drops. The fibre end is placed inside the pipe foot, in contact with the pipe metal surface, and will thereby detect early formation of water or ice on the actual pipe surface. The fibre is flexible and very thin (a small fraction of a millimetre) and because of this it can be easily inserted into an organ pipe and stay there without disturbing the sound or the instrument performances.

An organ contains different types of materials, e.g. different types of wood, leather, various types of lead–tin alloys, from pure lead to almost pure tin, brass and iron. It also contains a complex system of moving parts and the wind system needs to be airtight. This makes the organ to be an object very sensitive to microclimate changes and harmful environments. The organ becomes unplayable if just one of the materials or functions just mentioned is responding to the environment in a negative way. An organ may be affected by a number of deterioration mechanisms, generated by an aggressive environment, as follows.

The first key problem is dryness and mechanical damage. The *MC* inside wood leather and other hydrophilic materials is in equilibrium with the ambient *RH*. Changes in *RH* cause gain or loss of water and dimensional changes, i.e. expansion or shrinkage, respectively, followed by physical stress and risk of cracks in the wooden parts of the organ. This can make the instrument unplayable and a major and expensive repair work has to be performed. Continuous heating and warm-air stratification lower the *RH*. The problem is especially frequent with warm-air heating and organs located in the choir loft. Intermittent heating may start new cracks and in the long run the repetition of temperature/humidity cycles deepens the existing ones. Church dryness cannot be mitigated with humidifiers, because moisture blown in air eventually condenses on cold walls and generates moulds. The only remedy is to lower the indoor temperature or to use local heating.

The second key problem is the deposition of suspended particulate matter. Deposition of airborne dust, cloth fibres, aerosols and smoke affects not only the visual appearance but also the organ playing. Every church has important sources of airborne particles, e.g. candle smoke, external leakage, dust transported by churchgoer shoes and cloth fibres. In the long run, such particles tend to deposit on the internal artworks, including the organ. The deposition rate is highly affected by a number of physical mechanisms, most of them driven or enhanced by the church heating. These will be discussed in Chapter 8 and are due to air motions and inertial impaction, particularly efficient for medium-sized and coarse particles; warm air and cold surfaces (thermophoretic transport towards a cold surface, particularly efficient for small- and medium-sized particles); Brownian deposition speeded up by more efficient air molecular collisions, i.e. higher air temperature, particularly efficient for fine particles;

diffusiophoresis and Stefan flow transport in the case of moisture condensation, with high efficiency and irrespective of the particle size; as well as electrophoretic deposition, especially active in warm and dry environments, irrespective of the particle size.

We can easily see that most heating systems increase the internal air motions or the indoor air temperature creating gradients corresponding to surfaces that require some time to heat and to follow the air temperature level.

The third key problem is dampness. Dampness may be caused by a number of factors. The most frequent are water percolation through the roof, the walls or the windows; water-absorbing brick walls or walls deprived of the protective external plaster; frequent rains and humid local climate; crowded congregation; and use of gas or liquefied petroleum gas burners with combustion inside the church.

Another problem is the "tin pest" of tin rich pipes, frequently used in the organ façades because of the shiny appearance. At 13.2 °C and below, pure tin undergoes an allotropic transformation from white metallic tin (with tetragonal crystalline structure) to grey non-metallic tin (with cubic structure). This phase change causes the decomposition of this metal into a powdery form and induces the formation of corrosion-like stains, cracks and holes. As soon as the tin begins decomposing, the process speeds up. The damage is cumulative and progresses when thermal cycles are repeated around 13.2 °C, or the temperature drops below this threshold. The transformation is most rapid at about −40°C. The metal lattice must expand for pest formation and tensile (thermo-mechanical) stress may exacerbate it.

The temperature and especially the humidity influence the emission rate of organic acids (VOC), especially acetic and formic acid, from wood (Chiavari et al., 2008; Niklasson et al., 2008; Odlyha et al., 2010b). High *RH* and temperature levels will increase VOC emission and provide an active environment for the presence of absorbed water and the high energetic level of the water molecules (see Arrhenius equation). High *RH* speeds up the corrosion process itself and condensation onto the pipe foot certainly worsens the situation. The pipe metal is a lead–tin alloy and pipe corrosion has been studied in the EU-funded project Corrosion of Lead and Lead-Tin Alloys of Organ Pipes in Europe (COLLAPSE) (Bergsten, 2009). High-lead-content pipes are very sensitive to corrosion in the presence of organic acids but a small percentage of tin has a corrosion-protecting effect at low *RH*. However, at high *RH*, this protecting effect gradually disappears. High *RH* levels may create a favourable habitat for the growth, development and infestation of mould. Woodworms and insects can sometimes damage wooden or leather parts of the organ, especially in a humid, warm environment.

Some comments concerning organ preventive conservation, maintenance and restoration: the organ life cannot be conceived separately from the indoor climate and use of the church. This requires a general understanding of microclimatic factors that may create harmful environments for organs. Once harmful environments and practices have been detected, it is necessary to assess methods to mitigate or impede them, e.g. church (and maybe organ too) maintenance and restoration and installation of heating systems friendly to the church and the organ.

Organ builders' restoration practices should avoid the use of harmful products. Not adequately seasoned wood will release VOC. Long-term seasoned wood should be used, especially when the organ pipes are composed of alloys with high lead content. Some wooden species should be avoided, especially oak or nut, because of their high VOC emission rates. The temperature and especially *RH* will influence the VOC emission. High *RH* and temperature levels will increase emissions of organic acids.

Polyvinyl acetate is a synthetic polymer that is a basic component of some types of glue, which is popular for certain properties such as elasticity and flexibility at room temperature. However, it releases VOC and should be avoided in the organ restoration for the high corrosion potential.

When repairing a corroded pipe, the same alloy as the original pipe should be used for two reasons. One is to avoid galvanic corrosion on the contact area. The second is that repairing a corroded pipe using a more tin-rich alloy may also result in a heavier corrosion attack on the original pipe material in a more sensitive position higher up in the pipe due to the presence of organic acids. For instance, replacing a foot tip may increase the corrosion rate in the upper foot and closer to the sensitive mouth area. The reason is that the organic acids, entering through the toe hole, will not react with the new metal and thereby be consumed but will instead react with the original and more corrosion-sensitive metal higher up in the foot.

It is not convenient to remove the corrosion products from the metal pipes. Corrosion products are stable. The corrosion will continue until the environment will have acids and moisture in synergism. A nude metal surface, deprived of the corrosion products, will have faster reactions.

In order to mitigate corrosion, metal pipes should always be above the dew point and a good practice is to remember that for ventilation or other reasons, the metal pipe temperature should never drop below the dew point or below the temperature of other parts of the organ.

2B.6. WHAT IS THE BEST MICROCLIMATE FOR CONSERVATION? THE EUROPEAN STANDARD EN15757: 2010

We have seen that some humidity levels are associated with specific deterioration mechanisms or may accelerate them. The water molecules that are absorbed into the material may determine internal stress and some deformation to the structure; the amount of absorbed water (and therefore internal stress and deformation) is in equilibrium with the *RH* and also, to a lesser extent, with *T*. These deformations tend to be reversible in plastic, relatively young materials but irreversible and extremely dangerous in aged materials.

This is a good reason for suggesting some allowed or nonallowed microclimate conditions. For the materials most frequently used in creating works of art, some recommended values (or spans) of *T* and *RH* could be found in the literature, with the aim of determining the conditions of physical wellbeing, suitable for an appropriate conservation. These values are recommended with the aim of taking advantage from experience, avoiding the repetition of errors already made in the past, or keeping off from dangerous situations discovered with laboratory tests. However, a fundamental question arises: are microclimate norms and guidelines always useful and should they be always followed?

Precise recommendations exist to avoid or reduce chemical transformations, e.g. hydrolysis, oxidation, corrosion; to avoid metamorphism, hydration or phase transitions of unstable minerals or other materials; or to avoid microbiologic or other pest attacks. However, the problem is that a number of various and simultaneous types of risk exist and the best solution should be to find a compromise between the different needs and reduce as far as possible the various deterioration mechanisms. This requires a holistic approach and optimizing the synergism between the individual solutions. This always introduces a series of uncertainties to the multidisciplinary decision system.

In particular, it is useful to underline that recommended values of *T* and *RH* that are in principle suitable for *hygroscopic materials* are not always appropriate for individual *objects*. In fact, specific objects that have been kept for centuries under determined *T* and *RH*, have adapted to their environment reaching a new equilibrium of internal tensions with possible (permanent) deformation of their structure. The adaptation might imply plastic deformations or cracks that play the role of expansion joints for the object to respond to the ambient variability.

Generally speaking, an old hygroscopic object, which has adapted to its environment according to a complex internal system of tensions and has lost its initial elasticity, is unable to adapt once again in a reversible way to a novel microclimate and any change may be dangerous. For this reason, the knowledge of the past conditions is needed and the previous microclimate should be kept untouched. It can only be modified smoothing out *RH* fluctuations or reducing the range of variability.

Once a hygroscopic object has adapted to a certain *RH* variability, both in terms of average levels and fluctuations, it requires respecting the previous conditions as any change, or fluctuation exceeding the historic range, might affect durability. Abrupt climate changes are risky. We have seen that ASRAE (2007) have considered this point, have suggested not to exceed the historic *RH* variability, and that have introduced the concept of '*proofed fluctuation*' as an unplanned crash test to update the safety threshold. We will see that the European standard EN15757: 2010 recommends a more cautious limitation. Repeated cycles might accumulate stress and sooner or later they might cause mechanical damage (fatigue failure).

Seasonal changes, although they have a wider span, are less dangerous, as they occur more slowly, in a time much longer than the relaxation time of the object and with a rate slower than the penetration of heat and moisture in the material, so that there is no significant stress between the external layer and the interior. However, seasonal cycles might also be dangerous in the case of inhomogeneous materials. It is evident that furniture, or also a panel built with wood slabs constrained between them and having grains differently oriented, will undergo important internal stress when an element shrinks at an individual rate different from others.

It is obvious that vulnerable hygroscopic materials, e.g. aged wood or ivory, need a very constant microclimate and the risk associated with a rapid change of the physical environment is greater than for more elastic or deformable materials, e.g. hydrated paper, parchment and tapestry. However, the difference between *materials* and *objects* is often substantial: e.g. an old book with paper sheets and parchment cover, bound together with wire, is composed of parts that cannot freely swell or shrink, so that microclimate cycles that are harmless to the individual components may rapidly generate damage.

When a hygroscopic object is not irreversibly conditioned by its past environmental history and it is possible to choose the most appropriate climate for conservation, the best *RH* interval might be found by means of its adsorption isotherm (see Chapter 5). The goal is to find the best compromise among various problems, e.g. to avoid exceeding water for hydrolysis or microorganism colonization or/and exceeding dryness to avoid breaking polymers, or the *RH* intervals where a small change in *RH* causes a too large change of adsorbed water in order to keep the object in one of the intervals where *RH* changes do not generate strong internal tensions, and, finally, to respect the historic climate that often is the strongest condition, as we will see later.

However, it is not true that the best choice is always found in the laboratory by the physical and chemical analysis of the material. In fact, if the experimentally determined 'best' interval is far from the natural local conditions, it might be better to choose a naturally stable microclimate than an artificial one, which is conditioned by the good functioning of complex devices. On this ground, Thomson (1986) suggested 60% *RH* for museums in Great Britain and this value was deduced from the mean natural climatic values of that country. Similarly, for reasons of continuity, the National

Gallery, London, kept the 55% level, which was in equilibrium with the *MC* in wood before the installation of an air-conditioning (Padfield, 1994). Often, rules of thumbs, good examples and related recommendations have been improperly applied to other climatic contexts. Not always are choices adequately supported by scientific investigations.

In 2003, the European Committee for Standardization (CEN) established a Technical Committee, i.e. TC346, to produce standards for cultural heritage. The standard EN15757: 2010 '*Conservation of Cultural Property – Specifications for temperature and relative humidity to limit climate-induced mechanical damage in organic hygroscopic materials*' recommends priority of the historic climate for the class of materials affected by dangerous dimensional changes induced by humidity shrinkage or swelling. This novel standard abandons any list of suggested *RH* levels or intervals for materials such as wood and considers that a wooden object has been acclimatized and adapted to the most frequent *RH* average levels and fluctuations experienced over its lifespan, including the formation of yield and cracks. It might be risky to exceed the so-called safe band lying within the 7th and 93rd percentiles of fluctuations around the monthly variability of the historic climate. This safe band should be considered the target climate. The band is obtained excluding the 7% of the most extreme fluctuations in dryness and the same in dampness. Practically, the 7th and 93rd percentiles of the positive and negative *RH* fluctuations should be calculated and these two values should be algebraically added to the 30-day moving average of

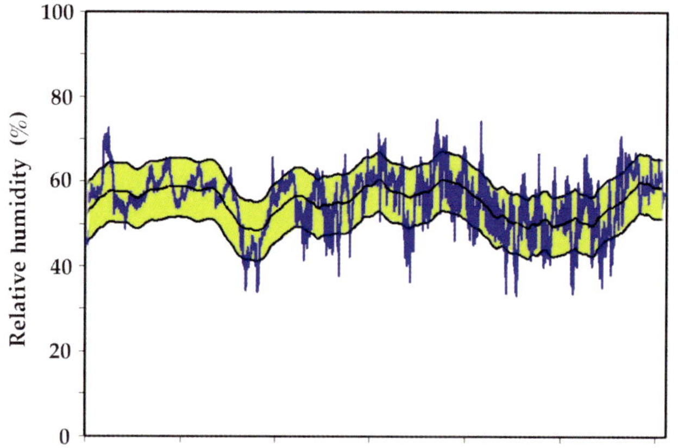

FIGURE 2B.19 Example of *RH* in an unheated chapel (blue line), the safe band (green) lying between the 7th and 93rd percentiles of the *RH* fluctuations and the 30-day moving average (thick black line in the middle of the band). Calculations performed according to EN15757: 2010.

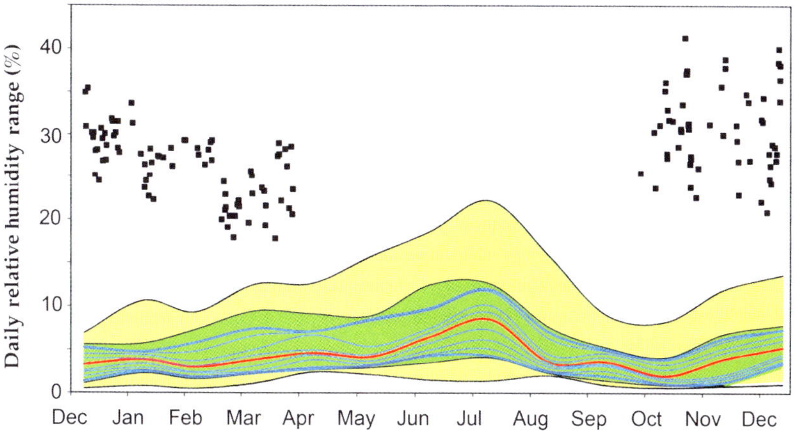

FIGURE 2B.20 Daily *RH* range in the case of a church with occasional warm-air heating. Yellow band, the daily *RH* ranges observed over the calendar year; green band, the safe band lying between the 7th and 93rd percentiles of the *RH* fluctuations calculated after 1-year monitoring, according the EN15757: 2010; red line, the 50th percentile; blue lines, other percentiles; and dots, heating episodes, fully outside the safe band and the historic climate as well.

the *RH* calculated over a net calendar year (Fig. 2B.19). This means that monitoring should be performed for 13 months because the 30-day moving average needs, but wastes, the two 15-day periods at the beginning and the end of the record.

In addition, EN15757: 2010 establishes that, in case an object should be relocated, or moved to a temporary exhibition, or an HVAC should be installed, the new conditions should be respectful of the safe band of the historic climate. For instance, if a heating system is operated continuously, it may lower too much the ambient *RH* and the supply of moisture might cause condensation, salt mobilization and biological colonization on cold walls. In such a case, the historic *RH* is preserved in the indoor air, which is irrelevant, but not at the interface with walls and artwork and this is crucial. On the other hand, if the heating system is operated occasionally, it might cause severe fluctuations or spot departures from the unheated regime. This is evident when the daily *RH* range is considered. The daily *RH* range is the numerical difference between the daily maximum and minimum *RH* values observed in a given location, as reported in Fig. 2B.20 for an occasional warm-air heating in a church. In such a case, the monthly variation (i.e. the 30-day mobile average) is removed and only the daily extremes are pointed out. In the figure, the occasional heating episodes are represented with black dots and all of them fall outside the safe band. This means that this specific heating system is potentially dangerous.

In practice, EN15757: 2010 imposes serious limitations to new HVAC installations. If a building is already heated, the ongoing heating might be kept for the future but any reduction of its impact is welcome. As opposed, if the building is unheated, only three possibilities exist:

1. to leave the building unheated;
2. to operate the HVAC with very gentle, but hardly comfortable, heating; and
3. to avoid central heating and only install local heating that is concentrated on warming people, leaving almost unchanged the temperature over the whole volume of the building and the fabric structures.

If the building has been recently heated, it should be controlled whether objects have been yet acclimatized or not. In the case they are not, and the novel climate has proved to be dangerous, it is convenient to return to the previous conditions.

2B.7. KEEPING CONSTANT RELATIVE HUMIDITY IN ROOMS AND SHOWCASES

RH is of primary importance in artwork conservation and in exhibition rooms, it should be kept both homogeneous in space and constant in time. However, in practice, several variations exist and dramatic changes occur, especially in the morning when windows and doors are opened during cleaning or when the heating system is switched on. In several cases, humidifiers are introduced to mitigate or compensate the effect of the

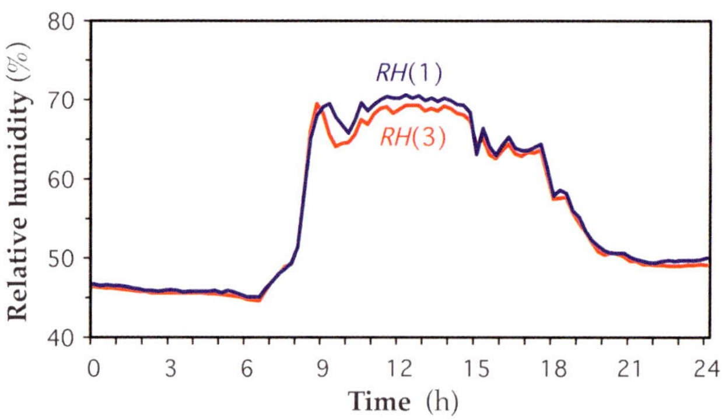

FIGURE 2B.21 Change in relative humidity at 1 and 3 m above the floor, i.e. *RH*(1) and *RH*(3), when the heating system and the compensating humidifiers are operating during the visiting time (for the temperature change, see Fig. 1.20(a)). The humidifiers were exceedingly powerful and, instead of mitigating the dry air, the net result was moistening. The Pollaiolo room, Uffizi Gallery, Florence, 18 February 1997. After the survey, the HVAC has been better tuned.

temperature change but this system is never well balanced and hardly successful (Fig. 2B.21).

Air-conditioning systems and humidifiers cause repeated potentially noxious local variations. Active devices create artificial microclimates but are unable to ensure a very homogeneous distribution of moisture and may be risky especially when they go out of order or the power supply is stopped for uncontrolled reasons. It is evident that clouds of moisture generated by humidifiers in exhibition rooms produce local perturbations also in terms of *RH*, with the effects that have been previously discussed.

Various systems are popularly used to control *RH*. A common problem is that the HVAC unit is controlled by temperature and *RH* sensors enclosed in a small plastic case in contact with the wall that has a large thermal inertia and a temperature generally different from that of the air. In addition, when moisture is blown inside a room to balance the temperature and obtain the selected *RH* level, walls, ceiling and furniture will absorb moisture, causing unbalances. For instance, in the Uffizi Gallery, Florence, a number of different systems have been used in the past and they give the opportunity of very interesting comparisons.

The Giotto room has a system that is based on controlled air introduced from grille diffusers sited in the ceiling above the *Madonna d'Ognissanti* by Giotto; as an equal flow is removed by intake slits on the wall near the floor level, this system generates a preferential path for the controlled air masses, which cross the room and partially mix with the room air. For this reason, this system is integrated with some traditional box humidifiers close to the walls. The result is a nonhomogeneous distribution of the

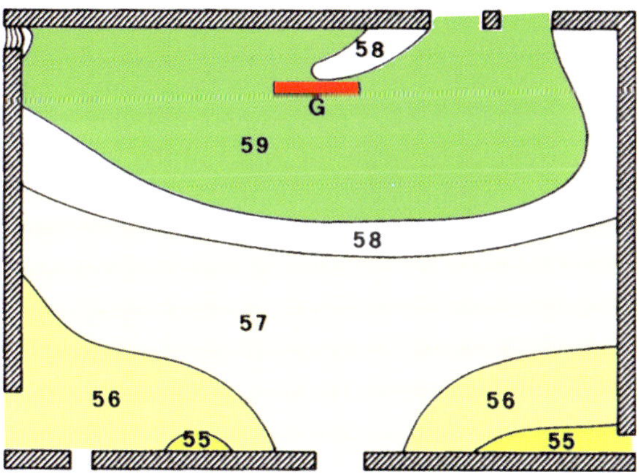

FIGURE 2B.22 Relative humidity in the Giotto Room, Uffizi Gallery, Florence. In the warm season, the cold air-condition released through diffusers in the ceiling falls, generating a perturbed zone; 12 August 1992 at 12.00. Label 'G' and the red line show where the Giotto's panel is located.

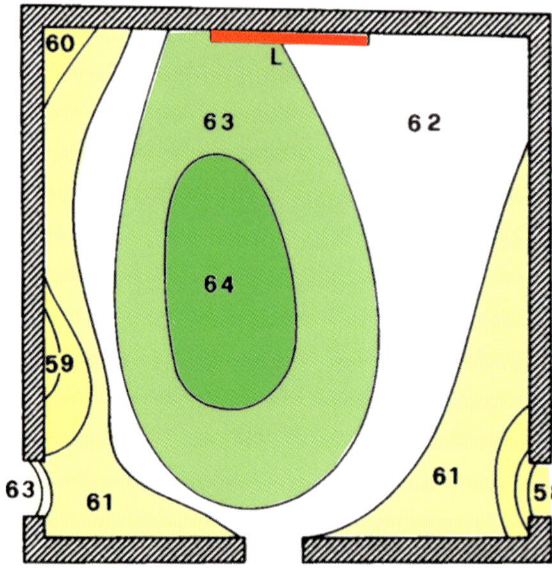

FIGURE 2B.23 Relative humidity in the Leonard Room, Uffizi Gallery, Florence. In the warm season, the sophisticated air-conditioning system releases treated air from an extended source below the ceiling and this system is unable to guarantee ideal conditions; 13 August 1992 at 18.20. Label 'L' and the red line show where the Leonard's panel is located.

humidity inside the room and, in particular, in the summertime, the cold flow of air-conditioner falls just on the Giotto's painting (Fig. 2B.22).

The Leonard room (Fig. 2B.23) in the Uffizi Gallery has a more sophisticated system that blows conditioned air from four pipe ducts hung to the ceiling, parallel to the walls, but at a certain distance from the walls, so that the cold air falls far from paintings.

In principle, this system is preferable, as the warm or cold air and the moisture are neither blown from a few points located on the walls, close to paintings, nor from the floor, with the risk of mixing and disperse dustiness. In this case, the conditioned air is blown from pipes aloft, far from exhibits. However, the room air is moved by the depression generated by intake slits on the wall near the floor level and is transported by air currents existing in the gallery when a door or a window is open.

Although the system is good in principle for a single room, the overall result in the context of buildings, where rooms are interconnected and air masses are transported from a room to another by internal temperature or pressure differences, is not fully satisfactory and practically not better than the simpler system in the Giotto room (Bernardi and Camuffo, 1995a).

A number of rooms have air-conditioner or humidifying fan coil units located close to walls and paintings, following an incorrect and often noxious practice very frequently found in museums and galleries.

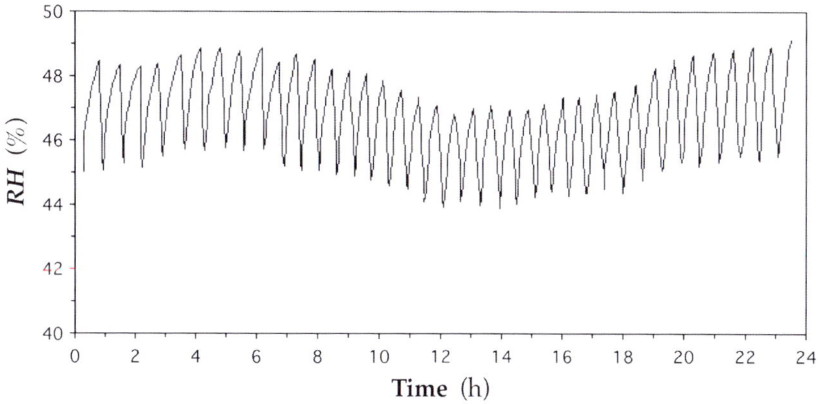

FIGURE 2B.24 Fluctuations in relative humidity (*RH*) generated by a sophisticated controlling system, which controls this variable within two stated limits, i.e. ±1% in the wider span 45% ≤ *RH* ≤ 50% determined by environmental variability on the larger scale (Private Gallery of Modern Paintings, Parma, February 1996).

More appropriately, the Botticelli room has some humidifiers and air-conditioner units located in the centre of the room, far from paintings.

Even sophisticated automatic systems that control the room climate keeping it to stay in a band between two selected levels cause continuous fluctuations between the lower and the upper thresholds (Fig. 2B.24). In such a case, the range of *RH* variability lies within a few percentage and the period is relatively short; as a consequence, the moisture perturbation cannot reach deep layers on artefacts and is extinguished in a very thin surface layer.

Repeated cycles may have negative impact on some thin objects or sensitive surface layers, e.g. tapestry, colour coatings, paintings on canvas, and lacquers that have a short relaxation time and are at risk for the forces that derive from the contact with other parts at longer relaxation time, e.g. the deeper layers in the case of colour coatings or lacquers.

One can easily imagine the consequences that warm-air heating systems may have in churches, where heating is operated once a week. In the above example of the church at Rocca Pietore, Italian Alps (Fig. 1.16), the temperature peak found at 1 m height, i.e. $\Delta T = 7$ °C, causes a humidity drop $\Delta RH = 10\%$; the temperature peak at 3 and 4 m, i.e. $\Delta T = 20$ °C, causes a sharper humidity drop $\Delta RH = 50\%$. The pipe organ, paintings on canvas, decorated panels and other wooden artefacts will shrink; on the other hand, the frescoed ceiling will mobilize internal moisture and evaporate, generating efflorescence. In the long run, the repetition of these cycles may have negative impact on artwork conservation.

Some porous or fibrous materials characterized by large heat capacity maintain constant both *T* and *RH*: e.g. a thick wall tends to keep a constant microclimate in its proximity and paintings hung to it or statues in wall niches benefit from these conditions; wooden boxes protect small objects against sudden changes of *RH*. One of the most effective materials, widely used in showcases, is the *silica gel* characterized by many fine pores and it is therefore extremely adsorbent. This material smoothes out abrupt changes of *RH* but does not ensure steady, predetermined conditions. Silica gel is often preconditioned at some desired *RH* levels by keeping it for a sufficient long time at a determined humidity level until equilibrium is reached. At this point, the silica gel behaves as a buffering agent, i.e. if the *RH* drops it will release moisture and if *RH* increases it will absorb moisture in order to offset changes. However, the silica gel slowly adapts to the new average environmental conditions and the buffering level changes in a rather uncontrolled way, so that it is necessary to change often the silica gel in the case with other silica gel, just conditioned at the desired level. This makes complex the management of the case. In addition, the amount of silica gel necessary for good conservation of exhibits in showcases has been calculated to be 25 kg m^{-3} of air to control (Thomson, 1986). This figure clearly shows that it is extremely difficult to preserve objects at constant *RH* avoiding fluctuations, disturbances or slow changes. Whenever possible, it is advisable to improve the natural microclimate with the help of all the passive systems that ensure the best stability and reliability.

Another method to maintain constant *RH* at a desired level inside an airtight case is to place in the bottom of the case a vessel containing a supersaturated solution of pure water with an excess of a certain solid substance that is in equilibrium with a specific value of *RH*, a number of these substances being listed in Chapter 12. These supersaturated solutions are characterized by a constant, typical vapour pressure and, consequently, by a constant value of *RH*. If the *RH* inside the case lowers, some water evaporates from the vessel to re-establish the equilibrium and some vapour condenses in the solution if the inside *RH* increases. The choice can be extended to a large number of buffered *RH* levels, of course by choosing chemical substances that are not noxious to the exhibits and controlling that salts do not migrate within the volume.

Sometimes showcases have a humidity level that is controlled by a microclimate generator that adjusts the *RH* by adding or removing moisture as required. The treated air is continually circulated between the case and the moisture-controlling unit. The *RH* sensor located inside the showcase drives a feedback of moisture exchanges and the operation stops when the target *RH* level is reached. If the whole system, i.e. the case and the control unit, have exactly the same temperature, there is a good probability that the desired *RH* is reached in average, i.e. with several dangerous fluctuations around the mean level. However, if temperature differences are found in the system, although the *MR* is exactly the same, the *RH* will present departures and the required homogeneity is lost. This happens normally for several reasons, e.g. the case and the control unit may have a different temperature, the exhibit is heated by the lighting system, the case has not exactly the same temperature everywhere because the most lighted panes overheat and warm the air, or the warm air is layering at the top of the box leading. These active systems are hardly fully reliable, need continuous control, are rather expensive and expose the works to the risk of dangerous departures in the case they lose tuning or go out of order.

2B.8. CONDENSATION ON COLD SURFACES

If a nonporous surface cools until it reaches the dew point, the environmental vapour begins to condense onto it, forming films of liquid water or droplets. A well-known example is the condensation on the window panes during winter, especially in rooms with an elevated *MR* due to the presence of people, when the glass conductivity lowers below the *DP* the pane temperature. Another example is the surface condensation that typically occurs in spring in the Mediterranean region, when the air becomes mild and rich in moisture (i.e. with an elevated *DP*) and especially when the warm and humid Sirocco wind blows. Historic buildings have thick walls with high heat capacity and a large thermal inertia and their temperature keeps a memory of the past cold season. The contrast between the elevated *DP* of the air and the low temperature of the thick walls, which remain below the *DP*, causes heavy condensation on the surface (Fig. 2B.25). If the surface is porous, the effects of the surface tension can favour condensation in the micropores even at temperatures above the *DP*, as we will see in Chapter 5. For this reason, in spring, thick walls of unheated buildings are very frequently damp.

A nice example for showing that the surface condensation is independent of the air temperature can be obtained with a stainless steel pot that is half filled with cold water and placed over fire obtained from the combustion of methane (CH_4), propane (C_3H_8), also known as liquefied petroleum gas (LPG), or other liquid fuels (Fig. 2B.26).

In the presence of excess oxygen, methane or propane burns to form carbon dioxide and water vapour, which are released into the room, i.e.

$$CH_4 + 2\,O_2 \rightarrow CO_2 + 2\,H_2O + \text{heat}$$

$$CH_3{-}CH_2{-}CH_3 + 5O_2 \rightarrow 3\,CO_2 + 4\,H_2O + \text{heat.}$$

In particular, propane produces twice the number of water molecules released by methane. The steam generated by combustion increases the

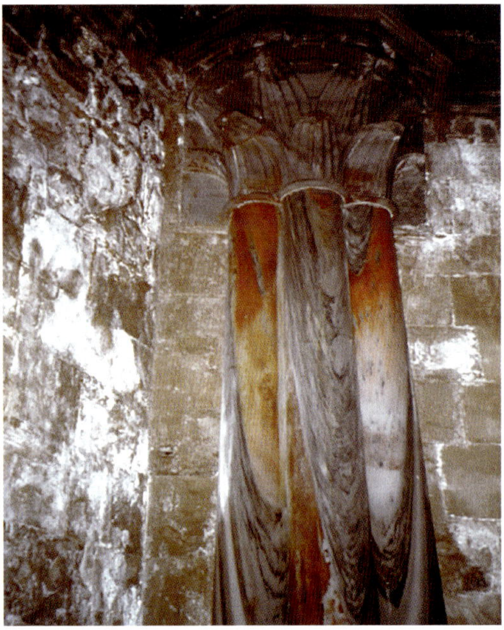

FIGURE 2B.25 In spring, the contrast between the air rich in moisture (i.e. with an elevated *DP*) and the low temperature (below the *DP*) of the thick walls causes heavy condensation on the surface, which is damp and fully covered with droplets. In this picture, taken in a room of Castel del Monte (Andria, Southern Italy), on 27 March 1996, when the surface is perpendicular to the flash light, the light crosses the droplets that are transparent and invisible but when the surface is nearly parallel, the light undergoes multiple reflection inside droplets that appear brilliant, with a silver-like appearance. The red colour on the white part of the marble columns is an algal biological patina.

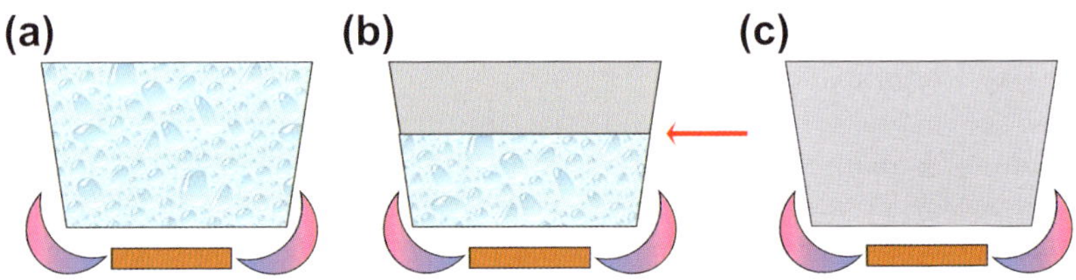

FIGURE 2B.26 (a) A pot of cold water placed above a gas flame is immediately covered with dew. (b) After a short time period, the upper part of the pot, above the internal water level (shown with the arrow), is warmed above the *DP* and the dew disappears. (c) When the internal water is above the *DP*, all the dew will disappear.

MR and, consequently, the *DP*, which may reach some 30−40°C. The pot temperature is lower than this *DP*: although flames lick the pot surface, the pot starts dewing immediately. The upper part of the pot, which is empty, has smaller heat capacity compared with the lower part, which is filled of water. For this reason, the metal surface overheats more rapidly in the upper part than in the lower part. After a few minutes, droplets will disappear from the upper part as soon as the surface temperature exceeds the *DP*. After few minutes, droplets will disappear from the lower part, when both the pot and the water temperatures rise above the *DP* level.

Following the above example, in some specific cases, condensation on limited surfaces can be eliminated either by increasing the surface temperature, e.g. with *IR* radiation or direct warming, or by diminishing the environmental *MR* with air that is poorer in moisture content.

Some positive results might be reached by impregnating the surface with hydrorepellent substances that increase the contact angle θ of water droplets, as we will see later (Chapter 5).

References

Adams, G.D.J., Irons, L.I., 1992. Practical aspects of formulation: the avoidance of product collapse. Pharmaceutical Journal 249, 442–443.

ASHRAE, 2007. Museums, galleries, archives and libraries, (Chapter 21). In: ASHRAE Handbook – HVAC Applications. ASHRAE – American society of heating, refrigerating, and air-conditioning engineers inc.

Baldini, F., Falciai, R., Mencaglia, A.A., Senesi, F., Camuffo, D., della Valle, A., Bergsten, C.J., 2008. Miniaturised optical fibre sensor for dew detection inside organ pipes. Journal of Sensors, Article ID 321065, 5. http://dx.doi.org/10.1155/2008/321065.

Bergsten, C.J., Odlyha, M., Jakiela, S., Slater, J., Cavicchioli, A., de Faria, D.L.A., Niklasson, A., Svensson, J.-E., Bratasz, L., Camuffo, D., della Valle, A., Baldini, F., Falciai, R., Mencaglia, A., Senesi, F., 2010. Sensor system for detection of harmful environments for pipe organs (SENSORGAN). e-Preservation Science 7, 116–125.

Bergsten, C.J., 2009. Collapse. In: European Commission, Chapuis, M. (Eds.), 2009. Preserving Our Heritage, Improving Our Environment, vol. 2. Publications Office of the European Union, Luxembourg, pp. 72–73, ISBN 978-92-79-09029-5.

Bernardi, A., Camuffo, D., 1995. Uffizi gallery in Florence: a comparison between two different air conditioning systems. Science and Technology for Cultural Heritage 4 (2), 11–22.

Bratasz, Ł., 2010. Acceptable and non-acceptable microclimate variability: the case of wood (Chapter 4). In: Camuffo, D., Fassina, V., Havermans, J. (Eds.), Basic Environmental Mechanisms Affecting Cultural Heritage – Understanding Deterioration Mechanisms for Conservation Purposes. COST Action D42 "Enviart". Nardini, Florence, pp. 49–58.

Bratasz, Ł., Kozłowski, R., Camuffo, D., Pagan, E., 2007. Impact of indoor heating on painted wood. Monitoring the altarpiece in the church of Santa Maria Maddalena in Rocca Pietore, Italy. Studies in Conservation 52, 199–210.

Bulska, E., Wagner, B., 2005. A study of ancient manuscripts exposed to iron-gall ink corrosion. In: Janssens, K., Van Grieken, R. (Eds.), Non-destructive Micro Analysis of Cultural Heritage Materials. Elsevier, Amsterdam, pp. 755–788.

Cailleteau, C., Angeli, F., Devreux, F., Gin, S., Jestin, J., Jollivet, P., Spalla, O., 2008. Insight into silicate-glass corrosion mechanisms. Nature Materials 7 (12), 978–983.

Camuffo, D., Bernardi, A., 1991. The microclimate of Leonardo's "Last Supper"; joint edition. European Cultural Heritage Newsletter on Research, and Bollettino Geofisico, Special Issue 14 (3), 1–123.

Camuffo, D., Bertolin, C., 2012. The earliest temperature observations in the World: the Medici Network (1654–1670). Climatic Change 111, 335–363.

Canadian Council of Archives, 2003. Basic Conservation of Archival Materials: Revised Edition, 2003 Chapter 3: Environment. Canadian Council of Archives, Ottawa.

Caneva, G., Nugari, M.P., Salvadori, O., 2009. Plant Biology for Cultural Heritage – Biodeterioration and Conservation. The Getty Conservation Institute, Los Angeles.

Chiavari, C., Martini, C., Prandstraller, D., Niklasson, A., Johansson, L.G., Svensson, J.E., Åslund, A., Bergsten, C.J., 2008. Atmospheric corrosion of historical organ pipes: the influence of environment and materials. Corrosion Science 50, 2444–2455.

Copedé, M., 1991. La carta e il suo degrado. Nardini, Florence, 165 pp.

Day, J., Stacey, G.N., 2007. Cryopreservation and Freeze-Drying Protocols. Methods in Molecular Biology, 368. Humana Press, Totowa, New Jersey.

Gelbrich, J., Mai, C., Militz, H., 2008. Chemical changes in wood degraded by bacteria. International Biodeterioration and Biodegradation 61 (1), 24–32.

Giordano, G., 1993. Tecnica Delle Costruzioni in Legno. Hoepli, Milano, 826 pp.

Havermans, J., 2010. Paper deterioration and the role of pollutants. In: Camuffo, D., Fassina, V., Havermans, J. (Eds.), Basic Environmental Mechanisms Affecting Cultural Heritage – Understanding Deterioration Mechanisms for Conservation Purposes. COST Action D42 "Enviart". Nardini, Florence, pp. 153–157.

Jakieła, S., Bratasz, Ł., Kozłowski, R., 2007. Acoustic emission for tracing the evolution of damage in wooden objects. Studies in Conservation 52, 101–109.

Jakieła, S., Bratasz, Ł., Kozłowski, R., 2008. Acoustic emission for tracing fracture intensity in lime wood due to climatic variations. Wood Science and Technology 42, 269–279.

Jones, D.A., 1996. Principles and Prevention of Corrosion. Prentice Hall, Upper Saddle River, N.J., 572 pp.

Kolar, J., Mozir, A., Strlic, M., de Bruin, G., Philar, B., Steemers, T., 2007. Stabilization of iron gall ink: aqueous treatment with magnesium phytate. e-Preservation Science 4, 19–24.

Kolar, J., Šala, M., Strlič, M., Šelih, V.S., 2005. Stabilisation of paper containing iron-gall ink with current aqueous processes. Restaurator 26, 181–189.

Kolar, J., Strlic, M., Havermans, J.B., 2004. Proceedings of the International Conference Durability of Paper and Writing. National and University Library, Ljubljana, Slovenia.

La Mantia, F.P., Morreale, M., 2008. Accelerated weathering of polypropylene/wood flour composites. Polymer Degradation and Stability 93 (7), 1252–1258.

Lattuati-Derieux, A., Bonnassies-Termes, S., Lavedrine, B., 2004. Identification of volatile organic compounds emitted by a naturally aged book using solid-phase micro-extraction/gas chromatography/mass spectrometry. Journal of Chromatography A1026, 9–18.

Laurenzi Tabasso, M., Marabelli, M., 1992. Il degrado dei monumenti in Roma in rapporto all'inquinamento atmosferico. Betagamma, Viterbo, 169 pp.

Leygraf, C., Graedel, T.E., 2000. Atmospheric Corrosion. Wiley-Interscience, New York.

Malešič, J., Kolar, J., Strlič, M., Polanc, S., 2005. The use of halides for stabilisation of iron gall ink containing paper – the pronounced effect of cation. e-Preservation Science 2, 13–18.

Mandrioli, P., Caneva, G., Sabbioni, C., 2003. Cultural Heritage and Aerobiology: Methods and Measurement Techniques for Biodeterioration Monitoring. Kluwer, Dordrecht, 245 pp.

McCleary, J.P., 1987. Vacuum freeze-drying, a method used to salvage water-damaged archival and library materials: a RAMP study with guidelines/prepared by John P. McCleary/for the/General Information Programme and UNISIST. UNESCO, Paris.

Mecklenburg, M.F., Tumosa, C.S., Erhardt, D., 1998. Structural response of painted wood surfaces to changes in ambient relative humidity. In: Dorge, V., Howlett, F.C. (Eds.), Painted Wood: History and Conservation. The Getty Conservation Institute, Los Angeles, pp. 464–483.

Michalski, S., 2000. Guidelines for Humidity and Temperature in Canadian Archives. Technical Bulletin No. 23. Canadian Conservation Institute, Ottawa.

Michalski, S., 2009. The ideal climate, risk management, the ASHRAE chapter, proofed fluctuations, and towards a full risk analysis model. 19 pp in: Contribution to the Experts' Roundtable on Sustainable Climate Management Strategies, held in April 2007, in Tenerife, Spain. The Paul Getty Conservation Institute, www.getty.edu/conservation/science/climate/climate_experts_roundtable.html.

Munier, I., Lefèvre, R., Losno, R., 2002. Atmospheric factors influencing the formation of neocrystallisations on low-durability glass exposed to urban atmosphere. Glass Technology 43C, 114–124.

Newton, R., Davison, S., 1989. Conservation of Glass. Butterworths, London, 322 pp.

Niklasson, A., Langer, S., Arrhenius, K., Rosell, L., Bergsten, C.J., Johansson, L.G., 2008. Air pollutant concentrations and atmospheric corrosion of organ pipes in European church environments. Studies in Conservation 53 (1), 24–40.

Odlyha, M., Bergsten, C.J., Thickett, D., 2010a. Volatile organic compounds released by wood. In: Camuffo, D., Fassina, V., Havermans, J. (Eds.), Basic Environmental Mechanisms Affecting Cultural Heritage – Understanding Deterioration Mechanisms for Conservation Purposes. COST Action D42 "Enviart". Nardini, Florence, pp. 107–114.

Odlyha, M., Jakiela, S., Bergsten, C.J., Slater, J.M., Niklasson, A., Svensson, J.E., Cavicchioli, A., de Faria, D.L.A., Thickett, D., Grøntoft, T., Dahlin, E., 2010b. Dosimetry for monitoring in organ pipes and in microclimate frames for paintings. Metals, 321–326 (Proceedings of the Interim meeting of ICOM-CC for Conservation Metal Working group October 11–15, 2010 Charleston, South Carolina, USA).

Padfield, T., 1994. Role of standards and guidelines. In: Krumbein, W.E., Brimblecombe, P., Cosgrove, D.E., Stainforth, S. (Eds.), Durability and Change. Wiley, New York, pp. 191–199, 307 pp.

Ramalho, O., Dupont, A.L., Egasse, C., Lattuati Derieux, A., 2009. Emission rates of volatile organic compounds from paper. e-Preservation Science 6, 53–59.

Richardson, B.A., 1993. Wood Preservation. E & FN Spon (Chapman & Hall), London, 226 pp.

Schoemaker, H.E., Piontek, K., 1996. On the interaction of lignin peroxidase with lignin. Pure and Applied Chemistry 68 (11), 2089–2096.

Schreiner, M., 2005. Corrosion of historic glass and enamels. In: Janssens, K., Van Grieken, R. (Eds.), Non-destructive Micro Analysis of Cultural Heritage Materials. Elsevier, Amsterdam, pp. 713–754.

Sebera, D.K., 1994. Isoperms: An Environmental Management. Tool Commission of Preservation and Access, Washington, DC.

Sedlbauer, K., 2002. Prediction of mould growth by hygrothermal calculation. Journal of Building Physics 25, 321–336.

Sedlbauer, K., Hofbauer, W., Krueger, N., Mayer, F., Breuer, K., 2011. Material specific isopleth-systems as valuable tools for the assessment of the durability of building materials against mould infestation – the "isopleth-traffic light". In Proceedings of the XII DBMC, International Conference on Durability of Building Materials and Components, Porto, Portugal, April 2011. p.8.

Sterflinger, K., 2010. Fungi: their role in deterioration of cultural heritage. Fungal Biology Reviews 24, 47–55.

Sterflinger, K., Pinzari, F., 2012. The revenge of time: fungal deterioration of cultural heritage with particular reference to books, paper and parchment. Environmental Microbiology 14 (3), 559–566.

Strlič, M., Kralj Cigić, I., Kolar, J., de Bruin, G., Pihlar, B., 2007. Non-destructive evaluation of historical paper based on pH estimation from VOC emissions. Sensors 7, 3136–3145.

Strlič, M., Thomas, J., Trafela, T., Cséfalvayová, L., Kralj Cigić, I., Kolar, J., Cassar, M., 2009. Material degradomics: on the smell of old books. Analytical Chemistry 81, 8617–8622.

Summit, R., Slicker, A., 1980. Handbook of Material Science, Vol. IV: Wood. CRC Press, Boca Raton, Florida, 459 pp.

Thomson, G., 1986. The Museum Environment. Butterworths, London, 293 pp.

Vanholme, R., Demedts, B., Morreel, K., Ralph, J., Boerjan, W., 2010. Lignin biosynthesis and structure. Plant Physiology 153 (3), 895–905.

Warscheid, T., Braams, J., 2000. Biodeterioration of stone; a review. International Biodeterioration and Biodegradation 46, 343–368.

Winandy, J.E., 1994. Wood properties. In: Arntzen, Charles J. (Ed.), 1994. Encyclopedia of Agricultural Science, vol. 4. Academic Press, Orlando, FL, pp. 549–561.

CHAPTER

3

Parameters to Describe Air Masses and Vertical Motions

OUTLINE

Outdoor monuments, archaeological sites and cultural or natural landscapes are particularly vulnerable to what is happening at a wide scale and the basic microclimate concepts that we have introduced for objects and collections kept in indoor environments are no longer sufficient and need further developments. Air pollution transport and diffusion, precipitation, gale winds, weather injuries and extreme meteorological events are further challenges to which monuments, sites and built environment are exposed and need a specific treatment.

So far, in a room or a street we have observed the air from the *Euler's* point of view, i.e. from a frame of reference fixed to the soil or a room (*fixed Euler frame*), where air changes temperature or humidity or is substituted with other air masses. All quantities are depicted as a function of position and time. We have observed the variation of a number of parameters in some selected points, single or forming a horizontal grid or a vertical

http://dx.doi.org/10.1016/B978-0-444-63296-8.00004-4

profile. It may be useful to change the point of view and consider the thermodynamic transformation of an air mass from a mobile reference frame, which is related to the air mass itself, so that the fixed reference is the air and the changeable parts are the locations where the air is moving (*mobile Lagrange frame*). In *Lagrange's* view, the observer follows an individual fluid parcel as it moves through space and time. Each of these systems presents advantages and disadvantages, and the choice of the most appropriate system depends on the specific problem. Each problem should be faced with an adequate choice of variables.

It is now necessary to introduce new variables for a better representation of the thermodynamic state of an air mass in case we are interested in a wide or large scale or in sites at different altitudes above the average sea level or when we need to interpret weather maps for some specific survey purposes. For instance, an air mass is considered while it is moving over space to be characterized by the three dimensions X, Y, and Z. In case the elevation increases, the air will adiabatically expand and cool, becoming supersaturated, forming clouds and causing precipitation over our sites. If water drops will precipitate, they will remove liquid water and thus the possibility of utilizing again the latent heat associated with the change of phase. All these thermodynamic situations can be described with new parameters that should be introduced at this point. Although at a first glance they might appear complex, they have the great advantage of being affected by some specific thermodynamic processes and not by others, so that they may characterize an air mass before and after the occurrence of some specific transformations to which they are invariant. All these parameters are popular in meteorology and air pollution and are useful to interpret specific environmental conditions within a wider context. In addition, they can be applied for a better understanding of the interactions between the air and the surface of sites, buildings and monuments.

3.1. EQUIVALENT TEMPERATURE

The *equivalent temperature* T_e (or t_e) or *isobaric equivalent temperature is the temperature that a moist air parcel would have if all water vapour in it were condensed out at constant pressure and adiabatically*, i.e. *all the latent heat released in the condensation is used to heat the air*. The adjective *equivalent* implies the sum of the latent and sensible heat contained inside an air parcel. It represents the molecular temperature that would be attained by an air parcel if all the latent heat $L_v w$ (where L_v is the latent heat of vaporization per unit mass of water and w the mixing ratio) potentially present inside the parcel were transformed into kinetic energy of the air molecules and liquid water were removed from the system, e.g. by precipitation. This process is only speculative, as all the forms of condensations that occur in nature do not involve the total amount of the moisture actually present.

This parameter is given by the equation

$$T_e = T + \frac{L_v}{c_{pd}} w \approx T + 2.5\,MR \tag{3.1}$$

where *MR* is the humidity mixing ratio in popular units (i.e. g kg^{-1}) and the right-hand side approximation (with constant L_v) shows that 1 g of water vapour raises by some 2.5 °C the temperature of 1 kg of air.

From this definition, this parameter is a conservative property of the air with respect to isenthalpic evaporation or condensation of droplets in the air, e.g. the formation or dissipation of fog or clouds, or condensation or evaporation of water in micropores of nonconductive materials.

At ground level, considering that at standard conditions, i.e. $p = 1013$ hPa and $t = 0$ °C, 1 m^3 of air has a mass of 1.275 kg, with the help of Eqn (2A.28), the previous equation can be approximated as

$$T_e \approx T + 1.95\ AH \approx T + 2\,AH \tag{3.2}$$

where *AH* is the absolute humidity in popular units (i.e. g m^{-3}) and 1.95 = 2.5/1.275. This equation shows that 1 g of vapour increases the temperature of 1 m^3 of air by some 2 °C.

This parameter, which is a function of both *T* and *MR* shows when and where the air exchanges heat (also called *sensible heat*) or moisture (i.e. *latent heat*) with an evaporating or absorbing surface or when two different air masses meet and mix together. In microclimatic studies, this parameter is useful to individuate exchanges between air and surfaces or the path of external air penetrating inside a building. When no interactions have occurred, this parameter is the same indoors and outdoors. However, it remains unchanged even if condensation, or evaporation, occurs in micropores of a nonconductive material, the heat being supplied or absorbed by the air. If the walls have low conductivity and very large heat capacity and all the heat is supplied or absorbed by them, the condensation in micropores does not affect the room temperature but lowers the room *MR* and, consequently, the T_e. The opposite occurs in the case of evaporation from micropores.

This parameter presents some advantages in environmental diagnostics and is the first step to define another parameter, namely, the equivalent-potential temperature, which is conservative with reference to both adiabatic and nonisobaric processes. For instance, the formation or dissipation of fog (or a cloud) cannot always be described with a combination of isobaric and adiabatic processes and it will be useful to introduce other parameters, which are conservative before, during and after uplift adiabatic expansions, cloud condensation and precipitation. The next step is to find invariance with respect to adiabatic vertical movements in atmosphere.

3.2. ADIABATIC GRADIENTS IN TROPOSPHERE

In the lower atmosphere, the trajectory of an air mass and its changes in heat and moisture content are of interest not only in weather forecast but also in transport and dispersion of airborne pollutants. This implies to know how the energy is distributed and transformed within an air mass, and the values that its temperature, moisture content and density will assume when time, space or level will change. When an air mass changes height, the ambient pressure changes and the work done to expand it is made using part of the thermal energy of the air molecules. The consequence is cooling. This is an example of *adiabatic* expansion (i.e. an expansion without exchange of energy with the external system), and the temperature changes are calculated using the Poisson equation, as we will see in the following. The convective mixing of the lower atmosphere is the cause of the decrease in air temperature with increasing height in the first 10–15 km of atmosphere, called *troposphere*. Above this layer, up to some 25 km, the air temperature increases again because in this second layer, called the *stratosphere*, the dominant mechanism is the absorption of ultraviolet radiation by oxygen and ozone molecules.

In a transformation that is reversible and slow, the internal pressure of an air parcel equals the external one and

$$\delta W = p\,\mathrm{d}V. \tag{3.3}$$

The condition of reversibility gives

$$\delta Q = p\,\mathrm{d}V + \mathrm{d}U \tag{3.4}$$

where the internal energy $U = U(T)$ is a function of T only, as well known for an ideal gas. If an amount of heat δQ is supplied to a gas parcel having mass m, all the heat is transformed into internal energy and the increase is

$$\mathrm{d}U = m\,c_v\,\mathrm{d}T, \tag{3.5}$$

and this equation is also valid if the volume is not kept constant. The first principle of thermodynamics can be written

$$\delta Q = p\,\mathrm{d}V + m\,c_v\,\mathrm{d}T. \tag{3.6}$$

The equation of state is preferably written in terms of p and T, as they are easily measurable parameters. The equation of state becomes

$$p\,\mathrm{d}V + V\,\mathrm{d}p = m\,\Re_a\,\mathrm{d}T. \tag{3.7}$$

By definition, for a thermodynamic transformation X the specific heat c_X is given by

$$c_X = \frac{(\delta Q)_X}{m\,\mathrm{d}T} = \frac{mc_v\,\mathrm{d}T + p(\mathrm{d}V)_X}{m\,\mathrm{d}T} = c_v + \frac{p}{m}\left(\frac{\mathrm{d}V}{\mathrm{d}T}\right)_X \tag{3.8}$$

In the case of an isobaric transformation, $dp = 0$ and the Eqn (3.7) becomes

$$p\,dV = m\,\Re_a dT \tag{3.9}$$

and

$$c_p = c_v + \frac{p}{m}\frac{m\,\Re_a}{p} = c_v + \Re_a \tag{3.10}$$

Substituting this result in the Eqn (3.6) gives

$$\delta Q = m\,c_v\,dT - V\ dp + m\,\Re_a\,dT = m\,c_p\,dT - V\ dp \tag{3.11}$$

and a further substitution in the equation of the state gives for the unit mass

$$(\delta Q)_{m=1} = c_p\,dT - \Re_a T\frac{dp}{p}. \tag{3.12}$$

For an adiabatic process $\delta Q = 0$; if the process is also reversible

$$c_p dT = \Re_a T\frac{dp}{p} \tag{3.13}$$

or

$$dT = k\,T\frac{dp}{p} \tag{3.14}$$

where $k = (c_p - c_v)/c_p = \Re_a/c_p = 287/1003 = 0.286$, which is valid for small variations of p. After integration, the *Poisson* equation is found for adiabatic and reversible processes, i.e.

$$\frac{T_2}{T_1} = \left(\frac{p_2}{p_1}\right)^k. \tag{3.15}$$

The atmosphere is called *adiabatic* when vertical motions occur without any external work and this happens when the density of the air parcels is always in equilibrium with that of the ambient air. The atmosphere is characterized by a *neutral* equilibrium with reference to the vertical adiabatic motions of dry air. It is possible to compute the vertical thermal gradient during neutrality. From the hypsometric equation of hydrostatics

$$dp = -\rho\,g\,dz \tag{3.16}$$

where p is the barometric pressure, ρ is the density, $g = 980.665$ cm s^{-2} the acceleration of gravity and z is the vertical coordinate. Remembering that from the equation of the state for perfect gases $R_a \rho T/p = 1$, Eqn (3.13) can be written

$$\frac{dT}{dz} = -\frac{\Re_a \rho g T}{c_p p} = -\frac{g}{c_p} = -\frac{9.8}{1003} \approx 0.01\ (\mathrm{K\,m^{-1}}) \tag{3.17}$$

which in practice corresponds to 1 °C cooling per 100 m adiabatic uplift. The vertical gradient $dT/dz = 1\ °C/100$ m is therefore called *adiabatic* and is represented with the symbol γ (especially in mathematical formulae) or Γ (especially in graphic representations).

3.3. POTENTIAL TEMPERATURE

We have seen that the temperature is representative of the average kinetic energy of the molecules constituting an air parcel but this value changes when the parcel is moved, changing its height above the Earth's surface. It is useful to find a parameter that allows for a comparison between the average kinetic energy of molecules belonging to two distinct air masses, independent of their height, i.e. referring them to a standard level. The *potential temperature* Θ *is the temperature a parcel of air would have if brought dry adiabatically from its initial state to the standard pressure of 1000 hPa*. Note that the round number 1000 is used instead of 1013, in order to make calculations easier. At sea level, the potential temperature is very close to the actual air temperature and it is exactly $\Theta = T$ when $p = 1000$ hPa. The potential temperature of an air parcel is independent of the height, similar to the temperature of mixed incompressible liquids.

Θ is computed with the help of the Poisson equation utilizing the initial values of T and p, i.e.

$$\Theta = T\left(\frac{1000}{p}\right)^k \tag{3.18}$$

where Θ is directly proportional to T but inversely related to p^k.

A region of the atmosphere in which Θ is constant along the vertical is named *dry adiabatic*. A representation in terms of Θ immediately shows where the atmosphere is adiabatic, i.e. $d\Theta/dz = 0$, with vertical isolines; where it is superadiabatic, $d\Theta/dz > 0$, with isolines tilted towards left; and where it is subadiabatic, $d\Theta/dz < 0$, with isolines tilted towards right (Fig. 3.1). Further details will be given in Chapter 7.

In meteorology, weather prediction and pollution transport, the basic equations are more easily solved if appropriate coordinates are used. For this reason, the geopotential, the pressure, the entropy and

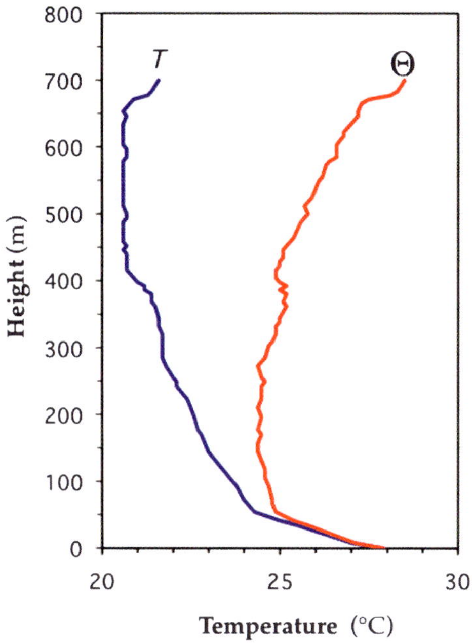

FIGURE 3.1 Vertical profile of the air temperature T (blue line) and potential temperature Θ (red line) in the Venice hinterland, on 25 June 1979 at 10 am. The first 150 m show a profile superadiabatic, i.e. with a gradient greater (in the absolute value) than the adiabatic one (Θ profile tilted towards left); from 150 to 280 m is the substantially adiabatic region (Θ profile vertical); and above the air is stable (Θ profile tilted towards right).

the potential temperature are used as a vertical coordinate instead of the geometric height (for a discussion see Kasahara, 1974). Potential temperature or isentropic coordinates are especially convenient for description of adiabatic motions. As in the definition of Θ, the atmosphere is characterized by adiabatic displacements in the whole region where $\Theta = \text{const}$, also $\delta Q = 0$, and trajectories are controlled by isentropic motions. Therefore, *all the surfaces with the same Θ are isentropic* and an air parcel having a given value of Θ will remain on the same surface characterized by a given potential temperature Θ or within the same layer bounded by the two isentropes Θ and $\Theta + d\Theta$, unless external work is supplied to it. Therefore, the isentropic representation of the atmosphere is a useful tool to forecast where a pollutant can or cannot be transported and whether it will be able to cross mountain chains or not (Fig. 3.2). In fact, except near the soil where heat is exchanged and in the absence of condensation, an air parcel tends to maintain the same entropy and potential temperature.

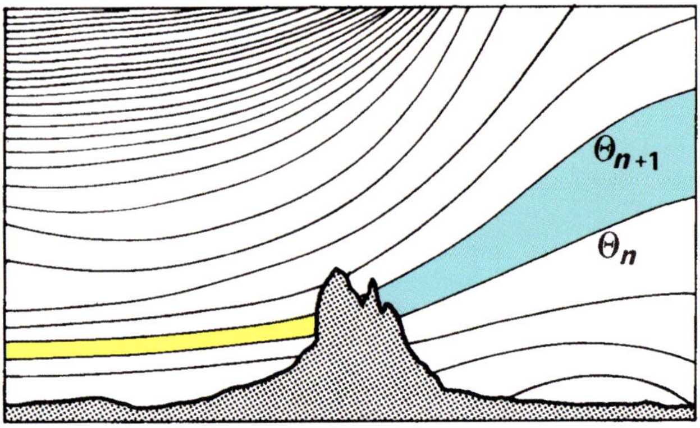

FIGURE 3.2 Vertical cross-section of potential temperature over a terrain with a mountain chain. The air layer between Θ_n and Θ_{n+1} will remain separated into two distinct parts (i.e. yellow and cyan) by the mountain chain.

It is also possible to find a relationship between potential temperature and entropy. Dividing by T both sides of Eqn (3.12) and integrating, one obtains the *entropy* $\mathfrak{S}$ of the air:

$$\mathfrak{S} = \int \frac{dQ}{T} = \int c_p \frac{dT}{T} - \int \Re_a \frac{dp}{p} = c_p \ln T - \Re_a \ln p \tag{3.19}$$

This equation holds for unsaturated air, assuming that the mixture of air and vapour behaves as an ideal gas (unsaturated vapour). For saturated vapour, the equation is formally identical, except for an additive constant (Goody, 1995). The relationship is easily obtained substituting in the previous equation the definition of Θ in Eqn (3.18) in logarithmic form, i.e.

$$\ln\Theta = \ln T + \frac{\Re_a}{c_p}(\ln 1000 - \ln p) \tag{3.20}$$

so that the entropy $\mathfrak{S}$ is immediately found, i.e.

$$\mathfrak{S} = c_p \ln\Theta \tag{3.21}$$

where the additive arbitrary reference constant has been omitted.

The Poisson equation can be compared with the assumption that the vertical temperature gradient is constant and given by Eqn (3.17), i.e. $dT/dz = \gamma$, so that

$$T = T_o + \gamma z \tag{3.22}$$

Dividing the hydrostatic equation (Eqn (3.16)) by the equation of state for perfect gases (Eqn (1.2)), and using the previous equation for T,

$$\frac{dp}{p} = -\frac{g dz}{\Re_a (T_o + \gamma z)} = -\frac{g}{\Re_a \gamma} \frac{d(T_o + \gamma z)}{T_o + \gamma z} \tag{3.23}$$

or

$$d\ln p = -\frac{g}{\gamma \Re_a} d\ln T. \tag{3.24}$$

Integration gives

$$\frac{p_2}{p_1} = \left(\frac{T_2}{T_1}\right)^{-g/\gamma \Re_a} \tag{3.25a}$$

$$\frac{T_2}{T_1} = \left(\frac{p_2}{p_1}\right)^{-\gamma \Re_a/g} \tag{3.25b}$$

where $\gamma \Re_a/g = k = 0.286$.

3.4. EQUIVALENT-POTENTIAL TEMPERATURE

The *equivalent-potential temperature* Θ_e *is the potential temperature corresponding to the equivalent temperature* T_e, i.e.

$$\Theta_e = T_e \left(\frac{1000}{p}\right)^k = \Theta + \frac{L_v}{c_{pd}} w \left(\frac{1000}{p}\right)^k \tag{3.26}$$

This parameter is conservative with respect to *dry adiabatic* processes, e.g. compression or expansion of air parcels due to change of height. The exponent k has been derived for dry air. Moist air, when reduced to the standard pressure of 1000 hPa, has a temperature somewhat lower than that of dry air and the departure increases with moisture content. However, this departure is in general very small, so that this parameter can be considered quasi-conservative for isenthalpic *pseudo-adiabatic* processes. The adjective '*pseudo-adiabatic*' refers to systems in which saturation occurs and the condensed water is removed from the system, e.g. by precipitation. The uplift of air masses, which form clouds and rainfall are an example of a pseudo-adiabatic process. Both the dry and the pseudo-adiabatic lines are represented in aerological diagrams, e.g. the Stüve diagram. This parameter varies for isobaric heating or cooling or gain of vapour supplied by oceanic waters relatively warmer than the air.

The above discussion on isentropic surfaces can be applied with a wider generality to Θ_e, as it also includes the possibility of formation of clouds and precipitation. When an air mass is forced to flow over a mountain chain, during the slope rise, it may arrive at a height above the *cloud condensation level*: the temperature drops below the dew point, the vapour is supersaturated and begins to condense forming cloud droplets and precipitation on the upwind slope. This upslope precipitation is known as *stau*. This mechanism continues during the cooling associated with the rise and stops when the air is heated by compression when it descends downwards along the other slope. Droplets that remain in the cloud will evaporate again. If the equivalent temperature of the uprising and descending streams is measured in the opposite slopes at the same height z_1, the value $T_e(z_1)$ is the same for both streams but is not conservative with changes of level, i.e. $T_e(z_1) \neq T_e(z_2)$. The equivalent-potential temperature Θ_e of an air parcel that moves adiabatically is independent of the height and is the same everywhere, so that an air mass can be always individuated by this parameter; when the parcel undergoes a nonadiabatic transformation, the change $\Delta\Theta_e$ gives the gain (or loss) of the thermal energy expressed as the sum of sensible plus latent heat.

In the previous example, the temperature T of the air has undergone important changes. During the upwind condensation, the latent heat has warmed the air mass and most of the liquid water has been removed from the system by precipitation. During the descent, the cloud droplets evaporate reabsorbing the heat that they have previously released during the condensation phase, so that this part of the process is reversible; however, the heat supplied by the precipitated water remains in the air, which blows warmer. This is the well-known downslope warm wind called *foen*. In all the stages of upslope cooling and precipitating and downslope warming, the equipotential temperature remains unchanged or quasi-unchanged.

3.5. VIRTUAL TEMPERATURE

A plume of smoke, a meteorological balloon or a cumulus cloud will rise while they have buoyancy, i.e. their density is lower than that of the surrounding air. However, the density of moist air is determined by two factors: the air temperature and the moisture content. The molecular weight of water vapour is 18.016 g mol^{-1}, whereas the average molecular weight of dry air is 28.966 g mol^{-1}. For this reason, the density of a moist air parcel at a given temperature and pressure is equalized by a warmer parcel of dry air at the same pressure. In order to simplify the description of atmospheric processes, it may be convenient to ignore the presence of moisture and make reference to an ideal dry air with an effective temperature that simulates the effect of moisture in terms of density.

The *virtual temperature* T_v (or t_v) *of a parcel of moist air is the temperature at which completely dry air would have the same density and pressure as moist air.* Also, this parameter is speculative but is very practical, as it allows applying to the moist air the equation of state for dry air. This parameter is used when comparing the buoyancy of two different air masses that will come into contact, or the vertical uplift of a puff of smoke, or a meteorological balloon.

Using the *Dalton* law, the pressure of the moist air is $p = p_a + e$ where p_a represents the partial pressure of dry air e the partial pressure of water vapour. Similarly, the mass of moist air is $m = m_a + m_v$ where the labels 'a' and 'v' are still used to distinguish between dry air and vapour. The equation of state for the moist air becomes:

$$\begin{aligned} pV &= (m_a \Re_a + m_v \Re_v)T = m\,\Re_a\left(\frac{m_a}{m} + \frac{m_v \Re_v}{m\,\Re_a}\right)T = m\,\Re_a\left(1 - \frac{m_v}{m} + 1.61\frac{m_v}{m}\right)T \\ &= m\,\Re_a\left(1 + 0.61\frac{m_v}{m}\right)T \approx m\,\Re_a\left(1 + 0.61\frac{m_v}{m_a}\right)T = m\,\Re_a(1 + 0.61w)T \\ &= m\,\Re_a T_v \end{aligned} \tag{3.27}$$

where T_v is defined as

$$T_v = (1 + 0.61w)T \tag{3.28}$$

but it is also often expressed as

$$T_v \approx \left(1 + \frac{3}{8}\frac{e}{p}\right)T \approx \frac{T}{1 - \frac{3}{8}\frac{e}{p}} \tag{3.29}$$

The derivation of this formula is simple. When w is small, as it is, the reciprocal of $(1 + 0.61w)$ is to a close approximation $1/(1 + 0.61w) \approx (1 - 0.61w)$. Substituting to w the approximation (2.19) and considering that $0.61 \times 0.622 = 0.379 \approx 3/8$, the latter equation is found.

From the above formulae it follows that $T < T_v < T_e$. Under usual meteorological conditions, the virtual temperature increment is $T_v - T \leq 5\,°\mathrm{C}$.

References

Theory and general applications

Goody, R., 1995. Principles of Atmospheric Physics and Chemistry. Oxford University, New York, pp. 324.

Kasahara, A., 1974. Various vertical coordinate systems used for numerical weather prediction. Monthly Weather Review 102, 509–522.

Further Reading

Belinski, V.A., 1948. Dynamic Meteorology. Ogiz, Moscow, pp. 591.

Brutsaert, W.H., 1982. Evaporation in the Atmosphere. Reidel, Dordrecht, pp. 299.

Byers, H.R., 1974. General Meteorology. McGraw-Hill, New York, pp. 461.

Haltiner, G.J., Martin, F.L., 1957. Dynamical and Physical Meteorology. McGraw-Hill, New York, pp. 454.

Houghton, D.D., 1985. Handbook of Applied Meteorology. Wiley, New York, pp. 1461.

Iribarne, J.V., Godson, W.L., 1981. Atmospheric Thermodynamics. Reidel, Dordrecht, pp. 259.

Matveev, L.T., 1967. Physics of the Atmosphere. Israel Program for Scientific Translations, Jerusalem, pp. 659.

Petterssen, S., 1956. Weather Analysis and Forecasting, vol. 2. McGraw-Hill, New York, pp. 265.

CHAPTER

4

Radiation and Light

OUTLINE

4.1. THE EMISSION OF RADIATION FROM BODIES AND THE EFFECTS OF THE ABSORBED ENERGY

According to the classical Maxwell theory of radiation (Maxwell, 1873), all bodies emit electromagnetic radiation, which propagates in space in the form of waves of the electric and magnetic fields. The radiation frequency ν is given by the ratio between the velocity of light c and the wavelength λ of the radiation, i.e.

$$\nu = \frac{c}{\lambda}. \tag{4.1}$$

http://dx.doi.org/10.1016/B978-0-444-63296-8.00005-6

The intensity I of the flux of energy emitted by a blackbody per unit time, unit wavelength and unit solid angle depends on ν and the absolute temperature T, according to the *Planck* formula of thermal emission (Planck, 1901)

$$I(\nu, T) = \frac{2h\nu^3c^2}{\exp(h\nu c/kT) - 1} \tag{4.2}$$

where $h = 6.626 \times 10^{-27}$ erg s $= 6.626 \times 10^{-34}$ J s is the Planck constant. The same phenomenon can be described in terms of corpuscular emission (Einstein, 1905), where the radiant energy is transported by photons having the individual energy

$$E = h\nu, \tag{4.3}$$

the intensity of a spectral component being given by the number of photons having a given wavelength. Although the two theories are equivalent in most cases, the classical wave representation is more convenient to describe some phenomena, e.g. diffraction, interference, polarization and the quantum theory is convenient for other phenomena, e.g. photoelectric effect, emission of cathode rays, and gas ionization by means of ultraviolet (UV) radiation. A synthesis of the two points of view has been made associating with light the dual nature of wave and corpuscle.

According to the Einstein theory, when a molecule absorbs a photon, its energy is used to excite an electron and this provokes a change of the energetic levels with a number of consequences. If the energy $h\nu$ exceeds the bonding energy of the electron, the latter may escape (photoelectric effect), or the exited molecule may dissociate, isomerize, transfer energy to another molecule, be transformed into a different compound with a photochemical reaction or undergo other degradation mechanisms (Wypych, 1995). Ionization in the upper atmosphere, ozone formation in the stratosphere, and molecular excitation of atmospheric H_2O, CO_2 and O_3 are well-known mechanisms that absorb some spectral bands of the solar radiation.

Solar radiation that reaches the Earth's surface is classified as follows (Table 4.1).

Insolation is a measure of solar radiation energy received on a given surface area and recorded during a given time. In the solar radiation, the shortest wavelengths (e.g. ultraviolet (UV)) are mainly absorbed by the ionosphere and in particular in the ozone layer; the longest wavelengths (e.g. infrared (IR)) are absorbed by atmospheric water vapor and carbon dioxide. The sun, whose photosphere is at some 5800 K, emits 99.9% of its energy within the wavelength interval 0.15–40 μm, with peak intensity in the visible light, around 0.47 μm. The UV radiation is composed of a small flux of photons. The total power associated with the UV band depends on the cumulative effect of the (small) number of photons and their (high)

TABLE 4.1 Wavelengths of the Solar Radiation Reaching the Earth's Surface

Wavelength (μm)	Radiation	Wavelength (μm)	Radiation
<0.001	γ rays and X-rays	0.72–1.5	Near infrared
0.001–0.2	Far ultraviolet	1.5–5.6	Middle infrared
0.2–0.315	Middle ultraviolet	5.6–1000	Far infrared
0.315–0.38	Near ultraviolet	>1000	Micro & radiowaves
0.38–0.72	Visible light		

individual energy, and is in general low. However, each individual photon has high energy and is potentially dangerous when a sensitive surface is reached. About half of the total power of the solar radiation lies in the visible light and most of the remaining in the near IR.

All bodies can absorb and emit electromagnetic radiation at all wavelengths, and the *'colour'*, i.e. the distribution of intensity of the spectral lines of the light, depends on the body temperature. This effect is quantitatively described by the *Wien* displacement law, which states that the spectral distribution of the energy density varies with the body temperature, the peak value being attenuated and displaced towards the IR with decreasing body temperature. Bodies at ambient temperature have emission peaks in the near IR and the emission is called thermal radiation.

More precisely, the *Wien* law states that the wavelength λ_p of the peak intensity for the emitted radiation is inversely proportional to the absolute temperature, i.e.

$$\lambda_p = \frac{\text{const}}{T} \tag{4.4}$$

where the proportionality constant is 2897 μm K (Born, 1952). This finding was obtained by differentiating the Planck equation with respect to the wavelength and equalizing the derivative to zero, i.e. $\partial I(\nu,T)/\partial\lambda = 0$. The higher the body temperature, the shorter the peak wavelength, and the lower the temperature, the longer the peak wavelength. The tungsten filament of an incandescent lamp has a temperature (or, better, the *colour temperature* i.e. the temperature of a blackbody having the same colour appearance) around 2800 K; its spectrum falls within the solar spectrum, with a lower spectral radiance, but with the maximum shifted to around 1 μm, in the near IR region. The candle flame reaches 1700 K; its spectrum has lower spectral radiance, beginning in the visible and extending into the IR region where the peak value is found at 1.7 μm. At 800 K, hot bodies are faint red and below this temperature they emit only invisible radiation in the IR region, with lower power; the peak is at 3.6 μm. At room temperature, the spectrum has a still lower power, ranging between 3 and 100 μm, and the peak is near 10 μm. As the power emitted by a body depends on T, a remote measure of this power can be used to detect the body temperature, which is called *radiometric temperature*.

4.2. RADIOMETRIC TEMPERATURE

The *radiometric temperature* T_{rad} *is the temperature deduced from the flux* Φ of radiant energy spontaneously emitted per unit surface and unit time, by a body with temperature T by means of the IR emission, according to the *Stefan–Boltzmann law*. This law establishes that the flux density emitted by a blackbody is proportional to the fourth power of its absolute temperature T, i.e.

$$\Phi = \sigma T^4 \tag{4.5}$$

where $T = T_{rad}$ so that the temperature can be written in explicit form

$$T_{rad} = \sqrt[4]{\frac{\Phi}{\sigma}} \tag{4.6}$$

and $\sigma = 0.286 \times 10^{-10}$ cal cm^{-2} min^{-1} K^{-4} = 5.67 erg cm^{-2} s^{-1} K^{-4} is the Stefan constant.

The flux of electromagnetic radiation emitted in the IR spectral window can be measured with a radiometer or a bolometer and furnishes the experimental observation of the surface temperature, without mechanical contact with the surface. This aspect is very important especially in the case of vulnerable surfaces that cannot be measured with contact thermometers (e.g. tapestry, paintings on canvas, frescoes, and polychrome wooden statues, where it is not possible to fasten sensors with glue). It is also convenient for remote monitoring (e.g. to measure ceiling temperature without scaffoldings) or whenever the contact thermometer may alter the surface temperature because the heat capacity of the item under investigation is smaller than the heat capacity of the sensor (e.g. a thin canvas, a sheet of paper). Similarly, when a surface is illuminated, a contact sensor will shield the contact area and will measure the temperature of the surface in the shade, which has a lower temperature compared with the illuminated area). It is also convenient for monitoring moving objects.

The radiometric temperature of a body is a function of several variables, some of which are intrinsic in nature, e.g. the albedo and the thermal diffusivity of the body, and some of which are due to atmospheric forcing, e.g. reflection of long wave radiation, air temperature, humidity, aerosols, cloud cover, and heat or moisture accumulated or lost during past meteorological conditions.

It is not easy to obtain accurate observations of T_{rad}. The radiometer (see Chapter 11) can be used to compare relative temperature levels and to find anomalies in subsurface conductivity (e.g. detachment of frescoes or the presence of heterogeneous materials) especially during natural or forced dynamic changes of temperature. Indoor measurements are easier and more reliable than outdoor IR observations.

The Stefan–Boltzmann law in the above form is valid for an ideal blackbody, when both the absorptivity and emissivity are equal to 1. In practice, real bodies have some departures from this theoretical formulation, and a '*grey*' correction should be added, which takes into account the *relative emissivity* of bodies. Therefore, the grey emissivity $\varepsilon < 1$ should be explicitly written in the Stefan–Boltzmann equation, i.e.

$$\Phi = \varepsilon\sigma T^4. \tag{4.7}$$

The values of ε are in general close to unity for most of the natural materials but are much lower for polished metals, as we will see in Chapter 11. As ε varies with the physical characteristics of the surface, e.g. rough or glossy and wet or dry, attention should be paid in order to distinguish apparent variations of T from changes of ε.

The radiation flux arriving from a surface A can be divided into two components: the radiation directly emitted by A and the diffuse radiation, i.e. emitted by third bodies B but impinging on A and being reflected by A. The radiation emitted by A is useful to know the surface temperature with the help of Eqn (4.7); the diffuse radiation is related to B and constitutes a disturbance to measurements. The sum of the two gives the total (i.e. emitted plus reflected) radiation Φ. Normalization is made dividing these quantities by the incident flux, and the following identity is found

$$\varepsilon + R = 1 \tag{4.8}$$

where R is the reflectivity of the surface. This equation shows that good reflectors are bad emitters, and vice versa; e.g. polished metals are good reflectors ($R \to 1$) but bad emitters. Candle smoke emissivity is close to 1 and the reflectivity is very low ($R \to 0$). A reflector behaves like a mirror: we cannot see it but we will see reflected what is in front of it. The reflectivity is strictly related to the '*albedo*', or reflection coefficient, except for the dependence on the incidence angle. Albedo is defined as the fraction of incident electromagnetic radiation that is globally reflected by a surface. While reflectance is defined as the directionally reflected fraction for a single incidence angle, albedo is the directional integration of reflectance over all directions. The albedo is 1 for a totally reflecting surface, e.g. a mirror, and 0 for a perfect blackbody. The name is derived from Latin and means 'whiteness' because a white surface reflects the luminous flux of solar radiation. Both the albedo and the reflectivity may change with the radiation frequency.

4.3. ANGULAR DISTRIBUTION OF RADIANT EMISSION OF BODIES

The radiant intensity $I(\theta)$ (i.e. the flux per unit solid angle) emitted in any direction from a unit radiating surface varies as the cosine of the angle θ between the normal to the surface and the direction of the radiation (Lambert law), i.e.

$$I(\theta) = I(0)\cos\theta. \tag{4.9}$$

At the same time, a radiometer sees the observed surface under a given solid angle, determined by the optical objective that characterizes the geometry of the instrument. When the instrument is normal to an extended surface, the objective sees the portion S_o of it; when it is rotated by the angle θ, it sees a larger portion of surface, i.e. $S(\theta) = S_o/\cos\theta$, although tilted by the angle θ (Fig. 4.1).

As a consequence, the flux of radiant energy entering the radiometer objective is

$$\Phi(\theta) = I(\theta)S(\theta) = I(0)\cos\theta\frac{S_o}{\cos\theta} = I(0)S_o = \Phi(0) \tag{4.10}$$

i.e. the *luminance (or radiance*, i.e. the amount of light leaving a surface) *of a radiating surface is independent of direction* and appears equally bright at all angles. A surface obeying this law is said to be *Lambertian*. An ideal blackbody is Lambertian. Most of the surfaces are Lambertian, at least in a first approximation. For this reason, the sun and the moon appear as flat discs, not as three-dimensional spheres; similarly, golden or painted domes appear flat. For this reason, Vittori and Mestitz (1975) suggested that in the four gilded horses of the Basilica of San Marco in Venice, the narrow scratches on the surface were deliberately made by a skilled engraver to gain the third dimension. However, other surfaces may present some departures from this law; e.g. the wings of butterflies that

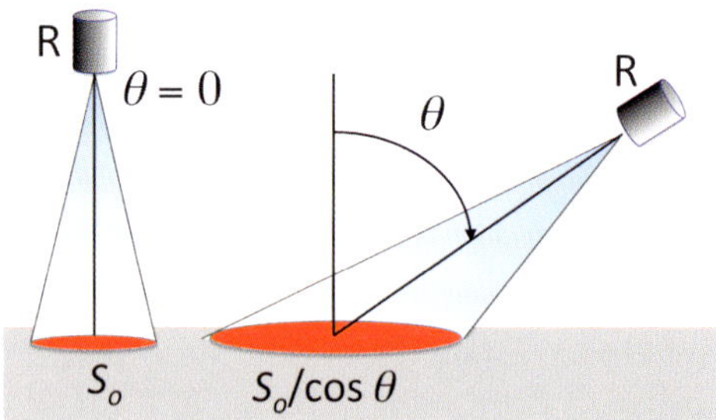

FIGURE 4.1 A radiometer R pointing to a surface sees a certain section of it (red disc). If the angle θ from the normal to the surface is increased, the radiometer will see a portion of surface increasing with θ or, more exactly, in the inverse proportion to cos θ.

change colour with the angle of sight and Persian carpets are well-known examples of non-Lambertian surfaces.

It might be useful to remember that *radiance* (W m^{-2} sr^{-1}) is a radiometric unit that measures the energy emitted in unit time (unit: Watt) from the unit surface (unit: square metre, m^2) in the unit solid angle (unit: steradian, sr) and *luminance* (unit: lm m^{-2} sr^{-1}) is a photometric unit that measures the intensity of the light (unit: lumen, lm) emitted from the unit surface in the unit solid angle in the frequency range in which the human eye is sensitive. Both are a measure of the power radiated by the unit area into the unit solid angle.

Another consequence of the Lambert law of emission is that surfaces differently exposed, even if into contact with air at the same temperature, may reach different temperatures as a result of the radiative balance. In fact, horizontal surfaces facing the clear sky emit IR radiation and receive back only a small fraction of the outgoing radiation. On the other hand, vertical surfaces face the horizon, which has an infinitely large optical air mass, so that the IR outcome nearly equals the income with almost null flux. For this reason, in clear sky evenings, dew forms first on the horizontal tops of cars, which face the open sky, and may cool down more rapidly than the vertical broadsides.

4.4. ATTENUATION OF LIGHT IN THE ATMOSPHERE

Corrections to the observed Φ should be made to take into account the attenuation of intensity caused by atmospheric absorption, and in particular by the water vapor, CO_2 and other gases, smoke and atmospheric aerosols. The atmospheric attenuation is exponential and varies with wavelength λ (*Bouguer–Lambert* law), i.e.

$$I(\lambda) = I_o \exp(-\alpha(\lambda)m) \tag{4.11}$$

where I_o is the intensity of incident radiation (coming from outside our atmosphere), $I(\lambda)$ is the intensity after having crossed the *optical air mass m*, and $\alpha(\lambda)$ is the absorption coefficient. This law can also be expressed in terms of the *optical path length x* through the absorbing substance, instead of the optical air mass m. This formula shows that a white solar beam, when crossing the atmosphere, has its spectral components $\alpha(\lambda)$ differently attenuated. As a consequence, the final mixture of colours depends on the optical air mass m, which is a function of the sun height $H_\odot$ or its zenith angle $Z_\odot = 90° - H_\odot$, i.e.

$$m = \sec Z_\odot . \tag{4.12}$$

This is the reason why daylight colour is continually changing, from sunrise to sunset.

When radiation is incident on the atmosphere, or a liquid or a solid body, part of it is transmitted, part is absorbed, and part is reflected. The ratio of each of these quantities to the incident flux, are, respectively, called *transmissivity Tr*, *absorptivity A* and *reflectivity R*. They add up to unity, i.e.

$$Tr + A + R = 1. \tag{4.13}$$

This formula has a general character and holds in average, either in bulk terms over the whole spectrum, or for any individual spectral wavelength λ, including the visible and thermal IR range.

4.5. DAILY AND SEASONAL CYCLES OF SOLAR RADIATION ON A SURFACE

Sunlight falls on outdoor monuments and penetrates through windows: it is composed of *direct radiation*, which forms the solar beams, and *diffuse radiation*, which is scattered by the sky in all directions. The diffuse component ranges from 20% to 30% of the total income in clear days to 100% in overcast ones. Scattering is a function of the wavelength (this is the reason why the sky appears blue), and the very energetic photons in the UV band are the most scattered ones, so that they may hit objects not exposed to direct light.

For any latitude ϕ and day of the year, defined by the solar declination $\delta_\odot$, direct light can be computed by means of the coordinates of the sun: the altitude $H_\odot$ over the horizon (i.e. solar height) and the azimuth $A_\odot$, i.e. the angular distance between the two vertical circles both containing the zenith and one containing the sun and the other the south point. The astronomical formulae to compute these coordinates are:

$$\sin H_\odot = \sin \delta_\odot \sin \phi + \cos \delta_\odot \cos \phi \cos \tau \tag{4.14}$$

$$\sin A_\odot = \frac{\cos \delta_\odot \sin \tau}{\sqrt{1 - (\sin \delta_\odot \sin \phi + \cos \delta_\odot \cos \phi \cos \tau)^2}} \tag{4.15}$$

where the *hour angle* $\tau = 180° \, t/12$ is computed from the time t, in hours and tenths of hour, from or to the culmination of the sun, i.e. from or to the true midday. This means that t is negative in the morning, vanishes at noon and is positive in the afternoon. The *solar declination* $\delta_\odot(j)$ for the jth day is found in astronomical ephemeris tables or, for our purposes, it can be computed with the simple approximation

$$\delta_\odot(j) = \delta_\odot(0)\cos\frac{2\pi j}{365} \tag{4.16}$$

where the jth day is computed since the winter solstice (i.e. 21 December) and $\delta_\odot(0) = -23° 27'$ is the declination at winter solstice. In this formula, the Earth's orbit around the sun is supposed to be circular. The mean of the Earth orbit *eccentricity* $e_\oplus$ is $e_\oplus = 0.0167$; the eccentricity is found by dividing the distance between the foci by the length of the major axis, and a circle is the limiting case as $e_\oplus$ approaches zero. The error in this approximation is small (some primes) and negligible except for precise astronomical calculations.

From the above formulae it follows that:

- at solar noon, when the hour angle τ is zero, $\cos\tau = 1$ at any latitude, and the *zenith angle* $Z_\odot$ (defined as $Z_\odot = 90° - H_\odot$) is $Z_\odot = \phi - \delta_\odot$;
- at *sunrise* and *sunset* (i.e. $H_\odot = 0$), at any latitude (except the poles), $\cos Z_\odot = 0$, and 2τ is the daytime length expressed in hour angle (h);
- the *daylength* (the time duration from the moment the upper limb of the sun's disk appears above the horizon at sunrise to the moment when the upper limb disappears below the horizon at sunset) $DL = 2\tau$ is computed under the condition $\cos\tau = -\tan\phi \tan\delta_\odot$, i.e. $DL = (24/\pi)$ $\arccos(\tan\phi \tan\delta_\odot)$ (h).

Solar beams are represented by straight lines originating from the sun and passing through windows; their intercepts on the opposite wall represent the light spot. The envelope of these spots describes the areas affected by the direct solar income during the course of the day as discussed below.

The flux density per unit time or *intensity* I_p of the solar radiation (also called *irradiation*) falling on horizontal, vertical or arbitrarily inclined planes (Robinson, 1966; Kondratyev, 1969; Bernardi and Vincenzi, 1994) facing the direction A_p (computed from the moment of the true noon, i.e. the south) and inclined by the angle β with the plane of the horizon is

$$\begin{aligned} I_p = I_o\{&\cos\beta(\sin\delta_\odot \sin\phi + \cos\delta_\odot \cos\phi \cos\tau) \\ &+ \sin\beta\{\cos A_p[\tan\phi(\sin\delta_\odot \sin\phi + \cos\delta_\odot \cos\phi\cos\tau) \\ &- \sin\delta_\odot \sec\phi] + \sin A_p \cos\delta_\odot \sin\tau\}\} \end{aligned} \tag{4.17}$$

where I_o is the intensity of the solar beam (near the surface) and the irradiation on a horizontal or vertical plane is obtained by setting $\beta = 0°$ or $\beta = 90°$, respectively.

An example of the hourly variation of the flux density of the solar radiation on the horizontal and vertical surfaces with various orientations is here discussed for the latitude $\phi = 45°$. In order to show the influence of the absorption due to the optical mass, first let us suppose that the atmosphere is perfectly transparent. At the winter solstice, the sun rises at $A_\odot = 124°$, i.e. close southeast (135°) and sets at $A_\odot = 236°$, i.e. close southwest (225°) and the vertical surfaces facing these two directions receive the maximum flux of energy; at noon, the solar height is low, i.e. $H_\odot = 90° - Z_\odot = 90° - (\phi - \delta_\odot) = 22°$ and the energy income on the vertical surface facing south is greater than that on the horizontal plane. At the equinoxes, sunrise and sunset occur just at east and west, and with a perfectly transparent atmosphere, the vertical surfaces facing these two directions receive the maximum flux of radiation. At this particular latitude, at noon the solar height is $H_\odot = 45°$, so that the solar energy that falls on the vertical surface facing south equals that on the horizontal plane. At the summer solstice, sunrise and sunset occur, respectively, at $A_\odot = 57°$, i.e. midway between northeast and east–northeast and $A_\odot = 303°$, i.e. midway between northwest and west–northwest. At noon the sun is high, i.e. $H_\odot = 68°$, so that the solar energy flux that falls on the vertical surface facing south is less than that on the horizontal plane, and at this particular latitude, the flux on the horizontal equals the flux at the winter solstice on the vertical surface facing south, and vice versa.

If the same calculations are made taking into account the attenuation of the solar radiation through the optical air mass, which increases with decreasing solar height, the results are quite different, as shown in Fig. 4.2. At the winter solstice (Fig. 4.2(a)), the declination $\delta_\odot(0) = -23° 27'$ is at its minimum and the optical thickness of the atmosphere is maximum. When the sun is very low on the horizon and the light beams are perpendicular to vertical surfaces facing the sun, the fraction of energy reaching the surfaces is small. The incoming intensity becomes relevant only when the sun is relatively high. At noon, the vertical surface facing south receives the maximum flux density and the horizontal plane receives much less.

At equinoxes (Fig. 4.2(b)), $\delta_\odot = 0$ and the atmospheric absorption is smaller. Near sunrise and sunset, the atmosphere reduces the solar energy (but less than in winter) and then the maximum flux density on the vertical surfaces facing the sun remains more or less the same, reaching the minimum at noon where the geometric effect of the slant beam dominates over the smaller atmospheric attenuation. On vertical surfaces not facing the solar motion (northern sector from west to east), the flux density is much smaller. The flux density on the vertical surface facing south equals the flux density on the horizontal plane. This is obvious at noon, when the solar altitude is $H_\odot = 45°$, forming the same angle with the horizontal plane and the vertical surface. On the other hours, for the vertical surface, the advantage of receiving the beam with an angle approaching the normal is substantially compensated for by atmospheric absorption.

At the summer solstice (Fig. 4.2(c)), the declination $\delta_\odot$ is at its maximum and the optical thickness of the atmosphere is further reduced. However, the height of the sun over the horizon is the dominant factor,

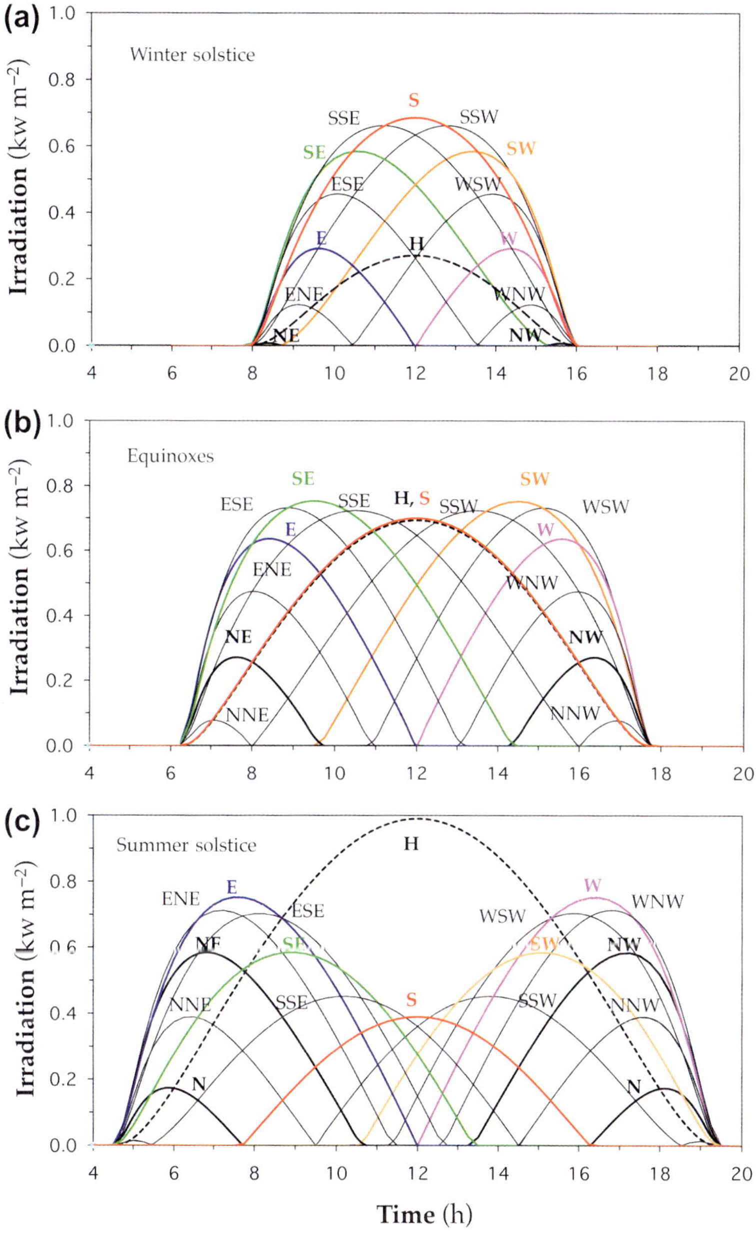

FIGURE 4.2 Hourly variation of the flux density of the solar irradiation I_p on the horizontal H (dotted line) and vertical surfaces (full lines) with various orientations (principal compass directions, thick colour lines; secondary, thin black lines) for latitude $\phi = 45°$. (a) Winter solstice, (b) equinoxes and (c) summer solstice.

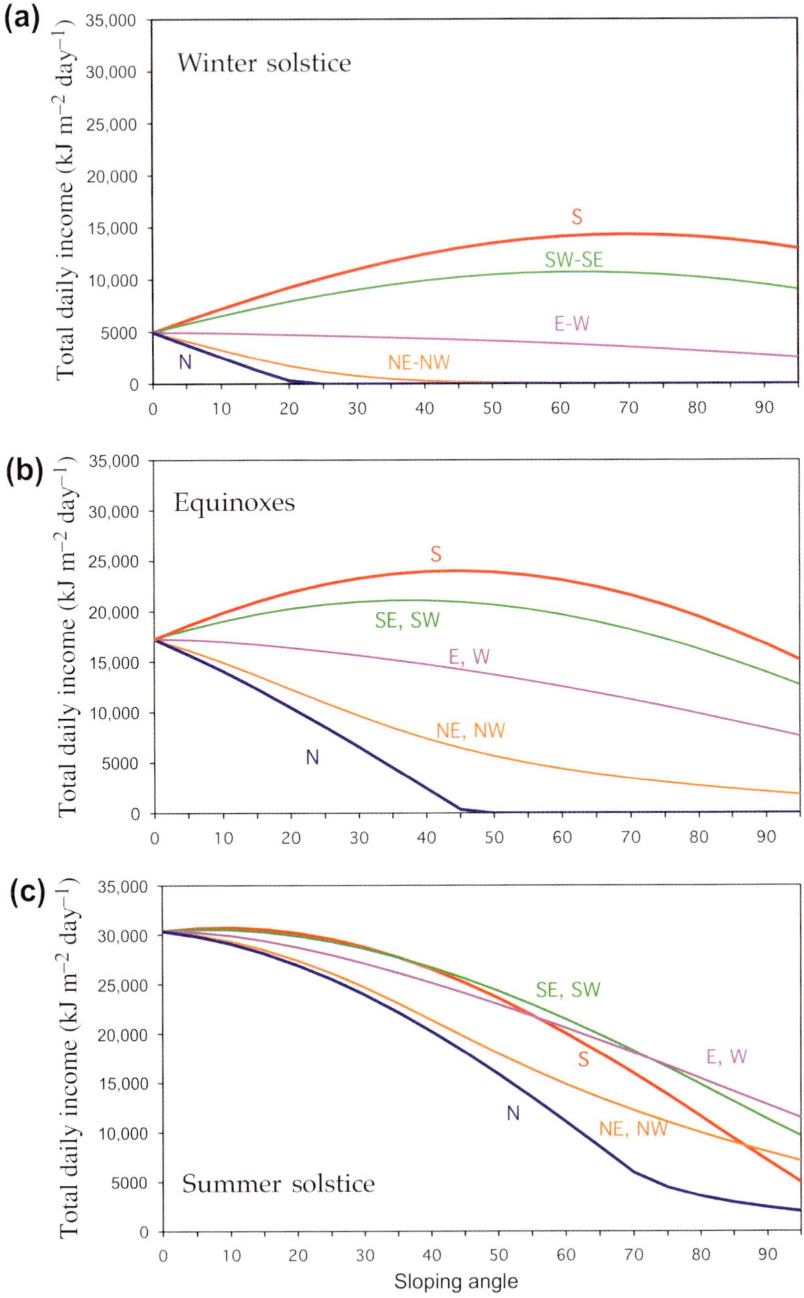

FIGURE 4.3 Daily total income I_t of the solar radiation on a slant surface oriented in the eight cardinal directions is represented against the sloping angle for the latitude $\phi = 45°$. (a) Winter solstice, (b) equinoxes and (c) summer solstice.

except near sunrise and sunset. In fact, the flux density increases from sunrise to east, then decreases and reaches the minimum at south and continues symmetrically till sunset. The elevated solar altitude makes the solar income maximum on the horizontal plane. The flux on the south surface is small and at noon it is close to the value reached at 6.30 a.m. on the north–northeast or at 17.30 p.m. on the north–northwest.

A comparison between the plots at the two solstices shows that at noon the summer flux on the vertical south surface is greater than the winter flux on the horizontal plane at the same hour and the summer flux on the horizontal plane at noon is even greater than the winter flux on the vertical surface facing south.

This example shows how the solar income is affected by the atmospheric optical length, although in clear sky. If the atmospheric attenuation is affected by the local weather (e.g. haze persistency, cloud cover) or pollution, the departures are even larger.

It is possible to integrate over the whole daytime the flux density of solar radiation for an arbitrarily oriented or slant surface (Fig. 4.3). At the winter solstice (Fig. 4.3(a)), the dominance of the southern sector is evident. At the equinoxes (Fig. 4.3(b)), the maximum instantaneous values of irradiation are similar for the whole southern sector from east–southeast to west–southwest, with a slight minimum in south, and the surfaces facing east and west having slightly less irradiation. At the summer solstice (Fig. 4.3(c)), all surfaces receive a similar flux, except for the northern sector. However, the daily total solar income on these vertical surfaces (i.e. $\beta = 90°$) shows a different situation: the south surface receives the maximum total and the east and west ones receive half of it. On the other hand, when β tends to zero, all the slant surfaces tend to the horizontal plane and the differences tend to vanish. For slant surfaces facing south and southeast or southwest, the total income reaches the maximum, respectively, at the slopes $\beta = 45°$ and $\beta = 40°$. The whole southern sector has convex plots with the maximum at a sloping angle that tends to vanish at the east and west orientations. The east and west plots are still convex but close to a straight line. The northern sector has concave plots, with the total income decreasing when the slope increases. The surface exposed to north is no more reached by direct solar radiation when $\beta \geq 45°$. At the summer solstice, all the integral values, except for those of the northern sector, are quite similar between them.

4.6. WHAT IS THE COLOUR OF NATURAL LIGHT?

Sunlight is optically white, as it is composed of a balanced mixture of all colours, i.e. all the wavelengths in the spectrum of incident light. However, when the solar radiation penetrates the atmosphere, the wavelengths corresponding to violet and blue resonate with the oxygen and nitrogen molecules that disperse them. Looking at the sky on a clear calm day, the characteristic interstellar blackness of empty space cannot be seen, but there is, instead, an intense blue, because the human eye is much more sensitive to this colour than it is to violet. The lighter blue or the whitish colour of the sky depends on the presence of minute drops of water of various diameters (which are even more evident when there are clouds), given that the dispersion is not selective but distributed over the whole spectrum. The band of solar radiation is initially white, but when part of the blue has been subtracted from direct sunlight and diffused in all directions by the air molecules, colours slightly change and the white objects that are far away, e.g. the snow caps of distant mountains, slightly take on a complementary yellow, orange or red.

This happens every day when the sun changes height and the light beams cross a continually variable optical thickness in the atmosphere. Colours change less when the sun is close to zenith (especially midday) and more when the sun approaches the horizon (sunrise and sunset) when rays are at a grazing angle and cross a thick atmospheric layer. Pollution may increase the opacity and false colours. The sun itself, which is white, becomes coloured for the progressive subtraction of blue: varying from brilliant yellow to purple with the time of day and the season.

In addition, vision is not objective and the eye has a perception that depends on the light intensity. The retina contains two types of photoreceptors, *rods* and *cones*, with different roles and activation thresholds. Rods function best at dim light but are not sensitive to colours; cones are useful for colour reception but need bright light. The light intensity required to observe an object varies with the individual person, his or her age, the luminance of the object, the background and the surroundings. In bright light objects appear with warm brilliant colours and tones shifted towards yellow; when the intensity is lowered, colours become colder and tend to vanish as soon as cones cease to operate. Under moonlight, all colours are replaced by shades of grey. This different response of the eye is called *Purkinje effect*.

4.7. EXHIBITION LIGHTING AND ELECTRIC LIGHT SOURCES

The human eye can visually discern a certain spectral portion of sunlight, and lamps should closely reproduce the solar spectrum, or at least the spectral bands corresponding to the colours existing in exhibits. In preventive conservation, exhibition lighting fundamentally implies three problems: correct vision and enjoyment of colours, direct and indirect effects of photons reaching exhibits and convective air movements and soiling generated by the light sources.

The variability of the daylight spectrum poses a question, which is very important when exhibition lighting is needed, i.e. '*What kind of lamp will give the same light as the sun?*' No lamp varies during the day, from red at early morning (sunrise) to the bright golden white at noon and red again at evening (sunset). No lamp gives the dim bluish grey of fog, darkens as the sky darkens with clouds, takes on the transparency of the drops of water when it rains, changes tone with the various seasonal changes in vegetation, or reflects the white of the snow. Every lamp can reproduce only one of these particular types of visions and never will it reproduce a dynamic sequence of colours, tones and shades as sunlight.

However, in front of a fresco, or in an exhibition room, it is necessary to choose the type of illumination to adopt. Would it be better to choose a light giving the same colour of a candle, a torch, or the sun at dawn or another precise time? This is a static sensation that has to be chosen, and once it has been realized, it remains unchanged. For exhibition lighting the first step is the choice of the most convenient light source that should ensure (1) a good reproduction of colours and (2) the minimum damage to exhibits. It is known that for most objects with photosensitive surface, the relative spectral damage sensitivity decreases exponentially from the UV to the IR, i.e. it rapidly decreases with the photon energy. On the other hand, the IR part of the spectrum is less efficient in discolouring or causing other chemical effects, but overheats surfaces. A good practice to reduce the above negative factors is to avoid excessive lighting intensity and especially direct exposure to UV and/or IR radiations that are noxious and not useful for colour vision.

All light sources have a characteristic emission spectrum, including in different proportions the UV band (which should, therefore, be filtered above all for organic materials) and the IR that forms a considerable part of the total spectrum. The harmful effects of UV and IR radiation may vary with the type of material and its sensitiveness to these wavelengths, and a universally valid quantitative comparison of the damage resulting from exhibition lighting intensity is difficult. If objects are photosensitive, then the best is to illuminate with the lowest intensity of visible light required for viewing and only for the strictly necessary duration.

Decay will depend on the specific material, the light spectrum and intensity. It should be remembered that the decay of a photosensitive surface is of cumulative nature, i.e. it depends on the total number of photons with given energy, and that this number is related to the light intensity and duration, i.e. the total dose of photons accumulated during the exhibit life. It may be useful to determine whether it is advisable to substitute a light source with another and whether, for example, the situation is partially or wholly improved by compensating sunlight with electric light sources.

Different light sources may have different effect on the perceived colours of exhibits depending on their *colour rendering index* (indicated CRI or, more precisely, Ra) that is determined by the spectral distribution of the light. For the best colour enjoyment, the lamp should produce a continuous spectrum, possibly close to solar light. Halogen lamps provide the best approximation and are used as a reference, with Ra = 100, although their spectrum is biased towards the red. The peak is in the near IR: as a consequence, they overheat the lighted surface and dissipate much heat in the room (Fig. 4.4).

A comparison of the overheating caused by different kinds of lamps is shown in Fig. 4.5 that points out the diverse contribution of the IR component. Four kinds of lamps are compared: a tungsten incandescent lamp, a halogen lamp, the early metal halides (i.e. before 1995 type, with an incandescent tungsten filament inside) and a fluorescent lamp that illuminated, all at the same *illuminance* (i.e. amount of luminous flux reaching a surface), an aluminium target that was previously blackened with candle smoke. The incandescent tungsten lamp caused the most intense overheating, followed by the halogen lamp and the early type of metal halides. The fluorescent lamp gave no measurable overheating.

Incandescent lamps (i.e. traditional tungsten filament and halogen lamps) have a continuous spectrum, CRI close to 100, with a spectral intensity that differs from sunlight for the lower emission temperature, which favours warm tones, i.e. orange and red. They have a warm tonality midway between solar light and candle; if they are dimmed, the colour appearances change and the light becomes even warmer.

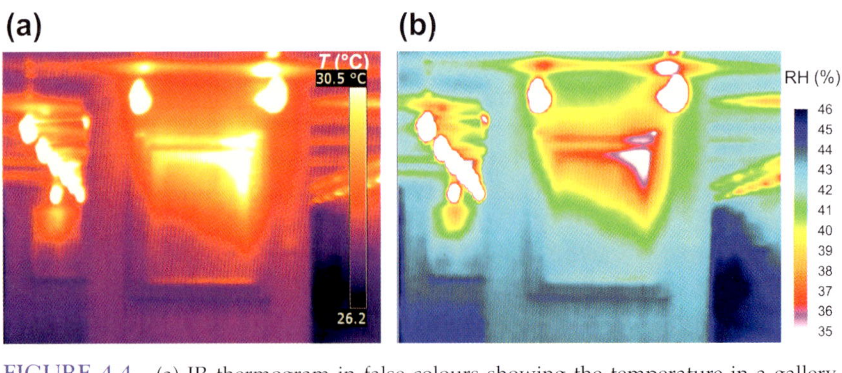

FIGURE 4.4 (a) IR thermogram in false colours showing the temperature in a gallery where paintings are lighted with fluorescent lamps. The upper part of the paintings is overheated. (b) Image in false colours of the *RH* at the interface between the painting and the air, calculated pixel by pixel from the IR thermogram and the moisture content in the air in contact with the surface.

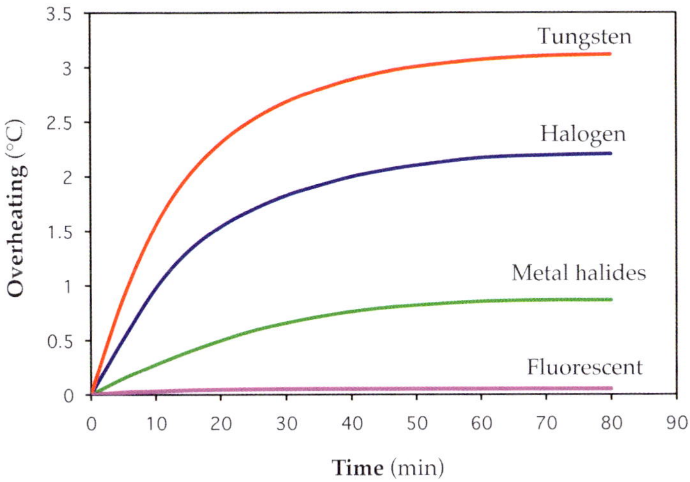

FIGURE 4.5 Comparison of overheating caused by different kinds of lamps providing the same illuminance (500 lx) to the same surface, i.e. a black aluminium target.

Halogen lamps have the tungsten filament temperature in the range 2700–3200 K and emit UV radiation in proportion of 2% or more of the visible light. However, nowadays they are provided with an anti-UV filter to stop this radiation. The main problems of incandescent lamps are large IR emission, heat dissipation and energy consumption.

The *efficiency,* i.e. the proportion between produced visible light and feeding electric power, varies with the power but is about 2% (i.e. 2% visible light and 98% heat dissipation) for a traditional tungsten filament lamp and 3.5% for a halogen lamp. The emitted light in lumen watt^{-1} is 10–15 for tungsten and 20 for a halogen.

When a spotlight is switched on, the energy reaching the illuminated area warms the surface. In Fig. 4.4, we have seen the effect on a painting on canvas; now we will see the effect on a painting on a wooden panel (Fig. 4.6), on a wooden statue and a wooden cupboard (Fig. 4.7) and on a mural (Fig. 4.8).

In 1993, the panel '*Madonna d'Ognissanti*' by Giotto (Uffizi Gallery, Florence), was illuminated with incandescent lamps. Surveys were made in the gallery to improve conservation. The panel temperature was measured and compared with the wall and floor temperature, as shown in Fig. 4.6. The wall close to the painting and the floor had a gradual temperature rise ΔT that reached some 1.5 °C after 6 h, with an average heating rate of 0.25 °C h^{-1}; the panel that has a lower thermal capacity was overheated by $\Delta T = 3$ °C after 4 h, with 0.75 °C h^{-1} heating rate, i.e. three times faster.

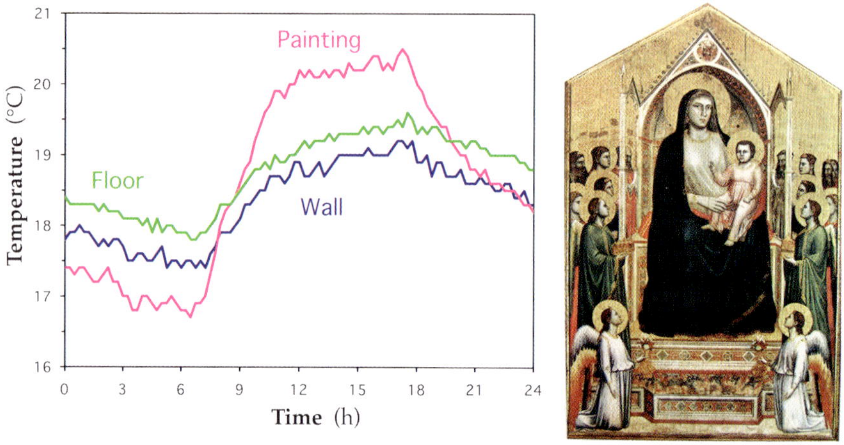

FIGURE 4.6 Overheating of the panel painting the *Madonna d'Ognissanti* by Giotto (Uffizi Gallery, Florence) compared with the wall and floor temperature (3 February 1993).

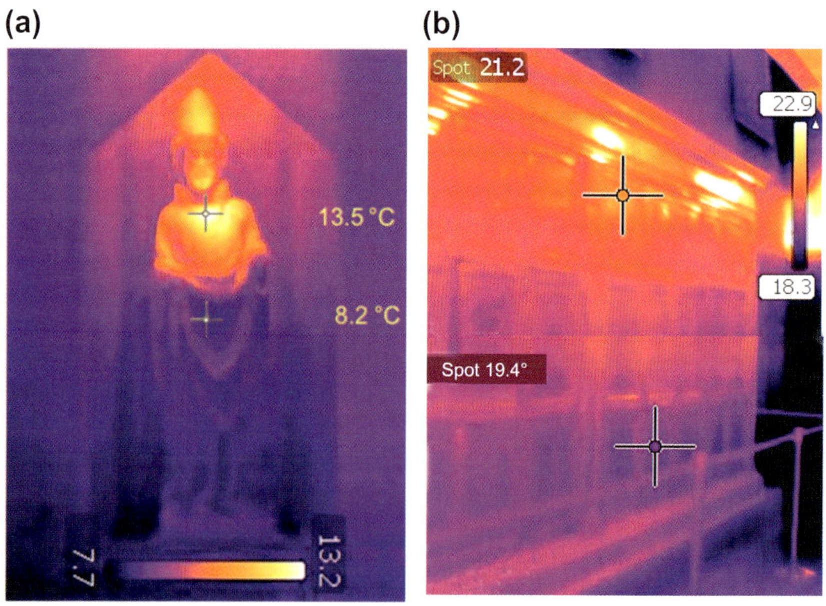

FIGURE 4.7 (a) A wooden statue illuminated with a halogen exhibition lighting placed over it. (b) A wooden cupboard illuminated with incandescent lamps placed aloft in front of the cupboard.

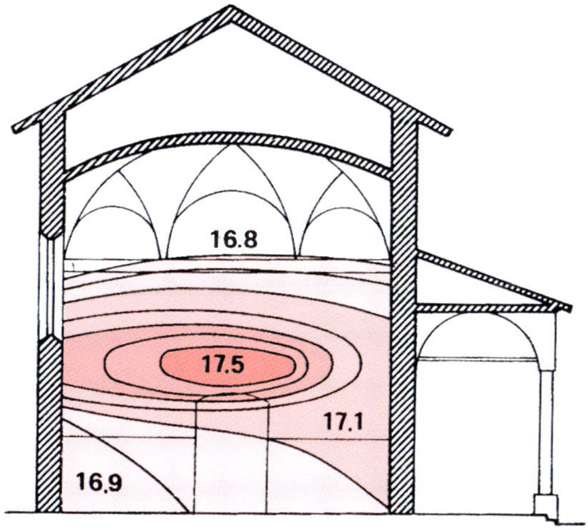

FIGURE 4.8 Warm spot generated by the incandescent lamps that several years ago were used to illuminate the Leonard's Last Supper, Milan. Today the lighting system has been changed to diffuse and cold light. Measurement taken on 9 May 1982. *Source: After Camuffo and Bernardi, 1991; by kind permission of European Commission, DG XII, and* Bollettino Geofisico. *See credits.*

The wooden statue in Fig. 4.7(a) is exhibited with a halogen light placed above it. The thermogram shows that the mitre, head and shoulder are hit by direct light and IR emitted by the exhibition lamp and are overheated by 5 °C, while the rest of the body illuminated with diffuse light remains colder.

A wooden cupboard cabinet is illuminated with some incandescent lamps placed aloft in front of it (Fig. 4.7(b)). A vertical temperature gradient is established on the front doors and the upper part has some hot spots (some 10 °C overheating) where the direct light beams hit the surface. A temperature and moisture gradient is also established across the wooden panels causing differential shrinkage, internal stress, cracks and warping.

In Leonard's Last Supper, Milan (Fig. 4.8), part of the radiant heat penetrates into the mural warming it, part is returned back to the air and part forces evaporation from the plaster, according to the energy balance (see Chapter 7). There is, subsequently, a considerable reduction in the relative humidity (*RH*) both in the air in contact with the object and that inside internal pores. The subsurface gradient and the differential expansion may cause cracking and flaking of the colour coating or the surface layer.

Finally, the convective cells that form above the lamps for the heat dissipation, will lead to the deposition of particles that will stick on surfaces, blackening them (Fig. 4.9). The problem of soiling will be extensively discussed in Chapter 8.

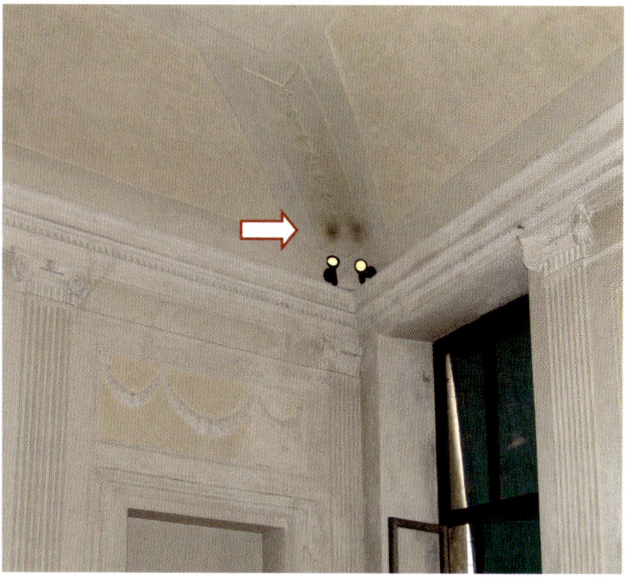

FIGURE 4.9 Black stains (arrow) formed over two spot lamps placed below the ceiling, at a short distance from it. The uprising air flow transports airborne particulate matter that will deposit on the surface, blackening it.

Fluorescent lamps have a low IR dissipation and for this reason they are said to give *cold light*. This is a clear advantage; however, their spectrum is characterized by a number of spectral bands, with some colours missing, while others may have exceedingly high intensity. The final effect is unbalanced colours. Sometimes these lamps are appreciated for the low IR emission; more often they are discarded because of the colour rendering that ranges from 80 to 98 for the best types.

Fluorescent lamps have higher UV emission in comparison with incandescent lamps, which may have negative effects in colour oxidation and fading. However, nowadays fluorescent lamps with anti-UV filters incorporated are commercially available so that this problem is eliminated.

Their efficiency is about 8–8.5% of the electric power. The emitted light efficiency is between 50 and 90 lumen watt^{-1}.

A combination of a fluorescent cold light with another fluorescent warm lamp is not a good solution. This combination will only give a superposition of spectral lines with excessively warm or cold tones with some gaps due to missing lines, so that the overall effect might, in the first instance, trick the human eye but will be unable to reproduce the appropriate chromatic appearance. It is, however, impossible to add a missing line, although it is always possible to balance a spectrum with an excessively dominant line, toning down the peaks with the complementary colour. For this reason, lamps with high density of spectral lines are preferable, while lamps with gaps between spectral lines are discarded. A combination of two lamps with opposed dominance of complementary colours is unadvisable because it is very difficult to balance two light sources and the balance will change with the lamp ageing. Such a combination creates double shades in two different colours and, in addition, the sum of the IR emitted by the two lamps, and the sum of the heat dissipated by the two lamps, are on the whole, always greater than the IR, or the heat dissipated from only one source of light having the same luminous efficiency.

Metal halide lamps produced after 1995 are gas discharge lamps, like fluorescent lamps, but under higher pressure. Their colour temperature is between 3000 K and 6000 K, apparently close to the solar 'white' light. High-quality lamps are available with a spectrum composed of many spectral bands simulating a continuous spectrum that reasonably reproduce the colours, although not as well as the halogen lamps. Their CRI is near 95. The emitted light efficiency is between 70 and 90 lumen watt^{-1}.

Early metal halides have been used to illuminate the Giotto and the Sistine Chapels because of the most acceptable compromise between heat dissipation and colour rendering, at least with regard to the colours existing in these two frescoes. In the case of Giotto, it was sufficient to apply a slightly coloured filter to improve the colour rendering. In the case of the Sistine Chapel, the producer tuned three times the mixture of the metal halides in order to obtain the best reproduction of the Michelangelo's frescoes.

Solid-state lighting is based on semiconducting materials that may generate light. The main types are light-emitting diodes (LED), organic light-emitting diodes, and polymer light-emitting diodes. LED is the most popular solution, particularly appreciated for the low energy consumption in a number of applications, including museum and galleries. In reality, any individual LED emits a single colour. However, by making an appropriate combination of coloured LEDs, e.g. red, green and blue (RGB), it is possible to obtain some white light.

Another methodology is to coat the LED surface with fluorescent dyes emitting in the green to red region to obtain white light. The colour rendering is not excellent, i.e. about 80 for the best types, but it can be improved with filters that reduce undesired colour peaks.

The emitted light efficiency is over 80 lumen watt^{-1}.

4.8. OPTICAL FILTERS AND OPTICAL FIBRES

The practice of keeping some areas highly or intensely illuminated in order to light up the furthermost parts of the room is not the best. Light should be reduced when it is too intense or sunlight should be integrated with electric light in the areas where it is too weak. The goal is to provide every place a homogeneous soft lighting.

Considering the damage that excessive lighting may cause in an exhibition room, direct sunlight should be avoided. The daily apparent solar motion induces sunlight spots to describe curved bands on the wall opposite to windows (or also in the wall perpendicular) and these bands move with the season: they are higher in winter and lower in summer (Fig. 4.10(a)).

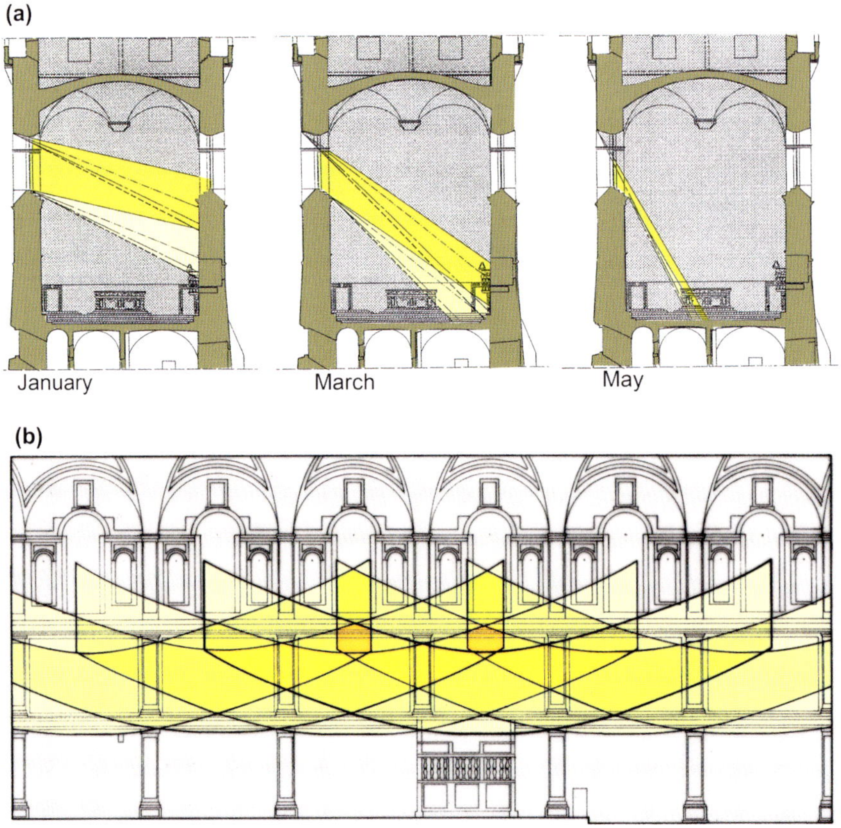

FIGURE 4.10 (a) Sunlight beams in the Sistine Chapel calculated for January, March and May. The seasonal solar cycle lowers the area hit by solar radiation and in July the beams are fully intercepted. Starting from July, the symmetrical trajectories are reproduced. (b) Spots of direct sunlight in the Sistine Chapel calculated for December 15. The windows are made of stained glass and the light passing through is partially dispersed. *Source: After Camuffo and Bernardi, 1986; by kind permission of European Commission, DG XII, and* Bollettino Geofisico. *See credits.*

The most popular method to stop light beams is to apply curtains to windows. However, when curtains are hit by solar radiation, their temperature rises and generates convective air cells. This airflow blackens the wall over the windows, or even the ceiling. Another problem with curtains is that dust and bacteria deposit on them; bacteria grow and when curtains are shaken off for closing windows, they may form dangerous clouds of dust and bacteria that will spread into the room.

The example of the Sistine Chapel, Rome, is reported in Fig. 4.10(b) for winter. The windows are made of stained glass and the light passing through is partially dispersed. However, as the effect of light is cumulative, i.e. is the sum of all the individual doses supplied at each exposure happened in the past. For this reason, the solar beams passing through the windows should be filtered out. In general, a second protective pane or a self-adhesive polyester film optically treated with the inclusion of metals, dies or ceramic nanoparticles. Ceramic is the most effective and stable technology. The former solution has been adopted for the Chapel.

Unfortunately, a number of paintings and wooden artefacts are daily reached by direct solar radiation and the impact is immediate. An example is reported for a painting on canvas (Fig. 4.11) and for a wooden stall (Fig. 4.12) in February morning in Italy.

Although the winter radiation is weak, the canvas overheats by 14 °C and the *RH* of air in contact with the painting drops by 30%. As opposed

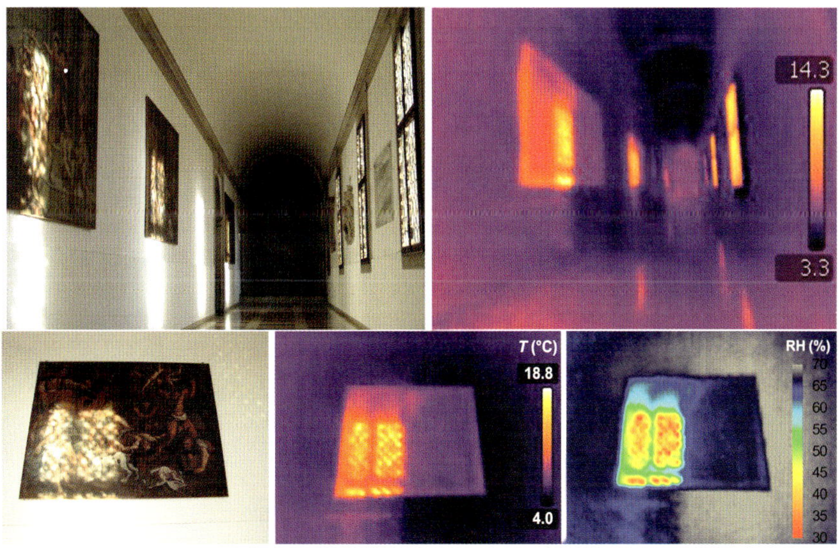

FIGURE 4.11 Top panel: solar beams, passing through stained windows, hit painting on canvas (February 2012, mid morning). Bottom panel: (a) a painting with solar spots, (b) the canvas temperature increases by 14 °C (IR thermogram) and (c) the RH at the interface between the air and the colour coating drops by 30%.

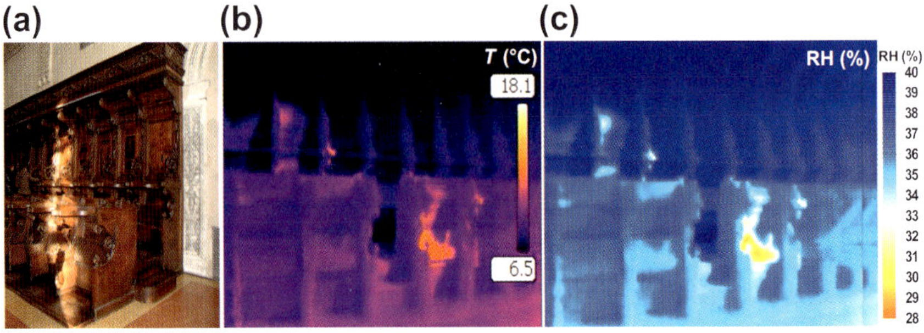

FIGURE 4.12 (a) A solar beam, passing through a window, hits choir stalls of the fifteenth century (February 2012, mid morning). (b) The wood temperature increases by 8 °C (IR thermogram). (c) The RH at the interface between the air and the wood surface drops by 10%.

to this, the thick wooden stall overheats by 8 °C and the *RH* drops by 10%. Everyday repeated cycles are noxious for conservation. Repeated cycles will cause yield or cracks as expansion joint for adaptation to the ambient variability.

In the Giotto Chapel, Padua (Fig. 4.13), sunlight was shaded for years with curtains. Now the problem has been solved by adding an antiglare glass pane externally to the old windows, with the aim of reducing the intensity of the solar beams. External panes have been applied at a distance of some 30 cm from the windows, leaving a free space on the bottom and the top, to allow free air circulation and mitigate the greenhouse effect between the two glasses. The external pane helps in protecting the stained glass against adverse meteorological factors and in reducing condensation on the internal side. However, rainfall cannot remove dust and particles that may accumulate.

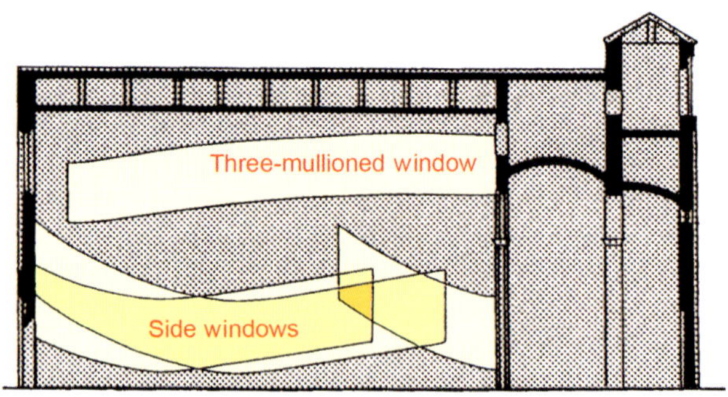

FIGURE 4.13 Spots of direct light beams in the Giotto Chapel, Padua, during the winter solstice. The upper band shows the daily displacement of sunlight through the three-mullioned window on the facade and the lower bands show the spots through the first, second and last side windows (the others are shielded by trees). The yellow levels indicate the areas with repeated sunshine. The Chapel is oriented with the facade facing 240° (west–southwest).

Glass is transparent to visible light but absorbs a considerable fraction of IR and UV radiation, leading to transformation of this energy into higher temperature of the pane. This fact is even more relevant for coloured stained glasses, which in addition absorb part of the visible band, depending on the colours used. As a result of overheating, the window will increase the temperature of the air that will come into contact with it and will act as a motor that generates an upward air stream flowing along the pane. This airflow will transport and deposit airborne particles on the glass and on the wall surface over the window. The result will be an increased soiling (see Chapter 8) and a favourable habitat for insects that appreciate light and milder environments (Fig. 4.14). Spiders will build their sticky nets that will further contribute to accumulate dust, spores and aerosols. In the case of condensation onto the glass surface, the deposited material may form an aggressive solution that will corrode the glass and the metal structure supporting it.

In sunlight, the harmful (and optically useless) parts of spectrum external to the visible optical range should be filtered out. Using optically selective glass, or by applying a self-adhesive filtering film on panes, can easily do this. However, it may also be necessary to dim the solar light, although this is a particularly delicate problem because it means that, in practice, a reflecting or coloured glass should be used. A reflecting glass is in principle similar to a mirror; however, if it is coloured, the light entering the room will not be white. Although everybody appreciates the gothic cathedrals that have large stained glass windows to make the austere interiors coloured, the introduction of coloured glasses is not accepted, especially in an ambience decorated with frescoes or pictures.

Several colours exist, but in practice the best choice of coloured panes (or optical filters on adhesive film) is one of the following two possibilities: grey and amber. White is not a colour but a perfect mixture of all colours of the solar spectrum. Black is not a colour but the total absence of all colours. Grey too is not a colour but a balanced toning down of the

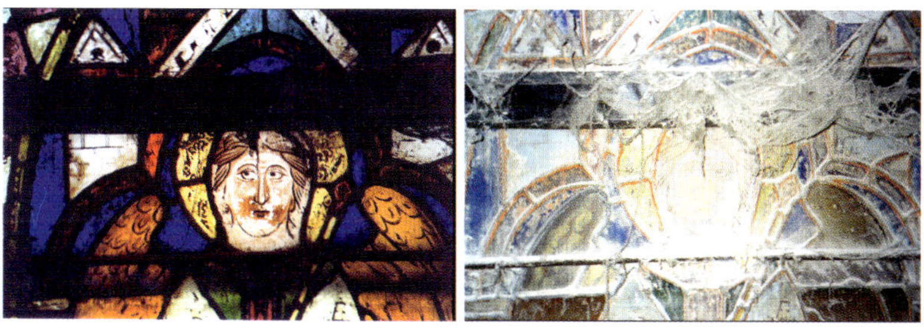

FIGURE 4.14 External and internal sides of a coloured stained glass window (Sainte Chapelle, Paris). The internal side is eventually covered with spider nets and heavy deposits of dust, spores and other particulate matter. Pitting corrosion is visible on the *grisaille* of the face.

whole solar spectrum so that none of the single colour dominates in that there is only a soft, general attenuation; grey is the first step from white to black. The choice of a grey glass is probably less popular, but it is certainly the most impartial. It does not alter the colours inside the room but only the intensity of sunlight coming from the exterior, as if the sky had a dense layer of clouds, almost a psychological trick played on the weather.

In contrast, using an amber pane of glass might seem, at first sight, absurd, because this really does alter colours, toning down all of them, except those of the same amber tone that are reinforced. However, an amber filter reproduces what happens to sunlight when crossing the atmosphere, i.e. the light we see a short time after dawn or before sunset. An amber glass might be considered, therefore, to be not responsible for falsifying colours, but rather, for falsifying time, so that in the morning the inside light is similar to the outdoor light as it was a few hours earlier, and vice versa in the afternoon. It falsifies the time of the day and, if it is raining outside, it also falsifies the weather.

Briefly, various types of filters exist to reach some specific goals:

- *UV absorbing*: to reduce the proportion of UV in the light beam.
- *IR absorbing*: to reduce the proportion of IR in the light beam.
- *Neutral density*: to reduce the intensity of light without changing the colour quality.
- *Colour correction*: to dim some exceedingly intense spectral band of a lamp.
- *Coloured*: to deliberately produce striking coloured effects.
- *Conversion*: to increase or decrease the colour temperature in the light.
- *Diffusion*: to diffuse or/and soften the light beam.
- Mixed: i.e. a combination of two or more of the above filters.

The choice of the filter depends on the special effect one would like to reach in order to get a safer light for conservation or a better light for colour enjoyment, to create special effects, or to adequately complement sunlight from windows with some electric light making the overall result pleasant.

Showcases with exhibition lamps inside will likely become small greenhouses because glass panes allow the passage of light in the visible range but less in the IR (Chapter 1). Given that the IR emission may be largely relevant, especially with incandescent lamps, heat is trapped inside the exhibition case, overheating objects and lowering the *RH* level. A good method, whenever possible, is to place lamps outside the case and transport light inside with *optical fibres*, which are made of a glass core surrounded by a cladding with lower refractive index. Light will be transmitted only if the glass is transparent to a specific wavelength and if the fibre is able to trap the beam in the core by total internal reflection, which is determined by the combination of the refractive indexes of the core and the cladding. Another good strategy is to use cold light from LEDs, if excellent colour rendering is not priority.

Most glasses are opaque to UV radiation for some strong absorption bands arising from electron excitation of two types: (1) when electrons are excited from the valence band to unoccupied states in an exciton (i.e. an empty electron state in a valence band) or in a conduction band and (2) when electrons are transferred from the shells of an ion to the shells of another one.

The absorption in the IR region especially occurs as a result of atomic vibrations in a molecule that has a dipole momentum, which has a fundamental vibration mode in the IR spectrum (Varshneya, 1994), as e.g. the H_2O molecule. If the optical waveguide is transparent to the optical bandwidth and is opaque to both the UV and IR bands, it acts as a filter that supplies light in the visible spectrum only.

Alternatively, the UV and IR bands can be cut out by inserting appropriate filters between the light source (e.g. a halogen lamp) and the fibre. Then the light is concentrated with a lens and supplied to the fibre. The filtered light beam, composed of pure light in the visible region, will be transported with the waveguide into the showcase and then the beam is opened again with another lens in order to illuminate the object within the desired light cone.

A popular practice is to use cold light from LED or fluorescent lamps. Both of these lamps have a band spectrum and not all colours are adequately reproduced, but the CRI for fluorescent lamps is generally higher than that for LED. However, fluorescent lamps have larger heat dissipation and this might be a problem when placed inside a closed showcase.

Note: According to CIE 157: 2004 and CEN/TS 16163: 2013 normative, materials of cultural property are grouped into four classes of sensitivity to luminous exposure. For each class, a limit in the cumulative annual luminous exposure (ALE, measured in lux-hours per year, i.e. lx h a^{-1}) has been established. This is calculated as a product of the allowed instantaneous illuminance level (measured in lux, i.e. lx) and the total annual exposure time (AET, measured in hours per year, i.e. h a^{-1}). As the cumulative effect is the main factor, ALE should be respected, whilst it is possible to increase the illuminance level if the AET is adequately reduced, leaving unchanged their product. Material sensitivity classes are as follows.

1. *Insensitive*, e.g. most metals, stone, most glass, ceramic, enamel, most minerals. This class has no limits in ALE, illuminance and AET.
2. *Slightly sensitive*, e.g. most oil and tempera painting, fresco, undyed leather and wood, horn, bone, ivory, lacquer and some plastics. Recommended ALE = 600,000 lx h a^{-1}. Reference values: illuminance = 200 lx and AET = 3000 h a^{-1}.

3. *Moderately sensitive*, e.g. most textiles, watercolours, pastels, prints and drawings, manuscripts, miniatures, paintings in distemper media, wallpaper and most natural history objects, including botanical specimens, fur and feathers. Recommended ALE = 150,000 lx h a^{-1}. Reference values: illuminance = 50 lx and AET = 3000 h a^{-1}.
4. *Highly sensitive*, e.g. silk, highly fugitive colourants, most graphic art and photographic documents. Recommended ALE = 15,000 lx h a^{-1}. Reference values: illuminance = 50 lx and AET = 300 h a^{-1}.

Note 1: the illuminance = 50 lx is the same for both class 3 and class 4 because 50 lx is the threshold limit for visibility. This fact has obliged to reduce the exposure time in class 4 compared with class 3.

Note 2: the reference values are typical vales whose product determines the recommended ALE. One may depart from the reference values provided that their product remains unchanged.

Source: CIE 157: 2004, Control of damage to museum objects by light, Wien, 2004.
CEN/TS 16163: 2013 Conservation of Cultural Heritage - Guidelines and procedures for choosing appropriate lighting for indoor exhibitions.

4.9. DETERIORATION OF WORKS OF ART CAUSED BY LIGHT

Various mechanisms of degradation exist, e.g. chemical reactions activated by photon energy, photooxidation of gaseous pollutants that may form an aggressive environment and colonization of phototrophic bio-organisms. Deterioration severity depends on several factors: photon energy (i.e. the wavelength of the spectral line), total number of photons as well as material response. The higher the photon energy, or the larger the number of photons, the more severe is the damage. For this reason, UV radiation is most dangerous in causing chemical transformations and decay. Incoming energy; changes in chemical composition and crystalline structure; type and concentration of molecules; optical properties of the surface and the subsurface layer; penetration depth; interactions with the molecular structure; and electronic state are all related factors that in general operate in synergy with the surface temperature and ambient humidity.

Light may alter crystalline or polymeric structures. Several organic molecules suffer because of this form of degradation, and damage occurs when the amount of energy absorbed by a molecule exceeds the bound energy. For this reason, the sensitiveness of a material increases with the photon energy, the most severe damage being caused by the shortwave UV light. The most evident effects are ageing of polymers, reduction of the plastic and elastic properties, colour fading, yellowing of textiles and wood degradation. An extensive presentation of the problem has been made by Michalski (1987) and Wypych (1995).

Colour changes might be related to changes in chemical or structural composition of dyes present in the surface layer. Colour fading is the main problem of watercolours, ink drawings, ancient manuscripts and coloured textiles. Oxidation of keratin exposed to UV radiation is responsible for yellowing of wool and decrease in tear and tensile strength and abrasion resistance. Wood is very sensitive to UV radiation, and the weathered outer layer increases sensitiveness to abrasion, having loosely connected fibres. Although UV penetration is limited to a depth of 75 μm, there is in-depth damage and after 1-year exposure degradation can reach 3 mm. However, one century is necessary to double the depth of the deteriorated layer. Surface coating may protect wood, and in the case of painted panels the problems are substantially two: the degradation of dyes and the mechanical stress experienced by wood after temperature and humidity changes.

Even in the absence of chemical degradation, the simple physical effect of light absorbed in the form of heat will affect the surface temperature and *RH* and may induce internal stress. Paintings are often intensely illuminated and overheated, as shown in Figs 4.4, 4.6, 4.7, and 4.10, either with electrical or with natural light.

In special exposition rooms, several attempts have been made to evaluate light levels and types, and some authorities have recommended lighting standards between 50 and 200 lx to limit fading, yellowing and breaking of binders. However, there is no known level of lighting that has been proven to be safe for colour changes. The real problem is that damage progresses with the cumulative number of photons that have reached the surface (and this involves light intensity, lighting duration and number of repeated cycles) and the energy of individual photons.

The energy is associated with the wavelength. Hence, the most obvious conclusion is that the whole visible spectrum should be allowed but harmful UV should be avoided with the use of appropriate filters. In addition, it should be useful that artwork which suffer fading (e.g. historic robes, tapestry and other coloured tissues, water coloured drawings, gouache painting) are usually kept in the dark with no light. They should be lighted with soft light, during short visiting times, as for the Leonard's *Virgin and Child with S. Anne and St. John the Baptist* in the National Gallery, London. Per each class of materials sensitive to light, recommended thresholds of light intensity and duration of luminous exposure are given by the standards CIE 157:2004, *Control of Damage to Museum Objects by Optical Radiation* ISBN: 9783901906275 and CEN/TS 16163: 2013 *Conservation of Cultural Heritage - Guidelines and procedures for choosing appropriate lighting for indoor exhibitions*. This principle can be usefully extended to preserving the whole museum or gallery in the dark and lighting exhibits only when people really want to watch them. This goal can be attained by placing in front of each painting or showcase a proximity sensor sensitive to IR emitted by visitors approaching the exhibit. This sensor temporarily switches on a soft lighting so that the lighting will follow the path and stops of visitors or even may guide the visit imposing selected watching times.

The intermittent use of soft light, when necessary, is different from the noxious practice of lighting panels or paintings for few minutes with powerful incandescent lamps that are switched on every time a visitor introduces a coin. Every painting is composed of a number of different layers (i.e. a colour coating, a preparation base, a canvas, a wood tablet or a wall) that are bad conductors so that the energy of the light beam is not transmitted or dissipated at the same rate in all the layers.

When a lamp is switched on, the pictorial coating is overheated and expands generating mechanical stress with the underlying layer that has not yet been warmed up, causing microfractures and partial detachment of the coating, which is transformed into tiny flakes, raised at the edges like the skin of a snake. After some delay, the heat reaches the preparation base and then the underlying frame, causing a dangerous surface stress every time the light is turned on/off, which tends, in the long run, to detach the coating from the support. This fatigue and ageing are important for painted panels and paintings on canvas, frescoes, wood, ivory, etc., where the mechanical stress is more dangerous than fading.

Flash light is usually forbidden in museums. The principle is theoretically sound, but in practice each flash has a very short duration so that the energy transferred to the objects is so small that its contribution is often negligible. For instance, considering that flash light may reach the intensity of daylight and usually has a duration ranging from 1/1000 s to 1/50,000 s, then, respectively, 1000 to 50,000 flash lights are equivalent to 1 s solar lighting as far as photon accumulation is concerned. Regarding heating, their cumulative effect is negligible, the energy being supplied by means of several very small contributions provided after time intervals that are extremely long compared with the short flashlight duration; consequently, the heat transferred is easily dissipated and overheating is never reached. Flashlight is much more dangerous for the UV, the guardians' eyes and might constitute a real risk only for extremely photovulnerable objects. If some paintings suffer due to light accumulation, a much more substantial advantage can be achieved closing curtains and window shutters as soon as the visiting time is up.

The position of electric light sources is important not only to avoid noisy reflections or glare. In a closed environment, the air tends to stratify into layers at different densities, corresponding to a gradient in temperature from the floor to the ceiling. The air heated by lamps rises forming convective vertical cells, which develop upwards until the warm air reaches another air layer with lower density or the ceiling. The bottom limit of the cell is determined by air with higher density, and for this reason, in a naturally stable environment, the convective cells cannot develop below the heat source. Therefore, lamps placed above paintings leave the paintings and the underlying space untouched by the convective cells, if we disregard the possible overheating generated by incandescent lamps. In the case of frescoed ceilings, it is better to keep the ceiling surface slightly warmer than the underlying air, in order to form a stable air cushion that reduces thermophoresis and inertial deposition. In addition, it is also useful to keep lamps (which should not dissipate too much energy) at a considerable distance from the ceiling, in order that the convective cells do not reach the ceiling.

The energy dissipated by incandescent lamps creates strong convective currents and turbulence above the light sources and air motions may be further increased by the presence of visitors. This favours inertial deposition of large particles. When the inertial force of these particles exceeds the viscous force of the air that transports them, the particles deviate from the fluid trajectory and impact on the surface sticking on it. Even in the event of a moderate laminar flow, the coarseness or roughness of the surface can locally perturb the trajectory of the particles, causing the same type of aerodynamic capture that, in the long run, is very effective. The difference in temperature between the warm air and the colder surfaces also increases the deposition rate of airborne particles (see Chapter 8).

It may be useful to remember that light is not the only factor responsible for colour fading: several pigments and traditional dyes are attacked by ozone or nitrogen oxides associated with photochemical smog. Photochemical reactions involve nitrogen oxides (NO and NO_2) and photons that generate ozone (O_3), the OH^- radical and other products including the irritant peroxyacetylnitrate (Andrews et al., 1996). Photochemical smog is important not only in the well-known cases of Los Angeles and Athens but also in many other sites, especially where the solar radiation is intense and in the proximity of an industrial area. In the Correr Museum, Venice, indoor O_3 peaks reaching 60 ppb were found in summer and simple measures against uncontrolled leakage, door and window opening were taken to stop its penetration. The indoor O_3 concentration was reduced by two-thirds simply by closing windows and doors during the hours with high concentration of photochemical smog. In general, ozone and other photooxidants show cycles that follow the solar radiation, with the maximum concentration during the strongest solar radiation. However, in coastal areas, ozone peaks have been found also during the nighttime: this is explained with the change of the breeze that transports back the photochemical smog formed inland during the daytime. The practice of filtering the air should be recommended to museums and galleries with vulnerable paintings or textiles.

A short comment about biological decay induced by light: the role of phototrophic algae and cyanobacteria, growing even under low light intensities, needs some comment. The hyphae release oxalic acid and cause mechanical damage when penetrating inside the rock or the masonry, or cause wood rotting. In addition, the photosynthetic microflora causes detrimental aesthetic effects because of the accumulation of various pigments (e.g. chlorophyll, carotinoids) and provides in certain cases the nutrient basis for subsequent microbial infections by bacteria and fungi. Phototrophic microorganisms, such as green or blue algae and cyanobacteria, use sunlight or electric light as the energy source for the photosynthetic process and release oxygen. The infestation is not limited to carbonatic materials as their carbon requirement is supplied by the fixation of atmospheric CO_2 (Ortega-Calvo et al., 1991).

In particular, green algae use light for phototropic synthesis and typically develop when underground humid rooms are illuminated. Their colour is green, typical of chlorophyll, and this means that for the synthesis of chlorophyll, they absorb and use all wavelengths except for green, which is partially or totally reflected or transmitted and, therefore, less relevant for their biological life. More specifically, two specific forms of chlorophyll used in oxygenic photosynthesis should be considered (Fig. 4.15(a)). Chlorophyll A, a photosynthetic pigment in photosynthesis in eukaryotes, cyanobacteria and prochlorophytes, absorbs most energy from blue–violet wavelengths (maximum at $\lambda = 430$ nm) and red (maximum at $\lambda = 662$ nm). Chlorophyll B acts indirectly in photosynthesis by absorbing light and transferring its energy to chlorophyll A. Chlorophyll B is yellow and primarily absorbs blue light (maximum at $\lambda = 453$ nm) and, secondarily, orange–red light (maximum at $\lambda = 642$ nm). In average, chlorophylls A and B primarily absorb blue–violet and orange–red light, which are the most effective colours for photosynthesis; green is mostly transmitted or reflected.

A study performed on marine algal growth, i.e. *nannochloropsis*, in both phototrophic (i.e. traking advantage of light) and mixotrophic (i.e. taking advantage of different environmental conditions) culture under three primary monochromatic light wavelengths, i.e. red LED, green LED and blue LED, and also in white LED, showed that the growth rate decreased using LEDs in the following order: blue, white, green and red (Das et al., 2011). Blue was the first, as expected, because it has the maximum photosynthetic efficiency and white was the second in relevance, because it is a mix of colours, including blue and red. However, the last two results were surprising because green was expected to have the lowest relevance and red a higher one. The finding is explained by considering that this alga lacks chlorophyll B and changes the absorption spectrum with photoacclimatation.

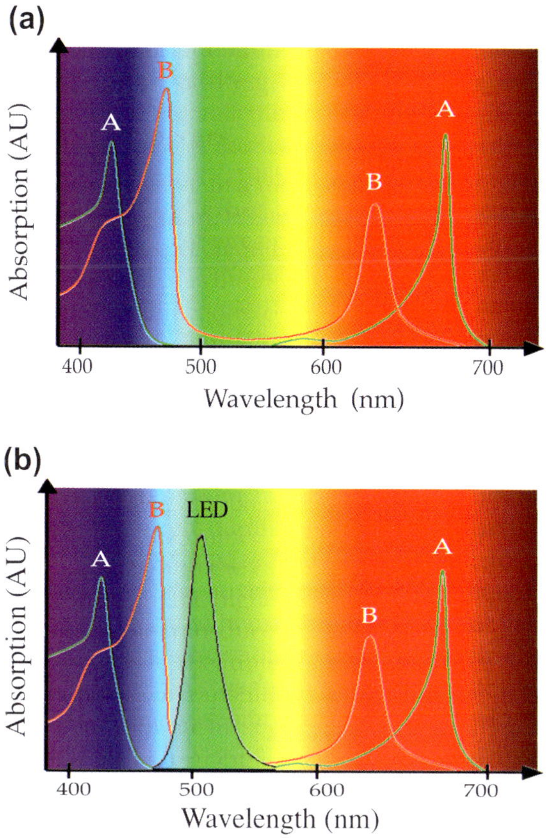

FIGURE 4.15 (a) Spectral absorption of chlorophyll A (A, green line) and chlorophyll B (B, red line). (Redrawn after Aushulz, Wikimedia Commons, see credits). (b) Spectral absorption of chlorophyll A and B and spectral emission of the green LED. The green LED produces cyan–green light peaked at 507 nm, in a band of no photosynthetic activity of chlorophyll A and chlorophyll B.

Therefore, one should conclude that, if a dark environment is lighted with cyan or green light, this light is less efficient than white light or other coloured light in performing the phototropic synthesis of chlorophyll and in developing algal colonization. Green LEDs emit cyan–green light, which shows peak at 507 nm (Fig. 4.15(b)). This light is not useful for the phototropic synthesis of algae. Green LEDs or the yellow light of sodium lamps might be safely used for mass tourism in catacombs, caves or other dark or underground environments that combine dampness with risk of algal colonization. Guided visits should be made using green or yellow light, sufficient to walk and distinguish architectural features. It might be possible, if strictly necessary, to use for a few seconds white light to rapidly show and appreciate colours in frescoes, paintings or other coloured decorations. In the most fortunate conditions, white light might be sustainable for a few seconds a day; certainly it is not for the whole visiting time.

References

Theory and general applications

Andrews, J.E., Brimblecombe, P., Jickells, T.D., Liss, P.S., 1996. An Introduction to Environmental Chemistry. Blackwell, Oxford, pp. 209.
Born, M., 1952. Atomic Physics. Blackie & Son, London, pp. 437.
Das, P., Lei, W., Aziz, S.S., Obbard, J.P., 2011. Enhanced algae growth in both phototrophic and mixotrophic culture under blue light. Bioresource Technology 102 (4), 3883–3887.
Einstein, A., 1905. Ueber einen die Erzegung und Verwandlung des Lichtes betreffenden heuristischen Gesichtspunkt. Annalen der Physik 17, 132–148.
Kondratyev, Ya, 1969. Radiation in the Atmosphere. Academic Press, New York, pp. 912.
Maxwell, J.C., 1873. A Treatise on Electricity and Magnetism. Dover, New York.
Planck, M., 1901. Über das Gesetz der Energieverteilung im Normalspektrum (On the Energy Distribution Law in the Normal Spectrum Radiation). Annalen der Physik Leipzig 4, 553–564.
Robinson, N., 1966. Solar Radiation. Elsevier, Amsterdam, pp. 347.
Varshneya, A.K., 1994. Fundamentals of Inorganic Glasses. Academic Press, Boston, pp. 570.

Applications to conservation

Bernardi, A., Vincenzi, S., 1994. Diurnal variation of solar radiation on differently orientated surfaces of monuments. Nuovo Cimento 17C (4), 431–442.
Camuffo, D., Bernardi, A., 1991. The microclimate of Leonardo's "Last Supper", joint edition European Cultural Heritage Newsletter on Research, and Bollettino Geofisico, special issue, 14 (3) 1–123.
Camuffo, D., Bernardi, A., 1986. Dinamica del microclima e scambi termoigrometrici tra pareti e atmosfera interna nella Cappella Sistina. Bollettino dei Monumenti, Musei e Gallerie Pontificie 6, 211–257.
Michalski, S., 1987. Damage to objects by visible and ultraviolet radiation. In: Lighting in Museums, Galleries and Historic Houses. Museums Association, London, pp. 3–16.
Ortega-Calvo, J.J., Hernandez-Marine, M., Saiz-Jimenez, C., 1991. Biodeterioration of building materials by cyanobacteria and algae. International Biodeterioration 28, 165–186.
Vittori, O., Mestitz, A., 1975. Artistic purposes of some features of corrosion on the golden horses of Venice. Burlington Magazine 864 (98), 132–139.
Wypych, G., 1995. Handbook of Material Weathering. Chem Tec, Toronto, pp. 564.

Further reading

Liou, K.N., 1980. An Introduction to Atmospheric Radiation. Academic Press, San Diego, pp. 392.
Liou, K.N., 1992. Radiation and Cloud Processes in the Atmosphere. Oxford University Press, New York, pp. 487.
List, R.J., 1971. Smithsonian Meteorological Tables. Smithsonian Institution, Washington DC, pp. 527.
Platridge, G.W., Platt, C.M.R., 1976. Radiative Processes in Meteorology and Climatology. Elsevier, Amsterdam, pp. 318.

CHAPTER

5

Physics of Drop Formation and Micropore Condensation

OUTLINE

5.1. HOW A CURVED WATER MENISCUS CHANGES THE EQUILIBRIUM VAPOUR TENSION

Microphysics of clouds can be usefully applied to cultural heritage conservation and the mechanisms for droplets formation are the starting point to discuss an extremely important phenomenon: the condensation in

Microclimate for Cultural Heritage, Second Edition
http://dx.doi.org/10.1016/B978-0-444-63296-8.00006-8

micropores. In reality, this mechanism is complex and rather unfamiliar, although the basic equation that controls the condensation in micropores also governs the formation of drops in the atmosphere, a subject well-known to meteorologists. However, the application to micropores is not immediate as it involves some relevant differences, e.g. the water meniscus with negative radius of curvature and the complex geometry of pores. For this reason, it will be useful to discuss the existing formulae and their degree of approximation. To this aim, their derivation is a necessary step.

In Chapter 2A we have seen that, in the atmosphere, water vapour becomes saturated at relative humidity (*RH*) of 100%, when the air temperature (and therefore the moisture dispersed in air) reaches the dew point (*DP*). However, it is useful to distinguish between the formation of droplets in the free atmosphere and the condensation onto a surface, or inside internal pores. In effect, the vapour tension in equilibrium with a curved liquid surface varies with the concavity or convexity (Fig. 5.1): it is higher at increasing convexity and lower at increasing concavity. For a concave surface (e.g. condensation into micropores and the meniscus in a wettable capillary), the radius of curvature r of the meniscus is $r < 0$; for a convex surface (e.g. droplets and meniscus in a nonwettable capillary) $r > 0$. For a plane surface, $r = \infty$ and this is the well-known physical condition of saturation in equilibrium with a plane surface of water that has been considered in the previous chapters.

The pressure of vapour $e(r)$ in equilibrium with the radius of curvature r of a given water meniscus is expressed as a function of r, and for a plane water surface $e(\infty) = e_{sat}(t)$ is calculated with the Magnus and Tetens equation. The greater the concavity or the convexity, the greater is the departure from the Eqn (2a.1).

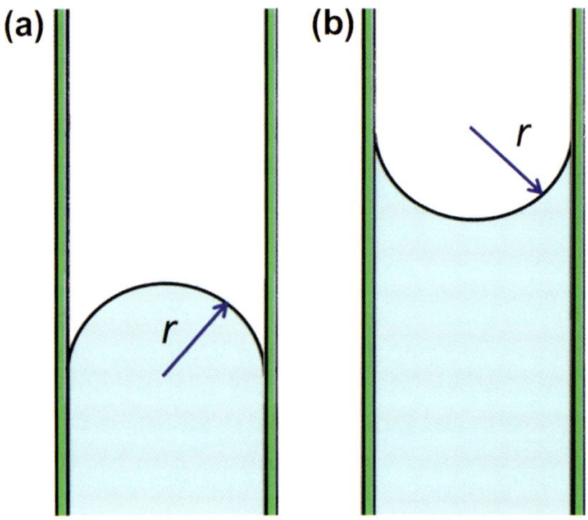

FIGURE 5.1 The meniscus in a nonwettable capillary (a) is convex, the centre of curvature is inside the liquid and the radius r is positive; in a wettable capillary (b) it is concave, the centre of curvature is outside the liquid and the radius r is negative.

In 1870, William Thomson, later Lord Kelvin, calculated the saturation pressure in equilibrium with a concave or a convex meniscus

$$\ln\left(\frac{e(r)}{e(\infty)}\right) = \frac{2\sigma V_m}{r\,\Re T} \tag{5.1}$$

where σ is the surface tension of water (represented in Fig. 5.2, e.g. $\sigma = 75.6$ erg cm^{-2} at $T = 273$ K and $\sigma = 72.2$ erg cm^{-2} at $T = 293$ K), V_m is the molar volume of the liquid (i.e. $V_m = 18$ cm^3 for pure water) and $\Re$ is the gas constant.

The surface tension derives from the different distributions of the polar H_2O molecules that are possible on the surface and in the interior of the liquid. Inside the liquid water, the H_2O molecules can be spatially organized in the three dimensions X,Y,Z forming flickering clusters (see Chapter 9). The most convenient organization of the charged polarities is thus achieved, and the lowest energy level is reached. In contrast, on the surface layer, a degree of freedom is lost and this constraint will require a higher energetic level. The H_2O molecules will distribute their H bonds limitedly to the XY plane, and the resulting bi-dimensional structure is stronger. A species of liquid membrane with some mechanical resistance is formed, able to support small insects. The strength of this membrane, i.e. the surface tension, will decrease at increasing temperatures because the increased kinetic energy of the molecular vibrations will tend to destabilize this strongly organized surface layer.

It is obvious that the ratio $e(r)/e(\infty)$, that is, the ratio between the pressure $e(r)$ and the vapour tension $e(\infty)$, represents by definition the *RH* close to the meniscus. The above formula can be rewritten in terms of the

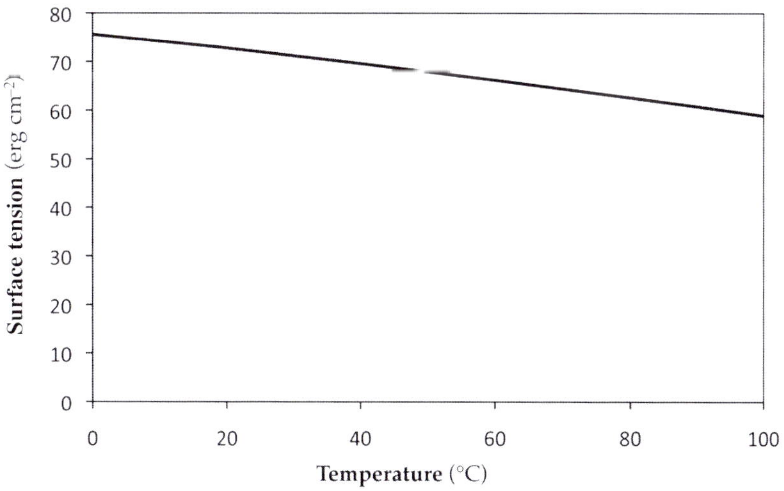

FIGURE 5.2 Surface tension of pure water in contact with air for water temperature ranging from 0° to 100 °C, i.e. from melting ice to ebullition.

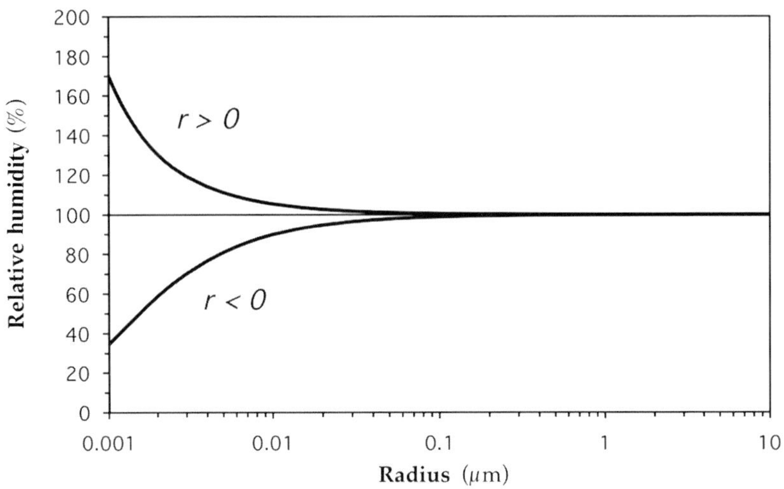

FIGURE 5.3 Relative humidity (%) in equilibrium with meniscus of water with radius r (μm), according to the Kelvin law. $r > 0$ for a convex meniscus, e.g. drop; $r < 0$ for a concave meniscus, e.g. wettable pore.

equilibrium for any specific radius of curvature, either positive or negative as:

$$RH(r) = 100 \exp\left(\frac{2\sigma V_m}{r\Re T}\right) \tag{5.2}$$

and the result is graphically shown in Fig. 5.3.

The departure from $RH = 100$ is relevant only for very small r, i.e. $|r| < 0.1$ μm. As the radius of curvature r is the main parameter, the Kelvin formula states that the logarithm of RH in equilibrium with the meniscus is inversely proportional to the radius of curvature of the water surface. RH levels higher than the above equilibrium will cause condensation; and if lower, evaporation. This equation clearly shows that for negative radii the argument of the logarithm is smaller than 1 and the equilibrium will be with $RH(r) < 100\%$; for positive radii, the argument is greater than 1 and $RH(r) > 100\%$.

The Kelvin equation was originally introduced to take into account the fact that in a capillary the tension of the vapour in equilibrium with menisci having different radii of curvature is not the same, and can be derived in two independent ways: one from the Stevin and Laplace laws and one based on the Gibbs and Helmholtz functions describing the free energy of a thermodynamic system.

A short comment about terminology, as the terms *micro*, *macro*, etc. are not always used to identify the same pore sizes and this may generate confusion: For example, Fitzner (1994) reports three classifications: the first one by De Quervain (1967) considers micropores for $r < 5$ μm, fine pores for

$r = 5\text{–}200$ μm, coarse pores for $r = 200\text{–}2000$ μm and large pores for $r > 2000$ μm. The second one follows the normative DIN 66131 (1973): micropores for $r < 0.002$ μm, mesopores for $r = 0.002\text{–}0.05$ μm and macropores for $r > 0.05$ μm. The third one by Klopfer (1985) considers micropores for $r < 0.1$ μm, capillary pores for $r = 0.1\text{–}1000$ μm and macropores for $r > 1000$ μm. Other definitions exist, based on various points of view. Generally speaking, the concept of small or large is related to the effects and the consequences of the phenomenon under investigation. In our case, we are concerned with (micro)pore condensation and two limiting sizes can be found. The lowest limit is the water molecule, which is of the order of 3×10^{-4} μm; pores with smaller diameter are not relevant for condensation. Another reference size is the interval 0.1–1 μm; above this the Kelvin effect becomes negligible and the pore size is irrelevant being unable to produce departures on condensation, and the term *macro* can apply. As a consequence, the term *micro* could apply to the small size interval 10^{-3} to 10^{-1} μm where the Kelvin effect is dominant.

5.2. DERIVATION OF THE KELVIN EQUATION FOR DROPLET FORMATION AND MICROPORE CONDENSATION

5.2.1. Derivation of the Kelvin Equation from Stevin and Laplace Laws

At the liquid–air interface, the surface layer is composed of a hydrogen-bonded network of H_2O molecules strictly bound between them and this fact can be expressed according to two different points of view: (1) the free liquid surface behaves as an elastic membrane and (2) some work is needed to change the surface area (or to stretch the surface). Both are described in terms of σ, which, respectively, represents: (1) the *surface tension* and is regarded as an elastic force per unit length and (2) the *specific free energy* of the surface and is considered an energy per unit area. The physical dimension of σ is expressed in the former case as force per unit displacement [dyne cm^{-1}] and in the latter case as energy per unit area [erg cm^{-2}]. Surface tension is a contractive tendency of the surface of a liquid in contact with air, or another specific gas, that allows it to resist an external force.

The surface tension does not include the whole surface energy, but represents the specific *Gibbs free energy of the surface*, i.e. the maximum work done for the formation of the unit surface area. If the surface area decreases, the energy of the system also decreases and for this reason free liquid surfaces have a natural tendency to contract, or to assume a spherical shape in the case of free droplets, because the sphere has the smallest surface to volume ratio.

The 'membrane' model is convenient and readily comprehensible, especially when describing a small insect that is walking on water and causes the surface to be deformed by its weight, or when tea leaves remain motionless on the water surface, apparently kept immobile by this 'membrane', while the water inside the pan is boiling with violent convective motions. The energy-oriented point of view is useful in a number of thermodynamic considerations. Both the above points of view play interesting roles, although the most appropriate model can be chosen whenever one or the other proves more to the point.

The Kelvin law can be derived in different ways (membrane, energetic and chemical potentials). Here it will be demonstrated following two different points of view, just to show how distinct approaches can converge. The membrane model has been preferred for the first demonstration; the energetic one will be presented in the next section.

The equilibrium of a spherical meniscus in a capillary (Byers, 1965; Mason, 1971; Sedunov, 1974; Kikoin and Kikoin, 1978; Matveev, 1984; Sivuchin, 1986) can be simply derived from the consideration that the pressure of a fluid in a horizontal plane is the same at each point. Consider a vessel filled with water with a glass capillary partially immersed in it and the entire system in thermodynamic equilibrium with the atmosphere; we will concentrate on what is happening to water vapour. The liquid surface will experience the saturation pressure of water vapour. At the horizontal plane crossing the meniscus, the vapour pressure will be exactly the same, both outside and inside the capillary. The same holds for the horizontal plane at the free liquid surface, so that the pressure exerted by the water column will be compensated by the forces exerted by the meniscus.

Kelvin first derived the equation considering the vapour density constant with height. Under this assumption, the *Stevin* law

$$p = \rho g h \tag{5.3}$$

(where ρ is the fluid density, g is the gravity acceleration and h is the height) gives the pressure $p(0)$ at the free liquid surface level $h = 0$ both for the vapour outside the capillary

$$p(0) = p(h) + \rho_v g h \tag{5.4}$$

and the water inside it

$$p(0) = p(h) + \frac{2\sigma}{r} + \rho_w g h \tag{5.5}$$

where ρ_v and ρ_w are densities of the saturated vapour and liquid water, respectively. Eqn (5.5) for liquid water includes the static contribution of the water column (i.e. Stevin law) and the term $2\sigma/r$ that represents the change in pressure (Laplace formula) due to the surface tension across the curved (spherical) meniscus into the capillary. Equating for $p(0)$ Eqns (5.4) and (5.5) the term gh is isolated, which, substituted into Eqn (5.4), gives the original Kelvin formula:

$$p(0) - p(h) = -2\sigma \frac{\rho_v}{(\rho_w - \rho_v)} \tag{5.6}$$

This formula is valid for a small capillary height in which the vapour density can be considered constant and there is no explicit mention of the wetting angles, i.e. surface wettability. A more general derivation can be similarly obtained using the barometric equation for the vapour pressure change with height and introducing the wetting angle.

In the capillary tube, the pressure originated by the surface tension σ of the meniscus at the top of the water column is caused by the component of a force tangential to the capillary F_t, generated by the surface tension all around the circumference $2\pi r$ of the capillary, i.e.

$$F_t = -2\pi\sigma\ \gamma \cos\theta \tag{5.7}$$

where θ is the contact angle between the water meniscus and the immersed part of the capillary; $\cos\theta$ is positive for wettable surfaces (also called hydrophilic) and negative for nonwettable ones (also called hydrophobic or hydrorepellent).

The *Laplace pressure* Π is obtained by dividing F_t by the section of the capillary πr^2,

$$\Pi = -\frac{2\sigma}{r}\cos\theta. \tag{5.8}$$

which considers the effect of the wettability of the surface by means of the correction factor $\cos\theta$. Equating the Laplace pressure Π to the Stevin hydrostatic pressure p of the liquid into the capillary, one obtains

$$\rho\, g\, h = -\frac{2\sigma}{r}\cos\theta. \tag{5.9}$$

This equation is useful in weather analysis because the term gh, called *geopotential height*, is obtained by dividing by ρ the two terms.

The next step is to consider the vapour pressure in a horizontal plane crossing the meniscus height. In equilibrium, it can be supposed that the lower pressure over the curved meniscus is in equilibrium with the natural barometric decrease of pressure of the vapour with height. The latter can be calculated by substituting into the state Eqn (1.1) for perfect gases the fundamental equation of the hydrostatics

$$dp = -\rho\, g\, dz \tag{5.10}$$

For a mole of gas, with molar mass $M = \rho\ V$, integration gives

$$\ln\left(\frac{p(h)}{p(0)}\right) = -\frac{Mgh}{\Re T} \tag{5.11}$$

which represents the *barometric* (or *hypsometric*) formula, which gives the vertical distribution of pressure for the increase of height $h = z_1 - z_2$ with vapour density decreasing with height. Following a similar approach to the derivation of the simpler Kelvin formula (5.6) and substituting the value formerly derived for the geopotential height, i.e. $-2\sigma\cos\theta/\rho r$, in place of gh in the barometric relation (Eqn (5.11)), the general Kelvin formula is obtained. This includes the interaction between water and solid (either wettable or nonwettable surfaces):

$$\ln\left(\frac{p(h)}{p(0)}\right) = -\frac{V_m}{\Re T}\left(p(0) - p(h) - \frac{2\sigma\cos\theta}{r}\right). \tag{5.12}$$

The curvature effect can be taken into account by substituting the equilibrium vapour pressure over the meniscus of radius r, i.e. $e(r)$, with $p(h)$ in the left-hand term, as the two pressures must be equal, being at the same horizontal level. The same holds for $e(\infty)$ and $p(0)$.

It should be noted that this more general Kelvin equation can be approximated in a practical form if the value of $p(0)-p(h)$ is very small compared to $p(0)$. More precisely, if the two following conditions are satisfied,

$$\begin{aligned} &\left|\frac{2\sigma\cos\theta}{r}\right| \gg |p(0) - p(h)| \\ &\frac{V_m}{\Re T}|p(0) - p(h)| \ll 1 \end{aligned} \tag{5.13}$$

then the Kelvin formula assumes the usual form

$$\ln\left(\frac{RH(r)}{100}\right) = \frac{2\sigma V_m \cos\theta}{r\,\Re\,T} \tag{5.14}$$

where $RH(r)/100$, i.e. the RH in equilibrium with a curved meniscus of radius r, has been substituted to the ratio $e(r)/e(\infty)$ in accordance with the definition given in Chapter 2A.

In the case of droplets, $r > 0$ and $\cos\theta = 1$; in the case of menisci of water condensed into micropores, $r < 0$ and θ is typical for each material, i.e. $\theta < \pi/2$ for wettable surfaces and $\theta > \pi/2$ for nonwettable ones.

An interesting ideal experiment was suggested by Kelvin (Thomson, 1870), based on the barometric formula. Let us consider two separate vessels (Fig. 5.4(a)) with water, posed at different heights z_1 and z_2 in a closed box in thermal equilibrium. At both heights the free water surfaces have equilibrium tension $e(\infty)$, but inside the box the pressure decreases according to the hypsometric formula. Therefore, the water molecules evaporating from the upper water surface will diffuse within the box. They will move from the level z_2 that has a lower pressure, and will

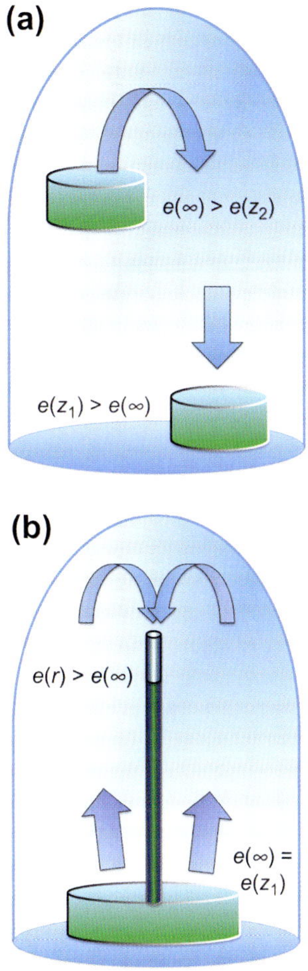

FIGURE 5.4 (a) Ideal experiment suggested by Kelvin, based on the barometric formula, i.e. pressure decreasing with height. However, at the flat water–air interface, the vapour pressure is the same, i.e. saturation. For the gradient that would develop at the interface, water will evaporate from the upper vessel and condense on the lower one. (b) Paradox that would arise if the external pressure were not distributed according to the barometric formula but were constant with height. In such a case, one should expect evaporation from the flat water surface (higher pressure) and condensation into the concave meniscus of the capillary (lower pressure). This is an impossible *perpetuum mobile*.

diffuse within the box, increasing the pressure at every level. At the lower level z_1, the pressure is also increased by a small amount ε, and condensation will occur in the vessel where the free water surface has the equilibrium pressure $e(\infty)$, lower than the newly created pressure at the same height, i.e. $e(\infty) + \varepsilon$. The consequence is that this thermodynamic system will cause evaporation from the upper vessel and condensation in the lower one, until all the liquid water will be transferred as in the case of two vessels communicating by means of a tube.

Another ideal experiment was suggested by Kelvin. If in a closed vessel a water reservoir and several vertical capillary tubes with different diameters, closed at the bottom and partially filled with water, were left for a sufficiently long time to reach equilibrium, the following situation is expected. All the water columns in the capillaries will be found in equilibrium with the barometric distribution of vapour, i.e. the excess water would have evaporated from the capillaries too much filled and would have condensed into those too less filled, and the height of the water columns would be the same as all the capillaries were open at both ends, partially immersed and communicated in the same water vessel.

If the external pressure were not distributed with the barometric formula, but were constant with height, the following paradox would arise (Fig. 5.4(b)). Let us consider a closed box having, at the bottom, a vessel containing water in equilibrium with its vapour; then a vertical capillary tube is partially immersed in the liquid, all at homogeneous temperature. In the capillary, a column of water will rise and stop at the level determined by equalizing the Stevin and Laplace pressures. The pressure of the saturated vapour in contact with the liquid is not the same over the flat surface of the water at the bottom and over the meniscus, being lower over the meniscus where the surface is concave. In the vessel, if the vapour pressure were constant with height, the equilibrium pressure would be $e(\infty)$ and the vapour would gradually migrate towards the meniscus into the capillary, where the pressure is $e(r) < e(\infty)$, establishing a gradient of concentration, and possibly of pressure, inside the box. At the top of the capillary, the vapour would become supersaturated with respect to the meniscus curvature and would condense, adding new liquid into the water column. The molecules that will condense into the meniscus will feed the water column, increasing its weight and generating in it a slow, continuous sinking motion. In practice, the condensation at the top of the capillary would be compensated by other molecules that would evaporate from the free water surface at the bottom where the equilibrium tension is $e(\infty)$. Thus, partially or totally neglecting the natural decrease of the vapour pressure with height, the above process would indefinitely continue with evaporation from the bottom, migration of vapour into the capillary, condensation onto the capillary meniscus, displacement of the column to reestablish the original level and so on, in a *perpetuum mobile* mechanism similar to a reversed fountain. With the approximate Kelvin formula one arrives to this absurd; if the rigorous formula (5.12) is considered, the system reaches its thermodynamic equilibrium as discussed.

5.2.2. Derivation of the Kelvin Equation from the Gibbs Potential

Gibbs derived independently the Kelvin equation, on the basis of chemical potentials. Here the main guideline will be followed; details can be found elsewhere (Byers, 1959, 1965; Mason, 1971; Pruppacher and Klett, 1980, 2003; Young, 1993). Gibbs considered the conditions in an isothermal and isobaric system of more than one component, such as a solution, where specific components might change the number of moles. To this aim, three main parameters should be defined: the *Gibbs surface free energy* $\mathcal{G}$

$$\mathcal{G} = U + p_s V_s - T\,\mathfrak{S} \tag{5.15}$$

the *free enthalpy* of the system

$$\mathcal{H} = U + p_s V_s \tag{5.16}$$

and the *Helmholtz free energy*

$$\mathcal{F} = U - T\,\mathfrak{S} \tag{5.17}$$

where U is the free energy of the system at temperature T, p_s is its pressure, V_s is its volume and $\mathfrak{S}$ is the entropy.

When the vapour molar fraction $\delta m/M$ is transported from the plane surface of the liquid (or from the atmosphere) to a droplet, the increase in free energy is

$$\Delta\mathcal{G} = \frac{\delta m}{M}\,\Re\, T\,\text{In}\left(\frac{e(r)}{e(\infty)}\right). \tag{5.18}$$

After some vapour has condensed, the increase $\Delta\mathcal{G}$ is due to the increase of surface energy, which is linked to the free surface S of the liquid under the action of surface tension σ, i.e. $\Delta\mathcal{G} = \sigma\,\delta\,S$. For a spherical droplet, the surface is $S = 4\pi\, r^2$ and the increase is $dS = 8\pi r\, dr$, the volume V is $V = (4/3)\pi r^3$ and $dV = 4\pi\, r^2\, dr$ and the mass that condenses is $dm = \rho\, dV = 4\pi\,\rho\, r^2\, dr$; hence $dS = 2\, dm/\rho\, r$ and $\Delta\mathcal{G} = 2\,\sigma\,\delta m\,\rho\, r$. Substituting this finding in Eqn (5.18) and considering that $M/\rho = V_m$, the Kelvin equation is obtained.

This derivation is very general and the Kelvin formula can be rewritten for any shape of meniscus, by using the ratio of the incremental values dS/dV of the meniscus surface and related volume

$$\ln\left(\frac{RH(r)}{100}\right) = \frac{\sigma V_m}{\Re T}\frac{dS}{dV} \tag{5.19}$$

An example of various geometric shapes is reported in Fig. 5.5.

After calculation, one obtains $dS/dV = 2/r$ for a sphere; $dS/dV = [(1/r_1) + (1/r_2)]$ for ellipsoids with principal radii r_1 and r_2; $dS/dV = [(1/r_1) - (1/r_2)]$ for saddles with principal radii r_1 and r_2;

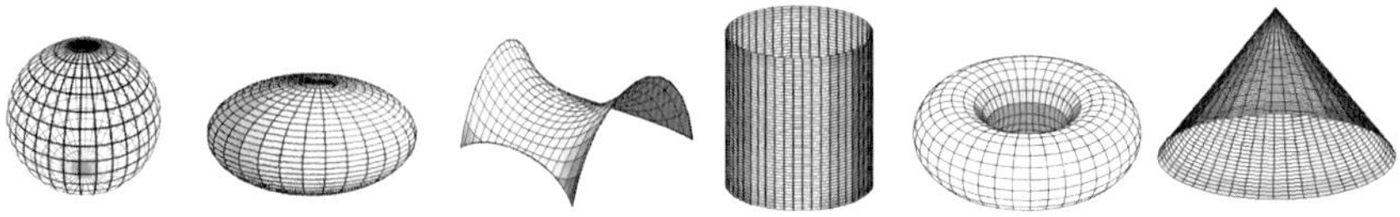

FIGURE 5.5 Geometrical shapes of sphere (SP), ellipsoid (EL), saddle (SA), circular cylinder (CI), circular torus (CT) and cone (CO). (b) Relative humidity (*RH*, %) in equilibrium with a concave meniscus of water with radius *r* (μm), according to the Kelvin law. C, cylindrical meniscus; S, spherical meniscus.

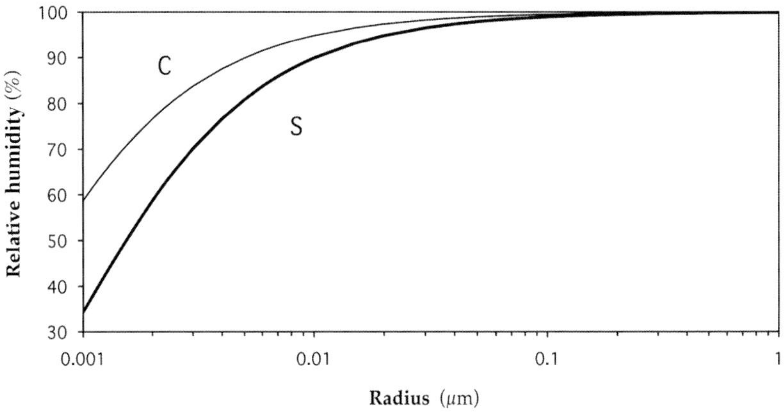

FIGURE 5.6 Relative humudity (*RH*, %) in equilibrium with a concave meniscus of water with radius *r* (μm), according to the Kelvin law. C: Cylindrical meniscus; S: Spherical meniscus

$dS/dV = 1/r$ for a circular cylinder or a circular torus (where r is the radius of the cylinder or the generating circle; note that the formula is independent of the cylinder height or torus radius); and finally, $dS/dV = 3(2r^2 + h^2)/(r^2h\sqrt{r^2 + h^2})$ for a cone of height h. However, the cone is unlikely for a small-sized liquid meniscus.

The equilibrium *RH* versus the radius of a spherical or a cylindrical meniscus is reported in Fig. 5.6.

5.3. THE FORMATION OF DROPLETS IN THE ATMOSPHERE: HOMOGENEOUS AND HETEROGENEOUS NUCLEATION

The first problem in the formation of droplets is that $r > 0$ and this needs supersaturation, i.e. $RH >> 100\%$, or the intervention of other phenomena that may counteract this physical effect. The Kelvin equation in the form (5.19) considers the energy balance derived from two counteracting factors: the positive work against the surface tension in the formation of the free surface of the meniscus (proportional to r^2) and the negative work derived from the energy released by the vapour–liquid change of phase of some water molecules (proportional to r^3), which comes from the tendency of water molecules to aggregate in the liquid state. In the formation of a droplet by condensation, the surface area is $4\pi r^2$ and the free surface energy is $4\pi r^2\sigma$ and considering in a similar way the contribution due to the increase of the volume, the elevation of the free energy due to the curvature of the surface is

$$\Delta G = 4\pi r^2\sigma - \frac{4}{3}\pi r^3 \rho \frac{\Re T}{m} \ln\left(\frac{e(r)}{e(\infty)}\right) \tag{5.20}$$

The algebraic sum of these two contributions, characterized by different powers of r, determines a graph whose maximum ΔG^* is a maximum of instability for the physical process and the corresponding critical radius is r^* (Fig. 5.7).

The critical radius can be determined under the condition $\mathrm{d}(\Delta G)/\mathrm{d}r = 0$, and is

$$r^* = \frac{2\sigma m}{\rho \Re T\left(\frac{e(r)}{e(\infty)}\right)} \tag{5.21}$$

All droplets with $r < r^*$, called *embryos*, are unstable and tend to dissipate; after the critical radius is surpassed, the embryo droplet grows in microseconds to a *'mature'* cloud or fog droplet. The growth may occur by accidental aggregation of other molecules, collision with other droplets or coalescence. The formation of a pure water droplet, without the

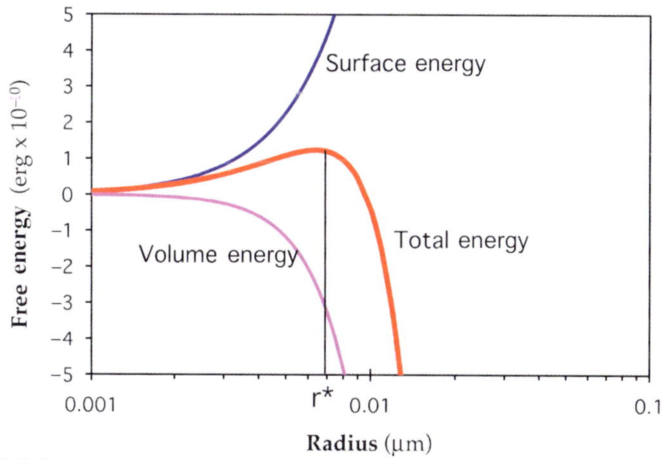

FIGURE 5.7 Free energy of a pure water droplet versus droplet radius and critical radius r^*. Lines refer to the surface free energy (blue), the volume free energy (pink) and the total energy (red line).

intervention of heterogeneous condensation nuclei, is called *homogeneous nucleation*. In this case, the system is composed of microspheres of water floating in the air, and the radius of the droplets is positive and a high supersaturation is required for equilibrium with the curved water surface, i.e. $RH >> 100\%$, $T << DP$. The smaller the droplet, the greater the supersaturation needed for the equilibrium.

When there is a population of differently sized droplets, the equilibrium pressure changes with the droplet size, the average pressure of the vapour characterizing the population being in defect for the smaller droplets and in excess for the larger ones. Therefore, the smaller droplets evaporate and the excess vapour condenses on the largest ones. In foggy days, monuments become wet due to this phenomenon, as the radius of curvature of each part of the monument surface is infinitely larger than the radius of fog droplets. There is no doubt that this is the dominant mechanism, as continental fog occurs in the absence of wind, therefore there is no impaction of droplets on the monument surface, and in the presence of fog the monument is generally in temperature equilibrium with the air, so that condensation is not accelerated by supersaturation due to monument temperature being below the *DP*. Eventually, droplets formed on the vertical surface of monuments grow until the larger ones, due to their weight, run downwards colliding other droplets, which supplies further water and increases their mass. These rivulets form small deposits of water at the base or on other horizontal surfaces; this water reservoir is then adsorbed by capillarity or furnishes the appropriate conditions for local weathering or biological deterioration.

Homogeneous nucleation can be generated in expansion cloud chambers, where very high supersaturation e.g. $RH \approx 400-800\%$ can be obtained for short times. Small droplets are formed on adiabatic cooling when the expansion ratio of the chamber exceeds 1.37 and $RH \approx 500\%$ (Mason, 1951) or also at the critical supersaturation $RH \approx 400\%$ (Madonna et al., 1961). In the early stage, the droplets are aggregates of water molecules around existing ions.

The water molecules have a large dipole moment, so that they can surround charged ions, and in this case the balance is made with three forms of energy: the surface energy $4\pi r^2\sigma$, the energy due to the latent heat of condensation $-(4/3)\pi r^3\rho\ L_v$ (where L_v is the latent heat of condensation or vaporization) as discussed before, and the electric energy of the Coulomb field of the droplet $q^2/(4\pi\varepsilon r)$, where q is the electric charge of the ion and ε is the dielectric constant. However, in this case high supersaturation is also needed, of the order of $RH \sim 400\%$. If a droplet carries a charge q, the equilibrium RH is

$$\ln\left(\frac{RH(r)}{100}\right) = \frac{2\sigma V_m}{r\Re T} - \frac{q^2}{8\pi\varepsilon\, r^4} \tag{5.22}$$

For this reason, in the free atmosphere droplets may only be formed by adsorption of water on some condensation nuclei (*heterogeneous nucleation*). When water vapour condenses on them, they form solutions whose equilibrium vapour pressure is relatively low. In other words, the depression formed by the chemical solution compensates for the Kelvin overpressure generated by the convex meniscus of the droplet, so that condensation will occur and the droplet will grow.

The most effective condensation nuclei reach equilibrium at $RH << 100\%$. A salt that turns into a solution by hygroscopic sorption from humid air is called '*deliquescent*' and many crystals with this property are found in the atmosphere; the most common one is NaCl generated by sea spray. The condensation nuclei begin to adsorb a number of water molecules and then grow in size becoming first deliquescent, then a very concentrated hygroscopic solution, and later a dilute solution, but the last step occurs when their radius is greater than r^*. However, the classical laws of thermodynamics cannot be applied to the first steps of this process because classical thermodynamics is only valid for infinitely diluted solutions. In particular, the *Van't Hoff* factor i that for soluble salts represents the effective number of moles after dissociation of 1 mol of the original substance, e.g. $i = 2$ for NaCl which transforms in Na^+ and Cl^-, varies as a function of the molality of the solution.

Solution effects are important not only because of the heterogeneous nucleation but also because they cause condensation on surfaces or micropores at relatively low RH. Chemical contamination causes departures from the Kelvin law and may be taken into account with some small changes to the original formula. The vapour tension of a solution is lower than that of pure water. The osmotic effect due to the ionic dissociation of the hygroscopic salts that form the condensation nuclei, that contaminate the surface of bodies or that are included in micropores may be very important. The solution effect is largely dominant in the presence of great concentrations of salts, as on the walls, or in the large pores with diameter D lying between 1 and 100 μm, when the Kelvin effect becomes negligible.

In the presence of hygroscopic salts, the Kelvin formula must be improved and two different approaches will be considered here. Mason (1971) considered the transfer of a small mass δm of water from a contaminated droplet to a plane surface of pure water. The solution will undergo a decrease of free energy

$$\Delta \mathcal{G} = \sigma_S \, \delta S - P\delta V_S = 2\sigma_S \frac{\delta m}{\rho_S r} - P\frac{\delta m}{\rho_S} \tag{5.23}$$

where σ_S and ρ_S are, respectively, the surface tension and the density of the solution; S and V_S are, respectively, the surface and volume of the droplet; and P is the osmotic pressure of the solution, variable with the molality. On the other hand, the decrease of the free energy is due to the work corresponding to the evaporation of the amount of water δm with initial equilibrium pressure $e_s(r)$, expanding the vapour to the lower pressure of the free surface of pure water $e(\infty)$ and, finally, condensing it at this pressure with a process that is reversible and isothermal, so that

$$\Delta G = \frac{\delta m}{M_w} \Re T \ln \left(\frac{e_s(r)}{e(\infty)} \right) \tag{5.24}$$

where M_w is the molar weight of water. Equating these two expressions of ΔG and considering that the osmotic pressure P is

$$P = \frac{\Re T \rho_w}{M_w} \ln \left(1 + \frac{i n_s}{n_w} \right) \tag{5.25}$$

where the labels s and w refer respectively to the solute and water, ρ_w is the water density, i is the parameter closely associated with the Van't Hoff coefficient for solution and depends on the molality, n is the molar fraction and m_s is the actual mass of the solute, the following formula is obtained:

$$\ln\left(\frac{RH(r)}{100}\right) = \frac{2\sigma_S M_w}{\rho_S r \Re T} - \frac{\rho_w}{\rho_S} \ln\left(1 + \frac{i n_s}{n_w}\right) \cos\theta \tag{5.26}$$

where $RH(r) = 100\ e_s(r)/e(\infty)$.

The other formula, by Wright (1936) is

$$\ln\left(\frac{RH(r)}{100}\right) = \left(\frac{2\sigma_S M_w}{\rho_S r \Re T} - \frac{i\ m_S M_w}{V_w \rho_w M_S}\right) \cos\theta \tag{5.27}$$

where m_s is the mass of solute, V_w is the volume of condensed water with dissolved salts, i.e. $(4/3)\pi r^3$ for droplets, and M_w/ρ_w is the molar volume V_m of water. The result is a function of the type of a hygroscopic salt, its solubility, the degree of dissolution and amount of water that has already condensed; as a consequence, it is continually variable, although the dynamics are strictly determined by the initial conditions. Of course, both the Mason and Wright formulae hold whatever is the sign of r and play an important role for the condensation in macropores, which always contain hygroscopic salts.

Both formulae show that the saturation pressure over a solution is lower than the saturation pressure over pure water. This happens because the saturation pressure is reached when the number of molecules of vapour that escape from the liquid equals the number of molecules that return into the liquid. In the case of a solution, the solute salt 'dilutes' and may aggregate to the water reducing the number of H_2O molecules that will be able to escape; therefore, the equilibrium is displaced towards a lower value.

5.4. BUBBLES

Water bubbles in air are well known: they are formed with a spherical layer of water with two air–water interfaces and therefore two membranes: one external and one internal. For this reason, the Laplace pressure exerted by them, and the energy needed to increase their surface, is twice the above values found for a meniscus with only one interface.

When the pressure inside a bubble increases by dp, its radius also increases by dr, and the external work $4\pi r^2 p\ \mathrm{d}r$ accomplished against the surface tension σ is transformed into free surface energy, i.e. $2\sigma \mathrm{d}S = 2\sigma \mathrm{d}(4\pi r^2) = 16\pi r \mathrm{d}r$, the factor 2 arising because of the contribution of both the inner and outer surfaces of the bubble. Conservation of energy requires that the external work equal the total increase of surface energy, and equating the two terms, a pressure twice the Laplace Π, given in Eqn (5.8), is found.

A bubble of air in water has only one interface and the pressure and the energy are as discussed in the previous sections. Although the surface tension tends to reduce the bubble size, as in the previous case but with only one surface, the pressure of the air entrapped inside counteracts it and the bubble can survive.

A completely different situation occurs for ephemeral bubbles of vapour in water. At first sight, vapour bubbles in water might appear complementary to water droplets in air but the change of sign of the radius of the meniscus makes ephemeral or impossible the life of vapour bubbles. The natural dynamics of the phenomenon, characterized by instability, leads the vapour bubbles (if any) to disappear in a very short time, and a comparison with embryos is more appropriate. A distinction should be made between bubbles determined by the presence of originally dissolved gases, e.g. CO_2 in mineral water, and those composed of vapour only. In the former case, bubbles form when a rise in temperature makes the gas insoluble and the pressure of the gas in the bubble opposes the contraction and vanishing of the bubble. Before boiling, vapour bubbles form on the hot pot surface. The external pressure, the unstabity of supersaturation and the above dynamics will determine the collapse of the bubbles. Each collapse is a water hammer against the pot that produces the typical buzzy noise.

The molecular forces acting in the surface layer of a liquid act as a membrane that does not allow the molecules to leave the volume of the liquid. Inside, the statistical distribution of the thermal motion shows that there are always some molecules that have greater speed and are less bound to the bulk liquid. These can be considered in the vapour state, although inside the liquid, and their pressure equals the tension of saturation of the external vapour. The equilibrium shows dynamics with perfect symmetry of behaviour, where the number of molecules leaving the liquid equals those passing from the atmosphere into it.

By definition, the boiling point (*BP*) is the temperature at which the saturated vapour pressure (inside the liquid) becomes equal to the external atmospheric pressure. The physics of this process is not clear if one thinks about what may happen when the pressure of the external vapour equals that of the atmospheric pressure but becomes clearer when thinking that equilibrium is reached with the *internal* vapour pressure and intense boiling starts as a consequence of the possibility of vaporization in the whole mass of liquid water (Matveev, 1985). However, also in this case, only a heterogeneous formation of bubbles is possible, as we will see later.

At temperatures below the *BP*, i.e. $T < BP$, the incipient bubbles have very small negative radii and therefore, according to the Kelvin equation, they are in equilibrium with low *RH*; however, the bulk liquid all around constitutes a very supersaturated environment and vapour bubbles cannot form or, if formed, they collapse immediately. Vapour bubbles can form only when $T \sim BP$. They form on the surface of the vessel for two reasons: (1) the vessel temperature is higher there, possibly $T >> BP$ and (2) bubbles develop not as small spheres with very small radius, but starting flat, adhering to the plane surface of the vessel where the radius of curvature is practically infinite. When a bubble grows being filled with vapour, and exceeds a critical dimension, it detaches from the vessel surface and comes up because of the buoyancy force. If a temperature fluctuation causes a temporary drop of temperature below the critical value, then the small unstable bubbles being formed in the vessel are immediately destroyed and the liquid water that replaces their volume impacts violently against the vessel, forming the characteristic buzz that precedes boiling.

5.5. MICROPORE CONDENSATION AND STONE WEATHERING

In the case of a pure, noncontaminated, nonreactive, nonporous surface, condensation occurs when the temperature of the surface T_S reaches, or drops below, the dew point (*DP*) of the air, i.e. $T_S \leq DP$, irrespective of both the degree of saturation of vapour (*RH*) and the temperature of the air at some distance from the surface. However, for simplicity, herewith it will be assumed that the air and material have the same temperature, at least in an air layer very close to the material, and the thermodynamic parameters will be referred to this layer in equilibrium with the surface.

By remembering the results in Chapter 2A for the *DP* spread, the same equation can be rewritten in terms of the Kelvin equation (Fig. 5.8)

$$\Delta DP = -\log_{10}\left(\frac{RH}{100}\right)\left(\frac{237.3+t}{7.5}\right) = -0.1158\,\sigma\,V_m\left(\frac{237.3+t}{r\Re T}\right)$$
$$= 2.057\times 10^{-8}\,\frac{\sigma}{r}\left(\frac{237.3+t}{273+t}\right) \qquad (5.28)$$

Evaporation occurs when $Ts > T_w$, where T_w is the wet bulb temperature of the air. Observation of the value of T_s allows us to distinguish between the two processes and determining, for example, whether a surface is wet because it is cold (condensation) or it is evaporating as a consequence of capillary rise. However, contamination with soluble salts lowers the equilibrium *RH* (and raises the *DP*), or the presence of micropores causes departures from this simple scheme and condensed water may be found in micropores at usual conditions, and also in relatively dry environments.

Inside droplets, all the water is in liquid state. In porous bodies, or in general, over a hydrophilic surface, the water molecules in contact with the material are adsorbed and strongly bound with the internal surface due to the presence of dipoles or image forces; the binding is so strong that this water is considered to be in the solid state. Beyond this solid layer, other water molecules are in the liquid phase and free. In pores with radius $r_p < 0.1\ \mu m$, the physical effect dominates; in pores with $r_p > 1\ \mu m$ and contaminated by soluble salts, the physicochemical effect that determines the equilibrium pressure for solutions may also cause condensation at lower *RH*. This typically occurs in a marine environment due to contamination with sea salts.

Water supplied by condensation causes dissolution of the material matrix; condensation–evaporation cycles cause capillary migration of dissolved salts and crystallization in other parts, e.g. efflorescence and subflorescence, thus weakening the material and causing both material and aesthetic value loss. This mechanism is clearly shown in thin sections of stones, where precipitated crystals are found in the pores (Fig. 5.9). Wet materials, e.g. rocks or mortars, may have their mechanical resistance diminished. In other cases, the presence of a film of water may decrease the free surface energy of the material, weakening it (Winkler, 1986).

In certain cases, e.g. argillaceous (containing clay minerals) limestone, water may alter the structure of the material, causing expansion, stress and fractures. In fact, the crystalline clay structure is composed of a series of wafers and positive ions are frequently trapped between the wafers. Water is able to penetrate the crystal because it is attracted by the hydroxyl groups and the inclusion of water molecules will cause hydrated clay to swell. However, the mechanism is complex at the molecular level and the swelling behaviour of clays cannot be explained

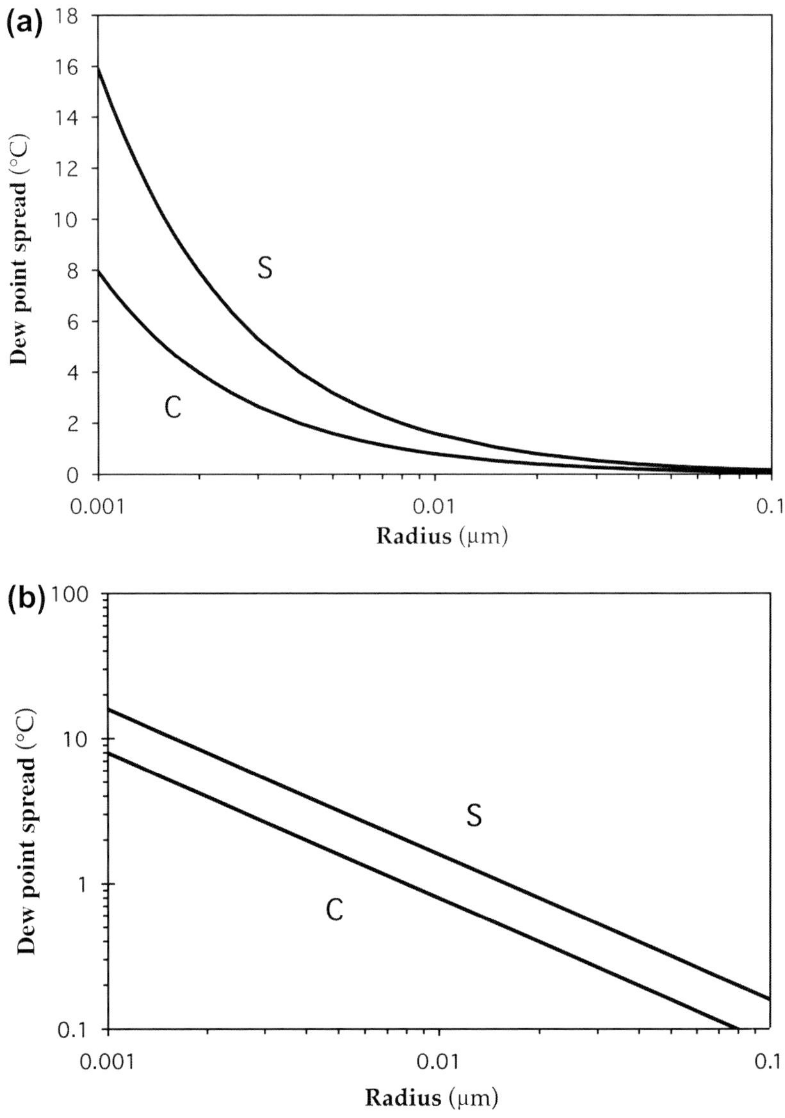

FIGURE 5.8 (a) Dew point spread (ΔDP, °C) in equilibrium with a concave meniscus of water with radius r (μm), according to the Kelvin law. C, cylindrical meniscus; S, spherical meniscus; linear scale. (b) Dew point spread (ΔDP, °C) in equilibrium with a concave meniscus of water with radius r (μm), according to Kelvin law. C, cylindrical meniscus; S, spherical meniscus; logarithmic scale.

only in terms of bulk liquid water properties but is an intricate interplay of the silicate surface, cation solvation effects and stress on the interlayer water (Bathija et al., 2009). This happens for clay minerals with expanding lattice, e.g. montmorillonite and vermiculite, but not for

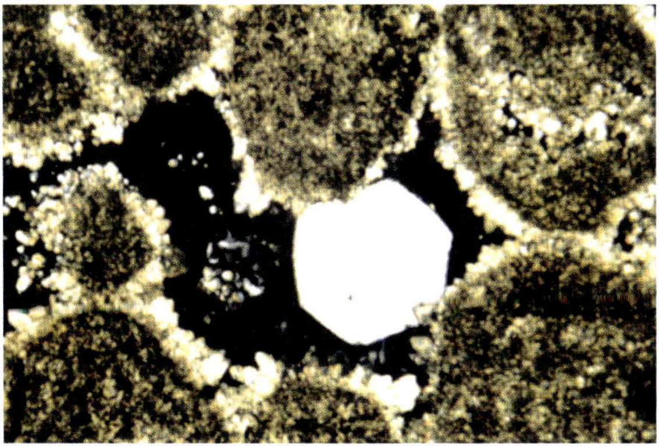

FIGURE 5.9 Thin section of oolitic limestone, where precipitated crystals are found in the pores. Authigenic dog-toothed calcite crystals cover the oolites; gypsum is more frequently found inside pores. A large authigenic hexagonal calcite crystal (white, in the middle) is growing inside the cavity. *Source: By courtesy of Prof. Marco Del Monte.*

nonexpanding ones, e.g. kaolinite, illite and chlorite. When the *RH* decreases, the adsorbed water evaporates but the structure between the wafers may have changed due to the formation of new crystals. The dimensional changes lead to hysteresis and in the long run, adsorption–evaporation cycles cause irreversible damage (Torraca, 1981, 1994). Dimensional changes for moisture exchanges may be of the order of 1 mm m^{-1} and water-repellent treatments cannot prevent moisture swelling and shrinking, although they may be delayed (Wendler, 1997). In some cases they can exacerbate damage.

Material dampness and air humidity favour biological life and weathering; this negative phenomenon becomes greater and greater when the duration, or the frequency, of the time of wetness (*TOW*) increases. *TOW* is an important parameter for the growth conditions of microorganisms on materials. It reflects the correlation between material structure (e.g. porosity, inner surface and cation exchange capacity) and the biosusceptibility of the material, i.e. the tendency for microbial contamination (Warscheid et al., 1993). Inside, the biological contamination is frequently restricted to certain physiological groups of microorganisms (e.g. fungi), whereas outside, microbial infections are mostly characterized by a complex network of algae, bacteria, fungi and lichens (Fig. 5.10). During the time in which a metal or a stone is wet, chemical reactions occur between the pollutants deposited and the material (Laurenzi Tabasso and Marabelli, 1992); the damage is linked to *TOW*, although no general formula has been yet found to link the damage to the *TOW*.

Condensation is also responsible for increasing the deposition rate of airborne pollutants. This fact is due to two different factors: (1) the particles and the hydrophilic gases that impact on a wet surface stick to it without

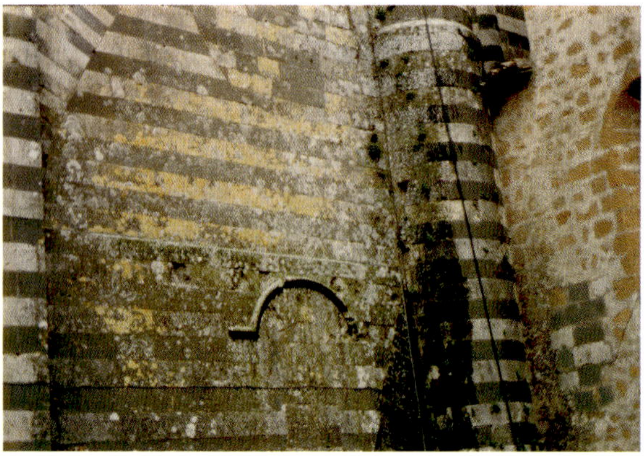

FIGURE 5.10 Lichens, fungi and algae colonizing the external wall of the Orvieto Cathedral, Italy, because they have found a favourable habitat for the frequent condensation and wetting on this surface.

bouncing, so that the capture efficiency of the surface is increased and (2) when condensation occurs, near the wet surface, several microphysical processes occur, the ultimate result of which is the increased transport of gases and particles towards the surface, as we will see in Chapter 8.

The problem of surface moisture and condensation is complex and depends on the physicochemical characteristics of both the atmosphere and the surface. The *RH* within a pore is a function of the temperature, mixing ratio, pore geometry, and presence and nature of soluble salts and can be considerably different from pore to pore as well as from atmospheric *RH*. Total porosity, total pore surface, spatial association of pores (that may form pockets and necks), pore size, pore form and pore radii distribution are important variables in monument weathering. Stones are characterized by a wide variety of pores and necks, with different shapes and sizes, which range from angstroms to millimetres, and can be grouped into several classes, depending on their properties based on laboratory analysis (Fitzner, 1994) and mineralogical characteristics (Jeannette, 1997). The porosity may change with time, especially in the subsurface layer where migration of salts, leaching, dissolution, erosion and other physical, chemical and biological attacks occur (Biscontin et al., 1993). However, although the situation is more complex, it is useful to introduce two basic types of pores: *open* and *internal pores* (Camuffo, 1984, 1988).

'Open pores', with very large outlets compared to the pore volume are found especially on the surface of bodies. The typical shape is that of a hemisphere, or a portion of a hemisphere (Fig. 5.11).

Open pores behave symmetrically with reference to droplets suspended in the atmosphere: the smaller the pore, the lower the *RH* required for equilibrium with the water meniscus. For each open pore, condensation

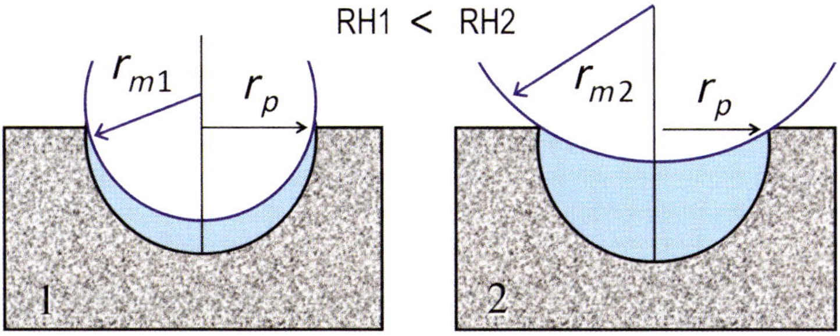

FIGURE 5.11 Reversible condensation in an open pore. When the ambient *RH* increases, the radius of the meniscus also increases, and vice versa. The figure shows two steps with $RH_1 < RH_2$; r_p is the radius of the pore and r_m is that of the meniscus. Cyan, condensed water. *(Source: By kind permission of Kluwer Academic Publishers. See credits.)*

begins at a low critical $RH(r_p)$ determined by the effective radius of curvature of the pore r_p. This is the geometric radius of the curvature of the pore minus the thickness of the mono- or bimolecular layer of water molecules adsorbed and strongly bound. Both condensation and evaporation are reversible and strictly follow any increase or decrease of the ambient *RH* level, always being in equilibrium. When the *RH* increases, condensation occurs and the radius of curvature of the meniscus (r_m) increases, i.e. the concavity of the meniscus is progressively lost, following the equilibrium with the variations in *RH* according to the equation

$$r_m = \frac{2\sigma V_m}{\Re T} \ln\left(\frac{RH}{100}\right) \tag{5.29}$$

An increase in *RH* corresponds to an increase in both condensed water and r_m, and vice versa. The hemisphere is completely filled at $RH = 100\%$. All the steps occur in equilibrium with *RH* and the process is reversible.

'Internal pores', with small outlets, are typically found inside bodies. They are connected to the atmosphere by a small hole across the surface layer. Alternatively, they may be in deeper layers, but connected to wider pores or capillaries (Fig. 5.12). They behave differently from open pores (Camuffo, 1984).

Condensation is characterized by an unstable equilibrium and is therefore irreversible, being triggered by a critical *RH* threshold level, and continues until the whole pore is filled with water. The condensation inside the pore begins at the low critical $RH = RH(r_p)$, which is in equilibrium with the radius of curvature of the pore r_p just wetted with a film of water. After a short time, the liquid water that is forming by condensation on the internal surface of the pore will reduce the free air pocket inside the pore. Consequently, the surface of the water meniscus too will be progressively reduced, as well as the radius of curvature of the meniscus r_m that will

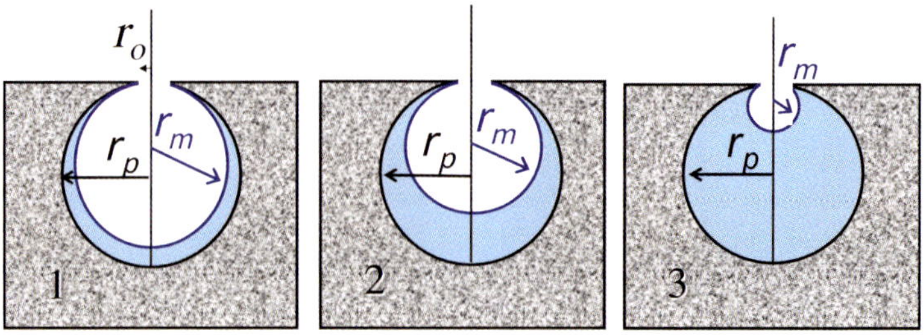

FIGURE 5.12 Irreversible condensation in an internal pore. When the ambient *RH* triggers condensation, the radius of the meniscus decreases until the whole pore is filled of water, and vice versa for evaporation. The figure shows three steps with the same *RH*; r_o (black short arrow) is the radius of the opening, r_p (black long arrow) is the radius of the pore and r_m (blue arrow) is the radius of the meniscus. Cyan, condensed water. *(Source: By kind permission of Kluwer Academic Publishers. See credits.)*

become smaller and smaller. The smaller the new radius r_m, the lower the equilibrium $RH(r_m)$. However, the actual *RH* inside the cavity, which initially was in equilibrium with the greater radius of curvature of the pore r_p, now corresponds to supersaturation for the smaller r_m, i.e. $RH(r_p) > RH(r_m)$ and the process is accelerated. Therefore, the initial level $RH(r_p)$ is no longer a neutral equilibrium value but a critical value of unstable equilibrium that triggers off the complete filling of the pore.

The reverse happens for evaporation. In order to reach evaporation, it is necessary that the ambient *RH* drop below a low threshold that is determined by the radius of the pore opening r_0. When *RH* drops below the critical level $RH(r_o)$, in the pore some water may evaporate from the free surface of the meniscus. While water is removed from the liquid reservoir inside the pore, the internal air pocket increases, and the meniscus too. This situation is unstable because the radius of the meniscus r_m is progressively increasing and the equilibrium $RH(r_m)$ level inside the pore becomes higher and higher, while the external *RH* level remains low, i.e. below the $RH(r_o)$ threshold. In conclusion, inside the pore the meniscus is increasing and could stay in equilibrium with higher and higher *RH* levels. However, outside the ambient *RH* is very low and this favours further evaporation and a continuous diffusive flow of water molecules from the interior to the exterior, until all the liquid water will be evaporated. Briefly, once triggered, the evaporation will be accelerated until the pore has been completely emptied.

However, it should be mentioned that although the condensation/evaporation inside the bulk pore is irreversible, when the whole pore is almost filled with water, the opening and the very small part of water underlying it, will behave as an open pore, because in this ultimate stage the meniscus varies without geometrical constraints and may change keeping equilibrium with the environment.

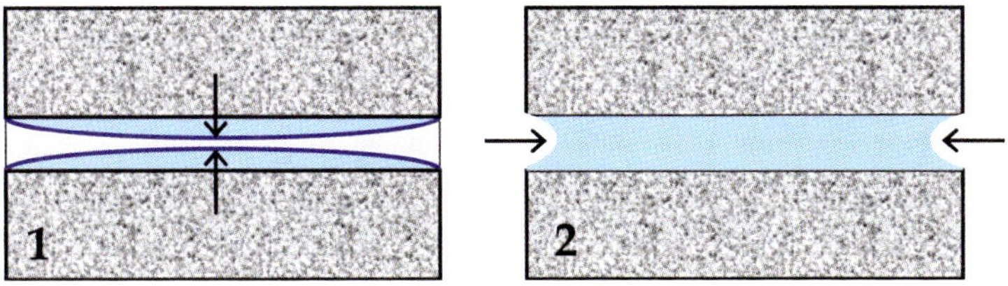

FIGURE 5.13 Condensation and evaporation in a capillary. (1) Condensation occurs with a cylindrical meniscus growing in thickness until the whole capillary is filled. (2) Evaporation occurs with a spherical or ellipsoidal meniscus deeply retreating inside the capillary. Cyan, condensed water.

In conclusion, open pores are characterized by reversible condensation/evaporation, are always in equilibrium with the ambient *RH* level and may contain a variable amount of water. Internal pores are characterized by irreversible condensation/evaporation, are in unstable equilibrium with the ambient *RH* level and may be found either empty or completely filled with water. They become empty after the *RH* drops below the critical level for evaporation and are filled if the *RH* has exceeded the critical level for condensation. If the ambient *RH* is at an intermediate level between the two, the pore will be empty or filled depending on which of the two thresholds came last.

Condensation in capillaries (Fig. 5.13) occurs similarly to that in internal pores, as condensed water makes smaller and smaller the radius of curvature of a cylindrical meniscus, determining a condition of unstable equilibrium and accelerating the condensation. When a capillary is full of water, a spherical meniscus forms at the beginning of the capillary and the evaporation starts in a reversible way, as in open pores.

Interesting is the case of the so-called *ink-bottle* pores (Fig. 5.14), i.e. cavities connected to other major cavities through smaller ducts, forming aggregates that can be compared to bottles having their neck in communication with the major cavity. According to the *Kelvin law*, in a porous material condensation occurs first into the smallest pores or long necks, which are the cavities with the smallest radii of curvature, whenever possible. When the necks are filled with water, air pockets remain entrapped in the pores and further condensation is impossible in steady-state conditions. Internal condensation may only continue when variations or fluctuations in temperature or atmospheric pressure cause the displacement of the water inside the pore necks. Nocturnal cooling of the body may reduce the pressure inside the pores and cause the water condensed in the neck to be sucked into the pore. According to the *Cantor law*, the required excess pressure is inversely proportional to the neck radius. When the excess pressure is sufficient, plugs of condensed water

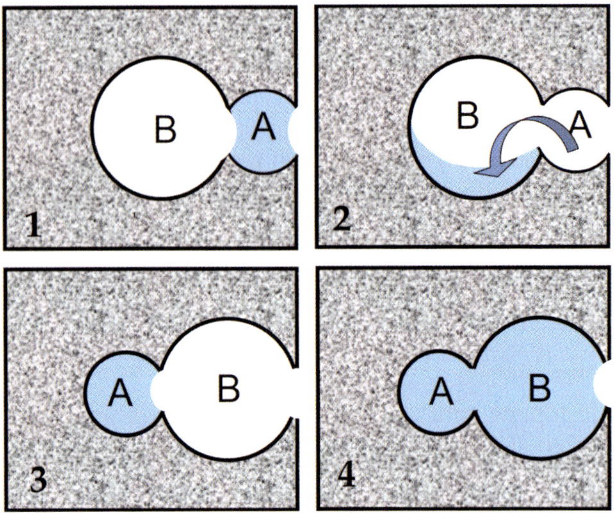

FIGURE 5.14 Condensation in a combination of pores, also called 'ink bottles'. Top line: a combination of a larger pore with a smaller one connected with the open air or a much larger cavity. (1) Condensation occurs inside the smaller pore A, closing access to B that remains filled with air. (2) If the temperature decreases, the air pressure in B will decrease and suck in the water that was condensed in A. Bottom line: a combination of a small pore with a larger one connected with the open air or a much larger cavity. (3) Condensation first occurs in the small pore A (4) and then continues in the larger pore B. Cyan, condensed water.

can be pushed out and trapped inside the pores or can be forced to migrate inside the material. Condensation may continue by means of these steps, so that the amount of adsorbed water is also a function of the pumping efficiency of these variations.

Similarly, evaporation from an open pore starts at its outlet, removing water from the pore and increasing the radius of the free meniscus. Evaporation is triggered off when the ambient *RH* drops below a critical value that is calculated according to the Kelvin formula, using the radius of curvature of the pore outlet (r_o), i.e. $RH(r_o)$. This critical value is lower than all the equilibrium values $RH(r_m)$, which can be calculated for all the increasing values of r_m, from r_o to r_p, following the loss of liquid water. Inside the pore, after some evaporation, the *RH* in equilibrium with the meniscus $RH(r_m)$ becomes relatively higher and higher compared with the external *RH*, that is $RH \leq RH(r_o)$, so that the external condition favours further evaporation. Consequently, the process is accelerated and is irreversible. Evaporation continues until all the liquid water (i.e. all the water inside the pore, except the first and second molecular layers in contact with the surface that are attached with a stronger ice-type bond) has evaporated.

The condensation–evaporation cycles being thermodynamically irreversible present noticeable hysteresis as condensation occurs when triggered by $RH(r_p)$ and evaporation by $RH(r_o)$. In practice, $RH(r_p)$ and $RH(r_o)$ are not just two precise levels of RH but two ranges, being determined by the actual distribution of r_p and r_o in the porous material.

5.6. ADSORPTION ISOTHERMS

The condensation–evaporation cycles can be represented by the so-called BET (Brunauer, Emmett and Teller) *adsorption isotherms* (Brunauer, 1945; Gregg and Sing, 1967; Mikhail and Robens, 1983), in which the amount of adsorbed water (*WA*) is plotted versus *RH*. In practice, this graph shows the change of weight due to the adsorption of water by an initially dry sample of material when the *RH* increases from 0% to 100% and then decreases again to 0%. The most common type (Fig. 5.15) is composed of branches forming a hysteresis loop.

The first branch AB, characteristic of the very low *RH*, occurs when the material surface, both external and internal (i.e. the visible surface and the surface in contact with the air but inside the pores), is progressively covered with a layer of water in the solid state, with a final thickness of two or three molecular diameters. In this branch, the graph of adsorption is coincident with the desorption one. The condensation–evaporation process is characterized by a one-to-one correspondence and is reversible depending linearly on changes in *RH*.

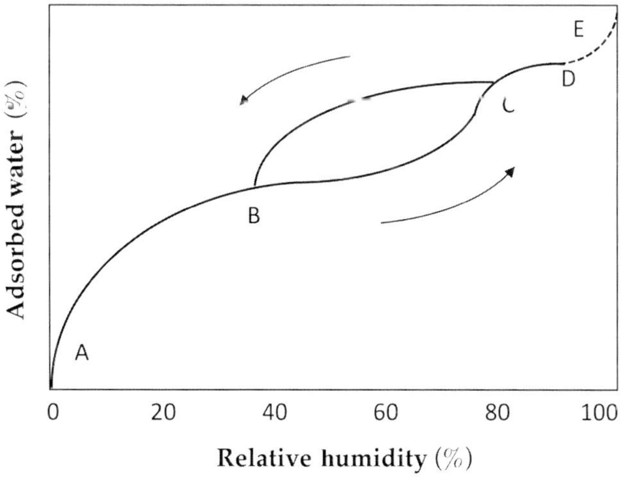

FIGURE 5.15 Adsorption isotherm for a condensation–evaporation cycle: absorbed water (*AW*) versus *RH*.

In the intermediate *RH* range, after the total surface has been covered with this solid layer, a further increase in the *RH* leads to the formation of bulk liquid water in the internal pores. This second branch (i.e. lower BC) becomes steeper and steeper as the pores are filled with water. In every pore with a radius r_p, condensation occurs when triggered by $RH(r_p)$ or by the chemical effect. If all the pores have the same size, the condensation branch is very steep. In this branch, the process is no longer reversible: if the *RH* decreases, some minor hysteresis loops start.

The third branch CD is characterized by very high *RH*, when all the internal pores are filled with water. In theory, this branch should be horizontal, as no further increase of *WA* is possible, also because when $RH = 100\%$ the surface is in thermodynamic equilibrium with the same number of molecules condensing and evaporating. In practice, this branch is tilted because the spectrum of the pores is wide and before reaching $RH = 100\%$ there are always some large pores that are empty and become progressively filled. In addition, the presence of contaminants or soluble salts causes the condensation to occur earlier, so that the vapour surrounding the sample is in a state of supersaturation forming liquid films or droplets on the external surface (i.e. the final rise DE). A similar condition of supersaturation occurs when the surface temperature drops below the *DP*, so that at the interface between the surface and the air the $RH > 100\%$. In any case, the amount of liquid water that can adhere to the external surface is modest and the final branch is slightly tilted, with a sharper rise near the upper extreme due to solution effect. If condensation stops before the droplets fall from the damp material, this branch is again characterized by a one-to-one correspondence and is reversible.

When *RH* decreases, the process first follows the last condensation–desorption branch ED and then DC, until the external water has evaporated. Continuing to lower the equilibrium *RH* after all the pores have been filled with water, evaporation may proceed, emptying the pores and this may occur only when triggered by $RH(r_o)$. As $RH(r_o) < RH(r_p)$, a hysteresis loop is generated, the condensation branch being related to the size of the internal pores r_p; the evaporation branch (upper CB) reflects the dimension of the outlet of the pores r_o. If all the pores have the same outlet, the evaporation branch is very sharp. A laboratory method to determine the pore size consists in measuring the derivative of the amount of water *AW* that is adsorbed in the stone by varying the *RH*. In fact, dAW/dRH, shows peaks corresponding to the dominant features of the pores, which are narrower or broader according to the distribution of the dimensions of the pores and their outlets. The Kelvin equation is the key to transform the values of *RH* that correspond to peaks in radii. After all the internal bulk water has evaporated, lowering the *RH*, layers of solid water (first branch BA) might start to evaporate, but they require a much greater latent heat (e.g. two or three times greater) than for the liquid–vapour transition.

Cycles involving the last branch DE mobilize the salts on the surface and subsurface layer; cycles involving the hysteresis loop dissolve and recrystallize the soluble salts at a deeper layer and cause outward migration with the formation of new efflorescence. The mechanical stress that follows thermohygrometric changes in a structure weakened by long-term dampness may have dramatic consequences. For an extensive review of the deterioration mechanisms generated by gypsum, soluble salts and dampness see Charola et al. (2007) Steiger et al. (2011). In addition to the chemical deterioration, the presence of water inside the pores favours biological epilithic and endolithic life and biological weathering. Saturated environments should be carefully avoided.

An experimental hysteresis loop similar to the above described one is presented in Fig. 5.16. It corresponds to a laboratory test of a brick sample immersed at low temperature (i.e. a bath of liquid nitrogen, temperature below 77 K) in a controlled atmosphere of nitrogen vapour, carried with a flow of helium. The desired partial pressures are obtained by varying the ratio between the nitrogen and the carrier gas, in order to obtain a controlled adsorption–desorption cycle. At that temperature, the nitrogen is adsorbed into the brick and the adsorption and desorption reflect in a change in the ratio of nitrogen to helium concentrations in the outflow. This ratio is determined by a thermal conductivity detector. The method is accurate and gives the pore size distribution of the sample. By expressing the data in volume (i.e. cm^3) of adsorbed nitrogen per unit mass of the sample, the same numerical values are obtained as for mass (i.e. g) of adsorbed water per unit mass of the sample, which are of more practical

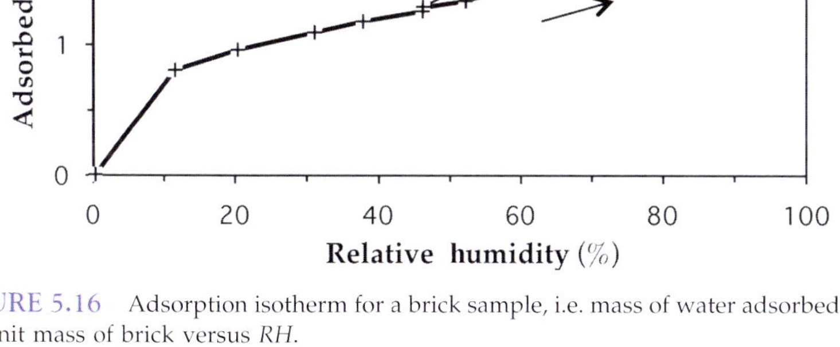

FIGURE 5.16 Adsorption isotherm for a brick sample, i.e. mass of water adsorbed (*WA*) per unit mass of brick versus *RH*.

use. In the figure, the hysteresis loop occurs in the *RH* range from 50% to 85% and the pores lie in the size range 18×10^{-4} μm < r < 75×10^{-4} μm. For *RH* above this hysteresis cycle, the isotherm continues to rise up to the last experimental value *RH* = 90%. This is due to the reversible filling of macropores with 'open' structure.

The surface of metals is covered with layers of adsorbed water molecules whose number is a function of the *RH* and the type of metal or alloy (see Chapter 2B). For all metals, the first monolayer starts to form at very low *RH* but it becomes completed only when *RH* = 30%; for higher *RH*, the specific metals differentiate, e.g. the second layer is completed at 40% for Co and Au and at 55% for Ni and Fe. At increasing *RH*, the deposition of water accelerates so that in equilibrium with moist atmosphere 10 or more monomolecular layers are deposited. The presence of heterogeneities (both structural and due to deposition of atmospheric pollutants) transforms the wet surface of the metal in an electrolyte for an electrochemical corrosion cell (Graedel, 1994).

From the previous section, it is evident that, when r_p approaches r_o, the area of the hysteresis loop also tends to vanish. When the pore population consists of aggregates of open pores, i.e. major spherical or irregular macrocavities including other smaller ones that have the appearance of hemispheres or microvalleys, the hysteresis loop disappears and the desorption curve coincides with the adsorption one.

In the case of materials characterized by *ink-bottle* pores, only the small cylindrical ducts (i.e. the *necks*) are filled with water and contribute to the hysteresis, as the condensation in a cylinder occurs with a cylindrical meniscus, and the evaporation occurs with a hemispherical one. Given the small amount of water allowable in the necks of the *ink bottles*, the hysteresis loop is modest.

Some very porous substances are used to buffer *RH*. This capacity derives from the fact that the moisture content of the substance varies greatly with the *RH* with which it is in equilibrium. *Silica gel* is widely used as it responds rapidly by absorbing or desorbing moisture when the ambient *RH* rises or falls. There are several types of silica gel with different buffering capacities. The greater the change in equilibrium moisture for unit change in *RH*, the greater the buffering capacity. Silica gel with regular density is more often indicated for the intermediate and low *RH*; silica gel with intermediate density is more appropriate for the high *RH*, and above 80% the buffering capacity becomes very effective. The silica gel can be conditioned to a specific level of *RH* by keeping it for a long time at the desired value of *RH* (e.g. in a closed environment conditioned by a saturated salt solution) until all the gel has absorbed the amount of moisture that is in equilibrium with the ambient *RH*. After this treatment, it can be used as a temporary buffer, as it will tend to absorb moisture if the *RH* will increase, or release moisture if the *RH* will fall. However, if it is put in a not previously conditioned case with objects in equilibrium with other values of *RH*, the moisture content of the silica gel will vary and the enforced buffering value will shift accordingly (Waller, 1992). Other porous or fibrous materials, e.g. wood, cork, wool and paper, have similar buffering properties. A most appropriate way to keep a constant level of *RH* in small closed volumes is the use of saturated solutions, as discussed in Chapters 2B and 12.

It might be useful to outline that when the adsorption is controlled by the Kelvin effect, loops in the adsorption isotherms are generated by the geometrical shape of pores and their communication outlets but that when another hydrophilic nature of the material is dominant, they may disappear. For instance, for the reasons previously discussed, paper is characterized by an equilibrium moisture content without loops, which increases with *RH*, not exactly linearly, but with a higher rate at the lower and especially the higher humidity levels, similar to a reversed 'S'. Of course, both the extremes of the natural humidity range (i.e. very elevated dryness and dampness) must be avoided in order to prevent the material from irreversible structural changes.

5.7. FREEZING–THAWING CYCLES

When the temperature drops below zero, freezing–thawing cycles develop and the pressure exerted by ice crystals in the pores of materials may have disruptive effects. At first sight, one would expect that the greater the size of the pore, the greater the force, so that stones characterized by large pores and high total porosity will be more exposed to risk. However, things are different and most European sandstones are quite immune to frost damage despite their low mechanical strength and coarse pore structure. The maximum pressure expected in sandstone has been evaluated to be some 10 times smaller than that in a poor limestone. Poor stones nonresistant to frost are characterized by a pore size distribution mainly peaking in the range 0.1–0.5 μm, which is rare in good stones (Everett, 1961). Similarly, Torraca (1981) noted that the damage by frost is more likely to occur in materials in which small pores are dominant, i.e. pore size between 0.1 and 1 μm, although the upper limit is not well defined as freezing damage has been found even in other stones with larger porosity. The problem of the damage caused by frost to porous materials is rather complex and depends on several factors: pore size distribution, geometrical combination of pores and capillaries and Kelvin effect for water and ice.

In a nonhumid climate, small pores have the highest probability of being filled with water at intermediate *RH* values, due to the combined action of Kelvin law and adsorption hysteresis, but, on the other hand, in pores with radius $r << 0.1$ μm, the Kelvin law causes a lowering of the freezing point for a curved meniscus. It is well known that freezing does not always occur at $T = 0$ °C; the smaller the radius of the meniscus, the lower the freezing point. In nature, supercooled droplets can be found in clouds as far as $\Delta T_f = 40$ K i.e. −40 °C; also, supercooled water can be found in pores and the theory of supercooling can be applied to determine the pore size of porous bodies by measuring the latent heat released at different temperatures (Fagerlund, 1973). When a porous material freezes, macroscopic ice crystals form in the larger pores and water is withdrawn from capillaries and finer pores where supercooled water remains longer. Thus, in principle, large crystals are expected to form in the larger cavities, whereas capillaries and smaller pores behave as a reservoir of supercooled water that may be displaced (either in the liquid phase or due to the lower equilibrium tension of ice) and feed the ice crystals that have already formed.

It is possible to calculate the change of the freezing point of water filling micropores with curved meniscus in the case of very simple pore geometry and the sphere is the simplest one. In addition, there is some evidence that the liquid−solid interface in a capillary is spherical, at least when the capillary is not too small. The Laplace equation holds not only for liquids in equilibrium with their vapour but also for solids in equilibrium with their liquid phase, i.e.

$$\Pi_{s1} = -\frac{2\sigma_{sl}}{r}\cos\theta \tag{5.30}$$

where the labels s and l refer to these two phases and $\sigma_{sl} = 17.2$ erg cm^{-2}. By proceeding as before with the derivation of the Kelvin equation, it is possible to compute how much the freezing temperature will depart from the usual level when we deal with particular ice shapes, as ice cristals of which snowflakes are composed etc. The most immediate similarity with the water droplets in Eqn (5.28) is the case of the equilibrium over (portions of) microspherules of ice, having curvature radius r. Similarly, the equilibrium with a (cylindric) needle is obtained dividing the result of a spherule by 2. It is known that a finely ground solid melts at a lower temperature and this physical law is largely used in foundries. Several equations exist with small empirical corrections (Fagerlund, 1973; Clifford, 1981; Iribarne and Godson, 1986); the chief equation for the Kelvin freezing point depression ΔT_f is

$$\Delta T_f = T_f\frac{2\sigma_{sl}M}{r\rho_s\Delta H} = T_f\frac{2\sigma_{sl}V_s}{r\,\Delta H} = T_f\frac{2\sigma_{sl}}{r\rho_s L_f} \tag{5.31}$$

where M is the molar mass of the substance, ρ_s is the solid density, ΔH is the molar heat of fusion and $L_f = \Delta H/M$ is the latent heat of fusion. For pure water $\Delta H = 18\,(80 - 0.5\,\Delta T_f)$ cal mol^{-1} where $0.5\,\Delta T_f = \int (c_w - c_i)\,dT$ where c_w and c_i are the specific heats of water and ice, respectively, at $p = \text{const.}$

Laboratory experiments on freezing in micropores show that the observed data lie between two curves calculated with the above equation, one computed with the surface tension σ_{sl} characteristic of the solid–liquid interface and the other with σ_{sv} i.e. the solid–vapour interface (Fig. 5.17). A possible interpretation for this is that, at usual meteorological temperatures, ice is always covered with a film of water, so that in reality, the solid–gas interface is the combined effect of two interfaces, i.e. solid–liquid plus liquid–gas interfaces, and the thickness of the liquid film between the two interfaces may change the observed values of the bulk surface tension.

It can be observed that in the above equation the sign of ΔT_f is linked to the sign of r and one might expect to find symmetrically warm ice at +40 °C in the same way as supercooled water is found at −40 °C. However, ice crystals are convex and the water in pores is characterized by a concave meniscus at the air–water interface, but the material–water interface is convex and is dominant in the case of icing, acting as a freezing nucleus; e.g. the method of determining the pore size of materials

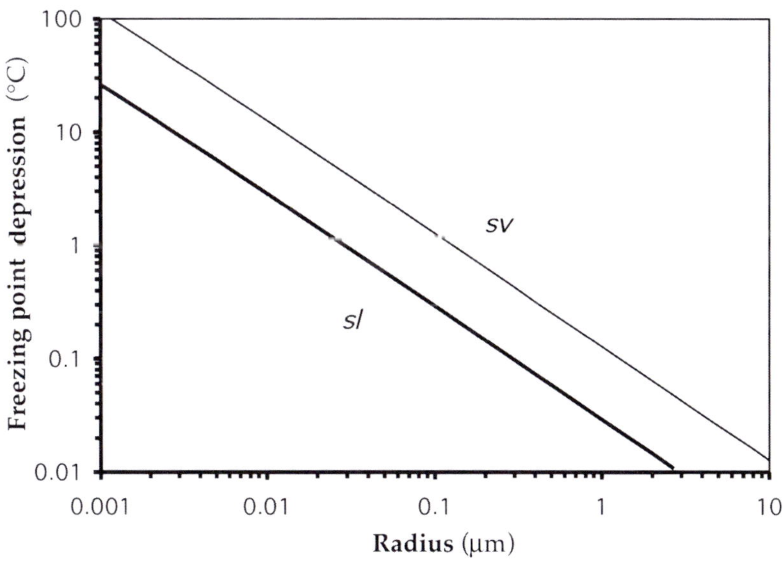

FIGURE 5.17 Freezing point depression (ΔT_f) calculated according the Kelvin equation for a spherical meniscus at the water–ice equilibrium (label *sl*, i.e. plot computed with the surface tension σ_{sl} characteristic of the solid–liquid interface), and the ice–vapour equilibrium (label *sv*, i.e. with σ_{sv} for the solid–vapour interface).

is based on supercooling. Probably, liquid microfoam might be similar to a structure with a dominance of negative radii of curvature to which these results might be applied; however, this phenomenon is theoretically possible but has never been observed.

Returning to the problem of the frost damage to monuments, Everett (1961) noted that the frost damage is associated not only with the existence of small pores in the already mentioned range but also with the simultaneous occurrence of pores of a critical size with larger pores. In his detailed study on the thermodynamics of frost damage, he found that the geometrical combination of cavities with different size plays a fundamental role and that during freezing, the excess pressure P_E that is exerted in a large pore of radius r_{lp}, connected to a supply of supercooled water at the reference pressure by a capillary or fine pores of radius r_{fp}, is proportional to the difference of the inverse of the pore radii, i.e.

$$P_{\mathrm{E}} = \beta \left(\frac{1}{r_{fp}} - \frac{1}{r_{lp}} \right) \tag{5.32}$$

where β is a proportionality coefficient. This difference may be small or large and is much more important than the individual value of each of the two terms. The contrast of large pores interconnected by fine pores gives the most dramatic results. Mechanical failure of the stone occurs if the excess of pressure determined by the above difference exceeds the mechanical strength of the porous material.

In addition to the Kelvin effect, the freezing point depression due to the solution of hygroscopic salts must be also considered. Internal evaporation occurs first in the largest cavities and other water with dissolved salts is withdrawn from the finest adjacent pores. Considerable amounts of salts are transported and precipitated in the coarsest pores and salt crystallization takes place preferably in large pores because of the energetically favourable lower chemical potential (Ginell, 1994). The crystals are more abundant in the lower part and at the bottom of giant pores, where the solution concentrates during evaporation, becoming saturated and precipitating salt crystals. The larger the pore, the greater is the possible amount of salt at the bottom. When the ambient RH increases, the hygroscopic salts hydrate, become deliquescent and a supersaturated solution forms and grows, until all the salt is dissolved; then the solution starts to dilute. Salts can be of atmospheric origin, can be leached from the soil, or due to constituents of the material, can be mobilized and transported by meteoric or ground water. In coastal regions, NaCl is a very important contaminant, which encourages condensation and lowers the freezing point. The same occurs in cold regions where deicing salts are used.

These salts lower the freezing point and prevent the mechanical stress due to the formation of ice crystals. The freezing point depression due to the presence of hygroscopic salts can be calculated with the simple formula valid for dilute solutions (for a derivation, see Adamson, 1986):

$$\Delta T_f = K_f\, m \tag{5.33}$$

where K_f is the freezing point depression constant and for water $K_f = 1.86$ K mol^{-1} and m is the molality of the solute. The formula shows that the value of ΔT_f depends only on the properties of the solvent and the molality of the solute. By the way, the same formula holds for the boiling point elevation, but with the coefficient K_b instead of K_f. For water, $K_b = 0.514$.

The freezing point depression ΔT_f is relevant for large contaminated pores, as shown in Fig. 5.18 where ΔT_f is plotted versus the percentage of anhydrous NaCl weight in water (g solute/100 g solution), which reaches −20 °C at saturation, i.e. 23% solution (Weast, 1985). This situation makes freezing in porous materials at Venice, and other coastal sites, a rare event. Apparently, this factor seems an advantage; as opposed, the long permanence of a film of liquid water on surfaces has several other negative aspects (e.g. faster chemical kinetics, increased pollutant deposition, dissolved salt migration and biological life). In addition, the contamination with soluble salts may lead to other important deterioration mechanisms, as we will see later.

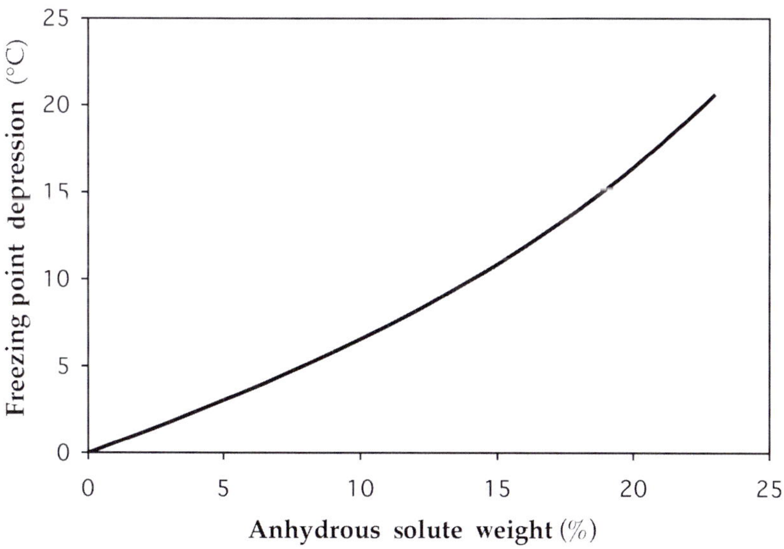

FIGURE 5.18 Freezing point depression (ΔT_f) due to the presence of dissolved NaCl at different concentrations (anhydrous solute weight percentage, i.e. g solute/100 g solution), up to saturation at 23%.

References

Theory and general applications

Adamson, A.W., 1986. A Textbook of Physical Chemistry. Academic Press, San Diego, 972 pp.

Bathija, A.P., Liang, H., Lu, N., Prasad, M., Batzle, M.L., 2009. Stressed swelling clay. Geophysics 70 (4), A47–A52.

Brunauer, S., 1945. The Adsorption of Gases and Vapors. Princeton University Press, Princeton, 511 pp.

Byers, H.R., 1959. General Meteorology. McGraw-Hill, New York, 461 pp.

Byers, H.R., 1965. Elements of Cloud Physics. University of Chicago Press, Chicago, 191 pp.

Camuffo, D., 1984. Condensation-evaporation cycles in pore and capillary systems according to the Kelvin model. Water, Air and Soil Pollution 21, 151–159.

Clifford, J., 1981. Properties of water in capillary and thin films. In: Franks, F. (Ed.), Water, a Comprehensive Treatise, Water in Disperse Systems, vol. 5. Plenum, New York, 366 pp.

Everett, D.H., 1961. The thermodynamics of frost damage to porous solids. Transactions of the Faraday Society 57, 1541–1551.

Fagerlund, G., 1973. Determinations of pore-size distribution from freezing point depression. Materiaux et Constructions 6, 215–225.

Gregg, S.J., Sing, K.S.W., 1967. Adsorption, Surface Area and Porosity. Academic Press, London, 371 pp.

Iribarne, J.V., Godson, W.L., 1986. Atmospheric Thermodynamics. Reidel, Dordrecht, 259 pp.

Kikoin, A., Kikoin, I., 1978. Molecular Physics. Mir, Moscow, 480 pp.

Madonna, L.A., Sciulli, C.M., Canjar, L.N., Pound, G.M., 1961. Low temperature cloud chamner studies on water vapour. Proceedings of the Physical Society 78, 1218–1222.

Mason, B.J., 1951. Spontaneous condensation of water vapour in expansion chamber experiments. Proceedings of the Physical Society B64, 733–779.

Mason, B.J., 1971. The Physics of Clouds. Clarendon Press, Oxford, 671 pp.

Matveev, L.T., 1984. Cloud Dynamics. Reidel, Dordrecht, 340 pp.

Matveev, A.N., 1985. Molecular Physics. Mir, Moscow, 448 pp.

Mikhail, R.S., Robens, E., 1983. Microstructure and Thermal Analysis of Solid Surfaces. Wiley, New York, 496 pp.

Pruppacher, H.R., Klett, J.D., 1980 and 2003. Microphysics of Clouds and Precipitation, first and second ed. Reidel, Dordrecht.

Sedunov, Yu.S., 1974. Physics of Drop Formation in the Atmosphere. Wiley, New York, 234 pp.

Sivuchin, D.V., 1986. Corso di Fisica Generale, vol. 2. Mir, Moscow, 583 pp.

Thomson, W., (Lord Kelvin), 1870. On the equilibrium of vapour at a curved surface of liquid. Proceedings of the Royal Society of Edinburgh 7, 63–69.

Weast, R.C., 1985. CRC Handbook of Chemistry and Physics 1985–86, sixty sixth ed. CRC Press, Boca Raton, Florida, pp. D213–D214.

Wright, H.L., 1936. The size of atmospheric nuclei: some deductions from measurements of the number of charged and uncharged nuclei at kew observatory. Proceedings of the Physical Society 48, 675–699.

Young, K.C., 1993. Microphysical Processes in Clouds. Oxford University Press, New York, 427 pp.

Applications to conservation

Biscontin, G., Driussi, G., Maravelaki, P., Zendri, E., 1993. Physico-chemical investigations of stone architectonic surfaces in Venice: the Scuola Grande dei Carmini. In: Biscontin, G., Graziano, L. (Eds.), Conservation of Architectural Surfaces: Stones and Wall Covering. Il Cardo, Venice, pp. 125–136.

Camuffo, D., 1988. Surface moisture and conservation. European Cultural Heritage Newsletter on Research 2 (5), 6–10.

Charola, A.E., Puhringer, J., Steiger, M., 2007. Gypsum: a review of its role in the deterioration of building materials. Environ. Geol. 52, 339–352.

De Quervain, F., 1967. Technische Gesteinskunde. 2 Aufl. Min. Geotechn. Reihe, Bd 1. Birkhäuser Verlag, Basel.

DIN 66131, 1973. Bestimmung der spezifischen Oberfläche von Festoffen durch Gasadsorption nach Brunauer, Emmett und Teller (BET), Grundlagen.

Fitzner, B., 1994. Porosity properties and weathering behaviour of natural stones, pp. 43–54. In: Zezza, F. (Ed.), Stone Material in Monuments: Diagnosis and Conservation, 2nd International Corurse on Monument Conservation, Adda, Bari, 222 pp.

Ginell, W.S., 1994. The nature of changes caused by physical factors, pp. 81–94. In: Krumbein, W.E., Brimblecombe, P., Cosgrove, D.E., Stainforth, S. (Eds.), Durability and Change. Wiley, New York, 307 pp.

Graedel, T.E., 1994. Mechanisms of chemical change in metals exposed to the atmosphere, pp. 95–105. In: Krumbein, W.E., Brimblecombe, P., Cosgrove, D.E., Stainforth, S. (Eds.), Durability and Change. Wiley, New York, 307 pp.

Klopfer, H., 1985. Feuchte. In: Lutz, et al. (Eds.), Lehrbuch der Bauphysik. Teubner, Stuttgart, pp. 265–434, pp. 628–635.

Jeannette, D., 1997. Structures de porosité, mécanismes de transfert des solution et principales altérations des monuments. In: Lefèvre, R.A. (Ed.), La Pietra Dei Monumenti in Ambiente Fisico e Culturale. European University Centre for Cultural Heritage, Ravello, pp. 49–77.

Laurenzi Tabasso, M., Marabelli, M., 1992. Il degrado dei monumenti a Roma in rapporto all'inquinamento atmosferico. Betagamma, Viterbo, 169 pp.

Steiger, M., Charola, A.E., Sterflinger, K., 2011. Weathering and deterioration, Chapter 4. In: Siegesmund, S., Snethlage, R. (Eds.), Stone in Architecture. Properties, Durability Springer Verlag, Berlin.

Torraca, G., 1981. Porous Building Materials. Iccrom, Rome, 141 pp.

Torraca, G., 1994. Physical condition. A Primary factor in the durability of stone, pp. 52–57. In: Zezza, F. (Ed.), Stone Material in Monuments: Diagnosis and Conservation, 222 pp. 2nd International Corurse on Monument Conservation, Adda, Bari.

Waller, R., 1992. Temperature- and humidity-sensitive mineralogical and petrological specimens. In: Howie, F.M. (Ed.), The Care and Conservation of Geological Material: Minerals, Rocks, Meteorites and Lunar Finds. Butterworth Heinemann, Oxford, pp. 25–50.

Warscheid, T., Becker, T.W., Braams, J., Gehrmann, C., Krumbein, W.E., Petersen, K., 1993. Studies on the temporal development of microbial infection of different types of sedimentary rocks and its effects on the alteration of the physico-chemical properties in building materials. In: Thiel, M.J. (Ed.), Proceedings of the International RILEM/UNESCO Congress "Conservation of Stone and Other Materials", Causes of Disorders and Diagnosis, vol. 1. E&FN Spon, London, pp. 303–310.

Wendler, F., 1997. New materials and approaches for conservation, pp. 181–196. In: Baer, N.S., Snethlage, R. (Eds.), Saving Our Architectural Heritage: The Conservation of Historic Stone Structures. Wiley, Chichester.

Winkler, E.M., 1986. A durability index for stone. Bull Assoc Eng Geol 23 (3), 344–347.

CHAPTER

6

Atmospheric Water and Stone Weathering

OUTLINE

6.1. ACID RAIN, RAINFALL AND CRUSTS

Atmospheric water causes stone to decay in several ways, which may be purely physical in nature or a combination of physical and chemical processes (Camuffo, 1994, 1995). Chemical weathering is not directly related to the concentration of pollutants in the air but to the synergistic action of meteoric water and pollutants that have been deposited on the

http://dx.doi.org/10.1016/B978-0-444-63296-8.00007-X

monument surface either in the dry or in the wet phase. Wet deposition is the result of the atmospheric scavenging occurring in clouds, when the droplets are forming and growing (*rainout*) and when they are freely falling (*washout*). Falling raindrops absorb hydrophilic gases and collect particles from the air by means of complex deposition mechanisms that will be described in Chapter 8. As in dry deposition, the capture rate is less efficient for particle sizes ranging from 0.05 to 2 μm and in this interval, the residence time in the atmosphere may reach 1 day.

The amount of pollutants that deposit in the dry phase is the result of the concentration of airborne pollutants, the velocities of the different microphysical transport mechanisms close to and at the interface between the air and the surface and the capture efficiency of the surface itself. It denotes the net fraction of the atmospheric pollutants that are ultimately deposited on the monument and that may react with it when environmental conditions will allow it. Measurements of the pollutant concentration in air, but not of the deposition velocity, are not sufficient.

Rainfall transports acidic substances from the atmosphere to the surface of monuments. Even the purely meteorological nature of the precipitation is relevant. In the Mediterranean climate, rain is not frequent in summer and a great quantity of pollutants is deposited in the dry phase. The same occurs in winter in the European regions that are under the influence of the high-pressure ridge that connects the Russian high with the Azores anticyclone. Under such conditions, in urban sites, dry deposition is much greater than wet deposition. If precipitation occurs in the form of heavy showers, the dry deposit is quickly washed away. Thus, the knowledge of the instantaneous precipitation intensity (i.e. amount per hour) is more useful than the monthly totals. The frequency of rainfall may be equally useful in giving the first indication about the possible dominance of wet or dry deposition.

In the 1970s and 1980s, rainwater had reached alarming levels of acidity. This phenomenon has been considered one of the most important challenges of the modern society, affected by industrial, traffic and domestic pollution. In the past century, several complaints were found in industrialized England where rainfall was severely contaminated and smog obscured the day (Brimblecombe, 1987, 1992, 1995).

Human activity is not the only factor responsible for the atmospheric acidification. A number of interesting documents were found that describe acid rain and its effects since 1670, i.e. before the industrialized era (Camuffo, 1992), as well as caustic, foul smelling 'dry' fogs, which were reddish and persistent, able to obscure the sun from one to several days and caused damage to vegetation and illness to animals (Camuffo and Enzi, 1995). These 'dry' (i.e. nonwetting) fogs were so frequent during the Little Ice Age (i.e. from 1430 to 1850) and particularly in the past century that the treatises of meteorology and agriculture distinguished between two kinds of fogs: the 'wet' one that is favourable for agriculture and the 'dry' one that is dangerous for it. The cause of this natural acidification was the activity of the Mediterranean volcanoes and more rarely Icelandic volcanoes, e.g. Laki in 1783. The 'dry' fogs were clouds of volcanic aerosols that were trapped in the planetary boundary layer in the warm season, when the atmosphere was particularly stable over the relatively cold sea waters and the Azores anticyclone favoured calm or light winds. Explosive volcanic emissions may reach the stratosphere and be trapped aloft for 2 or 3 years, possibly affecting Earth's climate. Natural acidification of rainfall or fog is typically caused by the less violent and continual emissions that remain trapped in the lower troposphere for days or weeks.

In southern Europe, rainwater acidity is often buffered by the presence of alkaline particles, i.e. dust from the Sahara desert, that the Sirocco wind picks up and blows from North Africa to Europe (Prodi and Fea, 1979; Camuffo, 1990). The frequency of occurrence of this kind of precipitation varies with the geographic location, reaching 50% of the cases in certain regions of Italy. A large part of southern Europe is composed of liming soil, mainly of calcareous nature. However, the soil of volcanic origin, or in some cases nonliming soil, is also naturally protected by another phenomenon. When precipitation with Sahara dust occurs, the droplets deposit on the ground a myriad of coarse particles that have not completely dissolved, thus leading to natural liming. When acid rainfall occurs, it will find the buffering deposit on the soil of the catchment basins and the water that reaches rivers and lakes has lost part of its original acidity. For this reason, the effect of acidification on the biological life of the water bodies over the Mediterranean area is less dramatic in comparison with that in northern Europe.

The situation is different for monuments, where the Sahara particles are easily removed by wind and rain, so that the surface is exposed to acid rain. However, another important difference exists with northern Europe. In northern Europe, the climate is rainy and most of the airborne pollutants are scavenged by rain and become acid rain. In southern Europe, rainfall is less frequent and deposition occurs mainly in the dry phase. The key factor is the dry deposit of particulate matter accumulated on monuments; on the other hand, rainwater makes possible particulate dissolution and chemical reaction.

Although the acidity of the precipitated rainwater as well as the turbulent character of the runoff attacks the stone, rainfall is not as damaging as drizzle that wets the surface making the dry deposit chemically reactive, without removing it. Under these circumstances, the pH of the rain may be of little importance in comparison with the acidity of the dry deposit, which generally contains elevated quantities of both acidic pollutants and catalysts, such as soot and oil-fired carbonaceous particles or carbonaceous fly ash due to the coal combustion (Amoroso and Fassina, 1983; Del Monte and Sabbioni, 1984; Del Monte et al., 1984; Camuffo et al., 1984; Camuffo, 1986; Chabas and Lefèvre, 2000), which have been dominant in several European countries in the past and now are mainly limited to some industrial areas.

Acid water acts by dissolving both the stone surface and the binding between granules. In the latter case, modest dissolution is sufficient to cause the loss of many granules, resulting in erosion. This rate of erosion may exceed the rate of dissolution by a factor of almost two for marble and three for limestone (Beadecker and Reddy, 1993).

Many sulphation and dissolution processes occur especially in stones that have a calcareous ($CaCO_3$) matrix; metals are corroded and external frescoes and murals are lost. However, even in those cases where the particles deposited on frescoes or murals are not subjected to direct wetting by raindrops or runoff, an opaque layer is still formed, resulting in soiling and darkening. This in turn means that the picture becomes less clear, so that each time a work of art is cleaned it is subject to a stress brought about by the very nature of the intervention.

The formation of crusts is related to the presence of water, but the necessary amount is generally supplied by rainwater, which is larger than that locally adsorbed by means of fog or condensation. These processes involve a very small amount of water, e.g. a thin film or droplets corresponding to a 0.1–0.2 mm thickness per night with clear sky (Camuffo and Giorio, 2003) i.e. much less compared with rainfall, which can reach some centimetres per day. However, indirect action is possible: condensation on nonabsorbing external surfaces may continue for some hours and form droplets that eventually run down in rivulets. The water collected at the base of the vertical surface is supplied by a relatively large area and may then be absorbed by capillary suction, which is much more effective (Bernardi et al., 1985).

The problem is manifold and removing the apparently 'main' cause does not necessarily imply that the negative consequences have been stopped. In the past years, there was unanimous consensus that SO_2 was mainly responsible for sulphation and stone degradation, so that much effort was put into reducing emissions. One of the most frequent remedies applied was to abandon oil and coal and pass to methane. The experience of northern Italy was that the concentration of SO_2 decreased and NO_x and O_3 increased after this change; no important benefits to monuments have been observed and it is too early to draw an exact balance of pros and cons. The oxidation potential has increased, as well as the photochemical smog. Nitrogen oxides are deliquescent at ordinary humidity levels and, consequently, they increase the time of wetness (TOW). Nitrogen oxides are generated by combustion, agricultural fertilizers, plants and microorganisms, i.e. in industrial, urban and country areas. As a consequence, the film of liquid water that develops on the contaminated surfaces favours the chemical reactions that increase in frequency of occurrence and duration. This example shows the complexity of the problem of the monument decay associated with atmospheric pollution and that any change in the emission strategy modifies the terms of the problem and may lead to unexpected results.

The visual features of the crusts of marble or compact limestone monuments are not only determined by the existing pollution deposited according to the prevailing micrometeorological conditions near the monument but also determined by the geometry of the surface and its exposure to rainwater. Similar monument shapes induce similar patterns of crusts. The pattern of the *white* and *black crusts* studied on monuments in Italy, and in many other European countries, is caused by the way meteoric water wets or washes the surface: where runoff dominates, dissolution and erosion occur and calcareous rocks form white areas; these areas are characterized by crystals of precipitated calcite and dolomite (Del Monte and Sabbioni, 1980) that form when the rainwater that contains the dissolved stone evaporates (Camuffo, 1994, Fig. 6.1(a), (b)); where the stone is wetted but not washed out, black gypsum crusts originate (Fig. 6.2(a), (b), (c), (d)), with embedded carbonaceous particles that seem to play an important role in this process.

Field observations (Camuffo et al., 1982, 1987; Camuffo, 1984a,b, 1986; 1990) have shown that in the zones exposed to runoff, gypsum crystals were only rarely found for their high solubility; as opposed, a thin layer of spatic calcite and dolomite crystals (Fig. 6.1(b)) due to dissolution in acidic rainwater, surface percolation, evaporation and crystallization was common. On the other hand, the maximum quantity of gypsum was found where runoff was prevented, i.e. in the *black crusts*. The acid rain hypothesis alone is not sufficient to explain the main phenomenon due to sulphation, i.e. the *black crusts*, which are one of the main problems affecting marble monuments (Fig. 6.2). Sometimes black crusts constitute an aesthetic problem only; more often they are noxious because they include pollutants that, in the presence of water, will interact with the underlying stone layer. Gypsum crystals formed by sulphation, i.e. after pure reaction between acidic rainwater and limestone, are white and the gypsum crusts are black because they include many black carbon particles.

In the case of marble and compact limestone, the visual features of deterioration patterns are related to the following three main types of interaction with water:

1. Runoff is associated with *washing white areas* in the zones where the surface has been dissolved and thinned. The eroded surface might be covered with precipitated crystals (Fig. 6.1).

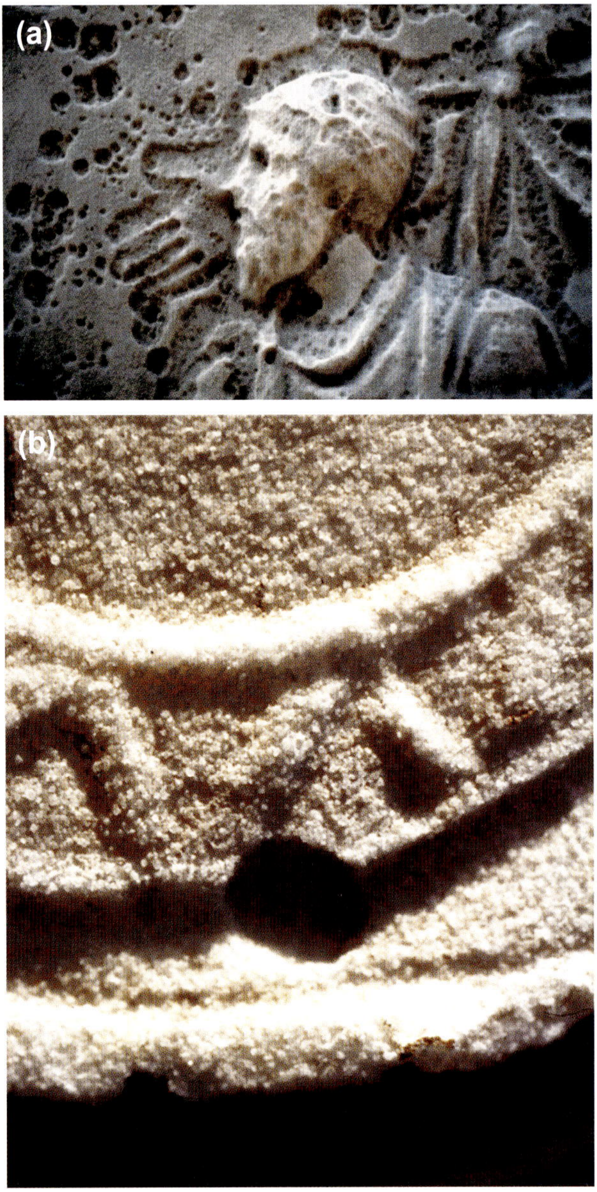

FIGURE 6.1 (a) *Washing white area* in a zone over which rainwater runs off. The surface has been thinned and is covered with crystals of spatic calcite. The pitting is due to bio-corrosion of microflora in the past (Trajan Column, Rome). *(Source: By kind permission of Butterworth Heinemann. See credits.)* (b) Calcite crystals originated from marble dissolution during rain and subsequent crystallization during rainwater evaporation on a shield in the Trajan Column, Rome. The hole is due to lichen colonization in the past. *(Source: By kind permission of Butterworth Heinemann. See credits.)*

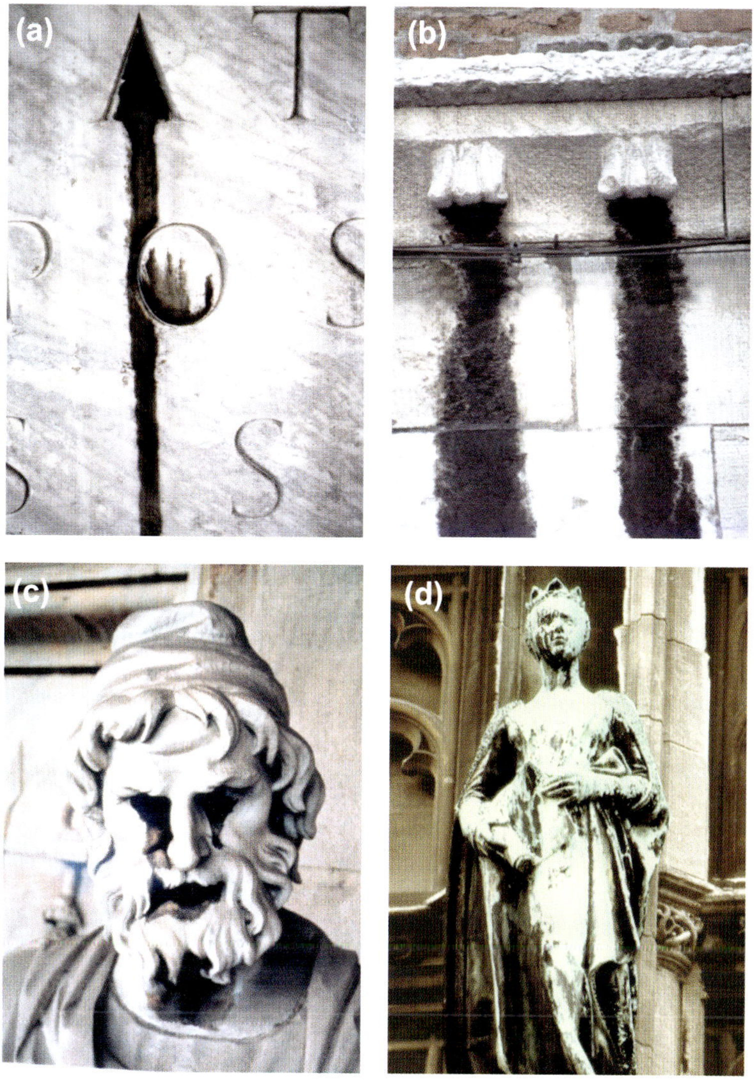

FIGURE 6.2 Examples of black crusts formed where rainwater wets (either splashing raindrops or percolation) but is unable to remove the dry deposit by means of runoff. (a) On marble, where the engraving of the letter A deflects the runoff (Aurelian Column, Rome; after Camuffo, 1984 a,b; by kind permission of Elsevier Science, see credits). (b) Below protruding decorations on limestone (*Palazzo dei Priori*, Padua). (c) In typical sheltered areas under the eyebrows and moustaches on marble (Constantine Arch, Rome). (d) Similar black deposits (mainly composed of gypsum and black soot) on a bronze statue (*Grand Place*, Brussels).

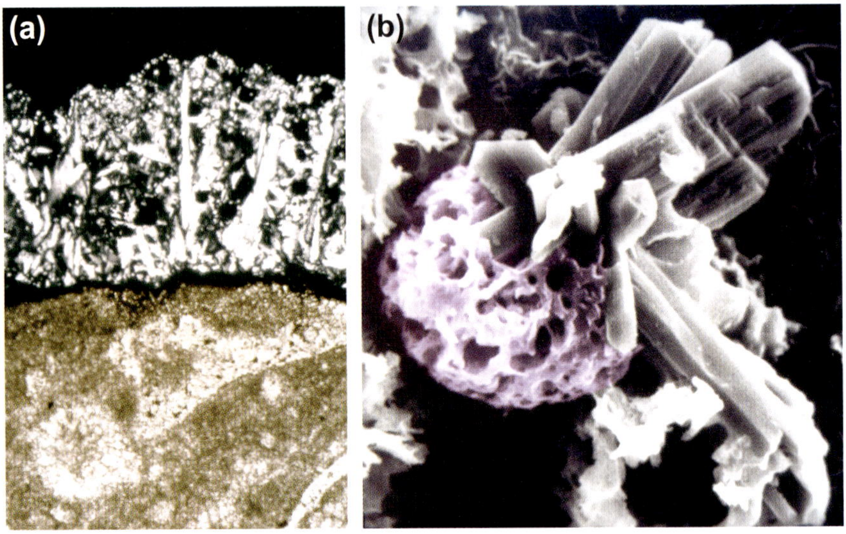

FIGURE 6.3 (a) Thin section of a black crust where white acicular gypsum crystals grow on the stone surface enclosing the black carbonaceous particles (black disks). (b) Gypsum crystals are growing on a carbonaceous particle (pink, false colour) after the particle has been wetted. *(Source: By courtesy of Prof. Marco Del Monte.)*

2. Dampness due to splashing droplets, or percolation near runoff, is associated with *black areas* and black crusts dominate. The crust is a buildup on the surface; a fraction of the constituting material originates from the rock and a part is airborne (Fig. 6.3(a), (b)).
3. Absence of both runoff and percolation, but possible condensation, is associated with *grey areas* where the stone is not chemically attacked but covered instead with a layer of dust, particles and debris. This causes a soft layer of airborne material deposited onto the surface (Fig. 6.4).

Although some specific mechanisms involved are still under discussion, fly ash, microsoot and single carbonaceous particles (formed during the combustion of oil) are active in forming *black crusts* when they are wetted by the percolation of rainwater. These combustion products contain sulphur compounds and catalysts and when wetted they will nucleate gypsum crystals and will remain embedded in a gypsum crust (Fig. 6.3(b)).

Gypsum partly results from the transformation of the underlying rock and partly from particle nucleation. Gypsum volume is 20% larger than that of calcite ($CaCO_3$) and this explains the mechanical damage that sulphation may cause when the reaction occurs inside pores or fractures. *Black crusts* have a *dendritic* appearance (i.e. they develop with a typical multibranching tree-like form) as a consequence of the mechanism whereby particles are deposited and bind together with gypsum crystals; the proportion between the gypsum and the particulate matter that has

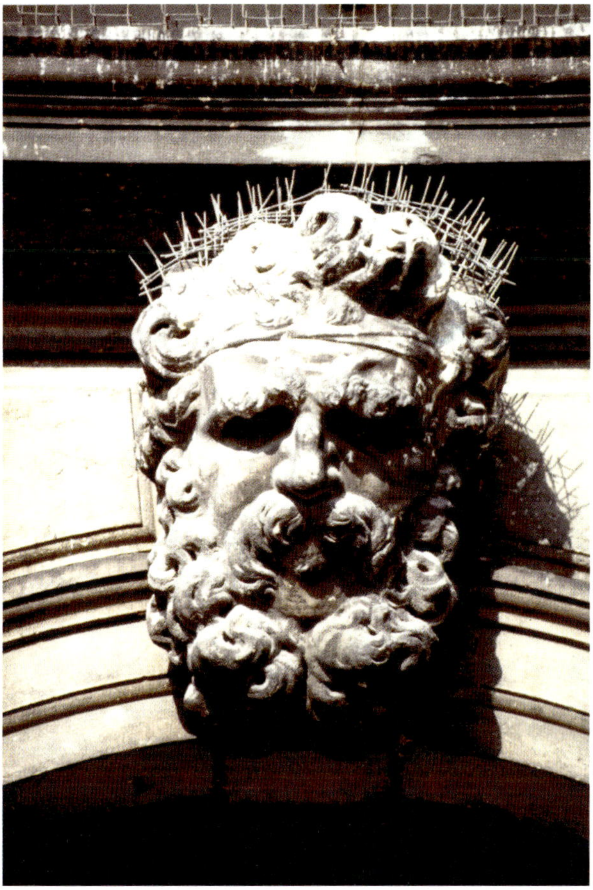

FIGURE 6.4 This face is an example of grey area under a protruding balcony that protects against rainfall. Grey areas are deposits of particles, dust and debris, accumulated on horizontal surfaces where there is neither runoff able to remove the deposit nor percolation able to trigger and feed chemical reactions between the deposit and the stone. The crown of thorns is an antibird device.

been deposited determines the mechanical strength of the crusts. Where the underlying substrate is not marble or limestone or the calcareous rock cannot be sulphated, gypsum is very scarce (i.e. only that originated by the particles) and the crust is very friable.

From the 1970s to the 1990s, when coal and oil were extensively burned, samples of crusts were taken over Europe and examined. All the *black crusts* contained carbonaceous particles and it was possible to reproduce this mechanism in the laboratory (Sabbioni et al., 1993). Of course, these dendritic crusts cannot be confused with the black patinas of biological origin (e.g. fungal melanin, lichens, algae, and cyanobacteria), with growth, combination and repetition of some elementary structures, possibly associated with biocorrosive pitting. In particular, the microflora also needs special microclimatic conditions to develop, chiefly a frequent supply of water. For this reason, black biological patinas can be found in places where the formation of crusts composed of soot and gypsum is impossible for frequent runoff (Fig. 6.5(a), (b)).

Sometimes, the situation is more complex due to an interaction between anthropogenic deposits and biogenic black materials, e.g. soot entrapped in microbial films. As melanin is made up of a mixture of polysaccharides, proteins, lipids and aromatic compounds, biological patinas can be identified by means of pyrolysis/methylation of crust samples (Saiz-Jimenez, 1995).

Green patinas typical of algal colonization may form on damp surfaces. In the windy and rainy climate of northern Europe, walls exposed to the dominant wind-borne raindrops are favoured. In the less rainy Mediterranean climate, algae can be found in the most humid microclimates (e.g. Venice) and prefer the horizontal surfaces in the shade, which are frequently reached by rainfall and where water may remain longer (Fig. 6.5(c)).

It should be noted, however, that the mechanisms of biodeterioration are not simply restricted to clearly evident patinas, pitting or other corrosion damages. The absorption of corrosive and nutritional atmospheric pollutants is even increased by the presence of sticky biofilms (Wittenburg, 1994). This way, the stone-colonizing microflora accelerates passively the reaction rate of biochemically induced corrosion processes and influences as a consequence the formation of crusts as their final result (Warscheid and Braams, 2000; Caneva et al., 2009). This phenomenon was documented by nearly unaffected stone surfaces below brass settings that act as a biocide and hinder the contamination of microbial biofilms (Becker et al., 1994). An interesting example of biocolonization of statues in the gardens of the National Palace of Queluz (Portugal) and comments about the conservation intervation has been published by Charola and Delgado Rodrigues (2012).

On stones exposed to low air pollution, biodeterioration might be more or less evident, e.g. for the presence of multicoloured photosynthetic algae, cyanobacteria or lichens. In highly polluted areas, it might be possible that microbial infections are not easy to detect, although bacterial or fungal microfilms are present.

Although the *washing white areas* are subject to dry deposition, the abundant runoff washes over the monument surface and in general removes most of the pollutants deposited on it. Falling rainwater will activate and continue their action. The result is that the surface is weathered, corroded, progressively leached and dissolved as a function of the total amount of rainwater that has flowed on it and the active acid contaminants.

When stagnant water evaporates, the calcite and dolomite dissolved in it will precipitate forming crystals.

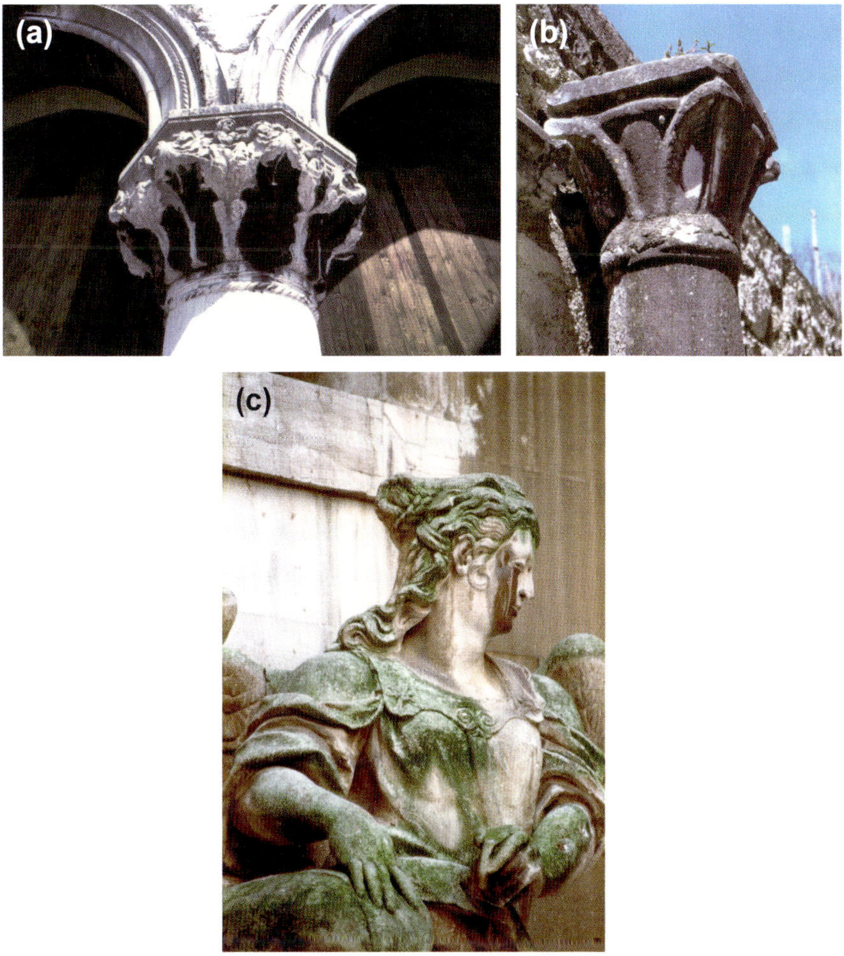

FIGURE 6.5 Comparison of crusts having different origin. (a) Black crusts originated by soot deposit, found in a capital: the parts where water runs are white, the parts wetted but not washed out are black. (Ducal Palace, Venice). (b) Black biological patches are found in places frequently wetted, where the formation of crusts composed of soot and gypsum is impossible due to the frequent presence of runoff, and white areas are found in dry, nonwashed areas where black deposits are expected in the case of heavy pollution. (Ravello, Italy). (c) Green patinas typical of algal colonization form on damp surfaces. At Venice, they prefer the horizontal surfaces on the shade, which remain damp longer (*Piazza S.Marco,* Venice).

Monuments are more or less severely damaged as a result of physico-chemical characteristics of the stone, type of pollutants and their deposition rate, local climate and geometric features of the monument. When one of these factors only slightly changes, the result too will change. For instance, in heavy polluted areas of Central Europe, the deposition rate is so fast that blackening is largely dominant over the washout, so that the white areas generated by washout practically disappear. In these regions, all the surfaces are covered with a thick black deposit and chemical reactions occur especially in the parts of the monument that are wetted more often. The outer layer of the limestone is transformed into a black gypsum crust with a thickness that may reach one or few centimetres. In the long run, the different physical and chemical characteristics of the transformed layer will cause its detachment and loss.

A further type of white area will appear: the internal clean surface where the fracture had occurred, which is generally close to the interface between the uncontaminated rock and the crust (Fig. 6.6). In zones where the atmospheric pollution has been drastically reduced, this *breaking white*

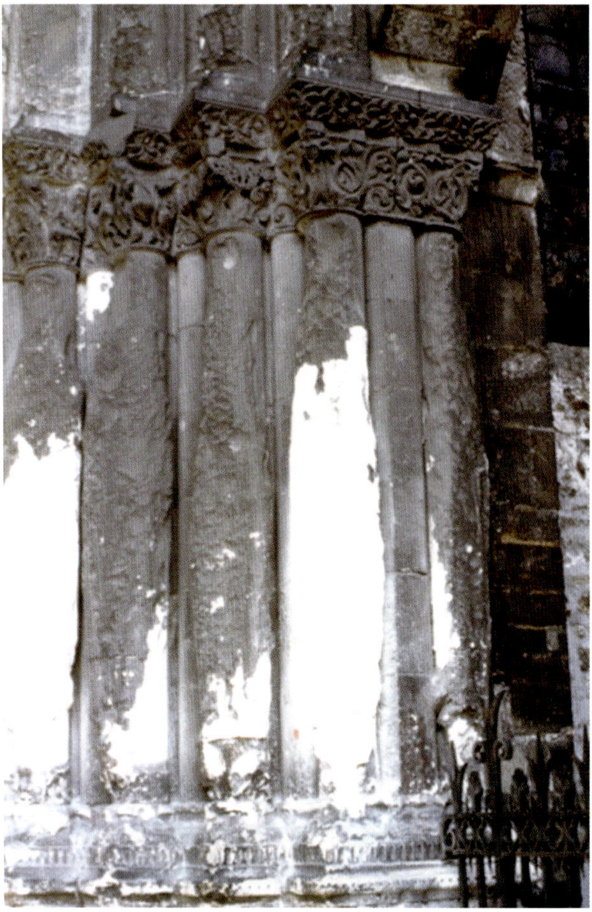

FIGURE 6.6 *Breaking white area* generated in a polluted environment: the white area is the internal surface where the fracture has occurred. Notre Dame en Vaux, Chalon sur Marne, France.

area remains clear for a long time; in zones where the pollution is still high, these white areas are blackened again in a short time.

When stones are characterized by large pores and are absorbent, the visual features are quite similar. Vertical surfaces absorb falling droplets and runoff cannot develop, so that the dry deposit is not removed, and may be transported inside pores. Washout and dissolution may only occur where great amounts of water are collected, e.g. under windows.

However, the porous stone eventually absorbs part of running water; the percolating water layer becomes thinner and thinner, the turbulence dampens, the flow becomes laminar and in some cases it might be extinguished too (Fig. 6.7). In turbulent regime, the surface is cleaned and eroded; in the laminar one, the material removed from the upper levels is released onto the underlying rock. When the regime is laminar, the first layer in contact with the surface becomes saturated and chemically

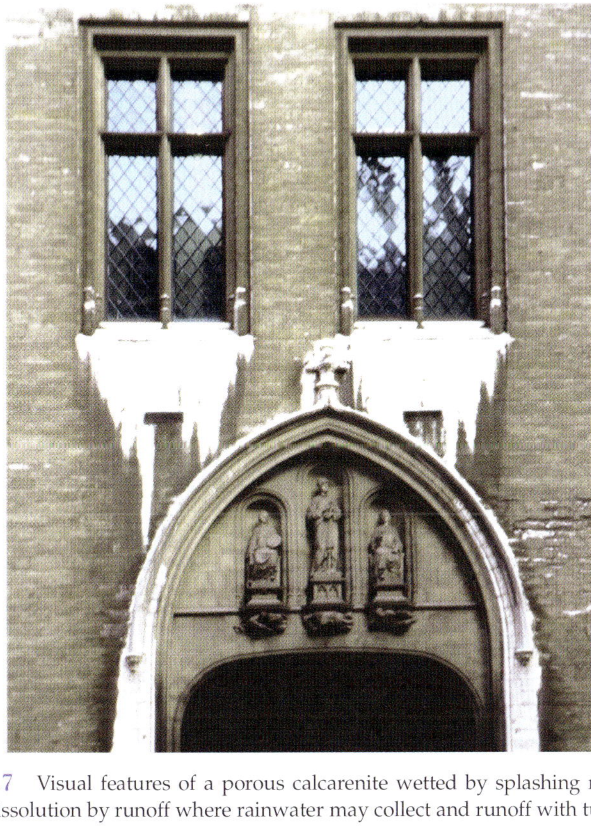

FIGURE 6.7 Visual features of a porous calcarenite wetted by splashing raindrops (the grey wall), dissolution by runoff where rainwater may collect and runoff with turbulence (the white area below the window) and in laminar regime after adsorption, releasing the material previously removed (the darker grey area below the dissolution area). The Mayor's Palace, Brussels.

inactive. When it is turbulent, the chemically active solution continually changes on the stone surface that is exposed to heavy dissolution and erosion due to loss of granules.

Generally speaking about limestone, sandstone, bronze or other materials, on vertical or sloping surfaces, beating raindrops and runoff cause material dissolution or corrosion and removal of leakage products, i.e. loss of mass. On horizontal surfaces where stagnant water remains and will disappear for evaporation, the leakage product may remain and precipitate, forming crusts on stones or patina on metals, i.e. gain of mass.

A short note about a popular mistake (not for chemists!) about averages of pH of rainfall: this parameter describes the hydrogen ion activity of rainwater and is defined as $-\log_{10}[H^+]$, where $[H^+]$ is the hydrogen ion concentration in moles per litre. Mixing two different volumes of different solutions with different pHs, the pH of the mixture is determined by the final H^+ concentration, which can be computed after the proportions of the initial concentrations and volumes. The same is to compute time averages of different pH values that have been found in successive samples of rainwater collected during one or more precipitation events. It is necessary to return to the original H^+ concentrations, average them (which is equivalent to mix together all the samples) and then compute the pH of the final mixture. Arithmetic averages of pHs do not make sense.

6.2. MECHANISMS OF PENETRATION OF RAINWATER AND EVAPORATION

While rainwater is flowing on the monument surface, part of it may penetrate inside, depending not only on the nature of the material (Dullen, 1979) but also on microclimatic factors. Penetration into capillaries and pores is possible only when the whole cavity surface (i.e. both the external and the internal one) is covered with a single or double layer of water molecules in the solid state and this process is not possible when the surface is completely dry, i.e. not covered with monolayers of solid water (Camuffo, 1991a).

When the surface of the material is still 'wet', i.e. covered with the monolayer of solid water, as generally occurs in the humid climates, liquid water penetrates (Fig. 6.8(a)) by simply progressing inwards with a concave meniscus without experiencing changes in curvature (no external work due to surface tension); in addition, the concave meniscus forms a lower vapour pressure in front of it, which favours penetration into the capillary (positive work).

In the case of a completely 'dry' surface, i.e. without the solid water monolayer, the meniscus of the water penetrating into the capillary is convex in the forward direction and must elongate gradually as the water

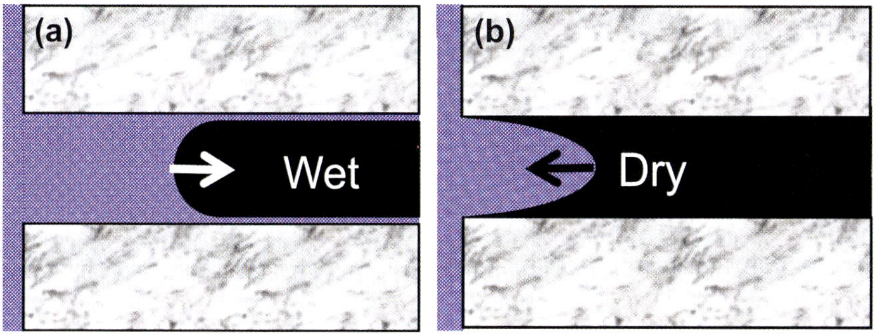

FIGURE 6.8 Penetration of water into a wet capillary when some water is percolating on the surface (blue layer on the left). (a) The capillary is initially filled with monolayers of water (i.e. wet), the water meniscus is concave and water can penetrate. (b) In a completely dry capillary, the meniscus is convex and water cannot penetrate. Arrows indicate the force exerted by the curved meniscus, i.e. onwards and backwards. *(Source: By kind permission of Maggioli Editore. See credits.)*

penetrates into the capillary (Fig. 6.8(b)). The work required against the surface tension and the higher vapour pressure over a convex meniscus is so great that water cannot penetrate under such conditions and runs over the monument without dampening it. For dry materials, a preliminary 'wetting' of the internal surface is a necessary condition for water to deeply penetrate and soak it.

In a warm, dry climate that is predominant during the summertime in southern Europe, the external surface of monuments and the surface of pores and capillaries in the subsurface layer are completely dried out because of the strong warming associated with the intense solar radiation. During a typical sudden shower in the afternoon, water cannot initially penetrate pores and capillaries, because their surface is not covered with monolayers of water in the solid state: lacking them, no liquid water can enter. A dry, or dehydrated monument needs some time to become wettable, i.e. to pass the surface of the capillary fringe from hydrophobic to hydrophilic, and then to be soaked.

Even microbial biofilms regulate the penetration of rainwater: first they reduce the capillary water uptake preventing rapid moistening of the uppermost material layer during showers, but later they conserve part of the water that has already penetrated into the material much longer than a noninfected material will do, especially when surface-active compounds are formed due to the biomineralization of hydrocarbons, frequently present on rock surfaces (Warscheid et al., 1991).

The penetration of water is favoured by pumping mechanisms and the wind is a very active generator of them. The first important mechanism is due to the interaction between the monument and the wind field. Airflows with speed u will develop local pressures p over the monument in accordance with the *Bernoulli* equation, which holds for an incompressible and *inviscid flow*, i.e. a flow of an ideal fluid that is assumed to have no viscosity (*inviscid fluid*):

$$\frac{1}{2}\rho\, u^2 + p = \text{const} \tag{6.1}$$

where ρ is the air density and the kinetic term $0.5\rho\, u^2$ has dimension of pressure and is called *dynamic pressure*. Each streamline is characterized by its own constant. The wind is unsteady in speed and direction and the monument causes continually local departures of the wind field and surface pressure, which are perceived as peaks and drops of a highly fluctuating variable. In addition to the Bernoulli effect at some distance from the monument, the departures are amplified in close proximity to the monument surface according to an aerodynamic coefficient, which varies point by point, and accounts for the interactions between monument shape and location, wind speed and angle of attack. Every change or fluctuation in the wind field is associated with a similar effect in pressure. The wind exerts over the surface a net force along the flow direction (*drag*) and across the flow (*lift*). In the frontal position, the drag drives the water into the monument and the lift displaces it along the surface. All these components depend on monument features and Reynolds number. The second mechanism is related to the formation of a boundary layer in proximity to the monument. The wind speed is affected by the presence of the monument and the aerodynamic disturbance forms a boundary layer where the air speed drops following a logarithmic profile with the distance from the surface until the air in contact with the surface adheres to it (no slip). This causes a retardation of the air motion and a pressure field that varies with the angle of attack. The third mechanism is intrinsic to the fluctuating character of the wind. Pressure fluctuations associated with wind gusts or lulls can amount to 2 or 3 hPa. They cause a fluctuating pressure difference between the interior and exterior of the monument, which may force the water flowing on the surface to penetrate the cavities and be trapped inside the pores. After many successive steps, water may deeply penetrate into the interior of the porous materials. External pollutants may also be trapped inside the material and continue the deterioration mechanisms each time microclimatic factors will reach convenient synergistic levels.

6.3. EVAPORATION FROM A DAMP MONUMENT

Drying is also controlled by several factors, i.e. atmospheric factors, e.g. specific humidity gradient and air turbulence near the surface, and intrinsic factors, e.g. pore and capillary size and distribution; another important factor is the type and concentration of dissolved salts. Gravity is also important for the downward transport of liquid water and determines a vertical gradient of dampness that is sometimes misinterpreted as capillary rise of ground water. When a wall or a stone is damp at its maximum capacity, the evaporation occurs first from the water film that covers the surface and the water that is lost is continually replaced by other water transported by capillarity from deeper layers. The drying rate $\partial W/\mathrm{d}t$ is controlled by the exchange with the atmosphere and mainly depends on wind speed, ambient relative humidity (RH) and sunshine. In steady conditions, initially the flow of moisture is constant and laboratory experiments show that the weight W of a damp sample decreases linearly and rapidly, so that $\partial W/\mathrm{d}t$ is constant. When the capillary transport of liquid water becomes unable to supply water at the same rate as evaporation occurs, the whole of the surface is no longer covered with a water film and the drying rate decreases until the mechanism is controlled by the diffusion of the moisture from the interior of the material. Laboratory experiments show a transition zone, in which $\partial W/\mathrm{d}t$ lowers and lowers until it reaches a very small value. Consequently, the weight of the sample decreases slowly and the plot of W versus t turns until it reaches another approximately linear trend, close to the horizontal, indication of very slow evaporation. In this step, the evaporation occurs inside pores and capillaries and then the vapour migrates outside through the pore and capillary fringe, until it reaches the surface where is quickly dispersed into the atmosphere. In the transition zone, the evaporation occurs first in pores close to the surface and where the free meniscus makes evaporation easier, and then in the deeper layers and in smaller pores where evaporation is more difficult. Eventually the vapour diffuses through longer paths and under weaker concentration gradients, reducing the exchange rate and the drying rate. In this last step, the loss of water occurs very slowly, reaching asymptotically the moisture content in equilibrium with the ambient. In practice, for a wall, this step may last for a very long time, e.g. months or years, depending on the thickness of the wall; the characteristics of bricks, mortar, plaster and colour coating; and the exchange with the atmosphere.

6.4. CAPILLARY SUCTION

Capillary suction occurs in all directions but a precise mathematical formulation is given only for the vertical ascent, which is the most important problem. The rise of water by capillarity is a well-known and important factor in monument weathering, especially in the case of historic brick buildings where the waterfront reaches considerable heights. The capillary rise can be simply calculated by means of the

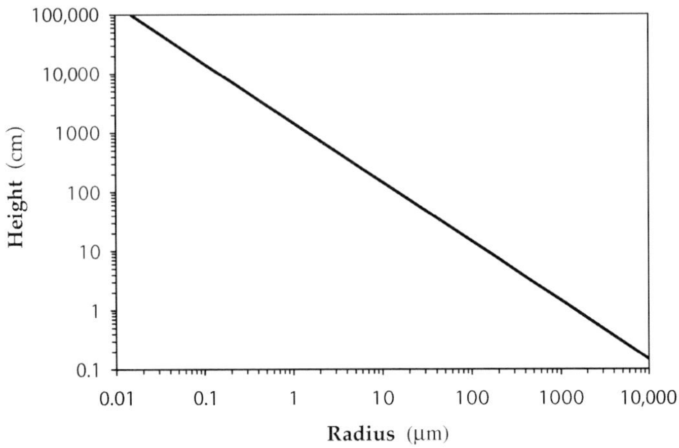

FIGURE 6.9 Height h of the capillary rise versus the capillary radius r.

Laplace formula, isolating h from the geopotential height in the *Stevin* formula

$$h = \frac{2\sigma}{r\rho g}\cos\theta \tag{6.2}$$

which states that the height h is inversely proportional to r and is affected by the wetting angle, which changes with the physical and chemical characteristics of the surface and the type and amount of dissolved salts. The formula is graphically reported in Fig. 6.9 where h has been computed for $\cos\theta = 1$; the plot represents the upper limit, as $\cos\theta \leq 1$. It can be observed that h reaches extremely high values for very thin capillaries (e.g. $h = 1.47$ km for capillary radius $r = 0.01$ μm at 20 °C) and decreases rapidly when the capillary size increases (e.g. $h = 1.47$ mm for $r = 1$ cm).

It is useful to comment on the above results, as for very thin capillaries it seems unbelievable. The volume occupied by one H_2O molecule in the liquid state can be obtained by dividing the molar volume of water, i.e. 18 cm^3, by the Avogadro number $N = 6.02 \times 10^{23}$, and the effective diameter is about 3 Å (1Å $= 10^{-4}$ μm). A capillary with radius $r = 0.01$ μm may contain 33 concentric layers, each one having the thickness of a water molecule; the first three molecular layers near the capillary surface are bound by molecular forces and cannot move but the water molecules that are in the inner part of the capillary, beyond this critical distance, can move freely and liquid water can flow. However, the cross-section of capillary is never constant, widening and thinning. In the case of these extremely thin capillaries, an irregular thinning of the cross-sections may be critical, hindering the free flow, and the probability of finding a capillary without any irregularity is extremely low. In practice, water suction is possible when changes in the capillary section never attain critical values and laboratory tests show that capillaries smaller than 0.1 μm are practically unable to absorb water (Winkler, 1994). However, if tests are made with materials with very thin capillaries, but with a well organized mineralogic structure, e.g. Carrara marble, capillaries may transport impressive amounts of water that may be misinterpreted for drops of water condensed on the monument surface. Capillary suction (in particular for ground water or rainfall percolation) is therefore much more efficient than condensation in causing dampness.

A simple but useful comment to Eqn (6.2): if the water supply is not limited, the level reached by rising water (i.e. *the front*) depends on two characteristics of the masonry, i.e. the capillary radius r and the contact angle θ. If a wall is homogeneous, the height reached by rising water should remain unchanged. If the front is not horizontal, the wall may have a non homogeneous internal composition or may be damp for another reason, different from capillary rise. An example is reported where a wall is damp in the lower part but the top levels of the waterfront are reached corresponding to gutters with some water leakage (Fig. 6.10).

It may also be useful to note that the most severe damage occurs on the top of the damp area where salts accumulate. This is apparently surprising because the wall is fed with water from the bottom and the flow of water, as well as the evaporation, will decrease with height until the last water reaches the front. If the lower part of the wall is affected by more water and evaporation, one could expect the most severe damage there.

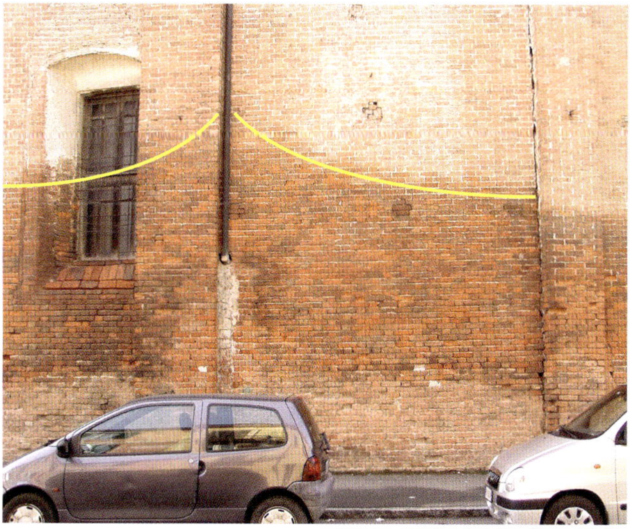

FIGURE 6.10 The upper border of the wet front peaks corresponding to a gutter. This is the real cause of wetting, not capillary rise (Cremona, Italy).

This is justified because it is true that most evaporation occurs in the lower part and that this leads to increased concentration or even precipitation of salts. However, the repeated or continuous flux of water dissolves the salt crystals and transports them upwards. On the front, the transport ends because there is no more water to continue this mechanism. The high concentration of salts is extremely noxious.

A common but misleading practice is to substitute deteriorated plasters in the frontal area with more resistant, waterproof mortars, e.g. with concrete. Before this intervention, the front had a given height that included, inter alia, a balance with the evaporation during rise. The stronger the evaporation, the smaller the capillary stream and less water would reach the theoretical height given by Eqn (6.2). In the case an impermeable layer is placed on the damp surface, in the absence of evaporation losses, rising water may reach higher levels as in Eqn (6.2). The result is that the front and the damage will be extended upwards, worsening the situation (Fig. 6.11).

Equation (6.2) was derived for the capillary rise (i.e. upward suction), equalizing the component of the force tangential to the capillary F_t, and the resulting Kelvin effect, to the action of the gravitational force on the water column. It is obvious that capillary suction, governed by the presence of curved meniscus and Kelvin law, occurs in all directions. In particular, no theoretical limit can be found for the displacement of water along a horizontal or downward path. In comparison with capillary rise (i.e. along the vertical), capillary suction is more active in the horizontal direction, being not counteracted by gravitation, and is favoured by gravitation in the downward direction. This can be experimentally proved by supplying water into a small hole made on a wall plaster.

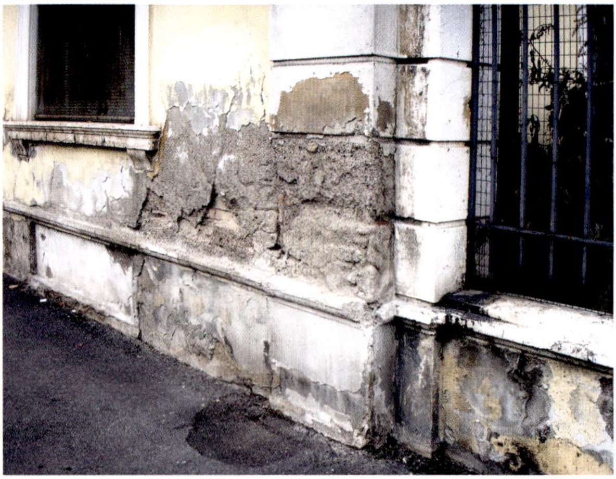

FIGURE 6.11 This waterproof band made to stop, or to mask, capillary rise had the opposite effect, i.e. it displaced the front upwards and increased the damage.

The stain of damp masonry develops similarly to a growing vertical ellipse that has the upper focus on the injection hole. The water supplied to a vertical porous surface will migrate upwards with minimum velocity, downwards with the maximum velocity and horizontally at an intermediate speed. In masonry, the vertical rise (where the suction force is opposed by the gravity) occurs at a rate that is half the horizontal spread.

The velocity at which water flows in a capillary is given by the *Washburn* equation

$$\frac{\mathrm{d}x}{\mathrm{d}t} = r\frac{2\sigma\cos\theta - r\,\rho\,g\,x\cos\phi}{8\eta x} \tag{6.3}$$

where x is the distance travelled by the water in the capillary, whatever is its direction, e.g. being inclined of angle ϕ with reference to the vertical so that $\cos\phi$ equals 1 for upward displacements, -1 for downward displacement and 0 for horizontal ones and η is the dynamic viscosity of water, e.g. 1307 μPa s at 10 °C and 1002 μPa s at 20 °C. From this equation it is possible to find the distance travelled by the water table after the time τ is elapsed, i.e.

$$x = \sqrt{\tau\frac{r\,\sigma\cos\theta}{2\,\eta}} \tag{6.4}$$

which is often concisely expressed as

$$x = A\sqrt{\tau} \tag{6.5}$$

where the coefficient A includes the dependence on the radius, the surface tension, the wettability and the viscosity. The penetration of water may continue for weeks or months. The velocity of penetration is faster at the beginning and is given by the above equations; later it decreases.

Similarly, the amount of adsorbed water (WA) varies linearly with the square root of the elapsed time τ i.e.

$$WA \approx B\sqrt{\tau} \tag{6.6}$$

where the coefficient B depends on the characteristics of the material, although some exceptions have been noted (Gummerson et al., 1980). If there is divergence in the water flow with lateral spreading of the waterfront, the cumulative water absorption increases more rapidly than $\sqrt{\tau}$ (Hall, 1981).

Capillary suction occurs with the progressive displacement of water, so that it cannot be stopped by pockets of air entrapped in the pore and capillary system, except for ducts without an outlet, in which the capillary pressure is counteracted by the entrapped air. Excesses of pressure that are generated by heating are important, e.g. on porous tiles where the cavities are topped by the glaze of the decorative ceramic surface, especially when thermal shocks are noticeable. A capillary closed at the upper extreme may be filled with water until the pressure of the air entrapped equals the Laplace pressure Π, determined by the meniscus curvature, minus the Stevin contribution due to gravity, as it has been discussed in the previous Chapter. Solar radiation falling on the material with air pockets embedded causes an increase in the pressure inside the cavities due to the temperature rise ΔT according to the law of gases

$$pV = (n_a + n_v)\Re T \tag{6.7}$$

where n_a and n_v are the number of moles of dry air and water vapour, respectively. This excess of pressure causes the water meniscus to withdraw and reach a new equilibrium due to the simultaneous presence of a gas and a saturated vapour, as follows: (1) n_a is constant and the partial pressure of dry air is inversely proportional to V and (2) n_v is not constant: in fact, ΔT causes a drop in *RH* and consequently other water molecules will evaporate to compensate the temperature effect until the original *RH* is reestablished, the *RH* being determined by the radius of the capillary (Camuffo, 1991b). The partial increase of pressure due to the vapour is shown in Fig. 6.12.

An efflorescence belt indicates the level reached by capillary rise: where water evaporates the salts will conclude their internal migration and will remain on the surface (or on the subsurface layer), impregnating the masonry with thick surface and subsurface deposits. Of course, damage is worsened in the presence of soluble substances, osmotic effects and salt crystallization. On plasters and frescoes, at high humidity, residual crystals of calcium oxide in the mortar may hydrolyse becoming calcium hydroxide, causing pitting as a consequence of crystal growth. Salt crystals have a

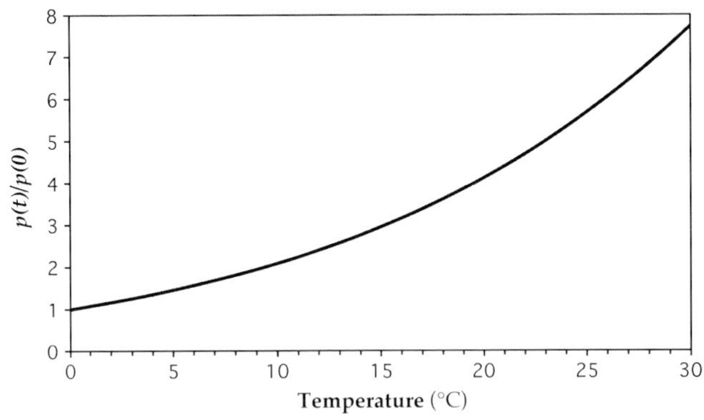

FIGURE 6.12 Increase in normalized partial pressure $p(t)/p(0)$ of the water vapour following a temperature rise t (°C) in a capillary due to both vapour heating and the increase of the number of water molecules that had evaporated to reestablish the original *RH* in the cavity. Note that $p(t)$ is not affected by retreat of water meniscus and increase in the volume of the cylindrical air pocket, the molecules being a saturated vapour and not a perfect gas.

disruptive effect as a consequence of the pressure generated by growing crystals, hydration and/or thermal expansion. The osmotic effect causes an upward rise of ions that displaces the wet front upwards. In the region where water evaporates, the transported salts will accumulate locally causing more and more damage (Arnold and Zehnder, 1990).

6.5. THE EQUILIBRIUM VAPOUR TENSION OVER A SOLUTION

In the previous section on heterogeneous nucleation, it has been anticipated that salts dissolved in the water will lower the water vapour pressure because the water molecules are diluted by the presence of the salt. It is useful to discuss some basic results of applied physical chemistry (for further details see Kireev (1977), Adamson (1986)) as this effect is often dominant and makes possible condensation in several circumstances. The *Raoult* law for ideal solutions states that the saturated vapour pressure p_A of the solvent A over a solution is always lower than that over the pure solvent. The greater the concentration of the solute, the lower the equilibrium vapour pressure, i.e. the saturation vapour pressure p_A of the component A over the solution is directly proportional to its mole fraction X_A, i.e.

$$p_A = p_o X_A \tag{6.8}$$

where p_o is the vapour pressure of the component A in the pure state, i.e. when $X_A = 1$. However, the concentrations of the solvent A and the solute B, X_B, are linked by the simple relation $X_A + X_B = 1$, so that the previous equation can be rewritten as

$$p_A = p_o(1 - X_B) \tag{6.9}$$

which can be rewritten in terms of the difference $\Delta p = p_o - p_A$. Applying the result to water vapour, the lowering of the partial pressure $\Delta e(X)$ of the water vapour in equilibrium with a diluted solution is directly proportional to the mole fraction X of the solute,

$$\Delta e(X) = e_o X \tag{6.10}$$

where, according to the common notation, e_o is the saturation pressure for pure water that has been indicated with $e(\infty)$ in the previous Kelvin formula, as the label 'o' is not anymore concerned with the curvature radius, but now indicates the pure solvent with 'zero' moles of dissolved solute. The Raoult law is an ideal limiting law for complete uniformity of intermolecular forces. A second ideal limiting law, i.e. the *Henry* law, for extremely diluted solutions, states that all the molecules of the solute B are completely surrounded by molecules of the solvent A and the partial pressure of the solute B becomes proportional to the mole fraction in the limit of zero concentration, i.e. when $X_B \rightarrow 0$

$$\lim p_B = k\, X_B \tag{6.11}$$

where k is a constant with the dimension of pressure and the behaviour of B is determined by the nature of the A–B interactions. In the case of deliquescent salts and concentrated solutions, the situation is better described by the Raoult law.

The lowering of the vapour pressure can be related to the change in free energy of the vapour by

$$\Delta \mathcal{G} = \Re T \ln\left(\frac{e}{e_o}\right) \tag{6.12}$$

where $\triangle \mathcal{G}$ is the change in free energy of the vapour in going from the vapour pressure e to the saturation pressure for pure water e_o. If the water in solution is in equilibrium with the vapour above it, $\triangle \mathcal{G}$ also represents the change in free energy of the liquid water.

6.6. CLIMATE CYCLES, SEA SPRAY AND SALT DAMAGE

Salts play an important role in decay because they can precipitate within an aqueous solution when the salt concentration becomes saturated from loss of water due to evaporation, Furthermore, some salts may change crystalline form or hydrate under some environmental conditions that may be changed by the presence of some contaminants. Deliquescent ones may dissolve and recrystallize when ambient RH drops below certain levels. Environmental cycles of temperature or humidity, or periodic heating due to direct solar radiation, may induce salt migration, precipitation, growth, hydration and expansion and may trigger disruptive cycles within the pores, causing fatigue or breakage of the internal material structure.

If a dilute salt solution within the pores is exposed to a low relative humidity, water will evaporate from the solution and the salt concentration will increase. The higher the concentration, the lower the vapour pressure in equilibrium with it, until a critical limit is reached, i.e. the lowest equilibrium *RH* at which the salt solution reaches saturation. If the ambient *RH* drops below this critical value, water molecules will continue to evaporate and some salt crystals will begin to precipitate and grow. The mechanism is in some aspects similar to what has been described for freezing. Evaporation begins in medium or large pores that are connected with the external environment and some precipitation begins.

Arnold and Zehnder (1991) noted that salt crystals form mainly in pores from 1 to 10 μm diameter and that they follow the following sequence: crystallization continues until the pore is filled, then crystals enter the neighbouring pores; successively the crystallization pressure generates some small cracks on the stone (or the mortar); and finally crystallization continues in the form of growing columns able to lift spalls and detach crusts. They also found that the crystal morphology is related to the substrate moisture: in the presence of liquid water, large crystals grow completely immersed into the solution; on a damp substrate, smaller and hygrometric crystals form a granular crust covered with a film of solution; in dryer environment, a humid substrate forms a fibrous crust; on less humid substrates, isolated columnar crystals grow; on nearly dry surfaces with localized supply of moisture, the crystals are very thin and acicular, hair-like and are called *'whiskers'*. New dampness may transform fresh whiskers into a more compact differently crystallized crust.

Every cycle of *RH* that crosses the critical equilibrium value causes a precipitation/dissolution cycle. In the general case, several soluble salts are present inside walls or stones determining a very complex situation for their interactions. In general, they precipitate in sequences according to their respective solubility, whereas the most soluble remains in solution, determining time sequences of crystallization or space patterns that reflect on vertical layering of the salt front when the source is ground water capillary rise. In this case, Arnold and Zehnder (1991) found that the salts are locally accumulated in a zone 0.5–3 m above the ground level, with mainly two distinct areas, the lower where most damage occurs where carbonates, sulphates crystallize, and the higher one characterized by its damp appearance where chlorides and nitrates are found. This vertical fractionation was explained considering that, during evaporation, salts with low solubility become supersaturated and are the first to precipitate, while the more soluble salts can continue their rise and will precipitate at upper levels. The more soluble the salts, the more hygroscopic they tend to be (Arnold, 1983). For this reason, the upper layer is characterized by a high concentration of very soluble and hygroscopic salts, which tend to form a humid belt on the top of the rising damp.

When a solution in equilibrium with its environment has reached saturation, if *RH* will decrease, it will cause evaporation, supersaturation and precipitation. Some solutions reach this critical condition at ordinary *RH*, and above the critical equilibrium levels of *RH*, only the precipitated salts can be found. Several deliquescent salts can be found on monuments, e.g. NaCl at $RH = 75\%$, $NaNO_3$ at $RH = 74\%$, $Mg(NO_3)_2 \cdot 6H_2O$ at $RH = 54\%$ and $Ca(NO_3)_2 \cdot 4H_2O$ at $RH = 53\%$, when the temperature is 20 °C, but the *RH* threshold changes with temperature. Daily microclimate cycles due, e.g. to solar heating or room heating may induce crystallization cycles in walls or monuments. For this reason, in the cold climate, week heating–cooling cycles in churches lead to important changes in the wall temperature, followed by severe damage due to crystallization cycles.

Some salts exist in two crystalline forms, i.e. anhydrous and hydrate, including some 'water of crystallization' as part of their structure, e.g. gypsum ($CaSO_4 \cdot 2H_2O$) and anhydrite (i.e. $CaSO_4$). This hydration is associated with 63% volume increase and swelling pressure, whereas dehydration with 39% volume decrease (Azam, 2007).

Breaking waves and whitecaps give a yearly production of sea salt aerosols ranging from 10^9 to 10^{10} tons that are transported inland by winds and clouds, forming cloud condensation nuclei, and will fall dissolved into raindrops. The production of whitecaps and the subsequent transport of sea spray depend on the wind speed. The sea salt concentration increases exponentially with the wind speed. With winds below 5 ms^{-1}, a variable threshold of sea spray is found, either due to higher variability of the wind field over the sea or due to the production of spray in the previous

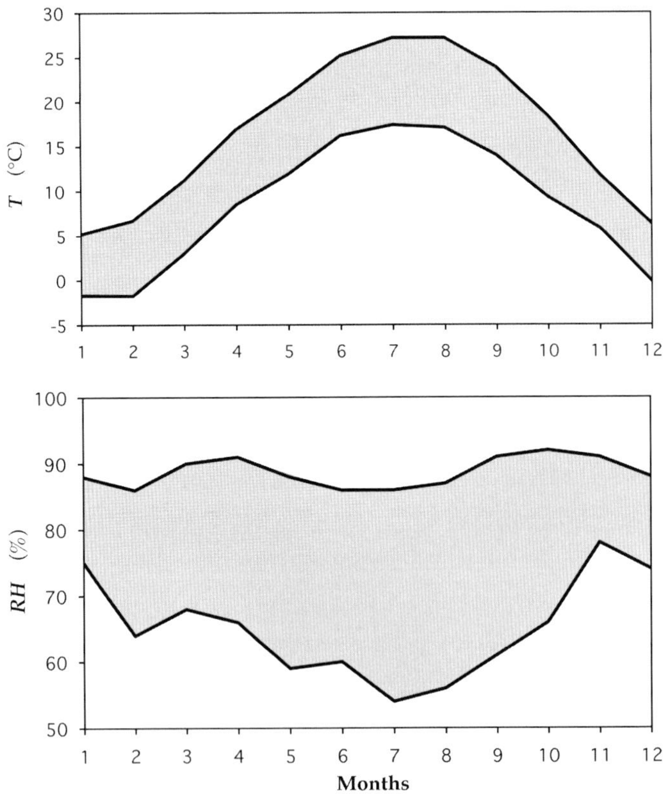

FIGURE 6.13 Seasonal variation of the average daily range of air temperature and relative humidity at Venice.

days. For wind speed $<10\ ms^{-1}$, the sea salt concentration increases with the first power of the wind speed; above this speed, the concentration increases again and reaches the third power at $20\ ms^{-1}$ and more than the fourth power as wind approaches the hurricane speed (Blanchard and Woodcock, 1980).

Some sites are very unfavourable for stone conservation, due to adverse natural conditions determined by climatic cycles and sea spray contamination, worsened by the presence of anthropogenic pollution. Venice is one of them and will be presented here to elucidate the problem with reference to the daily and seasonal values of relative humidity and air temperature (Fig. 6.13).

Sea spray contamination increases not only the TOW but also the solubility of some salts, causing many disruptive cycles. Initially, the solubility of gypsum increases with the sea spray concentration until it increases by a factor of four when it reaches the maximum, and then decreases again.

When a stone is contaminated with sea spray, the NaCl may be either in the crystalline form or in solution with adsorbed water vapour. At ambient RH below 75%, the NaCl is in the form of dry crystals and at higher RH values, it will adsorb water and dissolve in it. The solubility of calcium sulphate in an aqueous solution of NaCl will begin at $RH = 75\%$, will increase to $RH = 90\%$ and then will fall again as the NaCl concentration decreases, i.e. the RH increases. As the maximum solubility is reached at ambient $RH = 90\%$, at each transition towards increasing or decreasing RH values some gypsum crystal will form. This cycle studied by Price and Brimblecombe (1994) is extremely important in coastal areas. As every night the saturation is reached in Venice, every day this transition occurs twice.

In addition, in the midseasons and in summer, when the RH drops below 75% in the central part of the day, precipitation of the dissolved fraction of any soluble salt will occur every morning when the salt solution dries and a partial dissolution will occur every evening when the ambient RH rises above 75% and the contaminated stones begin to get wet. However, even in the absence of other soluble salts, NaCl alone is sufficient to cause severe damage. As not all stones have reached thermal equilibrium with the air, the number of cycles and the damage will vary from exposure to exposure, from building to building and from stone to stone.

The increase in gypsum solubility has another important effect. In white areas, rainfall dissolves in a short the gypsum crystals that have been precipitated on the surface, leaving unprotected the limestone and acid rainfall will eventually continue dissolution.

Hydration/dehydration cycle determined in coastal sites by temperature changes was determined by Gordon and MacDonald (1953). They found (Fig. 6.14) that at 40 °C and ordinary pressure, gypsum, bassanite, anhydrite and liquid water may coexist. Above 40 °C, anhydrite is the most stable form, below gypsum. However, the presence of sea spray lowers the transition temperature: the latter decreases with increasing NaCl concentration. When the solution contains 3 mol of NaCl, the transition temperature is 30 °C; at 4 mol, 25 °C; at 5 mol, 21 °C; and at saturation (i.e. 6.15 mol), the transition occurs at 14 °C. This suggests that for most part of the year, temperature and humidity cycles induce calcium sulphate crystals to undergo daily hydration/dehydration cycles with continuous volume expansion and contraction. The figure reports two bands corresponding to the range of temperature in January and June at Venice. In January, transitions are very rare and may occur for solutions with NaCl concentrations between 7 and 8 mol/l, whereas in summer, transitions are much more frequent and occur with concentrations between 3 and 5 mol/l. In the general case, however, a monument has not a homogeneous distribution of temperature, the surface is not homogeneously contaminated by salts, and it is not uniformly wet, and the concentration of dissolved NaCl is variable with height or orientation. Therefore, the most critical points may change with the day and the season.

Thenardite (Na_2SO_4) is a mineral impurity frequently found in cement-based mortars and concrete restorations made over the past 50 years (Haynes et al., 2008). Thenardite is an unstable mineral that under certain circumstances undergoes a transition to mirabilite, i.e. $Na_2SO_4 \cdot 10H_2O$, which is a hydrated crystal with the inclusion of 10 molecules of water. This happens under some critical temperature (T) and RH ranges,

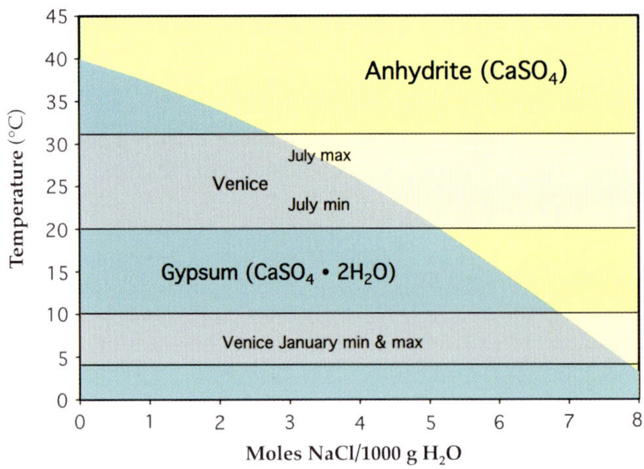

FIGURE 6.14 Transition between gypsum (cyan) and anhydrite (yellow) induced by synergism between temperature and NaCl contamination, in the presence of liquid water at ordinary pressure. The figure reports the bands of temperature variability in January and July at Venice, Italy. Transition is possible with medium NaCl concentration in summer and high concentration in winter. *(Source: By kind permission of* American Journal of Science. *See credits.)*

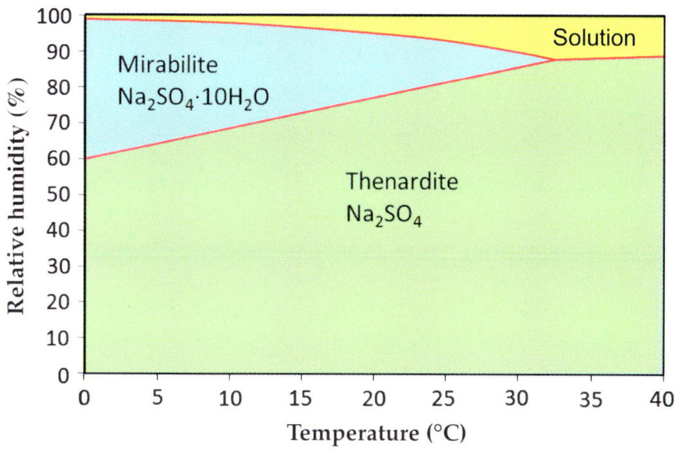

FIGURE 6.15 Temperature and relative humidity diagram for the mirabilite (cyan)–thenardite (green) transformations. The yellow area on the top indicates a solution of either mirabilite or thenardite, both deliquescent salts. All mineral transformations occur when the critical red line thresholds are passed.

i.e. $0 < T < 30\ ^\circ C$ and $60\% < RH < 88\%$, exactly when $RH = 0.8485\ T + 60$ (Fig. 6.15).

The inclusion of 10 molecules of water into each mirabilite crystal exerts a very high crystallization pressure inside pores, damaging the structure. Moreover, the transient stress can remain for a long period since the relaxation process is slow (Espinosa Marzal and Scherer, 2008). In contrast, the opposite transformation is associated with loss of water and volume reduction and does not exert mechanical forces. However, the 10 molecules of liquid water that will become free inside the concrete, or the mortar, will cause local dampness and iron bar rusting. Repeated cycles may accumulate stress and in the long run, they may cause severe decay.

The above examples have considered the presence of only one contaminant but the real situation is with much more complex synergistic actions. Other salts that are deliquescent at ordinary humidity levels, e.g. several minerals derived from nitrogen, extend the *TOW* and the range of ambient *RH* at which crystallization and hydration may take place; in addition, the chemical reactions, and the minerals that are generated, may be also influenced by the presence of sea spray.

Thermal expansion too may contribute to worsen the deterioration mechanism. The crystals formed will have mechanical characteristics different from the porous material. If the expansion coefficient of the salt is larger than that of the material, the newly formed crystals will reach the internal pore surface and will exert pressure when the temperature increases. If the coefficient is smaller, the same will occur, but when the temperature decreases.

6.7. SOME COMMON ERRORS THAT SHOULD BE AVOIDED

The risk of mistakes is particularly relevant in complex applications essentially of multidisciplinary character, especially when an expert (which is necessarily expert in only one or a few specific disciplines) gives a suggestion that may appear brilliant from his point of view but that may have negative consequences in other fields. Although human errors probably are one of the main, but less documented, cause of decay, nevertheless, their knowledge may help to avoid their repetition. This is especially true for the most frequent mistakes that are the consequence of erroneous popular beliefs or are the uncritical repetition of remedies that have proved useful in other apparently similar circumstances but without having analysed if they are equally convenient to each specific case. As a practical application, a short list of useful examples is here enclosed.

6.7.1. Condensation or Evaporation?

Several years ago, I read the report of a panel composed of experts in conservation who suggested some remedies for a famous church in Venice. The church was built in 1489 on the edge of a canal and was severely damaged by dampness. The report was amazing for a loop of contradictory statements that arrived at a misleading suggestion that, unfortunately, was applied.

The misleading loop of self-generating causes and effects was more or less as follows. *The wall is damp, and then evaporates. Evaporation forces the wall temperature to drop. When the wall temperature is dropped, condensation begins to occur. The suggestion is to warm the air in proximity of the wall. By warming the air, the relative humidity will decrease. When the air is dry, the condensation stops. The problem is solved.*

The wall was damp, from capillary rise and the presence of sea salt and evaporation increased the migration and accumulation of salts. Evaporation does not indefinitely lower the temperature of the interface between the air and the stone but tends to approach the wet bulb temperature. The temperature of evaporation (i.e. the wet bulb temperature) is well defined thermodynamically and is higher than the temperature of condensation (i.e. the dew point (*DP*)). As evaporation and condensation occur at two different temperatures, none of them can be the cause of the other one. Evaporation may never cause condensation, and when the two temperatures approach each other, the net flux of moisture stops and there is no practical exchange between the wet surface and the atmosphere.

The suggestion of warming the air in order to lower the relative humidity and stop condensation on a cold surface is scientifically unsound.

Condensation occurs on a cold surface whose temperature is below the *DP* irrespective of the relative humidity in air. In the Chapter 2A, we have seen that the *DP* is a function of the *humidity mixing ratio* or the *specific humidity* only and these (i.e. *MR* and *SH*) are lowered only if moisture is removed from the air, not if the air is heated. However, another mechanism may intervene, which eventually may stop the condensation. In the case of actual condensation on a cold wall, by heating the air, in the long run the wall too is heated. When its temperature becomes higher than the *DP*, condensation stops. For this reason, it is much more convenient to warm directly the wall and not the air (Camuffo and Valcher, 1986).

In addition, by heating the air, another negative mechanism was generated: the wall was damp from capillary rise and the water evaporated in the atmosphere; by lowering the relative humidity, the evaporation rate was increased as well as the migration of the soluble salts and the formation of efflorescence. Theoretically, this process will stop only when all the water in the canal in the canal evaporates by passing through the building walls. Evidently, the distinction between the elementary concept of condensation and evaporation is a basic one in meteorology but not equally familiar to other people.

A similar misinterpretation was recently repeated for a basilica in Venice. The diagnosis was again that the capillary rise caused evaporation and cooling and, consequently, the airborne pollutants were claimed to deposit on the mosaics by means of the Stefan flow (see Chapter 8). As it has been explained, evaporation cooling cannot generate condensation and the Stefan flow that is associated with capillary rise and evaporation works against deposition of airborne pollutants. In this case, the problem is capillary rise, mobilization and transport of internal salts and deterioration of walls, plasters, mosaics and so on.

Another misinterpretation that is commonly found is the analysis of the so-called *interstitial condensation* inside a wall. This problem is very important and some typical situations are discussed in Fig. 6.16 that shows some profiles of *T* and *DP* in a cross-section of a wall, the left-hand side being in contact with the exterior atmosphere and the right-hand side with the interior microclimate.

1. The first figure (Fig. 6.16(a)) shows the situation prospected in some textbooks without further specifications to illustrate the so-called *interstitial condensation*. This situation is claimed typical of wintertime, i.e. the inside temperature is high (domestic heating), outside is very cold and the air temperature is shown to be lower than the *DP*. In this figure, the following assumptions have been made:
   - the temperature and the *DP* profiles are linear inside the wall, which means that both *T* and *DP* were stationary outdoor and indoor for a very long time, e.g. several days;

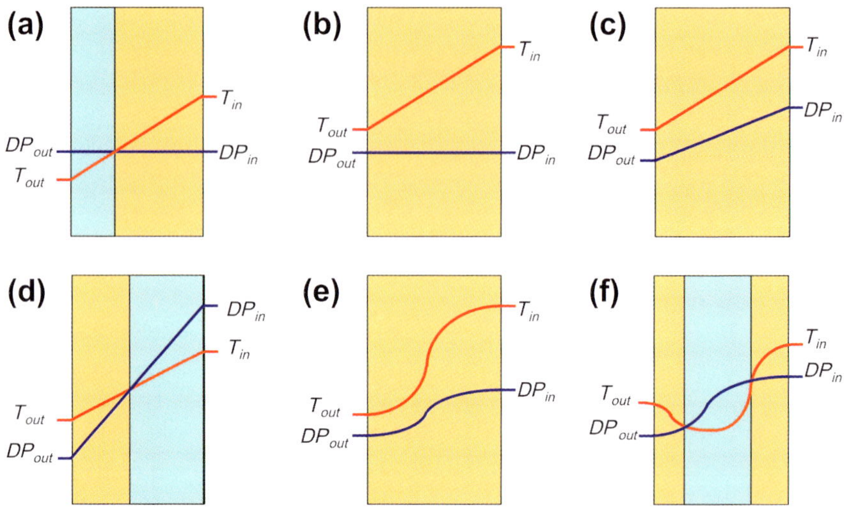

FIGURE 6.16 Winter temperature (T) and dew point (DP) profiles across the external wall of a heated building. (a) The situation is prospected in some textbooks under the following assumptions: T and DP are stationary both outdoors and indoors and the outdoor temperature is much lower than the indoor temperature and is also lower than the DP. Please note that the assumption $T < DP$ is impossible if T_{out} is the air temperature; however, it may be possible if T_{out} is the external wall surface temperature. In the left side of the wall (cyan) where $T < DP$, *interstitial condensation* occurs. (b) Under the same assumptions, but $T \geq DP$, no intersection is possible between the two profiles and there is not a slab of wall where condensation may occur. (c) In manned buildings, the vapour released by people increases the indoor DP. When the inside and outside temperatures are different and constant, the stationary profiles are as indicated and no intersection is possible. (d) If the building has a crowded room, the DP increases and may exceed the wall temperature T_{in} and condensation will occur in the cyan slab where the wall temperature falls below the DP profile. (e) Under nonstationary conditions, the temperature and the DP distributions inside the wall vary continually with time and depth, with curved, nonlinear profiles. (f) The situation when the outdoor temperature rises rapidly and the wall remains cold inside. The moisture can reach by diffusivity a cold slab (i.e. $T < DP$, cyan area) and saturation will occur inside the wall.

- the specific humidity, and consequently the DP, are the same outdoors and indoors as in the case of open widows or fast air exchange rate; and
- the outdoor temperature is much lower than the indoor temperature and is also lower than the DP. The illustrative sketch shows that the DP is reached somewhere at mid−wall thickness: the wall slab on the left, colder than the DP is wet and the slab on the right, warmer than the DP, is dry. We should comment on the third hypothesis of this illustrative example, i.e. $T_{out} < DP$. If T_{out} is referred to the air, this is thermodynamically impossible: the air temperature may never drop below the DP. As opposed, T_{out} should

be attributed to the wall, and it may temporarily happen that the wall temperature is found below the outdoor air temperature, e.g. change of blown air and/or large thermal inertia of the building envelope.

2. The second figure (Fig. 6.16(b)) is under the same boundary assumptions, but in the realistic case that the external air and the wall temperature are both above or equal to the *DP*. It is evident that no intersection is possible between the two profiles and there is not a slab of wall where condensation occurs (i.e. interstitial condensation), except for a possible contribution of the Kelvin (micropores) and Raoult (solution) mechanisms that are not considered in this example.
3. In manned buildings, the vapour released by people, for building use, or to compensate the *RH* level, raises the indoor *DP*. When the inside and outside temperatures are different and both remain unchanged, the stationary profiles are as indicated in Fig. 6.16(c) and no intersection is possible.
4. If the building has a crowded room, the *DP* increases and may exceed the wall temperature T_{in} (never the indoor air temperature) and condensation will occur in the wall slab where the wall temperature falls below the *DP* (Fig. 6.16(d)).
5. Under the effect of daily external and internal forcing, i.e. daily cycles of outside and/or inside temperatures and outside and/or inside specific humidities, the temperature and *DP* distributions inside the wall vary continually with time and depth, with curved, nonlinear profiles (Fig. 6.16(e)).
6. Due to the different velocities of heat conduction and moisture diffusion, it cannot be excluded that in some specific cases the two curves intersect (Fig. 6.16(f)) determining an internal damp slab. This especially occurs when warm and humid air is blown, while monuments and buildings are cold.

Without considering the Kelvin and Raoult laws, the two most frequent cases of wall condensation are as follows.

1. The indoor air temperature is higher than the outdoor air and wall surface temperature and the indoor specific humidity is much greater than the outdoor specific humidity (e.g. indoor humidifying systems, concentration of people), so that the *DP* of the indoor air is much higher than the *DP* of the outdoor air. Some vapour migrates from inside to outside through the pore and capillary fringe of the wall. When the air molecules meet a wall slab whose temperature is lower than the *DP*, condensation will occur. In some particular cases, walls have been protected against this problem by inserting a barrier against airflow through the wall, e.g. a polyethylene foil; in practice, however, failures

were unavoidable and the results very bad (Padfield, 1995–97). By the way, in this publication some other nice examples are reported.

2. According to the weather vicissitudes, the outdoor air may change temperature and specific humidity. On the other hand, buildings respond to the climatic forcing factors with a delay determined by the fabric heat capacity, heat conductivity, hygroscopicity, moisture diffusivity and so on. After a cold period, when building walls are cold, if wind blows warmer and moister air, the wall temperature may be below the *DP* and in this case condensation occurs immediately. The most effective situation is when the outdoor temperature rises rapidly, the interior of the wall remains cold, and *T* has a U-shaped distribution inside the wall. The moisture migrates by diffusivity more rapidly than the heat by conductivity and may reach a cold slab inside the wall (i.e. $T < DP$, shaded area) generating condensation (Fig. 6.16(e)). This situation is very typical when the warm and humid Sirocco wind blows from North Africa or in the transition between winter to summer, when the wall temperature follows with some delay the seasonal cycle of the air, whose moisture content increases with the growing season. For this reason, wall condensation is frequent in spring and rare in autumn.

The Kelvin and Raoult laws state that the vapour pressure lowers over a concave meniscus or when the rain or dew is contaminated by soluble salts. The immediate effect is that the equilibrium *RH* for saturation is lowered and the condensation point is raised above the *DP*. Let us consider the case of artwork exposed outdoors, but provided with preventive conservation heating, i.e. a device that supplies heat to the object to avoid condensation under normal circumstances. However, if the object surface is contaminated with soluble salts or pollutants, it may happen that condensation, or the deliquescence level, is reached well above the *DP*, when in the air $RH << 100\%$. This means that the protective device should be tuned to supply heat at a humidity threshold that should be calculated for the specific contaminants. If this problem is not adequately considered, the risk is that spurious wetting–drying cycles are forced with the effect of worsening the situation. This negative experience has unfortunately occurred to outdoor monuments and decorations, e.g. marble bas relieves or bronze panels nailed to wood doors, which suffered for a misleadingly increased number of condensation cycles, instead of receiving the expected protection.

6.7.2. Not Only Rising Damp

Dampness in the lower part of buildings is popularly attributed to capillary rise of ground water and the remedy is inevitably the same: to interrupt the supply of new water by cutting the walls and inserting an impermeable layer (e.g. lead, plastic) or impregnating brick and stones with polymers or other water-repellent substances. This intervention is absolutely correct when the actual cause is capillary rise, but exceptions to this rule are very frequent and the apparent analogy is that absorbed (or percolated) rainwater, under the influence of the gravity force, tends to descend downwards and is ultimately found to affect, for long periods, the basement of buildings.

An interesting example is given by the Romanesque Baptistery of the Cathedral of Padua, which the Florentine Giusto dei Menabuoi frescoed in 1375–1378. At present, the indoor frescoes survive but the exterior is in brickwork. The exterior too was originally protected by frescoed plaster, as some red traces of sinopite, i.e. the preparatory sketches of the frescoes, were visible. Today the bare bricks, the nude mortar joints and the presence of holes deliberately left to support timber scaffolds facilitate the penetration of water during rainfall. Having penetrated the thick walls, the water migrates in all directions, but the downward one is favoured by gravity, so that an accumulation in the lower part becomes evident. The drying rate is extremely low, depending on the hygroscopic nature of the walls and the exchanges with the ground (i.e. percolation, moisture migration) and the interior atmosphere (evaporation). The lower rows of frescoes (i.e. the first 2 m) were deteriorated by the dampness and the efflorescence that formed during the drying phase. The cause was erroneously attributed to capillary rise, and near the floor, a lead layer was inserted into the walls. This old intervention prevented rainwater from descending through the foundations and reach the ground; it was stopped at the lead layer level, and the only possibility was to evaporate into the atmosphere, increasing dramatically the formation of efflorescence. Suspecting this alternative possibility, it was easy to compare the moisture content in several points above and below the lead layer and the portion of wall between the ground and the lead was found dryer. The remedy was to remove the lead, open some ventilation holes in the wall and keep it ventilated for a few years and remove the efflorescence once the wall was dried. As the exterior was traditionally remembered in brickwork, it was not shielded with new plaster but was made water repellent (but perspiring) in order to avoid new penetration of rainwater. This happened more than 25 years ago and the restored frescoes are in very good conditions.

The Saint Rocco Oratory, Padua, built in 1542, with frescoes of the sixteenth century offers another case study. The frescoed wall for the first 2 m above the floor was severely damaged by dampness, and the owner, on the ground of the opinion of some experts based on some thermograph images that confirmed the lower temperature of the basement, decided to insert into the walls a physical barrier, i.e. a membrane that resists the capillary rise of moisture from the foundations of the building. This decision was not fully justified for two reasons: (1) the front of the salts was not horizontal but showed maxima in correspondence to gutter pipes that were embedded in the walls and (2) a small window in the lower part of the wall obliged the water to migrate all around: in the case of uprising flow, the part just above the lintel was expected to be dryer than that just below the window sill, and vice versa. A field survey that measured the specific humidity gradient close to the wall surface demonstrated that below the window the wall was dryer, and in addition, it was shown that the evaporation was greater in correspondence to gutters. This led to conclude that rainwater was released by the gutters and migrated along the walls wetting the lower part. An excavation along the foundation walls showed that the foundations were dry and confirmed the

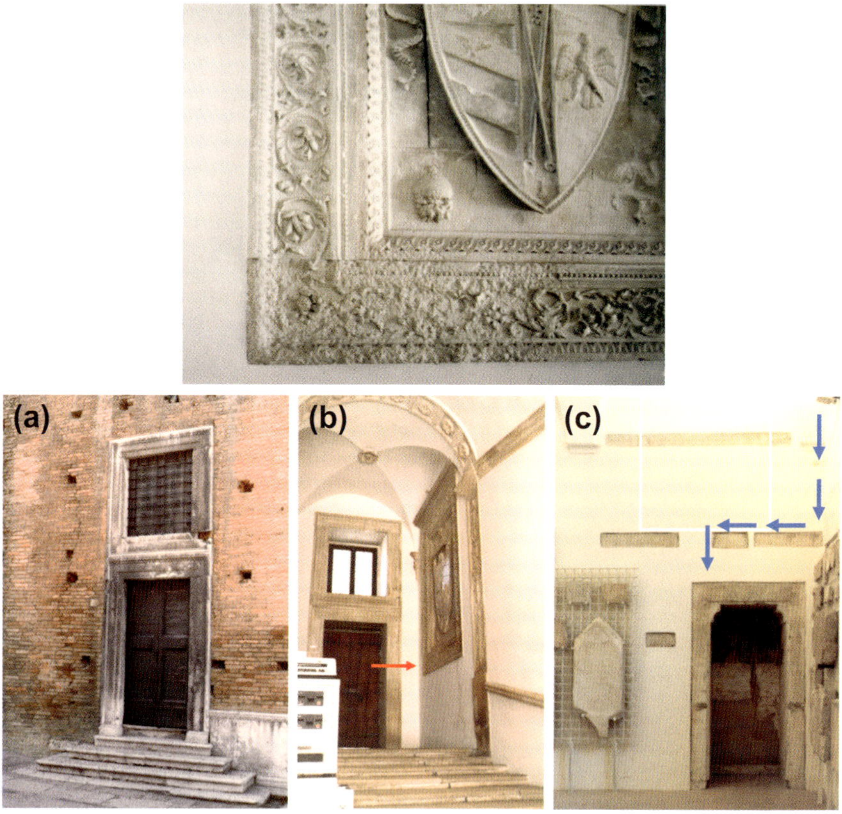

FIGURE 6.17 Top: The stone escutcheon with the edge of the lower corner severely deteriorated. Bottom: (a) The external wall, with the door and some holes left to support timber scaffolds in the case of needs. (b) The honour staircase seen from the open gate of the honour courtyard; the arrow points at the damaged corner of the escutcheon. (c) A view of the former salt depository on the back wall. The white rectangle corresponds to the escutcheon and the blue arrows show the path of internal water percolation and how water moves around the lintel. The Ducal Palace, Urbino, Italy.

hypothesis of deteriorated gutter percolation. After new gutters were installed external to the walls, walls began to dry and the problem was solved.

In the Ducal Palace, Urbino, built by Laurana in 1468, the lower corner of the main stone escutcheon in the honour staircase was severely deteriorated (Fig. 6.17). The staircase had a permanently open gate on the internal courtyard and acted as a chimney. A violent airflow from the courtyard continually blew in transporting cold air. It was suspected that this mechanism could cause condensation, but why just on that corner and not elsewhere? An additional hypothesis was that the room at the back was used in past centuries as a sea salt depot and that the sea salt might have contaminated the wall, making easier condensation at the deliquescence *RH* level, i.e. 75%. This required an investigation of both the staircase and the former salt depot, now used as museum room. Long-term monitoring of air and stone temperature and humidity excluded condensation, and in particular localized condensation.

The cause of deterioration was rainwater, because of some percolation that originated on the external walls, for some water penetrated through holes for timber putlogs, deliberately left to support timber scaffolds for temporary works. At each rain, the water percolation continued internally to the wall, until it reached an internal horizontal lintel. The lintel is composed of broken fragments, visible on the back wall surface, and was included inside the wall to support the monumental escutcheon. Water percolated along the lintel, wetted and damaged the escutcheon made of vulnerable stone and at a gap in the lintel, percolated down, leaving dry and untouched the rest of the escutcheon, as shown in the figure (Bernardi et al., 1985).

6.7.3. Drying Damp Murals

When a wall with murals has been incidentally soaked due to rainwater infiltration or other causes, it is exposed to high risk of damage from biological attack, chemical reactions or decreased resistance to physical injuries. In addition, two immediate problems arise: (1) the mobilization of soluble salts with formation of efflorescence and (2) soiling due to the increased deposition rate of airborne particulate matter, as all the particles that impact on a wet surface remain trapped there due to the adhesive force exerted by the film of water. A long TOW increases the probability that suspended particulate matter will be captured on the surface, making heavier soiling and blackening. The presence of water and the deposited particles constitute a favourable habitat for microorganisms, increasing the risk of biological and chemical deterioration. In addition, restoration is only possible when the wall is dry.

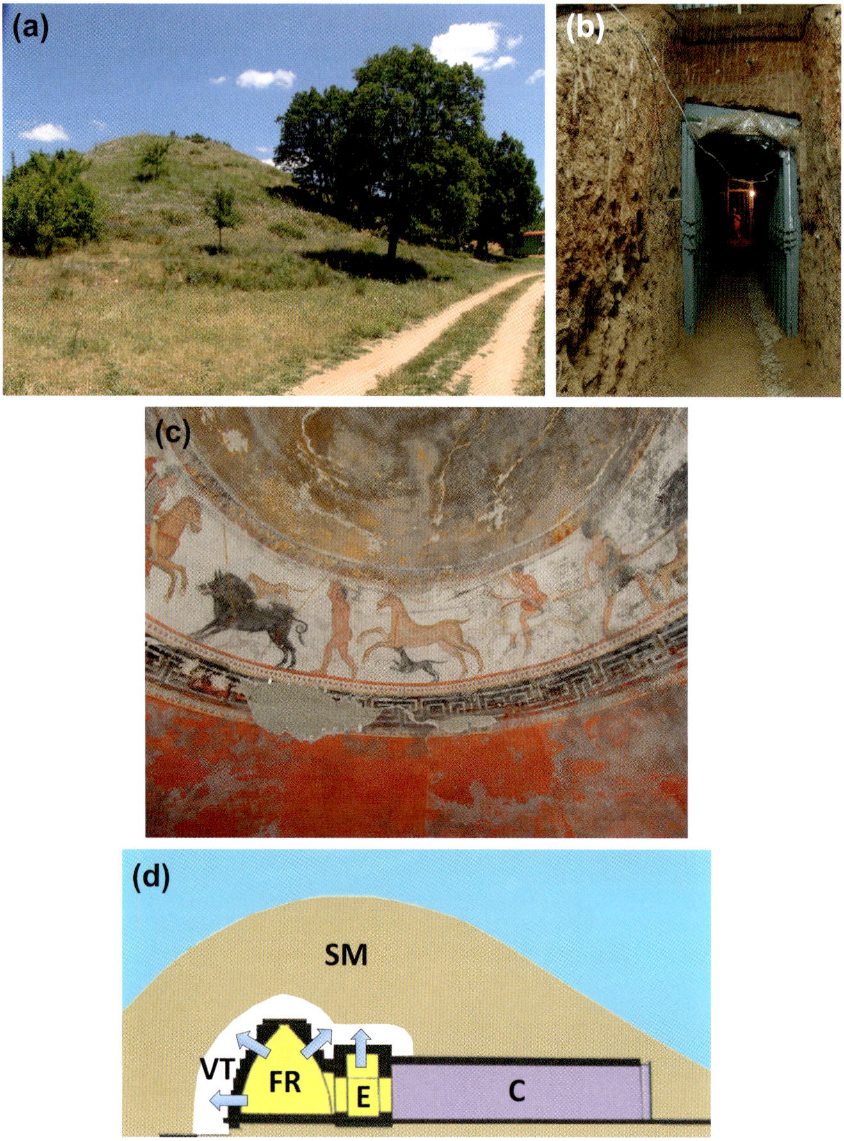

FIGURE 6.18 (a) View of the Thracian sepulchral mound near Alexandrovo, Bulgaria. (b) The ventilation tunnel excavated all around the funeral room. (c) Funeral room with hunting views. (d) Vertical cross-section of the tomb. Yellow: FR, funeral room; E, entrance room; violet: corridor (C) connecting the funeral room to the exterior; a closed door separates the corridor from the funeral room; White space VT: ventilation tunnel excavated into the earth (brown) constituting the sepulchral mound (SM). Arrows indicate the migration of water.

For the above reasons, it is desirable to dry the monument as soon as possible and in effect, it is not rare to see people ventilating and heating the room to increase the drying rate. Unfortunately, this practice has the negative consequence of increasing the damage due to the efflorescence. Better solutions are:

1. To favour wall evaporation from the opposite side of the fresco in order to diminish or stop the moisture flow across the fresco and, consequently, the total amount of dissolved salts that will deposit on the painted surface. In the case of the Padua Baptistery, some external bricks were removed from the thick wall, in order to favour back ventilation and internal wall drying.

   Two Thracian sepulchral mounds enclosing masonry rooms with paintings inside, one near Kazalunk and another near Alexandrovo, Bulgaria, were damp due to rainwater percolation and needed some drying (Fig. 6.18). It was decided to stop percolation and dry to the required level the painted masonry constituting the funeral room and the entrance. In order to avoid accumulation of soluble salt on paintings that in addition favour mould colonization, it was decided to make a forced migration of water and dissolved salts across the walls, but from the paintings to the exterior. To this aim, a tunnel was excavated in the earth of the mound all around, but externally to the funeral room in order to control the evaporation. The excavation had grilles for natural and forced air exchanges with the exterior. The methodology was successful in Kazalunk, and it was repeated in Alexandrovo.
2. Covering the painting with Japanese paper sheets, so that the water with the dissolved salt will cross the paper sheet and evaporate forming efflorescence that will encrust the paper, not the mural. Once the evaporation with the internal salt migration will be concluded, it is possible to remove the paper with the mineral encrustation from the dry wall.

References

Theory and general applications

Adamson, A.W., 1986. A Textbook of Physical Chemistry. Academic Press, San Diego, 972 pp.

Azam, S., 2007. Study on the geological and engineering aspects of anhydrite/gypsum transition in the Arabian Gulf coastal deposits. Bull Eng Geol Env 66, 177–185.

Blanchard, D.C., Woodcock, A.H., 1980. The production, concentration, and vertical distribution of the sea-salt aerosol. In: Kneip, T.J., Lioy, P.J. (Eds.), Aerosols: Anthropogenic and Natural, Sources and Transport, Annals of the New York Academy of Sciences, vol. 338, pp. 330–347. New York, 618 pp.

Brimblecombe, P., 1987. The Big Smoke. Methuen, London, 185 pp.

Brimblecombe, P., 1992. History of atmospheric acidity. In: Radojevic, M., Harrison, R.M. (Eds.), Atmospheric Acidity. Elsevier, London, pp. 267–304.

Brimblecombe, P., 1995. History of air pollution. In: Singh, H.B. (Ed.), Composition, Chemistry and Climate of the Atmosphere. Van Nostrand Rehinold, New York, pp. 1–18.

Camuffo, D., 1984a. Condensation-evaporation cycles in pore and capillary systems according to the Kelvin model. Water, Air and Soil Pollution 21, 151–159.

Camuffo, D., 1990. Acidic precipitation research in Italy. In: Bresser, A.H.M., Salomons, W. (Eds.), Acidic Precipitation, vol. 5. Springer Verlag, New York, pp. 229–265.

Camuffo, D., Enzi, S., 1995. Impact of clouds of volcanic aerosols in Italy in the past centuries. Natural Hazards 11, 135–161.

Camuffo, D., Giorio, R., 2003. Quantitative evaluation of water deposited by dew on monuments. Boundary Layer Meteorology 107, 665–672.

Dullen, F.A.L., 1979. Porous Media. Fluid Transport and Structure. Academic Press, New York, 396 pp.

Gordon, J., MacDonald, F., 1953. Anhydrite-gypsum equilibrium relations. American Journal of Science 251, 884–898.

Kireev, V., 1977. Physical Chemistry. Mir, Moscow, 572 pp.

Price, C., Brimblecombe, P., 1994. Preventing salt damage in porous materials. In: Proc. Preventive Conservation, pp. 90–93. (Ottawa).

Prodi, F., Fea, G., 1979. A case of transport and deposition of Saharan dust over the Italian peninsula and southern Europe. Journal of Geophysical Research 84, 6551–6960.

Applications to conservation

Amoroso, G.G., Fassina, V., 1983. Stone Decay and Conservation – Atmospheric Pollution, Cleaning, Consolidation and Protection. Elsevier, Amsterdam, 453 pp.

Arnold, A., 1983. Determination of mineral salts from monuments. Studies in Conservation 29, 129–138.

Arnold, A., Zehnder, K., 1990. Salt weathering on monuments. In: Zezza, F. (Ed.), The Conservation of Monuments in the Mediterranean Basin, pp. 31–58. (Grafo, Bari).

Arnold, A., Zehnder, K., 1991. Monitoring wall paintings affected by soluble salts. In: Cather, S. (Ed.), The Conservation of Wall Paintings, Paul Getty Trust. Thien Wah Press, Singapore, pp. 103–135.

Beadecker, P.A., Reddy, M.M., 1993. The erosion of carbonate stone by acid rain. Journal Chemical Education 66 (3), 104–108.

Becker, T.W., Krumbein, W.E., Warscheid, T., Resende, M.A., 1994. Investigations into microbiology. In: Bianchi, H.K. (Ed.), Ideas – Investigations into Devices against Environmental Attack on Stones. GKSS-Forschungszentrum, Geesthacht (F.R.G.), pp. 147–190.

Bernardi, A., Camuffo, D., Del Monte, M., Sabbioni, C., 1985. Microclimate and weathering of a historical building: the Ducal Palace in Urbino. Science Total Environment 46, 243–260.

Camuffo, D., 1984b. The influence of run-off in weathering of monuments. Atmospheric Environment 18, 2273–2275.

Camuffo, D., 1986. Deterioration processes of historical monuments. In: Schneider, T. (Ed.), Acidification and Its Policy Implication. Elsevier, Amsterdam, pp. 189–221.

Camuffo, D., 1991a. Aspetti microfisici delle precipitazioni acide in relazione al degrado dei monumenti. In: Morselli, L. (Ed.), Deposizioni acide, i precursori, l'interazione con l'ambiente e i materiali. Maggioli Editore, Rimini, pp. 339–350.

Camuffo, D., 1991b. Environment and microclimate. In: Baer, N.S., Sabbioni, C., Sors, A.I. (Eds.), Science, Technology and European Cultural Heritage. Butterworth- Heinemann, London, pp. 37–50.

Camuffo, D., 1992. Acid rain and deterioration of monuments: how old is the phenomenon? Atmospheric Environment 26B, 241–247.

Camuffo, D., 1994. Aspetti meteorologici e microclimatici nel degrado dei materiali lapidei, vol. 88. Accademia Nazionale dei Lincei, Contributi del Centro Linceo Interdisciplinare (Beniamino Segre). pp. 9–27.

Camuffo, D., 1995. Physical weathering of stones. Science Total Environment 167, 1–14.

Camuffo, D., Del Monte, M., Ongaro, A., 1984. The pH of atmospheric precipitation in Venice, related to both the dynamics of precipitation events and the weathering of monuments. Science Total Environment 40, 125–139.

Camuffo, D., Del Monte, M., Sabbioni, C., 1987. Influenza delle precipitazioni e della condensazione sul degrado superficiale dei monumenti in marmo e calcare. In: Ministero dei Beni Culturali ed Ambientali: "Materiali Lapidei", special issue of Bollettino d'Arte. Poligrafico dello Stato, Rome, pp. 15–36.

Camuffo, D., Del Monte, M., Sabbioni, C., Vittori, O., 1982. Wetting, deterioration and visual features of stone surfaces in an urban area. Atmospheric Environment 16, 2253–2259.

Camuffo, D., Valcher, S., 1986. A dew point signaller for conservation of works of art. Environmental Monitoring and Assessment 6, 165–170.

Caneva, G., Nugari, M.P., Salvadori, O., 2009. Plant Biology for Cultural Heritage – Biodeterioration and Conservation. The Getty Conservation Institute, Los Angeles.

Chabas, A., Lefevre, R.A., 2000. Chemistry and microscopy of atmospheric particulates at Delos (Cyclades-Greece). Atmospheric Environment 34, 225–238.

Charola, A.E.., Delgado Rodrigues,, M.J., 2012. The gardens of the National Palace of Queluz (Portugal) and comments about the conservation intervation. Associação World Monuments Fund Portugal and World Monuments Fund, Lisbon.

Del Monte, M., Sabbioni, C., 1980. Authigenic dolomite on marble surface. Nature 228, 350–351.

Del Monte, M., Sabbioni, C., 1984a. Morphology and mineralogy of fly ash from a coal-fuleded power plant. Archives for Meteorology, Geophysics, and Bioclimatology 35, 93–104.

Del Monte, M., Marcazzan, G.M., Sabbioni, C., Ventura, A., 1984b. Morphological, physical and chemical characterisation of particles emitted by a coal-fired power plant. Journal of Aerosol Science 15, 325–327.

Espinosa Marzal, R.M., Scherer, G.W., 2008. Crystallization of sodium sulfate salts in limestone. Journal Environmental Geology. 56 (3–4), 605–621. http://dx.doi.org/10.1007/s00254-008-1441-7.

Gummerson, R.J., Hall, C., Hoff, W.D., 1980. Water movement in porous building materials – ii hydraulic suction and sorptivity of brick and other masonry materials. Building and Environment 15, 101–108.

Hall, C., 1981. Water movement in porous building materials – IV the initial surface absorption and the sorptivity. Building and Environment 16, 201–207.

Haynes, H., O'Neill, R., Neff, M., Mehta, P.K., 2008. Salt weathering distress on concrete exposed to sodium sulfate environment. ACI Materials Journal 105 (1), 35–43.

Padfield, T., 1995–97. An Introduction to the Physics of the Museum Environment. Natural Museum, Denmark. http://www.natmus.min.dk/cons/tp (published in the website).

Sabbioni, C., Zappia, G., Gobbi, G., Pauri, M.G., 1993. Deterioration of ancient and modern buildings materials due to environmental factors. In: Brebbia, C.A., Frewer, R.J.B. (Eds.), Structural Repair and Maintenance of Historical Buildings. Computational Mechanics Publications, Southampton, pp. 235–242.

Saiz-Jimenez, C., 1995. Microbial melanins in stone monuments. Science Total Environment 167, 273–286.

Warscheid, T., Braams, 2000. Biodeterioration of stone; a review. International Biodeterioration & Biodegradation 46, 343–368.

Warscheid, T., Oelting, M., Krumbein, W.E., 1991. Physico-chemical aspects of biodeterioration processes on rocks with special regard to organic pollution. International Biodeterioration 28, 37–48.

Winkler, E.M., 1994. Stone in Architecture. Springer Verlag, Berlin, 313 pp.

Wittenburg, C., 1994. Trokene Schadgas- und Partkeldeposition auf vershiedenen Sandsteinvarietäten unter besonderer Berücksichtigung atmosphärischer Einftußgrößen. PhD thesis, Hamburg.

CHAPTER

7

Atmospheric Stability and Pollutant Dispersion

OUTLINE

http://dx.doi.org/10.1016/B978-0-444-63296-8.00008-1

7.1. INTRODUCTION

The way in which solar radiation is absorbed by a surface and the energy is partitioned between a solid surface and the atmosphere, the interactions between materials and the atmosphere or the mechanisms by which the atmosphere transports or disperses pollutants are problems common to several disciplines. The same findings can be applied to different scales: from the large scale of physical geography and long-range transport of pollutants, to the local scale of urban climate and pollution to the microscale of the monument surface and the internal boundary layers (*IBLs*) that develop in the air close to surfaces. The general results may be adapted to a number of specific cases. Application to microscale and, in particular, to indoor environments is characterized by the highest level of complexity. On the medium scale, the sun is the dominant source of heat, the soil is the dominant surface and the ground is the dominant medium. In a room, several sources or sinks of heat are found: lamps, heaters or air-conditioning devices and solar radiation through windows. There are six room surfaces, with different orientation with reference to the gravity force (air mixing is generated in the same way either by a warm floor or a cold ceiling); furniture and objects provide additional smaller surfaces. In addition, the surfaces are not homogeneous, having doors, windows and other peculiarities; and the material used for floor, ceiling and walls is not the same. For any selected monument, the same findings can be applied at least two times: on the medium scale to know the ambience where the monument lives (or dies) and on the microscale to interpret the interactions with the environment and the deterioration mechanisms. The local scale is particularly important in many cases, with some specific applications to the microscale. This chapter, which mainly focusses on medium or local scale, is preliminary to the next chapter, which specially focusses on the microscale.

The dispersion of effluents, the path and the concentration of gaseous pollutants are determined by the dynamics of the planetary boundary layer (*PBL*). The *PBL* consists of the first atmospheric layer particularly affected by the direct influence of the Earth's surface. It extends vertically to an altitude of 1000–1500 m.

The dispersion of pollutants is essentially related to the transport and mixing of air masses that are conditioned by the dynamic behaviour of the *PBL* with perturbations and asymmetric forcing. Several geophysical parameters are involved. None of them, taken alone, can simply and completely account for the various phases of this process. For this reason, attempts have been made to relate these parameters in various ways to obtain a novel holistic approach with bulk parameters that may be useful in specific applications.

Air turbulence may be of thermal or mechanical origin. Thermally generated turbulence, caused by convective movements above a warm soil surface, is particularly important and is responsible for phenomena such as *fumigation* and *looping*, which will be briefly described later. Mechanically induced turbulence, which is due to fluctuations of wind directions and intensity or due to friction and presence of obstacles, is also of considerable importance.

From a physical point of view, the key factors are:

- atmospheric stability generated by the vertical temperature gradient that follows the soil heating or cooling (thermal factor),
- wind and the turbulence characterized by weather factors or induced by the roughness of the Earth's surface (mechanical factor), and
- combination of mechanical and thermal factors and spatial distribution of topographic (or architectural) discontinuities or temperature changes.

It is clear that the wind intensity u is related to the dilution of effluents; the direction α from which the wind is blowing indicates the initial trajectory of a plume and the subsequent variations in wind intensity and direction will affect the dispersion. However, it is not easy to quantify this mechanism. The distance to which a pollutant is transported may be calculated in a simple manner only in the case that the wind is stationary, with no significant changes in direction or speed. In the nonstationary case, the path is not straight and the value of the average wind speed $<u>$ and the integral $\int u \, dt$ represent *the quantity of air passed through* the measuring sensor during the sampling interval. The transport is, in a first approximation represented by the average wind speed along the dominant direction and is correctly determined as the vector sum of the instantaneous wind vectors $\boldsymbol{u}_i$ observed at regular intervals, i.e. $\sum \boldsymbol{u}_i$. However, the wind vane always indicates a direction, even in the absence of wind. In the case of calm, or in the intervals between intermittent wind, i.e. when the speed is zero, the direction recorded by the wind vane is meaningless and monitoring is misleading. The majority of meteorological records is based on time averages and in the case of light wind, when the pollutant dispersion is critical, it is impossible to distinguish between real wind direction and vane left in some position after the wind has dropped.

In a Cartesian system of reference, the instantaneous wind vector $\boldsymbol{u}(t)$ is expressed by means of the components $u(t) \sin \alpha(t)$ and $u(t) \cos \alpha(t)$ and the displacement of a puff released by a point source is represented by $\int u(t) \sin \alpha(t)\, dt$ and $\int u(t) \cos \alpha(t)\, dt$. It is easily recognized that

$$<u \sin \alpha> \neq <u> <\sin \alpha> \neq <u> \sin <\alpha>, \tag{7.1}$$

and similarly for the cosine so that the popular practice of measuring $<u>$ and $<\alpha>$ separately may be useful for several meteorological purposes, but is not appropriate for the accurate determination of the transport of airborne pollutants.

Similarly, directional fluctuations indicate the horizontal fan in which dispersion occurs. The vertical dispersion is not only determined by vertical fluctuations but also by the soil roughness, slope and atmospheric stability. A simple treatment of the problem supposes that the atmospheric properties, during the diffusion process, are constant in time and homogeneous in space and, in addition, pollutants are not affected by decomposition, deposition or chemical or photochemical reactions.

The effect of the scale of the turbulence on pollutant dispersion is easily summarized. If the size of *eddies* (i.e. all the elements of the fluid that characterize the turbulent motion, each one with an identity and a life history of its own, e.g. rotors and vortexes) is much greater with reference to the size of a puff of smoke, the latter is moved (translated and rotated) by large eddies, without bringing about any variation in dilution or form. If the size of the eddies is much smaller than the puff size, the whole gaseous and particulate population within the puff is subject to translation and rotation but the shape and the position of the puff will remain unchanged. If, finally, eddies and puff have similar size, the dispersion is much more effective.

7.2. VERTICAL TEMPERATURE GRADIENTS AND PLUME BEHAVIOUR

The study of the atmospheric stability in relation to the vertical temperature gradient has led to many useful developments. In terms of vertical temperature gradients, the following definitions are used: *superadiabatic, adiabatic* or *neutral gradient, subadiabatic, isothermy* and *inversion gradient*. These gradients are associated with specific plume behaviour (Fig. 7.1). The temperature profile of the atmosphere (as measured by a

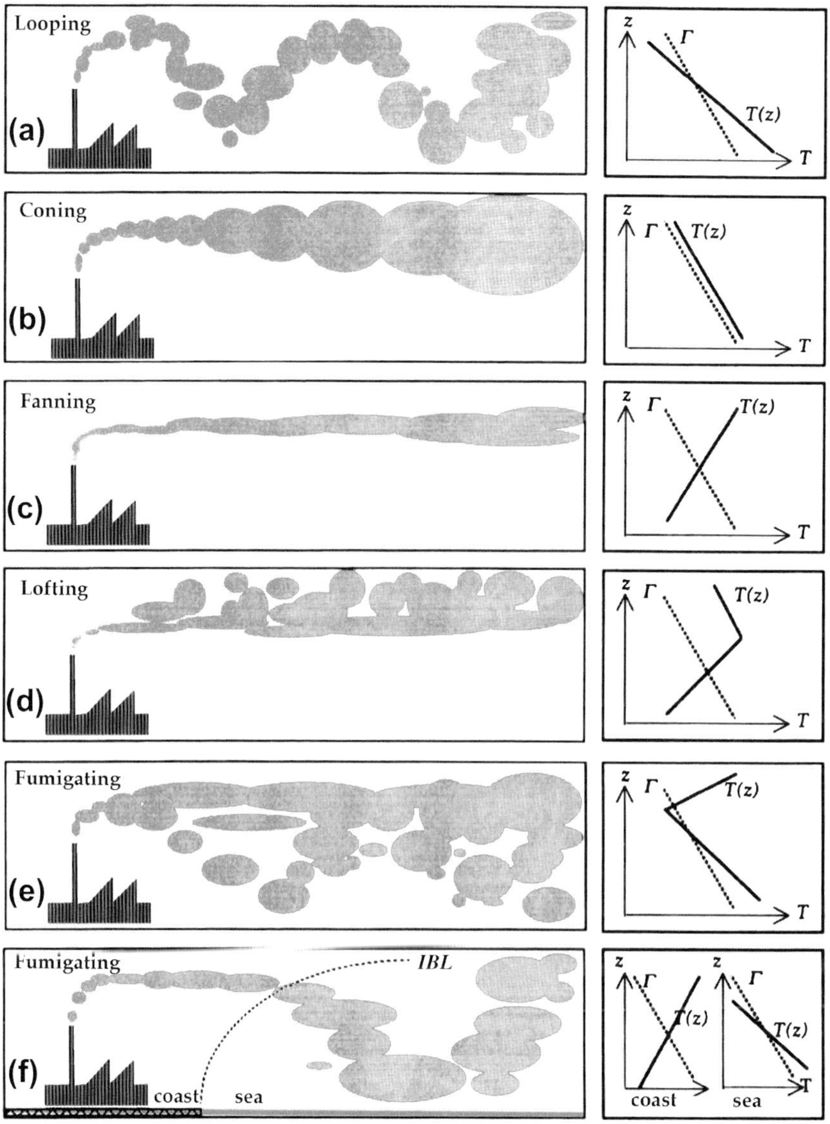

FIGURE 7.1 (a–e) Behaviour of a plume under different atmospheric stability.

radiosonde and indicated by the continuous line in the diagram on the right of the figure) is compared with the adiabatic gradient, conventionally indicated Γ, which is defined as $\Delta T/\Delta z = -1\ ^{\circ}\mathrm{C}/100\ \mathrm{m}$ (represented in Fig. 7.1 with a dotted line). It is also customary to speak in terms of *lapse rate* = *−gradient* in order to deal with positive figures instead of negative ones.

A temperature gradient is defined as being *superadiabatic* when it corresponds to the vertical cooling defined by the temperature gradient $\Delta T/\Delta z < -1°C/100$ m, or $\Delta\Theta/\Delta z < 0$ °C/100 m. An air parcel adiabatically raised becomes warmer than the surrounding air, gains buoyancy and is accelerated upwards. If it is lowered, it becomes colder and denser than the surrounding air and is accelerated downwards (unstable equilibrium, Fig. 7.2(a)). When the soil surface is considerably heated by solar radiation, this instability generates uprising warm flows (named *thermals*) as well as descending colder flows that operate a continuous *mixing* of the air masses in the *PBL*. The development of the superadiabatic layer is somewhat delayed with respect to the intensity of the solar radiation and in northern Italy it reaches a maximum altitude of about 250 m in winter and about 400 m in summer.

With regard to the dispersion of pollutants, the rising and descending currents of the convective cellar movements cause a characteristic continuous up and down movement of plumes released from a stack (Fig. 7.1(a)), called *looping*. In this condition, a source at the surface level is in a better situation to disperse pollutants, while an elevated source may reach the ground level, and is, therefore, worse for health.

The *adiabatic* gradient is equal to $\Delta T/\Delta z = -1°C/100$ m, or $\Delta\Theta/\Delta z = 0°C/100$ m, so that each vertical adiabatic motion changes the temperature of an air parcel exactly as the vertical actual distribution of temperature of the environmental air, thus leaving the parcel in *neutral equilibrium* with the surrounding air. For this reason, the atmosphere is said to be *neutral* (Fig. 7.2(b)). This type of gradient forms with overcast sky or when there is a strong wind. The overcast sky tends to reduce the radiative balance (i.e. solar income and infrared cooling) and keeps the soil surface and the atmosphere in neutral equilibrium; the turbulence induced by the strong wind leads to homogeneous mixing of the air masses and there is no stratification.

Under neutral atmospheric conditions, pollutants are symmetrically dispersed according to the Gaussian law with respect to the axis of the plume (or with respect to the trajectory, in the case of a punctual release, i.e. a *puff*). If the trajectory determined by the wind transport is a straight line, the dispersion gives rise to a cone of pollutant that is denser at the *apex* (beginning of dispersion) and then along the *axis* (or in the more general case along the *trajectory*, which is defined as the centre of each cross-section) and whose concentration decreases when radially departing from the axis. The conventional name of this dispersion is *coning* (Fig. 7.1(b)).

A *subadiabatic* gradient is $\Delta T/\Delta z > -1°C/100$ m, or $\Delta\Theta/\Delta z > 0°C/100$ m. In this case, an upward adiabatic movement makes the air parcel colder than the air at the level it reaches, so that the parcel is forced downwards, towards its original position. The opposite occurs when the

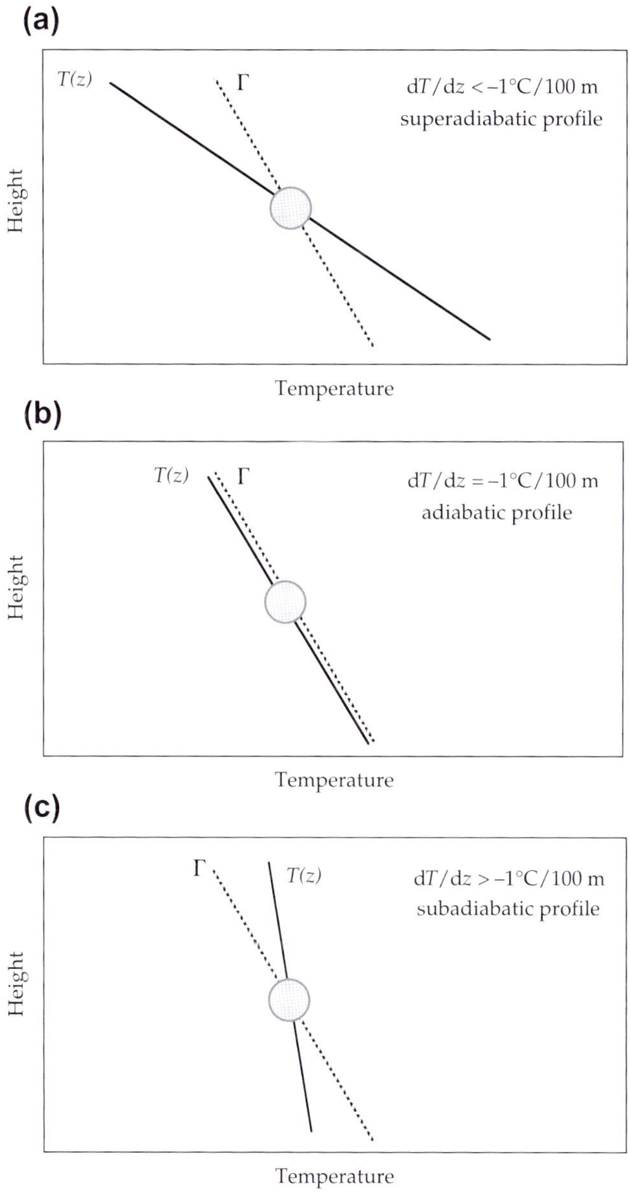

FIGURE 7.2 Vertical temperature profile $T(z)$ and stability of the air masses. (a) An air parcel, displaced upwards from its initial position, follows the adiabatic profile Γ and becomes warmer than the air at the same level and is accelerated upwards. A similar situation occurs for a downward displacement but in this case, the parcel becomes colder and is accelerated downwards. The equilibrium is unstable. (b) An air parcel, which is vertically displaced, always remains at the same temperature of the surrounding air and is in neutral equilibrium. (c) An air parcel displaced upwards becomes colder than the surrounding air and is forced to return to the original position. The same happens for downward displacements. The equilibrium is stable.

parcel is displaced downwards. Thus, the stable equilibrium tends to keep each air parcel at its original level and the atmosphere is said to be *stable* (Fig. 7.2(c)). Two important subclasses can be mentioned: the *isothermy*, i.e. $\Delta T/\Delta z = 0°C/100$ m, with the temperature T being constant along the vertical (moderate stability), and the *inversion*, i.e. $\Delta T/\Delta z > 0°C/100$ m, a positive gradient, with the temperature T increasing with the altitude (strong stability). Under these conditions, pollutants tend to lie at the same level and frequently they are found with moderate ventilation and weak sunlight. The inversion subclass is typical of clear windless nights, that is with infrared radiative loss (i.e. net soil heat loss) and no turbulent mixing. Isothermy may occur under weak radiative exchange (haze or thin stratus cover), especially in the transition between neutral atmosphere and inversion, or vice versa.

The larger the atmospheric stability, the greater the suppression of the turbulence and the smaller the mixing and the exchanges between atmospheric layers at different levels. A pollutant emitted at a certain height will be trapped at that level and will be transported and dispersed along that horizontal plane. Plumes emitted from tall chimneys remain aloft, while those emitted at ground level cannot uplift. As far as human health or monuments are concerned, this is the most favourable situation for elevated sources but is less favourable for low-level sources. A veering wind is one that turns clockwise with height; a backing wind turns counterclockwise. When the wind direction changes with height (*directional wind shear*), the plume will horizontally spread, the upper part departing from one side and the lower part on the opposite side, conferring a fan shape to the plume. This configuration is called *fanning* (Fig. 7.1(c)).

When a neutral layer is found above the stable one, as typically occurs above the nocturnal inversion layer, the pollutants released from tall chimneys cannot cross the inversion but once they have entered the neutral layer, they can be dispersed upwards. The process of rising of pollutants is called *lofting* (Fig. 7.1(d)).

An important phenomenon (opposite to the previous one) typically occurs in the morning that follows a calm night. After sunrise, when the solar radiation starts to warm the soil, a thin ground-based mixed layer develops, generated by the growing solar radiation. The nocturnal inversion layer begins to erode from below, the base of the inversion being the top of the ground-based superadiabatic layer under development. When the top of the superadiabatic layer reaches the altitude where the pollutants have been entrapped during the night, the pollutants are immediately transported up and down (looping) through the lower layer due to the violent mixing, which results in a sudden rise in the concentrations at ground level, and this process is known as *fumigation* (Fig. 7.1(e); the example (f) will be discussed later).

In the *PBL*, various sublayers may sometimes overlap one above the other, each with a different gradient. In the middle of the day, for example, because of the strong solar irradiation, a ground-based superadiabatic layer can be found. This is characterized by intense convective activity that mixes particles, gases and vapours dispersed within the layer. Above this, an adiabatic layer or a subadiabatic one can be generally found. At night, on the other hand, because of the radiant cooling of the ground, the first layer is frequently an inversion with a subadiabatic layer above it. In this case, the vertical mixing is suppressed and smoke stratification and fanning in the inversion layer and even cone-like dispersion in the upper neutral or slightly stable layer are formed.

Thermodynamic surveys made with radiosondes are useful to know the vertical temperature profile. Launching radiosondes at regular time intervals shows how the air temperature evolves during the course of the day. This is shown with a two-dimensional representation with the height in the ordinate and the time in the abscissa. There, the isotherms (as well as the isolines of any other thermodynamic parameter), by their reciprocal distance and position, indicate the intensity of the vertical gradients, the stratification and the dynamic evolution of the atmospheric layers. Each vertical intersection corresponds to a vertical profile. If in this plot the vertical distance d is measured, between an isoline and the next one, the layer is adiabatic when two isotherms, which differ from each other by 1 °C, are found at the exact distance $d = 100$ m; if $d < 100$ m, the layer is superadiabatic, while if $d > 100$ m, it is subadiabatic. The isolines are vertical and widely spaced when the gradients are of the isotherm type or vertical and close when there is a sudden change of air masses, e.g. the change of a breeze. Where there is an inversion, at the upper levels, the isolines are characterized by higher T values. When passing from stable to neutral or superadiabatic zones, the isolines show a sharp bending and the line connecting all these turning points indicates the base or the top of such layers. The base and top of such layers begin and end at the ground level or aloft at the point where they meet each other. Figure 7.3 is an illustrative example and corresponds to the final part of the erosion of the nocturnal inversion and the development of the mixed layer (Camuffo, 1980).

In order to forecast the course and dispersion of gaseous pollutants, it is necessary to know the exact level at which effluents have been emitted into the atmosphere, as well as the stability of the various layers into which the *PBL* is divided and also the horizontal discontinuities that are found. It follows, therefore, that urban and industrial pollutants will reach different levels and will have different destinies, with different paths and dispersion, and will ultimately affect different areas.

It might be useful to note that when T and z are represented in a diagram, the neutral atmosphere is defined by the slope of $\Gamma = 1°\text{C}/100$ m, where a steeper slope indicates instability and a flatter one indicates

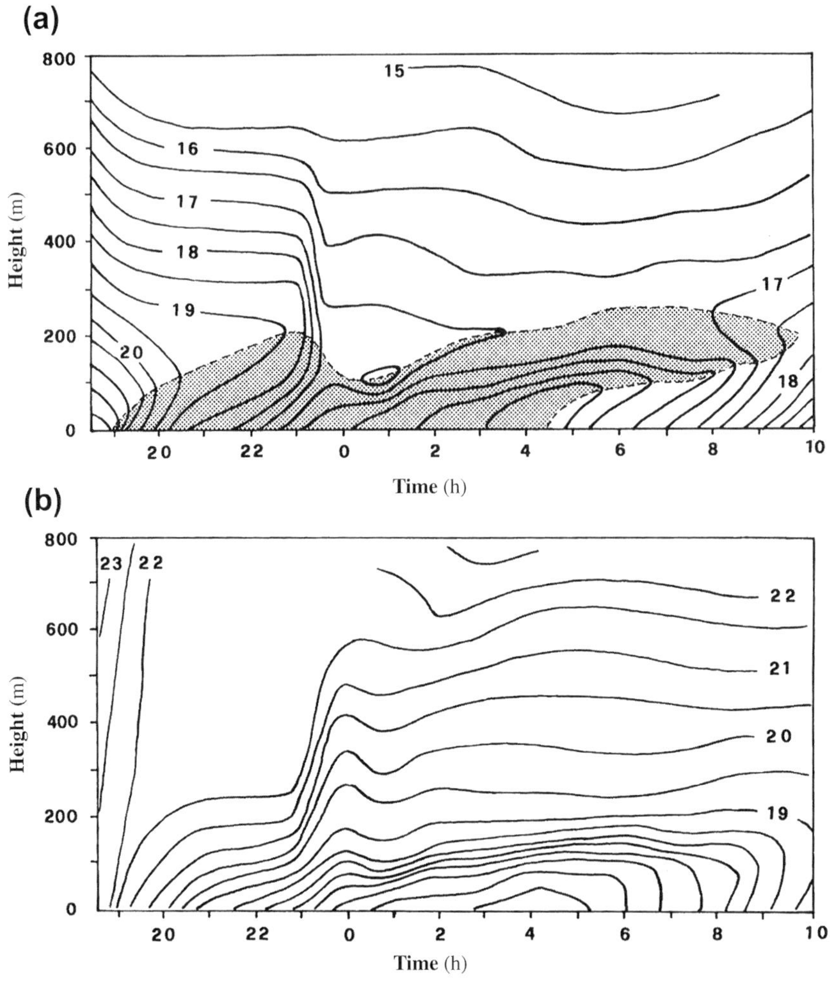

FIGURE 7.3 (a) Vertical versus time representation of the air temperature T; (b) vertical versus time representation of the potential temperature Θ. Padua, night from 8 July to 9 July 1978. In figure (a), the stable layer (i.e. inversion and isothermy) is shown with grey shading. Note that the inversion began to form at 7 pm (i.e. 1 h before sunset) and grew until 11 pm when the nocturnal breeze arrived. The dynamic effect lowered by some 50% the top of the inversion and the colder air masses caused a sudden drop of temperature. After the sunrise (4.30 am), the soil began to warm, forming a ground-based mixed layer and eroding the inversion from below. All soundings were made with the same tethered radiosonde, attached to a fishing nylon line. Sampling frequency: a sounding per hour.

stability. Very often, the eye is unable to exactly evaluate the slope and recognize the degree of stability from a diagram of this type.

When, instead, this is represented in terms of potential temperature Θ in a vertical versus time cross-section, the effect is much more immediate in that instability, neutrality or stability are represented, respectively, by $\Delta\Theta/\Delta z < 0°C/100$ m, $\Delta\Theta/\Delta z = 0°C/100$ m or $\Delta\Theta/\Delta z > 0°C/100$ m, where Γ corresponds to the intersection with the abscissa axis at an obtuse, right and acute angle, respectively, and this differentiation can be immediately seen with the naked eye. However, near the ground level and in the small scale, $T = \Theta$.

7.3. EFFECTS DUE TO TOPOGRAPHIC HORIZONTAL INHOMOGENEITY

Sometimes the *PBL* is not characterized by only one of the above characteristic layers but special effects may derive for the simultaneous presence of different layers, either in the vertical (multiple layering) or in the neighbouring regions, e.g. mainland and sea, with a frontal area where the two layers are in contact and overlap. In the event of topographic inhomogeneities, such as in a coastal region, the different temperatures of the sea and the mainland determine boundary layers with different stability. When these layers are pushed by the general circulation in the other region, an *IBL* develops along the frontal area.

Variations in the temperature gradients are found especially when the geophysical dissimilarities are extensive, such as along the sea coast. When the soil warms up or cools down, superadiabatic or inversion gradients are generated but the temperature of the water, and the temperature gradient above it, tend to remain unchanged. A plume of smoke, which during the night is borne by a land breeze, travels from the stable layer above the land to the unstable one at the marine surface that is relatively warmer (Fig. 7.1(f)). From then on, local convection causes looping. This is a characteristic type of fumigation and often occurs in Venice, Italy, not only during night-time but also during the whole winter. In fact, the Venetian hinterland during winter is very cold and is characterized by stability for most of the day, whereas the Adriatic Sea and the lagoon are relatively warm and generate a continuous mixing of the overlying air masses. The industrial area is located in the hinterland and the pollutants are entrapped within the stable layer at a certain height. When the air masses are transported over the warm lagoon, thermal mixing increases the pollutant concentration at the sea level, and especially at the city centre, where the urban heat island increases convective exchanges and turbulence.

Another *IBL* develops near the coastline. For example, during the daytime of the mid- and warm seasons, the relatively unstable mass of marine air transported by the sea breeze becomes completely unstable as it moves inland, so that a mixed layer develops. Inland, as one departs from the coast line, the height h of this *IBL* increases like a parabola, and for Italy, it is computed with the equation (Anfossi et al., 1976):

$$h = a\,\omega\sqrt{\frac{x}{<u>}} \tag{7.2}$$

which has been adapted from the original van der Hoven (1967) formula, where $a = 0.05\ \text{ms}^{-3/2}$, x is the distance from the coast, u is the wind speed and ω represents the *Brunt–Väisälä frequency* (considered a frequency in that it has the unit of time^{-1}) defined as

$$\omega^2 = \frac{g}{\Theta}\frac{\partial\Theta}{\partial z} = g\frac{\partial \ln\Theta}{\partial z} \tag{7.3}$$

where g represents gravitational acceleration, z is the height with respect to the soil and Θ is the potential temperature (in Kelvin). Please note that ω^2 is usually positive, but in proximity of the soil under superadiabatic conditions, it might be negative, depending on the sign of the vertical gradient of Θ. The Brunt–Väisälä frequency characterizes the atmospheric stability on the basis of the vertical temperature gradient $\partial\Theta/\partial z$.

Downwind of the coast, in the wind profile, another *IBL* develops because of the change in the surface roughness as the wind moves from the marine surface to the hinterland, or vice versa. This layer develops at a level that is proportional to the distance from the coast with a height h to shore distance x ratio that lies between $h/x = 1/10$ and $h/x = 1/20$ (Elliott, 1958; Panofsky and Townsend, 1964) and that varies according to wind intensity (Schlieting, 1955; Echols and Wagner, 1972).

The so-called *urban meteorology* studies similar phenomena, as well as those due to the formation of haze related to vapour emissions and to the condensation on airborne pollutants acting as condensation nuclei. When these aerosols are heated by solar radiation, they may generate a characteristic thermal inversion aloft.

7.4. URBAN CLIMATE: HEAT ISLAND AND AERODYNAMIC DISTURBANCE

The urban climate (Oke, 1978; Landsberg, 1981: Rosenzweig et al, 2011; Rohinton, 2012) is of the utmost importance as most of the population lives in cities and the majority of the cultural heritage is also concentrated there. The very fact that a town exists, the architecture of its buildings, the town use and the activities of the people alters, locally, the regional climate, leading to characteristic interactions with the atmosphere that may be more, or less, violent. Such interactions have led to a new definition of local climate: the urban climate, the main characteristics of which can be synthesized as follows.

In a town, the absorption of solar irradiation is elevated as a consequence of a number of factors: the main one is the presence of vertical surfaces with short distance between them (buildings and streets) that entrap the solar radiation after a multiple reflection and behave as a blackbody. The soil, being horizontal, during the day receives a certain amount of radiant energy depending on the solar height and cloud cover; during day and night, it looses thermal radiation towards the clear sky, which has a very low absolute temperature. The building walls, being vertical, will radiate horizontally and will receive back almost the same amount of energy emitted from the lower atmosphere or from nearby bodies at the same temperature. Other factors that contribute to the urban heat island include the storage of heat in materials such as terracotta tiles, bricks and asphalt; the low albedo of roofs and roads. Finally, we should consider the sensible and latent heat released by human activities, mainly domestic heating and traffic.

All the above factors determine higher temperature compared with the surrounding countryside. This is the so-called *heat island* effect that is particularly evident in the evening and the first hours of the night. The urban island also has contradictory aspects, e.g. the local rise of temperature tends to reduce relative humidity, time of wetness and the intensity and frequency of fog with respect to the surrounding rural area; on the other hand, the local increase of pollutant concentration and condensation nuclei tends to increase photochemical smog, haze, fog and thunderstorm frequency. For instance, London was a foggy town and the period of high pressure in 1952 caused a thick, persistent fog and many people died (Brimblecombe, 1987). After a severe regulation abated the emission of particulate matter that acts as condensation nuclei, the fog frequency was drastically reduced.

The heat released by the town generates buoyancy, thus leading to convective rise of the air masses. Under moderate stability conditions, the rural area is covered with low stratiform clouds or (high) fog, particularly frequent in the winter and midseasons, and the air that is convectively raised perforates the lower stable layer and a blue sky can be seen just above the town. This convection is even stronger in summer when the altitude attained by the ascending column of warm air reaches the cloud condensation levels, forming small cumulus clouds, typical of fine weather. If the cumulus clouds grow sufficiently, rain storms may occur over the town or the area downwind of it. Showers are more frequent in large towns, and the urban pollution, with its condensation nuclei, contributes to increasing the frequency of summer precipitation.

During clear nights, the soil cools due to the heat loss associated with the infrared emission and a ground-based nocturnal inversion develops. The air becomes colder, denser and still. In the case of breeze or moderate wind, these slip above the stable inversion layer; the inversion is attenuated or destroyed only in the case the wind aloft is intensely turbulent.

Tall buildings may emerge from the stable layer and protrude into the mixed layer with more intense wind. In this case, the turbulence induced in the cavity region and the wake downwind exchanges momentum, so that the upper layers of the wind field are slowed down and the lower ones are accelerated. Urban ventilation can, therefore, be greater than the rural one during the night. On an average, buildings tend to shield from strong wind. However, local distortions in the wind field are not homogeneous: in some areas, the wind speed is attenuated, while in other areas it is increased, because of induced turbulence (energy dissipation) or wind channelling (Venturi effect). Walking people might interpret such changes as wind gusts, but local people exactly know the precise locations where, for example, umbrellas get blown inside out. Tall buildings may exert important local departures to the wind field and form critical frontal situations, typically shielding, accelerating, deflecting and creating building channelling, corner streams, or inducing turbulent wakes or vortex flows. These effects cause well-known discomfort in the pedestrian areas and several consequences on monuments, especially those associated with the drag of falling raindrops, surface washout and deposited pollution removal.

Interesting studies of wind field perturbation within a built environment and wind tunnel experiments on urban models have been described by Plate (1982) and Simiu and Scanlan (1986).

7.5. DISPERSION AND TRANSPORTATION OF POLLUTANTS IN A CITY

Various mechanisms are involved in the dispersion and transportation of airborne pollutants. The activity of convective rising and divergence of the air masses above a town means that in the surrounding areas other air descends and converges, forming a large convective cell. This movement tends to lower the plumes emitted from chimneys near the town outskirts, thus fumigating into the town. The turbulence locally induced by buildings may lead to an increase in the pollutant concentration at ground level if the pollutants were originally released or transported at stack level. As opposed, pollutants emitted at the ground level are more widely dispersed; the impact on people who live on the ground floor is reduced but increased at upper levels.

The overall geometry of the town may lead to the so-called *canyon effect*. First of all, the wind nearly parallel to the main streets is channelled and the air flow remains entrapped in these urban valleys or canyons, increasing the concentration of the pollutants continually emitted by traffic. If the wind aloft crosses the street, a stationary horizontal vortex is formed in the urban canyon, partially entrapped between the two rows of buildings facing each other. The aerial flow at roof level, instead of crossing aloft the street, descends into the street and then rises again along the opposite side distributing to the windows the pollutants emitted at roof level (e.g. domestic heating and plumes from tall stacks) and on the street. If the angle of attack is intermediate, a helicoidal motion (i.e. screw motion) is generated in the street with similar effects (Munn et Rodhe, 1985).

The synergism between urban convective motion and pollution ultimately increases the local precipitation. Falling raindrops are very effective in removing particles suspended in the atmosphere as well as a good part of the gaseous substances, so that these are abated into the town in the form of the so-called *acid rains*. In the Mediterranean climate, the wet deposition of pollutants (i.e. acid rain) is only a very modest fraction when compared with the deposition during the dry phase, which causes much greater damage.

What is the difference between the urban climate and an undisturbed one? The answer is not easy (Cayan et Douglas, 1984; Jones et al., 1986; Lee, 1992) because there is no undisturbed climatic area, given that the regional climate is characterized by many microclimatic areas, the average of which could be considered, with extreme caution, only an approximate average. Many studies attempted to establish, in a general sense, the departure of the climate of an urban area from the surrounding countryside. This approach is based, obviously, on the comparison of environmental data, measured simultaneously at various locations, in order to show the local disturbance that emerges at the urban site. This problem has a clear implication when interpreting the long series of meteorological observations. In fact, the early instrumental observations began two or three centuries ago in the outskirts of small green towns, which have grown changing the local climate. However, the study of these long series is of uppermost importance for understanding the climatic changes. This is especially true because these observations are well documented and continued, without interruption, for centuries. An example is given by the instrumental series of Padua that began in 1725 and is composed of atmospheric pressure, air temperature, frequency and amount of precipitation, wind direction and other observations (Camuffo, 1984; Camuffo and Jones, 2002; Camuffo and Bertolin, 2012). These very important series, however, have a serious inconvenience in that towns have grown considerably over the past years and it is not easy to evaluate the urban disturbance from the recent times. In fact, towns have not only grown as a consequence of the number of inhabitants but have also undergone dramatic changes when gardens have been substituted with new buildings, when the streets have been asphalted, and when the building technology changed. Also, the rural climate all around is variable; its documentation is not as old as the urban one which, at one time, was only moderately affected by the presence of humans. Even the rural territory has, in many cases, been altered by land clearing, reclamation and transformation. This has affected the type of local vegetation, the whole energy balance at the ground level and, consequently, the heat exchanges with the atmosphere.

7.6. WIND FRICTION NEAR A SURFACE

Shearing stress and friction velocity are two fundamental parameters for the mathematical treatment of several processes in the *PBL*, especially those linked to the vertical transport of momentum, or turbulence generation. The friction strength of wind blowing tangentially to a surface per surface unit area is called *shearing stress* or *surface shearing stress*, is indicated by the symbol τ and is of the order of 1–10 dyn cm^{-2}. This shearing stress is due to the sum of a turbulent component and a viscous one. Near the ground, the viscous effect dominates and the Newton law of viscous friction applies, so that τ is proportional to the vertical gradient of the horizontal wind $\partial u/\partial z$, called *wind shear* or, simply *shear*, i.e.

$$\tau = \mu \frac{\partial u}{\partial z} \tag{7.4}$$

where μ is the *coefficient of molecular viscosity*, also called *dynamic viscosity* or, more simply, *viscosity*. For air, μ slightly depends on temperature, e.g. $\mu \approx 1.7 \times 10^{-4}$ P (i.e. poise = g cm^{-1} s^{-1}) at $T = 0$ °C and $\mu \approx 1.81 \times 10^{-4}$ P at $T = 20$ °C and is independent of pressure except for very low pressures.

The *PBL* is the lowest layer of the troposphere where wind is influenced by friction. The thickness (or depth, h) of the *PBL* is defined as the height at which the influence of friction is reduced to less than 20%. The surface stress τ is typically of the order 1–10 dyn cm^{-2} and $h \approx 2000\,\tau$ and ranges between 20 and 200 m (Lumley and Panofsky, 1964)

Moreover, the dynamic viscosity can be expressed in terms of the kinetic theory of perfect gases:

$$\mu = \frac{1}{3} \rho \, c \, \lambda \tag{7.5}$$

where ρ is the gas density, c is the average speed of the thermal motion of the gas particles and λ is the mean free path. For air, at $T = 0$ °C and under standard condition $\lambda \approx 5.5 \times 10^{-6}$ cm.

The viscous drag within the atmospheric viscous sublayer that is immediately adjacent to each surface can be determined with τ as above; however, beyond the viscous sublayer, at a considerable distance from the ground (or from any surface), the turbulent effect dominates and the drag can be expressed in terms of *Reynolds* stress, or the rate at which horizontal momentum of the air is being transferred vertically to the surface by means of turbulent transfer

$$\tau = -\rho <u'w'> \tag{7.6}$$

where u' *and* w' are the fluctuations of the wind (*eddy velocities*) along the mean wind direction and the vertical; physically, $<u'w'>$ represents the vertical transfer of momentum associated with the vertical component of the wind speed, i.e. the eddies present in the wind field or caused by the soil roughness. Both Eqn (7.4) and Eqn (7.6) show that the vertical transfer of the momentum is proportional to the vertical gradient of the wind speed, as will be discussed later. In the case of homogeneous wind field and isotropic turbulence, wind fluctuations are randomly distributed along the axes x, y and z so that $<u'w'> = 0$ and $\tau = 0$.

As the dimension of τ can also be interpreted as energy density, by analogy with the dissipation of the kinetic energy of the wind, which is dispersed by the eddy turbulence per unit of volume of air, it is possible to introduce a fictitious speed u_* that is homogeneous along the vertical and is called *friction velocity* as it is linked to friction, defined as

$$u_* = \sqrt{\frac{\tau}{\rho}} \tag{7.7}$$

Given that u_* has the dimension has the dimensions of speed, it derives from turning a complex phenomenon into a useful parameter that does not immediately correspond to any definite physical entity. In the turbulent layer, on the basis of the *Reynolds* stress, the friction velocity is defined by means of the eddies' contribution as

$$u_* = \sqrt{<u'w'>} \tag{7.8}$$

and in the viscous layer by the continuous, laminar increase of the wind speed as

$$u_* = \sqrt{\frac{\mu}{\rho}\frac{\partial u}{\partial z}}. \tag{7.9}$$

From these expressions, it can be seen that u_* is physically linked to the transport of momentum from one level to another.

In a general way, u_* can be expressed as a fraction of the average wind speed $<u>$, i.e.

$$u_* \approx \frac{<u>}{\mathcal{N}} \tag{7.10}$$

where the index of roughness $\mathcal{N}$ varies from 3 for perfectly smooth surfaces, such as snow or a calm lake surface, to 13 for grassy land.

7.7. VERTICAL FLUXES OF HEAT, MOISTURE AND MOMENTUM

In the atmosphere, a vertical profile of air temperature also implies a vertical transport of heat and a wind shear implies a vertical transport of momentum. The vertical fluxes of heat (also called *sensible heat*) $\mathcal{H}$, moisture (also called *latent heat*) $L_v\mathcal{E}$, and momentum τ are defined, respectively, as:

$$\mathcal{H} = \rho c_p {<}\Theta' w'{>} = \rho c_p \int \Theta'(t) \times w'(t) \mathrm{d}t = -\rho c_p K_H \frac{\partial {<}\Theta{>}}{\partial z} \tag{7.11}$$

$$L_v \mathcal{E} = \rho L_v {<}m'_v w'{>} = \rho L_v \int m'_v(t) \times w'(t) \mathrm{d}t = -\rho L_v K_E \frac{\partial {<}m_v{>}}{\partial z} \tag{7.12}$$

$$\tau = -\rho {<}u' w'{>} = -\rho \int u'(t) \times w'(t) \mathrm{d}t = \rho K_M \frac{\partial {<}u{>}}{\partial z} \tag{7.13}$$

where ρ is the density of the air, c_p is the specific heat at constant pressure, w is the vertical wind component, L_v is the latent heat of vaporization and $\mathcal{E}$ is the evaporation (or condensation) rate. As the vertical displacements involved are generally modest, the air temperature T can be used instead of the potential temperature Θ. As usual, the brackets $<>$ indicate the average value and the label $'$ indicates the fluctuating value, e.g. $u(t) = {<}u{>} + u'(t)$. The integrals of the correlation products between the fluctuating values of the vertical wind component w' and the corresponding fluctuations of the (potential) temperature Θ', moisture m' or speed u' represent the net transport of the related properties along the vertical. The sign minus has been introduced because in the surface boundary layer the fluxes are counted positive upwards.

The coefficients K_H, K_E, and K_M represent diffusivities, and several rough assumptions are often made in the surface *PBL* to make easier the mathematical treatment of this complex mechanism. The first assumption is that the vertical fluxes are constant with height (i.e. no accumulation or horizontal divergence); the second, often used quite successfully in engineering applications, is that all the coefficients are equal, although this has been verified only in near adiabatic conditions. This assumption allows predicting the distribution of a particular flow from measurements of another one. However, during strong inversions the radiative heat exchange and the pressure fluctuations may cause important departures for the heat and momentum transfer. For a discussion on this implication on diffusion models, see Lumley and Panofsky (1964).

The coefficient of heat exchange, K_H, also known as the *eddy diffusivity of heat*, refers to the heat flow due to the vertical transport of heat, because of either the convective motion or the vertical component of the wind fluctuations. It is defined as the coefficient of proportionality between the turbulent heat transfer and the vertical gradient of air temperature

$$K_H = -\frac{<\Theta' w'>}{\frac{\partial \Theta}{\partial z}} \tag{7.14}$$

This formulation represents the physical process that typically occurs near the soil surface in superadiabatic conditions. When the soil is warm, the air in the viscous sublayer above the soil is heated by conduction at the interface, gains in buoyancy and eventually escapes by convection out of the viscous sublayer. This warm air mixes with the environmental air above, i.e. in the superadiabatic layer, and forms bubbles of warm air that tend to rise by their buoyancy (*thermals*). At this point, the lower atmosphere is characterized by a positive temperature gradient and a convective motion develops. This motion is formed of many individual cell motions, with the warm cores formed by uprising thermals, associated with lateral descent of colder air that closes the cells. A net upward transport of heat is generated when the vertical uprising movements of the air w' are associated with the transport of warmer air (Θ') and downward movements are associated with colder air, i.e. $<\Theta' w'> > 0$, for both the upward and the downward transport, as both Θ' and w' change sign. The same result can also be found in another particular case, although in the absence of convective cell motion. This happens in the central part of windy days when the atmosphere is neutral but the soil is warm. In this case, the wind eddies coming from below are warmer than those coming downwards, leading to the same result. When the soil is colder than the air, in the presence of fresh wind, the eddy turbulence transports heat downward and $<\Theta' w'> < 0$. If the soil is cold and there is no wind, the air stratifies and tends to suppress turbulence, i.e. $w' \approx 0$ and also the product $<\Theta' w'> \approx 0$; in the case of stable atmosphere, the heat is not transported by eddies but flows downwards very slowly by means of conduction. This is another mechanism, characterized by a small efficiency and is not any more represented by Θ' or $<\Theta' w'>$.

Similarly, the coefficient of *eddy diffusivity of moisture* is defined as

$$K_E = -\frac{<m'_v w'>}{\frac{\partial <m_v>}{\partial z}} \tag{7.15}$$

The coefficient of mechanical exchange, K_M, also called *eddy diffusivity of momentum* or *kinematic coefficient of eddy viscosity*, represents the capacity of the atmosphere to exchange vertically the momentum as a consequence of the eddies induced by wind and is defined as the coefficient of proportionality between the vertical momentum transfer and the gradient of the wind speed:

$$K_M = \frac{<u'w'>}{\frac{\partial <u>}{\partial z}} \tag{7.16}$$

If the wind speed increases with height, the turbulence existing between two levels brings up some parcels of air with slower velocity and brings down some faster ones, exchanging momentum. The greater the turbulence (and the coefficient K_M), the greater the tendency for the distribution of the wind speed to become uniform.

7.8. HEAT BALANCE AT THE SOIL OR THE MONUMENT SURFACE

The heat balance at the soil governs the atmospheric stability, the dynamics of the *PBL* and many interactions between the soil (or a monument) and the atmosphere. Part of the balance considers how the net flux of radiant energy Φ is partitioned. The net flux Φ is the global solar income $I\downarrow$ minus the short-wave radiation $S\uparrow$ reflected from the soil and the long-wave component $L\uparrow$ reflected and emitted as blackbody infrared thermal emission, i.e.

$$\Phi = I\downarrow - (S\uparrow + L\uparrow). \tag{7.17}$$

Therefore, Φ is the radiant energy adsorbed or emitted by the soil. During the day, the solar income dominates and $\Phi > 0$; during the night, the radiative loss dominates and $\Phi < 0$; in practice, Φ is positive when the flow of energy is from the atmosphere to the soil, i.e. $\Phi\downarrow > 0$, and vice versa, i.e. $\Phi\uparrow < 0$. Everyday, the integral value $\int \Phi(t)dt \approx 0$; from winter to summer, these 24-h integrals are generally small and positive showing a daily gain of heat, whereas from summer to winter, they are negative.

At the soil (or monument) surface, the radiant flux of energy Φ is transformed into three heat fluxes: i.e. heat conduction G into the ground (or the material), sensible heat $\mathscr{H}$ into the atmosphere and latent heat $L_v\mathscr{E}$ into the atmosphere. This can be expressed by means of the equation of partitioning of the energy per unit surface and unit time, also called the heat balance equation:

$$\Phi = G + \mathscr{H} + L_v\mathscr{E}. \tag{7.18}$$

Given the convention adopted for the sign of Φ, the other fluxes are positive when the heat flow proceeds in the direction shown by the arrows, i.e. $G\downarrow$ heat from the soil surface to the deeper layers; $\mathscr{H}\uparrow$ and $L_v\mathscr{E}\uparrow$: sensible and latent heat from the soil surface to the atmosphere that is, respectively, heated and moistened. As opposed, $-G$ indicates the conduction of heat from the deeper layers to the soil surface; $-\mathscr{H}$, that the soil is cooling the air; and $-L_v\mathscr{E}$, that the soil is absorbing moisture or is dewing.

The rate at which heat flows through a building wall or a soil level at a depth z below the surface is directly proportional to the temperature gradient that is found at that depth, i.e.

$$G = -c_t \frac{\partial T}{\partial z} \tag{7.19}$$

where c_t, called *thermal conductivity*, is a coefficient of proportionality that is constant only for a homogeneous medium. In the ground, this is not strictly valid, as the moisture content in the soil, affected by rainfall, dew, evaporation and fringe diffusivity, changes with time and depth. For an infinitely thin layer, the heat transfer is regulated by the equations

$$\frac{\partial G}{\partial z} = -C\frac{\partial T}{\partial t} \tag{7.20}$$

$$\frac{\partial T}{\partial t} = \kappa \frac{\partial^2 T}{\partial z^2} \tag{7.21}$$

where C is the heat capacity of the medium and $\kappa = c_t/C$ is called *thermal diffusivity*. The latter coefficient physically expresses the speed of propagation of a thermal wave into a medium, which is proportional to the capability of transmitting heat in the presence of the unit temperature gradient (thermal conductivity) and inversely proportional to the capability of storing heat (heat capacity) (Munn, 1966).

When all the fluxes are positive, the heat balance equation shows how the radiant energy adsorbed by the soil surface is partitioned among the heating of the deeper layers of the ground, the heating of the atmosphere and the evaporation; when one or more of these fluxes change their direction and become negative, the equation shows the way in which each flux is transformed or supplies to or receives energy from the other fluxes (Figure 7.4). One or more components of this balance may be also zero, e.g. $\Phi = 0$ when the sky is completely overcast or slightly after the sunrise or before sunset; $G = 0$ twice a day when the heat flux into the ground inverts direction, or in the case of nonconductive surfaces; $\mathscr{H} = 0$ when air and soil surface have the same temperature and there is no evaporation or condensation; and $L_v\mathscr{E} = 0$ when the soil is dry and there is neither condensation nor evaporation (Camuffo et al., 1982).

Marble and bronze monuments have very little porosity in comparison with the soil, so that the amount of water adsorbed in the pores is extremely modest. For them $L_v\mathscr{E} \approx 0$ and the energy balance is practically reduced to $\Phi = G + \mathscr{H}$. The white Carrara marble of which the Trajan Column, Rome, is made becomes some 10 °C warmer than the air (Camuffo, 1993; Camuffo and Bernardi, 1993). Dark stones become some

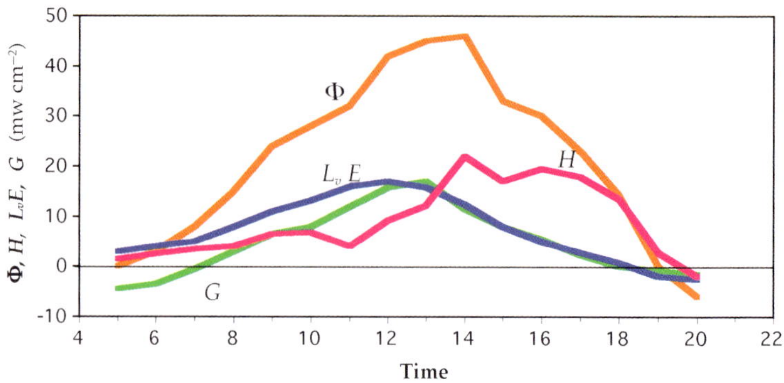

FIGURE 7.4 Soil energy balance $\Phi = G + \mathscr{H} + L_v\mathscr{E}$ during the daytime, at Padua, Italy (46° lat N), in August. Legend: Φ, orange line; G, green line; $\mathscr{H}$, red line; and $L_v\mathscr{E}$, blue line.

20 °C warmer. Bronze monuments are hollow and the bronze is a few millimetres thick. However, G is not negligible and the monument stores a huge amount of heat, reaching high temperatures.

For example, in clear summer days in Venice, the temperature of the gilded bronze horses of the St Mark Basilica sharply increased by some 15 °C at 11:30 when sunshine fell on the horses and then dropped by the same amount in 15 min when the horses entered into the shade of the bell tower (Figure 7.5). A few minutes later, when the shade passed away, they experienced a nearly symmetric increase. The dimensional changes of bronze, considered over the horse length and width, i.e. 250 cm and 72 cm, respectively, reached some 1.5 mm for the daily thermal cycle. As a consequence, huge tangential forces apply to the tree legs that are fixed to a basement made of white Istria stone. Fluctuations in monument temperature occur at temporary cloud shade. In addition, the bronze temperature fluctuates within a 6-min period. The thermal inertia of

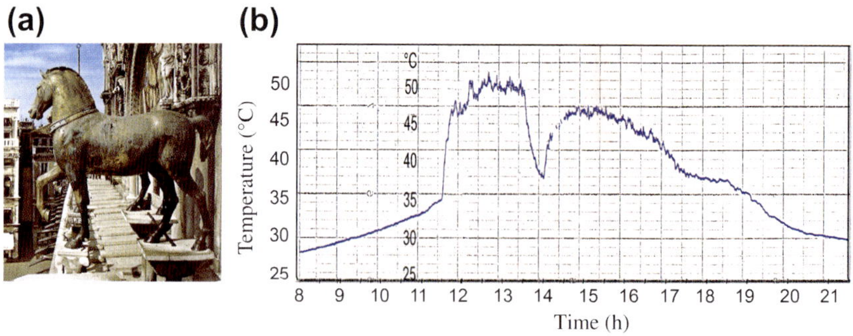

FIGURE 7.5 (a) One of the four horses in St Mark, Venice, and (b) a temperature record dated 18 August 1972, when the horses were exposed outdoors. The sharp temperature increase at 11:30 and the drop at 14:00 occur when the horse enters in the sunshine and in the shade of the bell tower.

monument cuts off the high-frequency fluctuations and only the longer periods are visible, which are induced by the sea breeze blowing over the city (Camuffo, 1981a; Camuffo and Vincenzi, 1985).

The daily cycles of Φ are affected by the cloudiness or abrupt changes of soil albedo, e.g. after showers, and vary with the seasonal change of the solar radiation and the vegetation. The moisture content of the soil and its vertical gradient affect the amplitude of G, $\mathscr{H}$ and $L_v\mathscr{E}$ and cause some asymmetry or delay (Camuffo et al., 1984). In fact, after a drizzle or abundant dew, the upper layer is moist, so that the evaporation rate is larger in the morning than in the evening, when the soil is dryer. The curve of $L_v\mathscr{E}$ is skew with the maximum in the morning and $\mathscr{H}$ has a similar skewed shape, but with the maximum in the afternoon, when the evaporation is reduced and more energy is employed to warm air. After some clear days, the upper layer of the soil will dry, so that the evaporation rate will increase in the afternoon, when the heat wave will reach the deeper, moister layers, mobilizing vapour. In this case, the curve of $L_v\mathscr{E}$ is skew showing a greater evaporation rate in the afternoon and, consequently, the maximum of $\mathscr{H}$ occurs in the morning. By plotting on a Cartesian reference frame the instantaneous values at time t of the fluxes $G(t)$, $\mathscr{H}(t)$, or $L_v\mathscr{E}(t)$ versus $\Phi(t)$ for a whole day, an ellipse is obtained, which is clockwise or counterclockwise, with the major axis more or less tilted and the minor axis more or less wide and the entire ellipse being slightly displaced upwards or downwards (Figure 7.6). Therefore, the daily cycles of G, $\mathscr{H}$ and $L_v\mathscr{E}$ can be calculated by means of the equations (Camuffo and Bernardi, 1982a,b):

$$G(t) = a_1\Phi(t) + a_2\frac{\partial\Phi(t)}{\partial t} + a_3 \tag{7.22}$$

$$\mathscr{H}(t) = b_1\Phi(t) + b_2\frac{\partial\Phi(t)}{\partial t} + b_3 \tag{7.23}$$

$$L_v\mathscr{E}(t) = c_1\Phi(t) + c_2\frac{\partial\Phi(t)}{\partial t} + c_3 \tag{7.24}$$

where the coefficients a_1, b_1 and c_1 indicate the first-order proportionality between each flux and the radiation income, i.e. the inclination of the major axis of the ellipse; a_2, b_2 and c_2 indicate the influence of the gradient of moisture into the ground, which causes positive or negative departures from linearity, i.e. the width of the ellipse, and the sign shows whether the ellipse is described clockwise or counterclockwise; and a_3, b_3 and c_3 indicate the background flux, independent of Φ, and the experimental problem of the divergence of the fluxes and/or the storage of energy, as the four fluxes that appear in the balance equation cannot be measured exactly at the same level as three fluxes are in the atmosphere and one is underground.

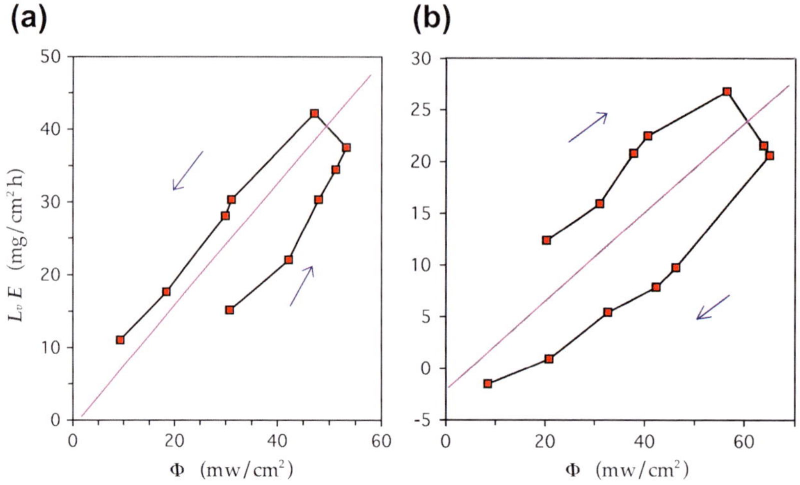

FIGURE 7.6 Plot of the evaporation rate $L_v\mathscr{E}$ versus the net radiation flux Φ over the soil in Padua (a) at the end of August, a few days after a shower, i.e. moist soil, and (b) in September, several days after a late summer shower, i.e. arid soil. Blue arrows indicate counterclockwise loop in (a) and clockwise loop in (b).

7.9. MAIN PARAMETERS USED IN MEASURING ATMOSPHERIC STABILITY AND TURBULENCE

'A precise definition of turbulence is difficult, if not impossible, to give' (Plate, 1982). For this reason, several parameters have been introduced, each of which may be, from time to time, useful or inappropriate.

However, the physical regimes ranging from turbulence to still air are fundamental in governing the mechanisms of deposition of pollutants and heat and mass exchanges. Monitoring atmospheric turbulence or stability in field surveys is at the same time important and extremely difficult. For this reason, it is necessary to become familiar with the most important definitions and their physical meaning and then to balance theory with the specific problem under consideration, as well as with the instrumental facilities and the experimental limits. In the absence of a well-focussed and well-planned observational plan, any environmental monitoring risks to be a mere collection of meaningless data.

Atmospheric stability may be defined as the tendency to mitigate (or enhance) vertical movements or existing turbulence. A number of parameters have been introduced to wholly and quantitatively describe certain atmospheric conditions, each of which illustrates a particular characteristic. These are not just limited to providing mathematical models but furnish new criteria for classification that may be particularly useful. It might be useful to remember some of the most famous parameters that have contributed to the development of this science. These parameters might be a bit abstract in that they have been introduced to give flexibility to the mathematical modelling of turbulence and their relevance is conditioned by their usefulness.

However, it is always possible to find a physical meaning – even though this might not always be immediately clear – and this section will focus on this discussion.

7.9.1. Kinematic Viscosity (ν)

The *kinematic viscosity* is an atmospheric variable defined as the ratio between the *dynamic viscosity* μ and the density ρ of the fluid, i.e.

$$\nu = \frac{\mu}{\rho} \tag{7.25}$$

and depends on both air temperature and pressure. For air at sea level pressure and 20 °C, $\rho = 1.205 \times 10^{-3}$ g cm^{-3} and $\nu = 0.15$ cm^2 s^{-1}. This parameter is used as a proportionality factor in the equation relating the accelerating (retarding) effects to the air motion, i.e. $\partial u/\partial t$, generated by fluid friction in a given wind speed profile:

$$\frac{\partial u}{\partial t} = \nu \frac{\partial^2 u}{\partial z^2}. \tag{7.26}$$

Equating Eqn (7.7) to Eqn (7.9), and operating with the help of Eqn (7.6) and Eqn (7.13), the kinematic viscosity ν equals the kinematic coefficient of eddy viscosity K_M. This is only a logically extrapolated similarity between the molecular regime and the free air turbulence. In fact, the eddy viscosity was introduced in 1877 by Bussinesq in analogy with the laminar flow and the relationship existing between the stress and the velocity shear. Physically, for reasons of continuity, the above equalization applies to the transition zone between the viscous layer and the external turbulent regime to which the above equations refer. In reality, although K_M has some formal analogy to ν, the mechanism is different and is expected to find K_M much larger than μ/ρ to account for the largely increased flux capabilities of the bulk turbulent flow in comparison with the molecular transport (Brown, 1991).

7.9.2. Reynolds Number (Re)

Re is the nondimensional ratio between inertial and viscous forces of a moving fluid:

$$Re = \frac{L\,u}{\nu} \tag{7.27}$$

where L and u are, respectively, the characteristic length and the characteristic speed of the system, while ν is the kinematic viscosity of the fluid. The physical meaning of Re can be deduced from the fact that the inertial forces tend to separate parcels of fluid that initially had different speeds. On the other hand, the viscous forces tend to lead to certain uniformity in the speeds at short distances and attenuate dissimilarities. At low Re values, when the viscous forces predominate over the inertial ones, the flow is laminar. A critical Re_C value is reached when the inertial forces become so great with respect to the viscous ones that turbulence is set up.

The Re number is often used in the field of hydrodynamic stability in order to evaluate the onset of turbulence. For example, in the case of a fluid that flows at a certain speed u over a surface, an *IBL* near the surface develops that is initially laminar and becomes turbulent after the fluid has covered the distance L until Re has reached the critical value Re_C, which generally lies between 10^5 and 3×10^6. In the atmosphere, Re is generally greater than Re_C, so that the air is generally found in turbulent regime. Outdoor Re_C is greater than in other closed systems; in pipes, for example, it is $2500 \le Re_C \le 5000$.

7.9.3. Richardson (Gradient) Number (*Ri*)

Ri is expressed in the form of a gradient (Richardson, 1920) and is the nondimensional ratio between the buoyancy forces (Archimedes) and the inertial ones due to the wind:

$$Ri = \frac{\frac{g}{\rho}\frac{\partial \rho}{\partial z}}{\left(\frac{\partial u}{\partial z}\right)^2} = \frac{\frac{g}{\Theta}\frac{\partial \Theta}{\partial z}}{\left(\frac{\partial u}{\partial z}\right)^2} = \frac{\pm\omega^2}{\left(\frac{\partial u}{\partial z}\right)^2} \tag{7.28}$$

where ω represents the Brunt–Väisälä frequency. The temperature gradient (and the related heat flow) that appears at the numerator is normally positive, except in sunny days near the soil, in the superadiabatic layer. The sign of this number is determined by the temperature gradient, while the denominator is always positive; the negative values are an index of instability and the positive ones are an index of stability. The numerator of Ri measures the density stratification or *static* atmospheric stability due to the temperature gradient. The denominator is of *dynamic* nature and measures the destabilizing effect, linked to the wind profile. In practice, the Ri represents the ratio of the work done against the gravitational stability and the energy transferred from the ensemble motion to the eddy turbulence.

When this and the following related parameters are measured in proximity to the soil, there is no difference between the actual temperature T and the potential temperature Θ. In practice, considering that T and Θ are equal between them at the soil level, at the height of 10 m the difference $\Theta - T$ equals 0.1 °C, which falls within the limits of experimental accuracy. For this reason, in microclimate studies, no difference between them is found and T is preferred.

7.9.4. McVehil Ratio (K_M/K_H)

This parameter is the ratio of two coefficients that define the vertical transport of momentum (K_M) and heat (K_H) between two adjacent atmospheric layers (McVehil, 1964).

When $(K_M/K_H) > 1$, the mechanical turbulence generated by the wind dominates over the thermal convection. When, however, $(K_M/K_H) < 1$, the convective mixing dominates over the eddy turbulence. Often, for reasons of simplicity and in the lack of observations, modellers assume $K_M/K_H = 1$; however, this assumption is valid in near-neutral and unstable conditions (Lumley and Panofsky, 1964).

7.9.5. Richardson Flux Number (Rf)

Rf is linked to Ri and to the previous ratio:

$$Rf = Ri \frac{K_H}{K_M} \tag{7.29}$$

which quantifies the role of turbulence in the vertical transport of heat and momentum, by means of the vertical fluxes of these properties. Rf can be written as a nondimensional ratio of two fluxes: the numerator is linked to the production (or destruction) of turbulent kinetic energy by means of the vertical heat flux $\mathscr{H}$ of thermal convective motions and the denominator with the shear production (or destruction) due to the dynamic action of wind, which involves the vertical transport of momentum and the gradient of the wind speed, i.e. the wind shear.

The Rf number can be rewritten in the following way:

$$Rf = -\frac{<\Theta' w'> \frac{g}{\Theta}}{<u' w'> \frac{\partial u}{\partial z}}. \tag{7.30}$$

This parameter has the sign minus as it was introduced to obtain positive numbers in the original studies on the onset of turbulence in a thermally stratified atmosphere. At $Rf \approx 0.2$, a balance is reached between the generation and destruction of turbulence and for this reason, this value is called the *critical Richardson number*.

7.9.6. Monin–Obukhov Length (L)

L is a dimensional ratio (it has the unit of a length) that characterizes a diabatic wind speed profile (i.e. with an exchange of heat) that involves both the sensible heat flow $\mathscr{H}$ and the friction velocity u_* (Monin and Obukov, 1953). The Monin–Obukhov length is defined as:

$$L = -u_*^3 \frac{c_p \rho}{k\, g \frac{\mathscr{H}}{\Theta}} = -u_*^3 \frac{c_p \rho\, \Theta}{k\, g \mathscr{H}} \tag{7.31}$$

where $k \approx 0.4$ is the von Karman constant. The term $u_*^3 \rho$ at the numerator represents a dynamic factor; the denominator involves the heat flow in entropic terms, i.e. $\mathscr{H}/\Theta$. As u_* represents the shearing stress, L is determined after the boundary conditions of drag and gain of entropy at the surface. When the heat flux vanishes, this length is infinite. It is negative during superadiabatic conditions and positive during inversions. In clear night conditions, a transition height is found, where eddies generated by the wind shear begin to be counteracted by buoyancy and $Rf \approx 1$. This transition height can be individuated as the Monin–Obukhov length.

7.9.7. Högström Ratio (*S*)

S compares the thermal stability given by the vertical gradient $d\Theta/dz$ with the wind destabilization pressure $1/2\, \rho <u>^2$ due to the wind kinetic energy (Högström, 1964) as follows:

$$S = \frac{\frac{\partial \Theta}{\partial z}}{<u>^2}. \tag{7.32}$$

All the above parameters require measurements that are either very difficult to realize or not particularly reliable. For this reason, the Högström parameter has been introduced, as it involves the static and dynamic coefficients in a simpler form.

7.9.8. Sutton Turbulence Index (*n*) and the Logarithmic Wind Profile

This and the following parameters only consider the wind profile. It is much simpler to measure the wind profile alone (or only the temperature profile), rather than complex measurements of the above-mentioned parameters. However, when only a single profile, representing approximately either the thermal or the dynamic aspects, is taken into account, the degree of approximation should be considered case by case, depending on the aims and the degree of accuracy required. The Sutton turbulence index has been defined for the theoretical study of wind turbulence (Sutton, 1947) and is based on the analysis of the vertical profile of the wind speed measured at two levels, z_1 and z_2:

$$\frac{u(z_1)}{u(z_2)} = \left(\frac{z_1}{z_2}\right)^{n/(2-n)} \tag{7.33}$$

The Sutton turbulence index n generally lies between 0 and 1 in cases of maximum and minimum turbulence, respectively, and is generally in accordance with other results but not always unequivocally. This index only considers the bulk effects of eddy turbulence and convective mixing on the wind profile and gives the degree of erosion on the basis of the logarithmic profiles, as suggested by the *theory of similarity*. This theory requires that, expressing the atmospheric variables in an appropriate dimensionless form, the profiles of these new variables must have a unique form when stated in terms of the basic independent parameters, also expressed in dimensionless form. Under this circumstance, the mathematical formulation is the same, whatever the atmospheric parameter involved. The similarity theory is very practical, not always rigorous.

The logarithmic wind profile was derived from the observation (in wind tunnel experiments) that in the turbulent regime the mean wind speed $<u>$ varies with the distance from the surface following the law

$$\frac{\partial <u>}{\partial z} = \frac{u_*}{kz} \tag{7.34}$$

where k is the von Karman constant and the shearing stress has been found constant throughout an air layer close to the ground, called *surface layer* or *constant stress layer*. By integrating the above equation, the logarithmic wind profile is obtained, i.e.

$$\frac{<u>}{u_*} = \frac{1}{k}\ln\frac{z}{z_0} \tag{7.35}$$

where the constant of integration z_0 is called *roughness length* and physically represents the height at which the average wind speed vanishes, i.e. $<u> = 0$.

7.9.9. Deacon Number (β)

Deacon number defined as

$$\beta = \frac{\partial \ln \frac{\partial u}{\partial z}}{\partial \ln z}, \tag{7.36}$$

is only related to the vertical profile of the wind speed, as the Sutton index, without discriminating the effects due to the transfer of heat and momentum (Deacon, 1949).

7.9.10. R Parameter

The *R parameter* only takes into account the *wind profile* or, more precisely, the wind attenuation at the ground due to friction and the exchange of momentum and is represented by

$$R = \frac{u_S}{u_g} \tag{7.37}$$

where u_s represents the wind at ground level and u_g the gradient wind (i.e. the wind determined by the pressure pattern, undisturbed by the soil roughness).

7.9.11. Wind Standard Deviation (σ)

The *standard deviation* σ of the wind is the wind turbulence statistically defined in terms of amplitude and frequency of departures from the average. Both the fluctuations in the wind direction θ, i.e. $\sigma\theta$, and speed u, i.e. σ_u are considered; the normalized value $\sigma_u/<u>$ is directly used in the equation to determine the concentration distribution in the Gaussian diffusion models, as we will see later in this chapter, as well as in Chapter 13.

7.10. PLUME DISPERSION

Several models have been developed to predict the concentrations downwind of a single source. The Gaussian model is popular for its simple mathematical apparatus and the agreement with the observed data for long-term averages. This model assumes that the dispersion is due to the random effect of eddies in the atmosphere that widen the plume while it is progressing along the downwind direction. In the case of a neutral atmosphere and steady wind direction, the maximum concentration is found along the plume centreline and the lateral diffusion is due to atmospheric turbulence; in the case of a wind direction continually variable around a prevalent direction, the plume meanders and the maximum concentration is again statistically found along the mean wind direction, downwind from the source. The crosswind distribution of concentrations is represented by a bell-shaped curve very narrow near the source and gradually broadening with increasing distance from it, i.e. as time elapses after the smoke release.

If a Cartesian reference is assumed, with the x-axis along the wind direction, the y perpendicular to it, but in the horizontal plane, showing the lateral displacement, and the z on the vertical, the atmospheric stability will differentiate the standard deviations of the wind fluctuations in these three directions, respectively, σ_x, σ_y and σ_z (also called diffusion coefficients) and the plume dispersion will be affected accordingly. When the atmosphere is unstable, vertical motions are favoured by convection and σ_z dominates; when the atmosphere is neutral, the diffusion coefficients are similar; when the atmosphere is stable, vertical turbulence is suppressed, i.e. $\sigma_z \to 0$, and σ_y describes the fanning or the meandering of the plume in the horizontal plane.

For an effective height h of an elevated point source, e.g. a stack, the solution for the plume concentration at ground level $\chi(x,y,z = 0,h)$ takes the Gaussian form

$$\chi(x, y, 0, h) = \frac{Q}{\pi \sigma_x \sigma_y <u>} \exp - \left(\frac{y^2}{2\sigma_y^2} + \frac{h^2}{2\sigma_z^2} \right) \tag{7.38}$$

where Q is the source strength and $<u>$ is the average wind speed at the height of the plume. Of course, the ground-level concentration directly downwind of the source is found by putting $y = 0$ in the previous equation, which reduces to

$$\chi(x, 0, 0, h) = \frac{Q}{\pi \sigma_x \sigma_y <u>} \exp - \left(\frac{h^2}{2\sigma_z^2} \right). \tag{7.39}$$

The maximum ground-level concentration χ_{max} is obtained by equalling to zero the time derivative of the previous equation, i.e.

$$\chi_{max} = \frac{2Q}{e\pi <\mathrm{u}> h^2} \frac{\sigma_z}{\sigma_y} \tag{7.40}$$

(where e is the Neper number) and occurs at the distance x where $\sigma_z = h/\sqrt{7.2}$.

The *effective height* h of the source is the height at which the plume stabilizes after an initial rise. Therefore, h is given by the geometrical stack height h_{st} plus the plume rise Δh, which is due to the momentum (determined by the vertical speed of the smoke into the chimney and the interaction between the vertical stack jet and the horizontal wind flow) and the buoyancy (determined by the emission temperature, i.e. the low density of the smoke, which is warmer than the surrounding air). Several formulae exist, which depend on the jet speed and emission temperature as well as the environmental air speed, temperature and stability.

7.11. STABILITY CLASSES TO EVALUATE ATMOSPHERIC STABILITY

From a practical point of view, the use of complex parameters to describe the atmospheric stability is limited, substantially, to a few cases where detailed measurements can be carried out. Even under ideal conditions without any discontinuity at ground level and in stationary regimes, where it could be assumed that Gaussian diffusion prevails, forecasting the wind turbulence or simply deducing it from other meteorological parameters without calculating it from wind fluctuations and the vertical temperature gradient is still a problem. Many scientists have dedicated much of their time to search for a reliable method that supplies reasonable values of wind variance and plume dispersion from other simple observations. Practically, the method consists in determining classes of stability and linking them to typical values of σ. The stability classes are determined on the basis of observations and very simple considerations that are, however, valid only in a general sense. These criteria tend to focus on the link between the dynamic evolution of the *PBL* and bulk classes of turbulence that summarize the situation. This practical point of view leads to the definition of classes of stability, as follows.

7.11.1. Brookhaven

The first fundamental contribution is due to the Brookhaven National Laboratory (Singer and Smith, 1953), which defined five classes of wind turbulence and tried to correlate each with temperature gradients, wind intensity, seasonal and diurnal cycles, solar radiation, cloud cover and the Sutton turbulence index. The Brookhaven turbulence classes, referring to wind records taken over a period of 1 h, are defined in Table 7.1.

TABLE 7.1 Brookhaven Turbulence Classes

Class	Wind Vane Trace
A	Fluctuations in wind direction exceed 90°
B_2	Fluctuations in wind direction between 90° and 45°
B_1	Fluctuations in wind direction between 45° and 15°
C	The anemograph trace, because of the continuous and regular fluctuations, appears like a wide, uniform band
D	The anemograph trace appears like a continuous line and existing fluctuations do not exceed 15°

The Brookhaven classes can be closely related to the temperature gradient, i.e. A, B_2, B_1, C and D, in order of decreasing instability. Class D is characterized by dispersion in a stratified atmosphere and the stability is stronger with respect to the other classes and is often associated with marked inversions, but sometimes with unstable conditions too.

The correlation with the wind intensity is less strict. Classes A and B_2 are generally associated with weak winds, while class C is associated with strong winds. Classes B_1 and D appear under considerably variable conditions. In particular, under stable conditions, winds associated with D are light to gentle at ground level and more intense aloft.

By combining the above results, it is possible to obtain the class by looking at the record of the wind wane because fluctuations are generally related to thermally convective or mechanical wind stress. Classes A and B_2

characterized by weak winds under unstable conditions are, essentially, of a convective nature. Class C is mainly of a mechanical nature, being associated with strong winds and neutral gradients. Class B_1 is of a mixed nature, while class D is characterized by very modest turbulence.

The seasonal character is visible but not very marked. The hourly distribution is, on the other hand, evident. Classes A and B_2 make very little contribution and only occur in the middle of the day. Similarly, class B_1 varies seasonally from 10–20% to 85%, while class D is almost complementary to class B_1; the sum of these two represents approximately 85% of the total cases. Class C, in qualitative terms, has the same but less-marked trend as B_1.

The correlation with direct solar radiation confirms what has already been said about classes A and B_2 but class C is mostly associated with weak sunlight or cloud cover, while class D is linked to clear night-time sky or weak daylight sunshine or extensive cloud cover and isothermy or transitory phenomena late in the afternoon.

Singer and Smith also underlined that their results, when supported by a clear correlation, could be extended to other sites in open countryside at any latitude and climate similar to that at Brookhaven, as long as an anemometer with similar characteristics was used at an altitude of about 100 m. Their study, in practice, shows which parameters are strictly related to certain classes of turbulence at a given site, confirming some precise dynamic relationships. At the same time, the study showed that any correspondence is of statistic nature, with large spread of values and lacking a one-to-one correspondence. It should be used for crude estimations only.

7.11.2. Pasquill

Pasquill (1961, 1962, 1974) made a further contribution to this problem, because he combined, in a more flexible way, the theoretical and experimental results obtained by various research teams. The basic simplifying hypotheses were stationary wind conditions; wind profile constant with height and dependence on air–soil interactions. The chimney smoke is dispersed following a Gaussian distribution both vertically and horizontally; the smoke dispersion may be represented either from a real or a virtual source. The real source is the chimney exhaust level H and the virtual source is the real source H to which an extra height ΔH should be added to take into account the effect of the plume buoyancy and keep the virtual source and the plume at the same level.

Pasquill stated that even under these conditions the effective height reached by a plume and the lateral spread should be calculated by measuring the variance of the wind fluctuations. However, in the absence of any direct measurement, in the case of short emissions (lasting a few

TABLE 7.2 Pasquill Stability Classes

Unstable	Neutral	Stable
A: very unstable B: moderately unstable C: slightly unstable	D: neutrality	E: slightly stable F: moderately stable G: most stable

minutes) and in case of emissions fairly close to the ground in open, flat countryside, Pasquill proposed using approximate evaluations of the plume spread and height for six stability classes that can be characterized in terms of wind intensity, sunshine (referred to England) and cloud cover. The advantage is that the classes can be attributed on the ground of the knowledge of prevalent local climatological characteristic. Under conditions of strong stability, he suggested no values, because the results could be somewhat erroneous. It should be quite clear that such values cannot be used in urban or industrial areas because of the additional dynamic turbulence generated by buildings.

The six stability classes determined by Pasquill mirror the results obtained by the Brookhaven climatic correlations and are defined in Tables 7.2 and 7.3, where the sunshine intensities defined 'strong' or 'weak' relate to the values measured at midday in summer or winter, respectively, in England. Night is defined as the period from 1 h before sunset to 1 h after sunrise, when the balance of the radiative exchange between the Earth and the sky vanishes (in England) passing from a positive flux during the day to a negative one at night.

It was suggested to use the D class for the first and last hour of the day, as defined above, and for the periods, night or day, characterized by completely overcast sky.

The conditions of strong stability were introduced successively.

TABLE 7.3 Key to Pasquill Stability Classes

Surface Wind Speed (m/s at 10 m)	Daytime: Strong Insolation	Daytime: Moderate Insolation	Daytime: Slight Insolation	Night-Time: Thiny Overcast ≥4/8 Low Cloud	Night-Time: Clear Sky or Cover ≤3/8 Low Cloud
<2	A	A-B	B	–	–
2–3	A-B	B	C	E	F
3–5	B	B-C	C	D	E
5–6	C	C-D	D	D	D
>6	C	D	D	D	D

These classes became very popular when Gifford (1961, 1976) plotted the dispersion coefficients σ_y, σ_z versus the downwind distance. These fundamental plots, known as Pasquill–Gifford diagrams (Figure 7.7), were obtained from field observations of the plume standard deviations derived from wind data and pollutant concentration levels measured under different atmospheric stability. The plots of dispersion versus downwind distance form, on a double logarithmic paper, a set of parallel

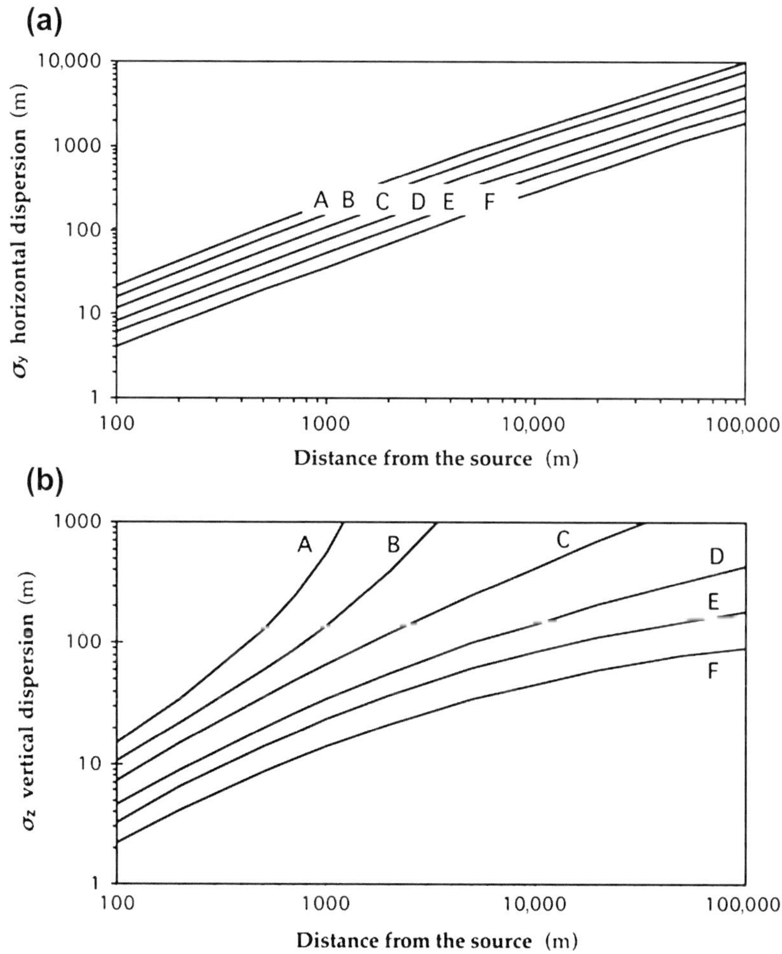

FIGURE 7.7 The Pasquill–Gifford dispersion coefficients σ_y and σ_z showing the horizontal and vertical standard deviations of the plume concentration for the different stability classes A to F. (a) The lateral dispersion coefficient σ_y and (b) the vertical dispersion coefficient σ_z are plotted versus the downwind distance from the source.

lines, one per each stability class, when σ_y is represented and a set of divergent lines for σ_z. The simplicity of using this method for estimating the dispersion coefficients, in a time when computer and suitable data recorders were not yet invented, made soon very popular this classification.

7.11.3. Subsequent Extensions

Turner (1964) extended the classification to urban areas, while Smith (1970; 1972) developed a scheme that took into account the roughness of the ground, the development of the mixing layer, theoretical and experimental data resulting from a further 10 years of field research. Pasquill classes, however, because of the simple application, are in practice the most popular and utilized tool when direct observations of turbulence and radiation are missing or when it is more convenient to use crude approximations than measuring the required values. Consequently, the terminology *'equivalent to a given Pasquill class'* (McElroy, 1969) was introduced to extend borders regarding the application of the original classes.

In reality, it is necessary to be very careful when applying criteria of stability to different sites, especially if these have topographic anomalies such as, mountain or coastal sites (Dobbins, 1979; Camuffo, 1981b; Berlyland, 1991). The fundamental consideration is that each site should be defined in terms of local climate by looking for the various dynamic phases that the *PBL* is subject to. In the most favourable case, whereby the operator is sufficiently familiar with the criteria, it would be possible to correlate, in a first instance, and then to deduce the local turbulence from the type of parameters that proved to be the most efficient ones.

References

Theory and general applications

Anfossi, D., Bacci, P., Giraud, C., Longhetto, A., Piano, A., 1976. Meteorological surveys at La Spezia site. In: Longhetto, A. (Ed.), Atmospheric Pollution. Elsevier, Amsterdam, pp. 531–554.

Berlyland, M.E., 1991. Prediction and Regulation of Air Pollution. Kluwer, Dordrecht, pp. 312.

Brimblecombe, P., 1987. The Big Smoke. Methuen, London, pp. 185.

Brown, R.A., 1991. Fluid Mechanics of the Atmosphere. Academic Press, San Diego, pp. 486.

Camuffo, D., 1980. Fog and related diffusion potential at Venice: two case studies. Boundary Layer Meteorology 18, 453–471.

Camuffo, D., 1981b. Fluctuations in wind direction at Venice, related to the origin of the air masses. Atmospheric Environment 15, 1543–1551.

Camuffo, D., 1984. Analysis of the series of precipitation at Padova, Italy. Climatic Change 6, 57–77.

Camuffo, D., Bernardi, A., 1982a. The diurnal trend in surface mixing ratio at Padova, Italy. Boundary-Layer Meteorology 22, 273–282.

Camuffo, D., Bernardi, A., 1982b. An observational study of heat fluxes and their relationships with net radiation. Boundary-Layer Meteorology 23, 359–368.

Camuffo, D., Bernardi, A., Bacci, P., 1982. Computing the flux of moisture from net radiation and soil wetness. Boundary-Layer Meteorology 22, 503–510.

Camuffo, D., Bertolin, C., 2012. Recovery of the early period of long instrumental time series of air temperature in Padua, Italy (1716–2007). Physics and Chemistry of the Earth 40–41, 23–31.

Camuffo, D., Jones, P. (Eds.), 2002. Improved Understanding of Past Climatic Variability from Early Daily European Instrumental Sources. Kluwer Academic Publishers, Dordrecht, Boston, London, pp. 392.

Cayan, D.R., Douglas, A.V., 1984. Urban influences on surface temperatures in southwestern United States during recent decades. Journal of Applied Meteorology 23, 1520–1530.

Deacon, E.L., 1949. Vertical diffusion in the lowest layers of the atmosphere. Quarterly Journal of the Royal Meteorological Society 71, 89–103.

Dobbins, R.A., 1979. Atmospheric Motion and Air Pollution. Wiley, New York, pp. 323.

Echols, W.T., Wagner, N.K., 1972. Surface roughness and internal boundary layer near a coastline. Journal of Applied Meteorology 11, 658–662.

Elliot, W.P., 1958. The growth of the atmospheric internal boundary layer. Transactions, American Geophysical Union 39, 1948–1954.

Gifford, F.A., 1961. Uses of routine meteorological observations for estimations of atmospheric dispersion. Nuclear Safety 2, 47–51.

Gifford, F.A., 1976. Turbulent diffusion typing schemes: a review. Nuclear Safety 17, 64–86.

Högström, U., 1964. An experimental study on atmospheric diffusion. Tellus 16 (2), 205–251.

Jones, P.D., Raper, S.C.B., Bradley, R.S., Diaz, H.F., Kelly, P.M., Wigley, T.M.L., 1986. Northern hemisphere surface air temperature variations: 1851–1984. Journal of Applied Meteorology 25, 161–179.

Landsberg, H.E., 1981. The Urban Climate. Academic Press, New York, pp. 275.

Lee, D.O., 1992. Urban warming? – an analysis of recent trends in London's heat island. Weather 47 (2), 50–56.

Lumley, J.L., Panofsky, H.A., 1964. The Structure of Atmospheric Turbulence. Interscience, New York, pp. 239.

Mc Elroy, J.L., 1969. A comparative study of urban and rural diffusion. Journal of Applied Meteorology 8, 19–31.

Mc Vehil, G.E., 1964. Wind and temperature profiles near the ground in stable stratification. Quarterly Journal of the Royal Meteorological Society 90, 136–146.

Monin, A.S., Obukov, A.M., 1953. Dimensionless characteristics of turbulence in the layer of atmosphere near the ground. Doklady Akademii Nauk SSSR 93, 257–267.

Munn, R.E., 1966. Descriptive Micrometeorology. Academic Press, New York, pp. 245.

Munn, R.E., Rodhe, H., 1985. Compendium of Meteorology. In: Part 6-Air Chemistry and Air Pollution Meteorology, vol. II. World Meteorological Organisation, WMO No. 364, Geneva, pp. 209.

Oke, T.R., 1981. Boundary layer climates. Methuen, London.

Panofsky, H.A., Townsend, A.A., 1964. Change of terrain roughness and the wind profile. Quarterly Journal of the Royal Meteorological Society 90, 147–155.

Pasquill, F., 1961. The estimation of the dispersion of windborne material. Meteorological Magazine 90, 33–49.

Pasquill, F., 1962. Atmospheric Diffusion, first ed. Van Nostrand, London, pp. 297.

Pasquill, F., 1974. Atmospheric Diffusion, second ed. Wiley, New York, pp. 429.

Plate, E.J., 1982. Engineering Meteorology. Elsevier, Amsterdam, pp. 700.

Richardson, L.F., 1920. Some measurements of atmospheric turbulence. Philosophical Transactions of the Royal Society, London, Ser A 221, 1–28.

Rohinton, E., 2012. An Urban Approach To Climate Sensitive Design. Strategies for the Tropics. Spon Press, New York.

Rosenzweig, C., Solecki, W.D.., Hammer, S.A., 2011. Climate Change and Cities. First Assessment Report of the Urban Climate Change Research Network. Cambridge University Press, Cambridge.

Schlicting, H., 1955. Boundary Layer Theory. Mc Graw-Hill, NewYork, pp. 535.

Simiu, E., Scanlan, R.H., 1986. Wind Effects on Structures. Wiley, New York, pp. 589.

Singer, I.A., Smith, M.E., 1953. Relation of gustiness to other meteorological parameters. Journal of Meteorology 10, 121–126.

Smith, M., 1970. Recommended Guide for the Prediction of the Dispersion of Airborne Effluents. American Soc. Mechanical Engineers, New York, pp. 85.

Smith, F.B., 1972. A schema for estimating the vertical dispersion of a plume from a source near ground level. In: NATO/CCMS Proceedings of the 3rd Meeting of the Expert Panel on Air Pollution Modeling. Tech. Report N. 14, XVII, 1–14.

Sutton, O.G., 1947. The theoretical distribution of airborne pollution from factory chimneys. Quarterly Journal of the Royal Meteorological Society 69, 426–436.

Turner, D.B., 1964. A diffusion model for an urban area. Journal of Applied Meteorology 3 (1), 83–91.

Van Der Hoven, I., 1967. Atmospheric transport and diffusion at coastal sites. Nuclear Safety, 490–499.

Applications to conservation

Camuffo, D., 1981a. Hot-horse Anemometry. Atmospheric Environment 15, 1767.

Camuffo, D., 1993. Reconstructing the climate and the air pollution of ancient Rome during the life of the Trajan column. Science of the Total Environment 128, 205–226.

Camuffo, D., Bernardi, A., 1993. Microclimatic factors affecting the Trajan column. Science of the Total Environment 128, 227–255.

Camuffo, D., Vincenzi, S., Pilan, L., 1984. A first-order analysis of the heat wave in the soil. Water, Air, & Soil Pollution 23, 441–454.

Camuffo, D., Vincenzi, S., 1985. Computing the energy balance of a statue of bronze: the San Marco's horses as a case study. Science of the Total Environment 44, 147–158.

CHAPTER

8

Dry Deposition of Airborne Particulate Matter: Mechanisms and Effects

OUTLINE

Microclimate for Cultural Heritage, Second Edition
http://dx.doi.org/10.1016/B978-0-444-63296-8.00009-3

8.1. INTRODUCTION

Outdoors, the deposition in dry phase of airborne pollutants is responsible for the formation of crusts on buildings and monuments, as already discussed. Indoors, the deposition causes soiling on paintings, frescoes and all surfaces, with a bad visual perception that reduces artwork enjoyment. However, visual aesthetics is not the only one affected, because deposited particulate matter constitutes a potential risk for chemical reactions or microbiological infestation. The aim of this chapter is to review the physics behind this mechanism to understand when, how and where deposition occurs in view of mitigation remedies.

The dry deposition of pollutants involves all particles suspended in air, from the finest ones with molecular size to the largest ones, as coarse as grains of sand. Naturally, the particle size is of fundamental importance, in that for the smallest particles, the air is perceived as a discontinuous medium of suspension that continually hits the particle with random molecular bombardment; for the largest particles, air is perceived as a continuous hydrostatic or hydrodynamic medium. The former is called '*molecular regime*', where individual molecular exchanges are dominant; the latter is called '*continuum regime*' where the carrier fluid is perceived as a continuum medium.

It is therefore necessary to make a first subdivision, based on the particle size, determined by its diameter D. However, it is not simple to define a diameter for nonspherical or rough particles and a number of empirical definitions in terms of effective diameter exist. One definition is the distance between opposite sides of a particle, measured crosswise of the particle and on a line bisecting the projected area. The second definition is the distance between two tangents on opposite sides of the particle profile. Another popular definition is the diameter of a circle whose effective area equals the projected area of the surface. However, the most practical definition is the diameter of a sphere having the same properties (e.g. aerodynamic, optical) of the particle. The list of definitions might continue but in this context, it is sufficient to know that the 'diameter' of nonspherical particles may have a definition that is not the only one existing and should be adapted to the problem under consideration. In the following discussion, this term will be used in a very general sense.

The parameter used for determining the molecular or the continuum character of the medium is the so-called *Knudsen number*, Kn, defined as

$$Kn = \frac{2\lambda}{D} \tag{8.1}$$

i.e. the ratio between the mean free path λ of the molecules of the carrier gas and the particle radius $D/2$. In air, at standard conditions, $\lambda = 0.065\ \mu m$.

The *molecular regime* is determined by the condition that $Kn >> 1$, i.e. D is much smaller than λ and molecular impacts occur individually and punctually, each with their own singularity after random time intervals.

In the *continuum regime*, $Kn << 1$, i.e. the size of the particle is much larger than the average mean free path of the molecules in the air; there are so many simultaneous impacts that their individual contribution is averaged out. The single impacts of air molecules loose their identity and the air can be considered a continuum medium.

Between these two extreme approximations, a *transition regime* whereby $Kn \approx 1$ ($D \approx \lambda$) is considered. Here, the resulting net effect of the impacts of the air molecules decreases rapidly as the volume of the particle increases.

The mobility of particles decreases with decreasing Kn, i.e. increasing D, in that, during its movement, a large particle will necessarily impact on a larger number of air molecules, so that the viscosity of the air becomes more and more important.

Deposition occurs as a result of various causes whereby microphysical phenomena and the local microclimate can reduce or increase the deposition rate, with efficiency depending on the particle diameter. Some deposition mechanisms vary their efficiency according to the size of the particles and may have a simultaneous, synergistic action on the same type of particle (with positive or negative synergism), resulting in a rather complex process.

For each deposition mechanism, and diameter, the number of particles that will deposit on a surface depends on their concentration in the atmosphere. In other words, if all particles have the same probability of sticking on a surface, the higher their concentration, the greater the number of particles that will deposit on it. However, the abundance is not the only factor that governs the deposition. The probability of deposition varies with the probability of the particle to cross the viscous layer that envelops every surface and the frequency of impacts for the specific type of particle. This depends on particle size and the microphysical state of the air in close proximity to the surface. In fact, a huge number of fine to medium particles are found suspended in the atmosphere because they have a long *residence time* and a small number of coarse particles are found, because they deposit in a very short time. Therefore, the statement concerning deposition should be refined as follows: any kind of particle, determined by its diameter, will deposit at a certain rate. The total number of particles that will deposit over a given time will depend on both the individual deposition rate and the relative abundance of each species.

It is possible to describe this problem in terms of *speed* instead of *time*. In fact, the statement that the number of particles $N(D)$ with a specific size (i.e. expressed with the diameter D), which will reach the unit surface per unit time, is proportional to their concentration $C(D)$ in ambient air is formulated as

$$N(D) = v_{dep} \times C(D) \tag{8.2}$$

where the coefficient of proportionality v_{dep} has the dimension of space $\times$ time^{-1} and is called *deposition velocity* or *deposition rate*. It is important to remember that the coefficient v_{dep} is not a constant value, but a complex function of the particle diameter, the air temperature, the gradients of temperature and moisture, the relative humidity (RH), the electric field, the air turbulence, the surface characteristics and other factors that will be discussed later. For this reason, different values of v_{dep} will cause different numbers N of particles to deposit on the surface under the same concentration C of airborne particles. In practice, N can be reduced either by reducing v_{dep} or C; the former method requires appropriate microphysical conditions obtained on the ground of an accurate knowledge of the theory of deposition; the latter needs a simple filter. All the commercial devices are based on air filtering but they do not consider that the air movements they generate will increase v_{dep}. This negative consequence, and the nonlinearity of this phenomenon, may nullify the benefit, or even increase the deposition. This apparent paradox is justified by the consideration that N is proportional to C, whereas v_{dep} has a higher order dependence on some variables, e.g. air turbulence. For instance, if the air in a room is naturally still and thermally stratified, as usual, and if one wants to purify it by forcing the air to pass through a filter, the ventilation and the turbulence that will be induced might cause more deposition than when keeping motionless dusty air. It is possible to decide the measures to adopt in order to reduce N by lowering v_{dep} on the ground of the theory and the experimental observation, based on special field surveys carried out to know the deposition mechanisms that are active in the room.

However, not all particles that impact on a surface will deposit on it and some of them will bounce back. In order to know whether a particle will deposit or not, it is necessary to consider the balance of all forces acting on the particle, the surface and its capture efficiency. The forces in play are the molecular forces of contact and adhesion, those of an electric nature or due to the presence of a film of water, and the inelastic deformation of the impacting particles and the surface.

In the following discussion, the main deposition processes will be described; further details can be found in more general studies (Friedlander, 1977; Pruppacher and Klett, 1980, 2003; Sehmel, 1980; Hinds, 1982; Hidy, 1984; Seinfeld, 1986; Buffle and van Leeuwen, 1992) or other specific papers mentioned later.

8.2. RANDOM WALK AND BROWNIAN DIFFUSIVITY

Brownian motion, also called *random walk* or the *drunk's walk*, is the zigzag motion of fine particles, the irregular trajectory being composed of a number of casually oriented short rectilinear branches, where the start and end of each branch is determined by molecular impacts. The sharp changes in the path occur when the particle is hit more on one side than another due to the thermal bombardment with molecules of the carrier fluid, which are in a state of violent agitation corresponding to the thermal level. This motion was discovered in 1827 by the botanist Robert Brown, but the formula for the process was calculated by Einstein (1905, 1956) and widely discussed by other scientists (Chadrasekhar, 1943; Ming Chen Wang and Uhlenbeck, 1945; Wax, 1954). A particle is defined as being *Brownian* when its size is small ($Kn >> 1$), or not too large, compared with the mean free path of the fluid and is sensitive to changes in momentum at each impact.

In the molecular regime, as a consequence of each individual casual impact with the air molecules, the kinetic energy tends to level out around their average value, while the motion of each single particle is described in terms of casualty.

Let us consider the individual displacement vector $\mathbf{s}_i$, which represents the rectilinear displacement of the particle from the ith collision to the next; then the position vector $r_n = \sum \mathbf{s}_i$, which joins the initial position of the particle to the final one after n collisions. The bold, here and in the following, is used for vector notation. It is possible to calculate the mean square displacement from the initial position undergone by the particle after a certain time t, i.e.

$$< r_n^2 > = < \sum_{ij} \mathbf{s}_i \mathbf{s}_j > = < \sum_i \mathbf{s}_i^2 > + < \sum_{i \neq j} \mathbf{s}_i \cdot \mathbf{s}_j > \tag{8.3a}$$

where $< \sum_i \mathbf{s}_i^2 >$ represents the mean square displacement after each impact and the algebraic sum of the scalar products of the casual displacements is $< \sum_{i \neq j} \mathbf{s}_i \cdot \mathbf{s}_j > = 0$, as the contribution of random positive factors equals the contribution of negative ones. The previous expression reduces to

$$< r_{\mathrm{n}}^2 > = n \; < a_i^2 > = < a_i^2 > \frac{t}{\Delta t} = \alpha t \tag{8.3b}$$

where $<a_i>$ represents the mean displacement after each collision, Δt the mean interval between two consecutive impacts and $\alpha = <a_i^2> / \Delta t$. The motion of a particle in a viscous fluid is described by the *Langevin* equation:

$$m\frac{d^2x}{dt^2} = -f\frac{dx}{dt} + F_x \tag{8.4}$$

where m is the particle mass, f is the friction coefficient, and F_x is the casual force acting on the particle and the random character of the impacts gives $<F_x> = 0$.

The friction coefficient f for large particles is based on *Stokes* law $f = -3\pi\mu D$ in laminar regime, where μ is the dynamic viscosity of the fluid. For very small particles, i.e. $Kn >> 1, f$ is proportional to the second power of D, according to the *Epstein equation* (1924)

$$f = \frac{2}{3}\left(1 + \frac{\pi\alpha_o}{8}\right)D^2\rho\sqrt{\frac{2\pi kT}{m}} \tag{8.5}$$

where $\alpha_o \approx 0.9$ is an empirical correction coefficient, ρ is the air density and m is the mass of the air molecules.

The Brownian motion is governed by two acting forces that are of different nature: the first one is the viscous drag, proportional to the velocity in accordance with the Stokes law in the continuum regime ($Kn << 1$), and the second is the random bombardment, typical of the molecular regime ($Kn >> 1$). Dividing both terms of Eqn (8.4) by the mass m, the same equation can be rewritten in the classical form

$$\frac{dv}{dt} = -\beta v + a_c(t) \tag{8.6}$$

where the unit of the coefficient β is time^{-1} and can be interpreted as the inverse of the viscous relaxation time of the particle and $a_c(t)$ is the random acceleration.

By multiplying both sides of Eqn (8.4) by x and transforming it according to the identities

$$x\frac{d^2x}{dt^2} = \frac{d^2\frac{x^2}{2}}{dt^2} - \left(\frac{dx}{dt}\right)^2; \quad x\frac{dx}{dt} = \frac{d\frac{x^2}{2}}{dt} \tag{8.7}$$

and considering that the displacements made along the axes x, y, z are equally probable, i.e. $<x^2> = <y^2> = <z^2> = <r^2> /3$, one obtains

$$<x^2> = \alpha\frac{t}{3}; \quad \frac{d<x^2>}{dt} = \frac{\alpha}{3}; \quad \frac{d^2<x^2>}{dt^2} = 0; \quad <m\left(\frac{dx}{dt}\right)^2> = \frac{\alpha f}{6}. \tag{8.8}$$

The principle of *equipartition of energy* states that molecules in thermal equilibrium have the same average energy associated with each independent degree of freedom of their motion and that this energy is

$$<\frac{1}{2}m\left(\frac{\mathrm{d}x}{\mathrm{d}t}\right)^2> = \frac{1}{2}kT \tag{8.9}$$

where k is the *Boltzmann* constant, allows the calculation of the value $\alpha = 6\ k\ T/f$, and the mean square displacement $<r_\mathrm{n}^2>$ can be found:

$$<r_\mathrm{n}^2> = \frac{6kTt}{f}. \tag{8.10}$$

This equation, first derived by *Einstein* (1905), constitutes a probabilistic representation of the expected displacement. The mean square displacement of the diffusing particles increases linearly with elapsed time and fluid temperature T; it decreases inversely with the friction coefficient f that is related to the (dynamic) fluid viscosity μ.

The dispersion of particles that were originally close to each other is the most probable result of the natural reorganization of the internal distribution of a system as a consequence of individual and casual molecular impacts. This happens without the intervention of collective organized motions that characterize *hydrodynamics*, which holds in the continuum regime and is governed by fields of force or pressure gradients.

The diffusion coefficient can be derived in terms of random walk (Seinfeld, 1986) by integrating the equation that governs the concentration C of Brownian particles

$$\frac{\partial C}{\partial t} = \mathscr{D}_B \nabla^2 C \tag{8.11}$$

that gives $\partial <x^2> /\partial t = 2\mathscr{D}_\mathrm{B}$ and $<x^2> = 2\mathscr{D}_\mathrm{B}\, t$. Equating the value of $<x^2> = (1/3)<r_\mathrm{n}^2>$ given by Eqn (8.10) with the result obtained integrating Eqn (8.11), the Brownian diffusivity $\mathscr{D}_\mathrm{B}$ is obtained (Fig. 8.1):

$$\mathscr{D}_B = \frac{kTC_C}{3\pi\mu D} \tag{8.12}$$

This is the fundamental *Stokes–Einstein* relation that has been improved with the addition of the empirical *Cunningham* slip correction factor C_c (Fig. 8.2). In the continuum regime, $C_c \to 1$ and $\mathscr{D}_B$ varies as D^{-1}; however, in the molecular regime, when the aerodynamic mobility of the particle increases, C_c is given by the equation

$$C_c \approx 1 + 1.657\frac{2\lambda}{D} \tag{8.13}$$

and $\mathscr{D}_\mathrm{B}$ varies as D^{-2}. In practice, C_c increases the efficiency of the Brownian deposition of fine particles, where the number of simultaneous

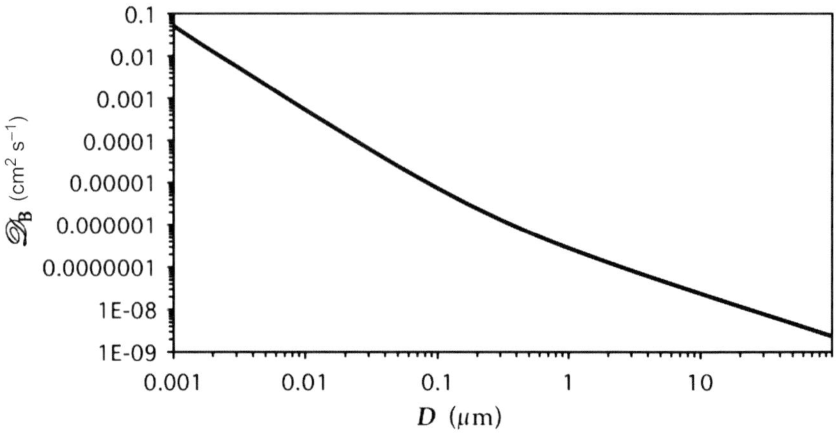

FIGURE 8.1 Brownian diffusivity $\mathscr{D}_B$ (cm^2 s^{-1}) of particles with size $0.001 \leq D \leq 100$ μm.

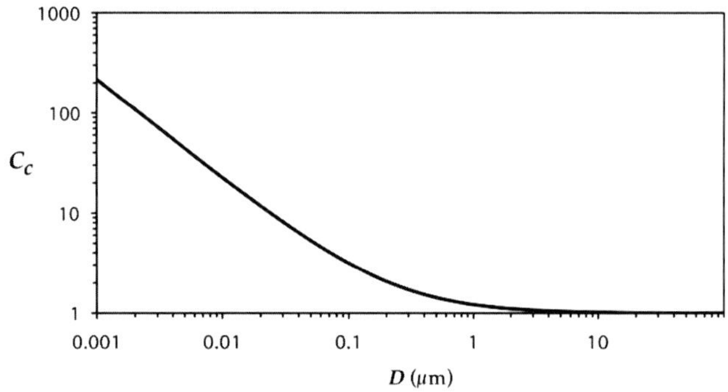

FIGURE 8.2 Cunningham slip factor C_c versus particle diameter D (μm).

molecular impacts becomes smaller and smaller and the net resultant is less and less attenuated.

The Stokes–Einstein formula shows that the diffusivity is proportional to the air temperature that determines the molecular excitation but is inversely proportional to the fluid viscosity and particle diameter. In the case of large particles, a very large number of random impacts are averaged and the net resultant tends to 0, as well as the diffusivity. The diffusivity is independent of the particle density, unlike the velocity, which arises from the thermal excitation.

Einstein (1905) derived the diffusion coefficient considering a solute, with nonhomogeneous concentration C, dispersed into a fluid, contained in a cylinder with unit cross-section. S_1 S_2 are two parallel cross-sections normal to the cylinder axis with a space of $2\Delta x$ between them, where Δx

represents the mean displacement undergone by the solute molecule due to the impacts that have occurred during the time interval Δt. S is the cross-section just in the middle of S_1 and S_2. C_1 and C_2 are the concentrations in proximity of the two sections S_1 S_2. Only half of the solute particles initially enclosed in the portion of cylinder between the sections S_1 S and S S_2 will move in the direction of the other section, passing through S, i.e. $(1/2)\ C_1\ \Delta x$ and $(1/2)\ C_2\ \Delta x$. The net amount of solute that diffuses across S in the interval Δt, i.e. $\Phi_{\Delta t}$, is given by the difference between C_1 and C_2, i.e. $\Phi_{\Delta t} = (1/2)\ (C_1 - C_2)\Delta x$. Comparing the differential and finite terms, we get

$$\frac{C_1 - C_2}{\Delta x} = -\frac{dC}{d\boldsymbol{n}} \tag{8.14}$$

where $\boldsymbol{n}$ is the direction of the cylinder axis, and substituting in the previous expression,

$$\Phi_{\Delta t} = -\frac{1}{2}\Delta x^2 \frac{dC}{d\boldsymbol{n}}. \tag{8.15}$$

The net flux per unit time is

$$\Phi = \frac{\Phi \Delta t}{\Delta t} = -\frac{\Delta x^2}{2\Delta t}\frac{dC}{d\boldsymbol{n}} \tag{8.16}$$

By definition, the diffusion coefficient is the coefficient of proportionality between Φ and the concentration gradient $-dC/d\boldsymbol{n}$ and Equation (8.16) gives:

$$\mathscr{D}_B = \frac{\Delta x^2}{2\Delta t} \tag{8.17}$$

However, although the final displacement Δx is the result of many random steps, Δx can be substituted with the Brownian mean free path λ_B, so that $\Delta t = \lambda_B/<v>$, where $<v>$ is the mean speed of the Brownian particle, i.e.

$$\mathscr{D}_B \approx 0.5\lambda_B <v>. \tag{8.18}$$

8.3. BROWNIAN DEPOSITION

The *Schmidt number Sc* is defined as

$$Sc = \frac{\nu}{\mathscr{D}_B} \tag{8.19}$$

and represents the ratio between the kinematic air viscosity ν and the diffusivity $\mathscr{D}_B$ of particles. In practice, Sc represents the ratio between the momentum diffusivity (linked to ν) and the mass diffusivity (linked to $\mathscr{D}_B$). In the case of a low *Sc*, the particles have a great diffusivity, are very small and are scarcely conditioned by the viscosity of the medium, so that they will easily cross the laminar layer that envelops smooth surfaces when the turbulence is moderate and impact on the surface. At high *Sc*, particles are giant, with small diffusivity, and this kind of deposition becomes less and less relevant. In practice, Brownian deposition is very important for fine particles with diameter ranging between 0.001 and 0.01 μm; it is still important for *D* of the order of 0.1 μm, but is negligible for micrometric particles.

Following the gas theory, the average *thermal speed* v_{th} (or root mean square velocity) of airborne particles in equilibrium with air can be calculated from the translation kinetic energy $E_c = (3/2)\,kT$,

$$v_{th} = \sqrt{\frac{3kT}{m}} = \sqrt{\frac{18kT}{\pi \rho_p D^3}} \tag{8.20}$$

where m is the mass and ρ_p is the density of the particle. The *mean velocity* of the flying particles v_{MB} (Figure 8.3) can be calculated by integrating the product of each individual velocity (given by the Maxwell–Boltzmann distribution) and its probability, i.e.

$$v_{MB} = \sqrt{\frac{8}{3\pi}}\,v_{th} = \sqrt{\frac{8kT}{\pi m}} = \sqrt{\frac{48kT}{\pi^2 \rho_p D^3}}. \tag{8.21}$$

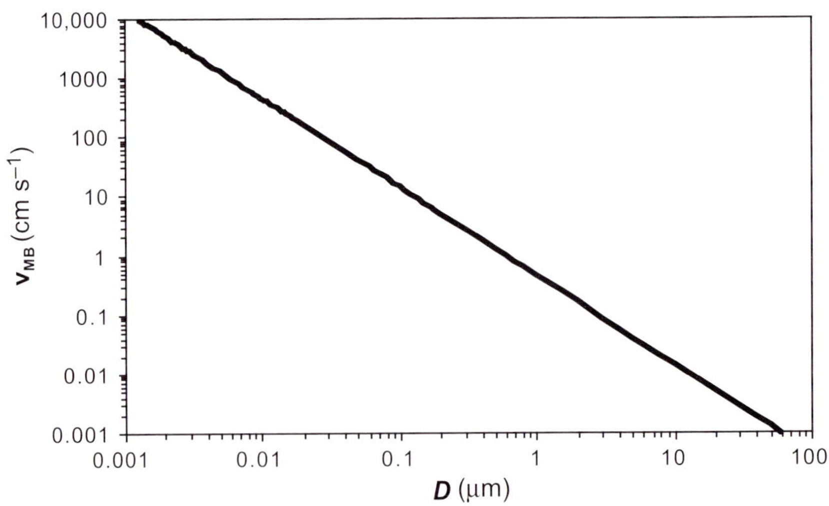

FIGURE 8.3 Maxwell–Boltzmann velocity v_{MB} distribution of suspended particles versus particle diameter *D* (μm).

It can be assumed that the Brownian deposition velocity v_B is equal to the mean velocity of the flying particles at the beginning of the process but, after some time, the interfacial layer will be progressively depleted of the particles that stick on the surface and new particles will arrive from the atmosphere. After time Δt, the number N of particles that will have deposited on the surface is

$$N(\Delta t) = 2\sqrt{\frac{\mathscr{D}_B \Delta t}{\pi}} \tag{8.22}$$

A new concentration layer originates near the surface, and the flux of particles that impact on the surface is now conditioned by the development of the concentration layer. In this layer, C varies over time and with the distance z from the surface. In stagnant air, the concentration layer evolves regularly and C is homogeneous and constant at the interface and beyond a certain distance from it.

The deposition velocity is conditioned by the particle diffusivity $\mathscr{D}_B$ and by the factors that determine the development of a very thin diffusive layer based on the surface, having thickness $\delta(t)$, in which it can be assumed that $C(z)$ varies linearly with z. Under this assumption, the *Brownian deposition velocity* $v_B(t)$ is

$$v_B(t) = \frac{\mathscr{D}_B}{\delta(t)} \tag{8.23}$$

and numerically coincides with the values of $\mathscr{D}_B$ (Fig. 8.1) but is expressed in cm s^{-1}, for the case of a layer having thickness $\delta = 1$ (Fig. 8.3). The value of $\delta(t)$ is characterized by the dynamic regime of the fluid, the thickness of the air layer in the near surface region, the particle size and the time t elapsed after the formation of the concentration layer.

8.4. THERMOPHORESIS

Thermophoresis (literally: transport due to temperature) is due to the fact that in the presence of a temperature gradient in the atmosphere, a particle is pushed towards the lower temperature zone because of the less-energetic molecular impacts. The phenomenon was already known in the past century in that scientists had recognized that no particles were found near hot bodies and Einstein (1905, 1956) gave the first mathematical explanation of this phenomenon.

In the molecular regime, the particle is hit by air molecules with stronger individual momentum $(p_i)_h$ on the warmer air side and weaker individual momentum $(p_i)_c$ in the colder side. The result is that the particle will experience a net momentum $\sum_i (p_i)_h - \sum_i (p_i)_c$ that will push it towards the colder part. This explanation holds for small particles in molecular regime, not for large ones that are in the continuum regime. An explanation in molecular terms leads to wrong results. In fact, suppose a cubic particle with side Δl_x, oriented with the temperature gradient. When the particle size increases, the impacting surface $(\Delta l_x)^2$ too increases, as well as the difference in temperature ΔT_x between the two faces of the particle normal to the gradient. As the thermophoretic force is proportional to the temperature gradient $\Delta T_x/\Delta l_x$, ΔT_x will also be proportional to Δl_x and the total gain will be proportional to $(\Delta l_x)^3$. However, this is not true, because on increasing the particle size the particle mobility will be reduced. The problem cannot be dealt with statistical thermodynamics because the mobility is of hydrodynamic nature and holds in the continuum regime. In practice, the mechanism depends on a number of factors: the temperature difference at the interface between the air and the particle, which is determined by the thermal conductivity of the gas and the particle; the regime of flow behind the particle; and the resistance of the fluid, which is proportional to the velocity in laminar flows and varies with quadratic law in nonlaminar flows. Epstein (1924) made a careful mathematical treatment of the exchange of momentum between the gas and a sphere in order to determine the resistance experienced by the sphere while moving through a gas, critically reviewing all the previous theoretical and experimental results by Langevin, Cunningham, Lenard and Millikan. He considered the two cases of spherical particles (droplets) very small and large compared with the mean free path of the gas. Introducing the ratio of the heat transferred inside the particle to the heat received in unit time from the molecular impacts, i.e. the ratio of the internal conductivity of the particle to the conductivity of the gas, he found a complete agreement between theory and experiment. In the case of comparatively large spheres, he started with the consideration that the particle exerts a relevant influence on the bulk motion of the gas molecules, so that the gas is perturbed by the onset of hydrodynamic stress, dependent on the velocities. The physical meaning of the stress components is the transportation of a given component of momentum in a given direction. He then applied the principles of conservation of both mass and energy, i.e. the number of molecules impinging on a surface element of the particle equals the number of those reflected by it and the energy brought by the impinging molecules equals the energy of the reflected ones. Under these hypotheses, the force of resistance experienced by a large particle (i.e. when $Kn^2 \to 0$) in the presence of a surface slip is the momentum conferred on the sphere by the molecules of the fluid in which the sphere is immersed, as reported in all the hydrodynamics textbooks. Similarly, Talbot et al. (1980) considered the case of molecules colliding obliquely on the surface of the particle: the normal component of the momentum is compensated, but not the tangential one, especially in the case of skin friction. On average, air molecules coming from a warmer gas region deliver more tangential momentum than those coming from the colder region and the net result is a tangential momentum transfer.

Various formulae have been proposed to calculate of the thermophoretic velocity $\mathbf{v}_T$. In the molecular regime ($Kn >> 1$):

$$\mathbf{v}_T = -\frac{\nu}{8\pi T}\frac{dT}{dn} \tag{8.24}$$

$$\mathbf{v}_T = -\frac{3\nu\frac{dT}{dn}}{4\left(1+\frac{\pi\varepsilon}{8}\right)T} \approx -0.55\frac{\nu}{T}\frac{dT}{dn} \tag{8.25}$$

$$\mathbf{v}_T = -0.33\frac{\mathscr{D}_B}{T}\frac{dT}{dn} \tag{8.26}$$

where ε represents a correction factor, generally $\varepsilon = 0.9$. Eqn (8.24) was the first approach by Epstein (1924) and Eqn (8.25) by Waldmann and Schmitt (1966). Equations (8.24) and (8.25) are independent of particle size; Eqn (8.26) is inversely proportional to D for the presence of $\mathscr{D}_B$ and it better describes experimental results.

In the continuum regime ($Kn << 1$):

$$\mathbf{v}_T = -2\frac{\sigma k_a^2}{p(k_p+2k_a)}\frac{dT}{dn} \tag{8.27}$$

$$\mathbf{v}_T = -0.15\frac{\mathscr{D}_B}{T}\frac{dT}{dn} \tag{8.28}$$

where ν is the kinematic viscosity of the air; n is the distance from the surface; σ is the aerodynamic slip factor, generally $\sigma = 0.2$; and k_p and k_a are the thermal conductivities of the particle and the air, respectively. Equation (8.27) is the Epstein equation as reported by Friedlander (1977).

Two formulae where proposed for the entire Knudsen range. The formula by Pruppacher and Klett (1980, 2003) is

$$\mathbf{v}_T = -B\frac{k_a}{p}\frac{dT}{dn} \tag{8.29}$$

where the particle size, expressed in terms of Kn, and k_a and k_p are included in the coefficient

$$B = -\frac{0.4(1+\alpha Kn)(k_a+2.5k_pKn)}{(1+3Kn)(k_p+2k_a+5k_pKn)} \tag{8.30}$$

where $\alpha = 1.257 + 0.400\ \exp(-1.10Kn)$. The formula by Talbot and al. (1980) is

$$\mathbf{v}_\mathrm{T} = -\frac{\nu\, K}{T}\frac{\mathrm{d}T}{\mathrm{d}\boldsymbol{n}} \tag{8.31}$$

where the parameter K

$$K = \frac{2C(R + C'Kn)\{1 + Kn[1.2 + 0.41\exp(-0.22Kn)]\}}{(1 + 3C''Kn)(1 + 2R + 2C'Kn)} \tag{8.32}$$

is expressed in terms of the ratio $R = k_a/k_p$, Kn and the constants $C = 1.147$, $C' = 2.20$ and $C'' = 1.146$. In general, $0.1 < K < 0.6$. The graphs of Eqns (8.29) and (8.30) are reported in Fig. 8.4, which has been computed for $R = k_a/k_p = 0.2$. The two equations are substantially coincident for fine particles; after $D = 0.01\ \mu\mathrm{m}$ the departure keeps increasing, and the equation by Talbot et al. overestimates excessively the velocity of large and giant particles. Other theoretical studies and comparisons with experimental data are reported by Hidy and Brock (1970, 1972), Slinn and Hales (1970) and Prodi and Tampieri (1982).

In general, the thermophoretic velocity of fine or large particles is proportional to the intensity of the temperature gradient in front of the

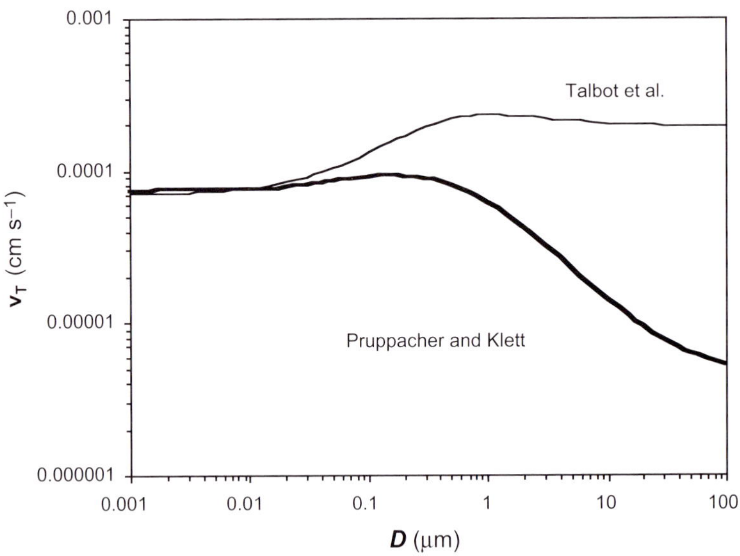

FIGURE 8.4 Comparison of the thermophoretic velocity v_T computed by means of the formulae suggested by Pruppacher and Klett (1980, 2003) and Talbot et al. (1980) for the unit temperature gradient $\mathrm{d}T/\mathrm{d}\boldsymbol{n} = 1\ ^\circ\mathrm{C\ cm}^{-1}$. The former is more close to the experimental data, whereas the latter departs at the medium size and the error increases with the larger particles.

surface where deposition occurs (the surface is colder than air). If, however, the surface is warmer than the surrounding air, the temperature gradient is inverted so that the thermophoretic force tends to protect the surface against deposition. It should be remembered that in front of the surface, the gradient of the air temperature diminishes exponentially as the distance n from the surface increases. Thus, at the interface where thermophoretic deposition occurs, $\mathbf{grad}T$ is much more intense than the average measurable value $\Delta T/\Delta n$ even at the minimum distance Δn allowed by the instruments.

8.5. DIFFUSIOPHORESIS

Initially, the molecules are close to each other and move in a *random walk*. This causes diffusion, passing from a nonhomogeneous situation to a homogeneous one. Pure diffusion occurs when some molecules of a species 'A' are immersed into a fluid 'B' and because of molecular agitation, the A molecules gradually move away from their original position, so that after a sufficient long time, there will be an equal distance between the A molecules that are thus uniformly distributed throughout the B fluid. This is the pure diffusion of A within the medium B, while the carrier B remains apparently unaffected, except for having participated in the dispersion of A. If the concentration of A is so high that the concentration of B is also initially affected, showing a minimum where A is maximum, then A diffuses diverging in B and B diffuses converging in A by mutual diffusion. If the concentration of A is so great that the partial pressures are also involved, the problem cannot be treated any longer in terms of statistics of random motions and diffusion but should be treated in terms of partial pressures and mass convection, using the laws of the hydrodynamics. A transition regime can also be considered, in which both these approaches, i.e. statistics and hydrodynamics, are simultaneously involved. When dealing with atmospheric pollution, the concentrations are usually so low that the first case is appropriate; condensation on a surface and evaporation in the free air are examples of the second case. Forced evaporation may be an example where the partial pressures are affected. There are various processes in the atmosphere that can give rise to nonhomogeneous concentrations of various gases. We will concentrate here on water vapour because of the high or low levels of concentration that may be locally reached and the effects that may follow on the deposition of pollutants. There are many sources or sinks of water vapour, such as natural or human combustion, transpiration and respiration by living beings, condensation, evaporation and sublimation.

Gradients in the concentrations of both water vapour C_w and dry air C_a give rise to two *diffusive flows*, but in opposite directions, and the number of these molecules N_w and N_a that cross unit surface in unit time interval is given by the *Fick* equation

$$N_w = \mathscr{D}_{wa}\frac{\mathrm{d}C_w}{\mathrm{d}\boldsymbol{n}} \qquad N_a = \mathscr{D}_{aw}\frac{\mathrm{d}C_a}{\mathrm{d}\boldsymbol{n}} \tag{8.33}$$

that has already been seen deriving the Einstein formula, where $\mathscr{D}_{wa} = \mathscr{D}_{aw}$ is the coefficient of mutual diffusion between dry air and water vapour; experimental observations give $\mathscr{D}_{aw} = 0.239\ \mathrm{cm^2\ s^{-1}}$ at $T = 8\,°\mathrm{C}$. This equation was derived in the heat transfer theory and is based on the assumption that diffusivity is constant in space. Of course,

$$\frac{\mathrm{d}C_w}{\mathrm{d}\boldsymbol{n}} = -\frac{\mathrm{d}C_a}{\mathrm{d}\boldsymbol{n}} \tag{8.34}$$

Any particle that is suspended in a region where dry air and vapour molecules are mutually diffusing is subject to two opposing phoretic (i.e. transport) forces and is drawn by the resultant, in the direction of the flow that exerts the preponderant action, i.e. the dry air flow given that the mass of the air molecules ($m_a \approx 29$) is greater than that of vapour ($m_w = 18$).

The resulting diffusiophoretic velocity $\mathbf{v}_\mathrm{d}$ is given by the equation:

$$\mathbf{v}_\mathrm{d} = -\sigma_{aw}\mathscr{D}_{aw}\frac{\mathrm{d}\gamma_a}{\mathrm{d}\boldsymbol{n}} \tag{8.35}$$

where σ_{aw} is the diffusion slip factor

$$\sigma_{aw} = \frac{m_w - m_a}{(\gamma_a m_a + \gamma_w m_w) - (\gamma_a + \gamma_w)\sqrt{m_w m_a}} \tag{8.36}$$

calculated on the basis of the molar fractions of dry air γ_a and vapour γ_w, respectively

$$\gamma_a = \frac{C_a}{(C_a + C_w)}; \quad \gamma_w = 1 - \gamma_a = \frac{C_w}{(C_a + C_w)} \tag{8.37}$$

A vapour concentration gradient thus leads the particle to follow the movement of the air molecule, going in the direction opposite to that of the vapour. In the event of a wall that releases vapour for surface or subsurface evaporation (e.g. in the micropores or the capillary fringe), if only the effect of pure diffusiophoresis is considered, all the particles tend to approach the wall and deposit there; on the other hand, if a surface absorbs water vapour, the opposite occurs. However, diffusiophoresis will generate transport by another phenomenon: the Stefan flow, which will radically change this situation.

8.6. STEFAN FLOW

In the case of an initial nonhomogeneous distribution of dry air and vapour, the process that leads to a uniform distribution of both the constituents leads to pure diffusiophoresis as discussed above. If, however, the quantity of dispersed vapour does not remain constant (as in the case of condensation on a surface, or evaporation), then an additional phenomenon is generated, that is, the Stefan flow (Goldsmith et al., 1963; Slinn and Hales, 1970; Pruppacher and Klett, 1980, 2003; Prodi and Tampieri, 1982; Vittori, 1984). This is the result of some microphysical processes that all take place at the same time and, therefore, are rather complex. None of these steps can take place individually and a partial look at the phenomenon easily leads to misunderstanding. For this reason, it is often misinterpreted.

To describe this phenomenon, it is useful to consider a cold surface on which condensation occurs, at temperature T that is only slightly below the dew point (DP) i.e. $T = DP - \delta T$. In a cloud droplet, δT is really an infinitesimal value, but in the case of a monument, measurable values of ΔT can be reached. At the interface between the air and the surface where the water molecules deposit, the vapour pressure diminishes by a quantity δp_w and assumes the vapour saturation value $e_{sat}(T)$, which is given by the Magnus formula. This determines a local minimum, even though very modest, with respect to the undisturbed value p_w, so that $p_w - \delta p_w = e_{sat}(T)$. It might be useful to remember that in practice, the value of p_w is a small fraction (i.e. 0–3%) of the total air pressure and δp_w is a small fraction of p_w, so that δp_w is a very small fraction of the total pressure and depends on the temperature difference between the condensing surface and the free air. In the case of a cloud droplet that is always in close equilibrium with the surrounding air, δp_w is zero or nearly zero; in the case of a relatively cold monument and warm, humid air, δp_w may reach higher values, up to one or a few hectopascals. Therefore, in the air (i.e. dry air and water vapour) close to a condensing surface, a weak convective *hydrodynamic* transport of the whole air mass is generated in order to compensate for the drop of pressure δp_w. As at the interface a number of vapour molecules are continuously subtracted from the atmosphere by condensation and, moreover, the convective flow that is incoming has fewer vapour molecules due to the concentration gradient of the vapour established near the condensing surface, the concentration of dry air molecules increases near the surface. Therefore, the partial pressure p_a increases by $\delta p_a = -\delta p_w$, so that the total atmospheric pressure $p = p_a + p_w$ remains homogeneous, in that this is an open system in equilibrium. Briefly, the system at the interface reaches a dynamic equilibrium with a lower concentration of water vapour and a higher concentration of dry air, which determine two opposite diffusive flows, one of vapour towards the surface and one retrograde of dry air, with the consequences discussed in the previous section.

The key problem is now to see how this equilibrium may be kept for a long time, because the higher concentration of dry air molecules that has formed near the surface generates a retrograde *diffusive* transport of the dry air molecules since the surface is a sink of vapour but not a source of dry air. Therefore, the departure of the molecules of dry air leads to a local reduction of the total pressure and, as a consequence, a convective mass transport of *hydrodynamic* nature develops, which exactly compensates the system keeping it in dynamic equilibrium, i.e. with stationary p as well as the localized maximum $p_a + \delta p_a$.

It should be noted that, in order to describe as clearly as possible this phenomenon in all its details, it has been divided into logical subsequent steps, which are not separated in the nature. The reality is summarized saying that *the surface condensation generates two opposite gradients of water vapour and dry air, with opposite diffusiophoresis. This situation generates a compensating hydrodynamic flow*, which is known as the *Stefan flow*.

An accurate definition of the Stefan flow is based on the balance of one component, as follows. The condition that the net transport of dry air at the interface of the condensing surface has a null balance (neither accumulation nor loss of dry air) means that *the diffusive transport of dry air is exactly compensated for, at each point, by the hydrodynamic transport of the surrounding air*, the latter being called the Stefan flow.

From the above definition, the speed $\mathbf{v_S}$ of the Stefan flow can be calculated from the balance of the number of the dry air molecules at the interface, i.e. equating $\mathbf{v_S}C_a$, which represents the hydrodynamic flow due to the convective transport of dry air described in Eqn (8.2), with the Fick diffusive flow of dry air due to the concentration gradient described in Eqn (8.33) and taking into account the opposite signs of these two flows, i.e.

$$\mathbf{v_S}C_a = \mathscr{D}_{aw}\frac{\mathrm{d}C_a}{\mathrm{d}\mathbf{n}}. \tag{8.38}$$

Isolating $\mathbf{v_S}$, using the definition of γ_a in Eqn (8.37), and remembering that the sum of the concentrations $C_a + C_w$ remains constant, one obtains

$$\mathbf{v_S} = \frac{\mathscr{D}_{aw}}{\gamma_a}\frac{d\gamma_a}{d\mathbf{n}}. \tag{8.39}$$

A particle in the proximity of a condensing surface is subjected to two counteracting mechanisms: the Stefan flow and the diffusiophoresis.

However, the diffusiophoresis is much weaker and the net result in vector terms is

$$\mathbf{v_R} = \mathbf{v_S} + \mathbf{v_d} = -(1 + \sigma_{aw}\gamma_w)\frac{\mathscr{D}_{aw}}{\gamma_w}\frac{\mathrm{d}\gamma_a}{\mathrm{d}n} \approx 0.8\mathbf{v_S} \tag{8.40}$$

as $\sigma_{aw}\ \gamma_w = 0.21$; this means that the Stefan velocity is largely dominant over the diffusiophoretic one and the net result is that the Stefan velocity is reduced by 20%. It can be noted that $\mathbf{v_S}$ is, moreover, the only one responsible for the transport of large particles. Like thermophoresis, in fact, diffusiophoresis tends to vanish for the larger particles (i.e. for $Kn \to 0$, $\mathbf{v_R} \to \mathbf{v_S}$), while active transport due to the hydrodynamic Stefan flow has the same efficiency for every type of particle, even those that are some micrometres in diameter.

This very efficient transport mechanism develops during either condensation or evaporation, and it transports, indiscriminately, all the particles near the surface during condensation or opposes all the other deposition processes during evaporation. Vittori (1973) measured the dust-free space around an evaporating droplet, i.e. the distance over which the Stefan flow is dominant and pushes away all the suspended particles, and in his experimental work, this dust-free layer was found to be some 360 μm thick.

At first sight, it would seem that condensation (that is, a surface temperature equal or inferior to the *DP* of air) is reached in an internal environment only in exceptional cases. It should be remembered, however, that this phenomenon occurs in the micropores very frequently, as can be seen from the Kelvin equation.

From experimental data, Goldsmith et al. (1963) related the deposition velocity $\mathbf{v_R}$ to weak pressure gradients $\mathrm{d}p/\mathrm{d}n$

$$\mathbf{v_R} \approx 1.9 \times 10^{-4}\frac{\mathrm{d}p}{\mathrm{d}n} \tag{8.41}$$

with $\mathbf{v_R}$ in cm s^{-1} and $\mathrm{d}p/\mathrm{d}n$ in hPa cm^{-1}. Unfortunately, there are no direct measurements for the case of monuments and laboratory experiments are very difficult to carry out. Applying this formula to the typical value for heavy condensation on monuments reported above, i.e. $\delta p_w = 1$ hPa, and the dust-free space found by Vittori for $\mathrm{d}n$, the order of magnitude of the deposition onto the monument surface is found, i.e. $\mathbf{v_R} \approx 5 \times 10^{-3}$ cm s^{-1}. This is very slow, but the effect cannot be ignored, as at this speed the equivalent of an air layer of some 18 mm is swept in an hour, and in a clear night, condensation may continue for several hours.

Another empirical formula derives from the observation that the Stefan flow is a consequence of the average displacement of the gas that substitutes the water molecules that condense on the surface, or vice versa in the case of evaporation. Dimensional considerations lead to relate the molecular masses of dry air m_a and water vapour m_w to the condensation (or evaporation) rate $\mathscr{C}$ (g cm^{-2} s^{-1}):

$$\mathbf{v_R} \approx \frac{m_a}{m_w} \frac{\mathscr{C}}{\rho} \approx 1.6\,\mathscr{C} \tag{8.42}$$

The number of captured particles depends on the amount of condensed vapour and the concentration of particles in the atmosphere. The $\mathbf{v_R}$ is about 0.005 cm s^{-1}/100 W m^{-2} of latent heat transferred by condensation, which is equivalent to the deposition of a water film with a growing thickness of 0.15 mm h^{-1}. On a monument, the condensation occurs at a lower rate than on grassy land, which can reach 45 W m^{-2} so that $\mathbf{v_R} < 3 \times 10^{-3}$ cm s^{-1}. This indicative value is very similar to the one found above with the Goldsmith formula.

The opposite occurs with evaporation, keeping in mind, however, that while condensation is a slow but continuous process, evaporation can occur very quickly under certain circumstances if a great deal of energy (e.g. direct sunshine) is supplied and there is sufficient ventilation. Therefore, in the evaporation phase, this hydrodynamic flow, even for relatively brief periods, can efficiently protect a surface against the deposition of pollutants, in that the $\mathbf{v_R}$ in the evaporation phase can reach values that are 10 times greater than in the condensation phase.

To avoid confusion, it is necessary to clarify that this complex effect that enables the calculation of resulting deposition velocity $\mathbf{v_R}$ has several names. Some people have called $\mathbf{v_R}$ the '*Stefan*' only, as in a first-order approximation, because the hydrodynamic Stefan flow is largely dominant with respect to mutual diffusion. Others, instead, call the resulting effect only diffusiophoresis, placing the accent on the initial phase that starts the whole process.

8.7. GRAVITATIONAL SETTLING

In the general case of particles immersed in a viscous fluid, forces do not cause acceleration but lead to a final velocity $\mathbf{v}$ so that the intensity of the force $\boldsymbol{F}$ is balanced by the resistance R (i.e. the *viscous drag*) exerted by the fluid on the particle and $\mathbf{v} = \boldsymbol{F}/f$, where f is the friction coefficient that characterizes R. Gravity-attractive force generates buoyancy of bodies immersed in a fluid, as a consequence of the net *Archimedes'* hydrostatic balance $\boldsymbol{F}_\mathbf{g}$

$$\boldsymbol{F}_\mathbf{g} = \frac{\pi}{6} D^3(\rho_p - \rho_a)g \tag{8.43}$$

where ρ_p and ρ_a are the densities of the particle and the air, respectively, and g is the acceleration of gravity. The viscous drag R of air is derived from the *Stoke* law (for spherical particles) and corrected for aerodynamic departures linked to Kn:

$$\mathscr{R} = \frac{3\pi\mu D\mathbf{v}}{(1 + \alpha Kn)} \tag{8.44}$$

where α is the same coefficient defined for Eqn (8.30). In laminar conditions, the terminal settling velocity $\mathbf{v_g}$ is:

$$\mathbf{v_g} = D^2 \frac{\rho_p g}{18\mu} C_c \left(1 - \frac{\rho_a}{\rho_p}\right) \approx D^2 C_c \frac{\rho_p g}{18\mu} \tag{8.45}$$

where ρ_a/ρ_p and is of the order of 10^{-3} and can be neglected compared to 1. The formula clearly shows that in still air, the sedimentation velocity is proportional to the density of the particle and to the square of its diameter D and becomes proportional to D for fine particles, when $Kn >> 1$ and the Cunningham C_c varies as D^{-1}. The particle density usually ranges between $\rho_p = 1\ \mathrm{g\ cm^{-3}}$, e.g. water droplets, and $\rho_p = 2.65\ \mathrm{g\ cm^{-3}}$, e.g. quartz granules. In the case of nonspherical particles, the aerodynamic effect due to the shape, expressed in terms of equivalent aerodynamic diameter, can be calculated. Note that dust and desert sand granules acquire a spherical shape because of constant rolling.

Sedimentation deposits large particles on horizontal surfaces and is particularly efficient in calm air, especially for particles having a diameter greater than 2 μm. The settling velocity $\mathbf{v_g}$ is shown in Fig. 8.5.

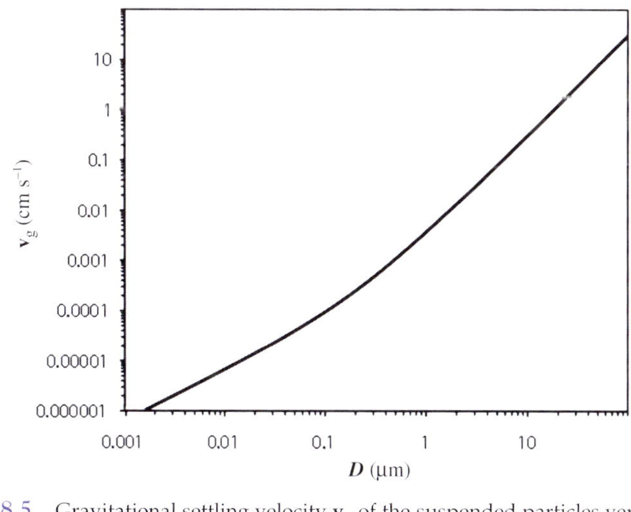

FIGURE 8.5 Gravitational settling velocity $\mathbf{v_g}$ of the suspended particles versus particle diameter D (μm).

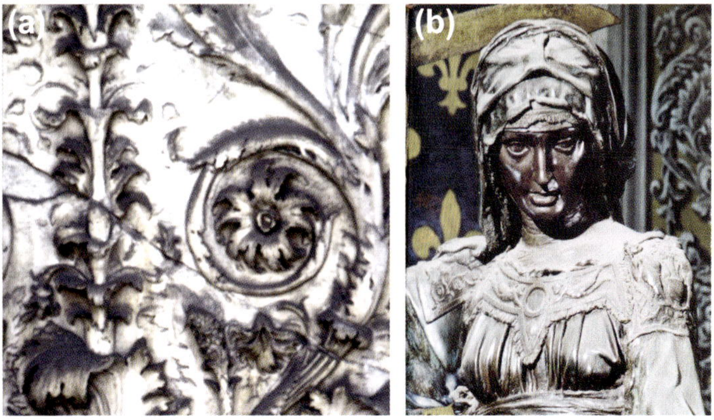

FIGURE 8.6 Particles deposited by gravitational settling on the upper part of decorations (bas-relief with flower decorations on white marble, Ara Pacis, Rome) or (b) statues (dark bronze by Donatello, Giuditta, in Palazzo Vecchio, Florence).

Coarse particles deposited by gravitational settling are typically visible dust that makes dirty the upper part of horizontal surfaces, e.g. decorations, statues and furniture. In general, it has a grey tone that appears black on white surfaces, and vice versa. It generally forms a soft, incoherent layer that can be easily removed with blown air or gentle brushing (Fig. 8.6).

When convective motions occur, the descending and uprising airflows apparently compensate and the average effect of downwards and upwards motions leaves unaffected the average values. However, the presence of convection generates mixing and turbulence, so that Stokes law cannot be applied. The turbulent drag is much greater and particles remain much longer in suspension following microscale air movements. Therefore, during the daytime, when heating, ventilationg and air conditioning (HVAC) systems and people movements generate turbulence and resuspension, the concentration of coarse particles is maximum during the middle of the day, when rooms are crowded with visitors. During the night, the concentration of fine- and medium-mode particles remains unchanged but the concentration in the coarse mode has considerably decreased for the gravitational settling especially active on largest particles. At the end of the night, the fine and medium particle concentration is still unchanged but almost all large particles have disappeared being deposited on the floor. They will be resuspended by the new day visitors (Fig. 8.7).

In the past, the most visible particle deposition was due to smoke and gravitational settling. This was particularly evident where walls were not perfectly vertical but offered some slightly protruding parts that were rapidly covered with a dust layer. In order to avoid this soiling mechanism, artists used the strategy of inward tilted vertical surfaces. An

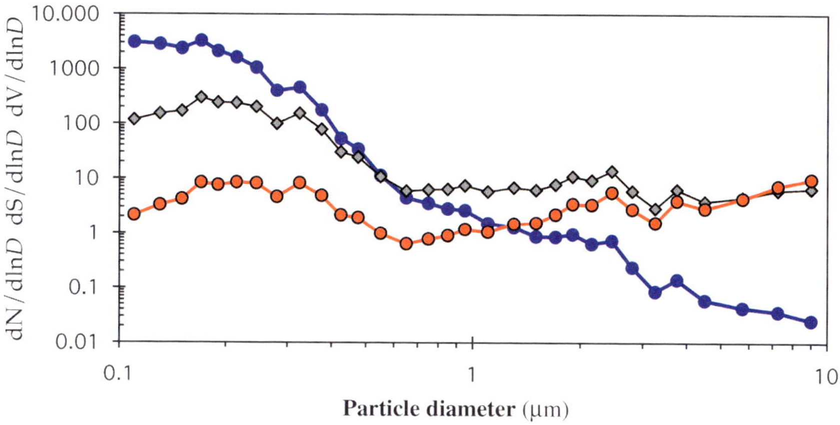

FIGURE 8.7 Changes in spectral distribution of the concentration of suspended particles during the day. During the visiting time, the sources, resuspension and turbulence, determine a variable concentration in the coarse range of suspended particles. At noon (red line, label 12), coarse particles reach the maximum; at midnight (blue line, label 24), many coarse particles have settled; the next morning at 6:00 a.m. (pink line, label 6), most of the coarse particles have disappeared by sedimentation. Museum Correr, Venice, measurements at 12.00, 24.00, and 6.00 of the next day, 12 and 13 February 1996.

example is given in Fig. 8.8 where Giotto inclined the gilded nimbus surface. Dust deposited on the upper part of the protruding border of the nimbus but left gold clean and bright. In addition, the tilted, curved and wavy surface augmented reflection of light towards the observer, increasing brightness. As another example, in the Sistine Chapel, Rome, Michelangelo built a second wall to paint the Last Judgment. The wall was 20 m tall and 20 cm inclined, with 1% tilting angle. This is totally irrelevant for improving perspective, however, but useful against settling particles because their distance from the wall will progressively increase

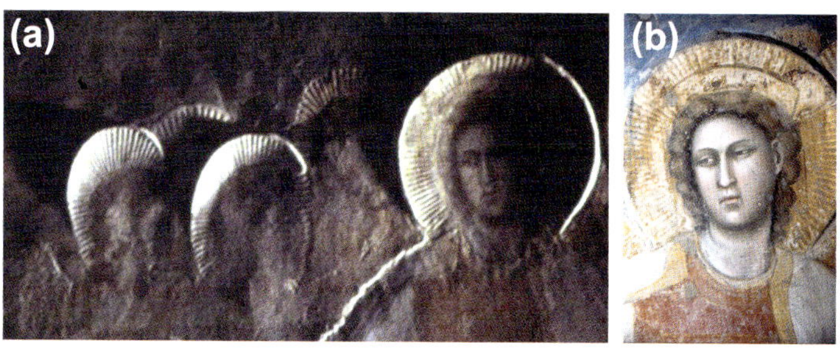

FIGURE 8.8 Giotto at the Scrovegni Chapel, Padua. Giotto inclined the gilded nimbus surface to avoid dust depositing on it and to keep it clean and bright. A protruding nimbus appears evident at glazing light (a) and disappears at normal daylight illumination (b).

8.8. ELECTROPHORESIS

Deposition due to electrostatic forces does not differ very much from gravitational sedimentation, since both cases involve movement generated by a force field. There are, however, some differences that should be pointed out. The first – and the most obvious – is that electrostatic Coulomb forces are either positive or negative, are inversely proportional to the square of the distance and are integer multiples of the elementary charge e of an electron, corresponding to the degree of ionization of the particle or the surface on which the particles are attracted. This may be caused, for example, by ultraviolet radiation or violent impacts of gas molecules or particles (colliding with each other or foreign bodies).

Atmospheric charges are especially frequent at low RH that prevents the dispersion of charges. The worst situation is in winter, in heated buildings, when there is not an adequate compensation for the adverse combination of low moisture content in air and indoor temperature, making the indoor RH level too low. The principal mechanism of charging is the so-called *diffusion charging*, i.e. the development of charge through random collisions of airborne ions and particles. Dust is highly charged. Smoke resulting from combustion is highly charged when first formed, e.g. 74% of particles in tobacco smoke were found to be charged (Cadle, 1965). Smoking cigarettes in a closed environment is extremely dangerous as the electrostatic charges lead particles to stick on surfaces in a short time.

It is not only dryness that is relevant to electric charges. Particles formed by condensation at room temperatures are initially uncharged but they rapidly become charged due to the gaseous ions that diffuse between them. Static electric charges are generated when the curvature of the water meniscus is changed, e.g. splashing raindrops form a myriad of fine droplets carrying electric charges that will be attracted by and will stick on surfaces. This is why lightening is generated in cumulonimbus clouds, or shower curtains after having collected charged droplets become charged and tend to envelop the person inside. Water is sensitive to electric charges that induce separation and redistribution of the charges inside the liquid, as it occurs in a trickle of water pouring out of a tap when a comb is brought near it (Fig. 8.9).

Electrostatic forces, however, arise not only when both the particle and the surface are charged but also when a charged particle generates an equal and opposite charge (image charge) when approaching an uncharged surface or when a charged surface generates a dipole on a conductive particle, such as water droplets. Furthermore, electric charges

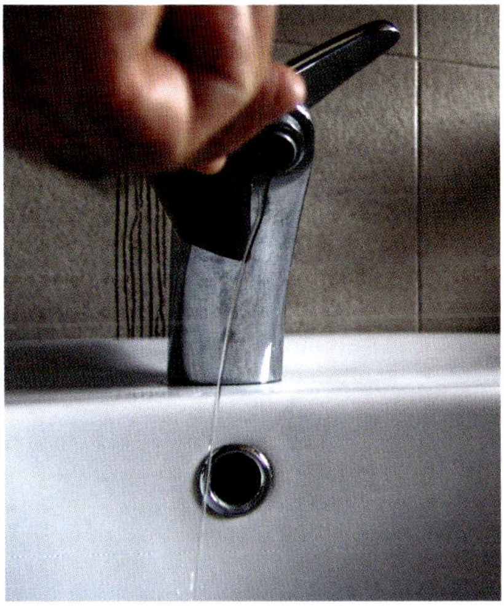

FIGURE 8.9 Water trickle being attracted and bent by electric charges of a comb previously put through hair.

may be generated even when neither of the two bodies is initially charged (*contact charging*). Electric forces may arise because of the difference of potential that is generated when two different bodies meet because of the local interaction of the permanent charges on the surface of the two bodies. Contact charging occurs during the separation of nonmetallic (semiconductor) particles from solid surfaces. Dust and paint, for example, are both semiconductors and the electric interaction depends on the relative donor–acceptor properties of the surfaces that come into contact with each other.

Electrostatic deposition becomes important only in the presence of low *RH*, while in damp environments, the conductivity of the air increases and the surface is covered with a film of water molecules so that the phenomenon is greatly reduced; on the other hand, humidity increases the adhesion of particles on moist surfaces, enhancing the capture efficiency of the surface (Phenix and Burnstock, 1990; Buffle and van Leeuwen, 1992).

It should be noted that the 'particle' is not only an agglomerate of inorganic and/or organic molecules, or a fragment of a mineral, but also implies biological structures such as fungal and bacterial spores, 'dwarf cells' or even vegetative cells. Due to their specific surface charge, depending on their nutritional status and related growth pattern as well as on the composition of the surrounding atmosphere and surface solution, they are even attracted or repelled by the material surface (Marshall, 1984).

8.9. PHOTOPHORESIS

When a suspension of particles is subjected to an intense beam of light, a motion is generated, called *photophoresis*. This phenomenon is measurable only with black particles containing elemental carbon, e.g. candle smoke, soot or fly ash generated by the combustion of wood, charcoal, oil. There is not a universally accepted theory to explain this mechanism. The first idea is to consider photons having momentum q

$$q = \frac{h\nu}{c} = \frac{h}{\lambda} \tag{8.46}$$

(where h is the Planck constant, ν is the frequency and λ is the wavelength associated with the photon and c is the light speed), and therefore, a light beam may exert a force on particles, which absorb or reflect photons. The pressure p exerted by a light beam is

$$p = \frac{E}{c}(1 + R) \tag{8.47}$$

where E is the energy impinging on the unit surface per unit time and R is the reflection coefficient, i.e. $R = 0$ for a totally absorbing black body and $R = 1$ for a totally reflecting mirror. A mirror experiences a pressure that is twice the pressure exerted on a black surface.

However, this simple explanation based on the so-called radiation pressure is not realistic, as suspended particles do not straightforwardly advance as they were pushed by a pressure but proceed either linearly or in the form of spirals, loops and so on. This behaviour is more commonly interpreted in terms of radiometer effect, i.e. the absorbed photons generate temperature unbalances on the particles and the surrounding gas, so that the resulting effect is similar to that for particles in a temperature gradient. Spiralling motion is therefore generated by particle rotation that changes the orientation of the temperature gradient and the associated thermophoretic effect. In fact, laboratory experiences on the radiometer effect in a gas (Landsberg, 1979) have shown that the effect of light pressure is largely overwhelmed by the effect of the convective motions in the gas, generated by the temperature unbalances due to the light absorption. The radiometer theory (Cadle, 1965) predicts that the maximum pressure $p_{\max}$ is exerted when the gas mean free path equals the effective dimensions of the particle, i.e. $Kn \approx 1$, and is given by the equation

$$p_{\max} = \frac{3\mu}{2D}\sqrt{\frac{\Re T}{M}} \tag{8.48}$$

(where μ is the dynamic viscosity and M is the molecular weight of the gas) so that the photophoretic force F_{ph} is described by equations that change with the gas pressure p, i.e. it varies as the cube of the particle diameter at low pressure, as the square at intermediate pressure, and linearly at high pressure, as follows:

$$F_{ph} = \frac{\pi a p D^3 G_p}{24T} \quad \text{for } kn >> 1 \tag{8.49}$$

which is the formula derived by Rubinowitz in 1920 for the low-pressure region, where a is an accommodation coefficient and G_p is the thermal gradient within the particle; in the transition region where the particle size is about equal to the gas mean free path, Hettner derived in 1926 the following formula:

$$F_{ph} = 0.25\pi\mu D^2 G_p \frac{\sqrt{\frac{a\Re}{MT}}}{\frac{p}{p_{\max}} + \frac{p_{\max}}{p}} \quad \text{for } Kn \approx 1 \tag{8.50}$$

and the following one for the high-pressure region

$$F_{ph} = \frac{3\mu^2 D \Re G_p}{2pM} \quad \text{for } Kn << 1. \tag{8.51}$$

This mechanism suggests once more that works of art should not be lighted with intense radiation and that smoke or soot particles not only produce the worst blackening but also are the most respondent to this effect.

8.10. AERODYNAMIC DEPOSITION: INERTIAL IMPACTION AND INTERCEPTION

When a fluid motion undergoes a sudden change, all the airborne particles are subjected to two forces: the drag of the fluid and their inertia. The latter tends to maintain the initial momentum so that individual trajectories depart from the airstream and a number of particles may impact on a collecting surface, where they can adhere or (more rarely) bounce. The larger and heavier the particles, the greater the conservation of momentum and the departures; for this reason, this deposition mechanism is very efficient for micrometric particles and is most efficient when the particles measure about 4–5 μm and the air velocity is high. This first kind of aerodynamic deposition, called ***inertial impaction***, occurs when *the inertia force acting on a particle is greater than the viscous force of the fluid so that the particle cannot exactly follow the path of the airstream and impacts on a collecting surface.*

On the other hand, smaller and lighter particles are dragged around the obstacle by the deflected airstream before contact can be made. However, some of them may have their *trajectory approaching the surface and they come into grazing contact with the surface just because the flow brings them within touching distance*. This second mechanism, called ***inertial interception***, is less efficient, especially because the trajectory progresses in the viscous layer near the surface and the particle is slowed down and becomes important only in the presence of surface roughness.

These two forms of aerodynamic deposition occur whenever air flows near a surface as the friction generates turbulence; the airstream may be part of a more general circulation or may be generated by pressure differences (e.g. open windows and/or doors) or may be due to thermal imbalances that generate convective motions over or along surfaces. Also in the case of laminar motion, uneven areas may cause local particle trajectory departures and interception on a surface that is close to the trajectory.

As the *Reynolds number* (7.27) can be used as a measure of the ratio of the inertial forces to the viscous forces and is also an index linked to the onset of turbulence, it is evident that the efficiency of this deposition can be related to this parameter as well as to the *Schmidt number* (8.19), which is the ratio between the momentum and the mass diffusivities. Wind tunnel experiments over an even, flat surface have shown (Hicks, 1982) that the velocity of aerodynamic deposition $\mathbf{v_a}$ for small particles can be expressed in terms of *Sc* and the friction velocity of the airstream u_*

$$\mathbf{v_a} = Au_*\, Sc^{-2/3} \tag{8.52}$$

where the coefficient A is $0.06 < A < 0.6$. From the definition of *Sc*, it follows that $\mathbf{v_a}$ decreases when D increases (as shown in Fig. 8.10), but only

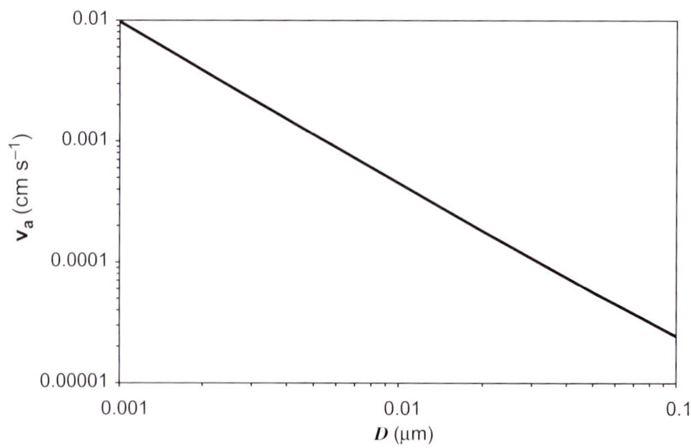

FIGURE 8.10 Aerodynamic deposition velocity $\mathbf{v_a}$ of small particles, i.e. with $D < 0.1$ μm.

when $D < 0.1$ μm and this formula is valid in that range. After having reached a minimum that generally lies in the $0.1 < D < 1$ μm interval and depends on atmospheric turbulence, the efficiency of this mechanisms increases again (Caporaloni et al., 1975; Sehmel, 1980).

However, when the airspeed becomes too fast, the airborne particles do not stick anymore and some deposited particles will be removed and airborne. However, there is not a precise correlation between turbulence and speed, and it is convenient to associate the deposition rate with the particle mass and the degree of turbulence rather than with the average air velocity. The inertial contribution dominates with large particles and rough surfaces but it is difficult to express a mathematical formula because of the departures due to the microdistortions in the fluid and the local coarseness.

Inertial deposition typically occurs in the presence of air currents or when walls are not in thermal equilibrium with air. Air coming into contact with cold walls cools, becomes denser and sinks, forming a descending stream along the surface; in contrast, warm walls form uprising currents. When a free convective motion develops along a wall, an internal boundary layer (*IBL*) is formed near the surface (Schlichting, 1979). The movement begins at very low speed and in laminar regime; while the path progresses, the speed as well as the *IBL* thickness increase. After a critical distance L_{crit} has been progressed, when the speed has reached the critical value U_{crit}, a sharp transition occurs to the turbulent regime. In the case of a wall warmer than the surrounding air, i.e. uprising flow, the critical condition can be expressed in terms of the *Reynolds number*, *Re*, measuring the distance L from the base of the wall and the airspeed U close to the surface at the level L; in the case of cold walls and descending currents, the process is the same but starting from the top of the wall. In a room, the onset of the turbulence is governed by the atmospheric stability, the difference of temperature ΔT between the air and surface and the surface roughness. The dynamic system may reach stationary conditions or may oscillate between some limiting configurations that are determined by boundary conditions or are forced by external perturbations. In a turbulent regime, the inertial deposition is greatly increased, with the effect of surface blackening in a short time.

In the case of smooth surfaces, the transition between laminar and turbulent regime occurs when the *Rayleigh number*, *Ra*, is $Ra \sim 10^9$. This number is the product of other two numbers, i.e. the Prandtl (*Pr*) and Grashof (*Gr*) numbers.

The *Prandtl number* is defined as

$$Pr = \frac{\nu}{\kappa} \tag{8.53}$$

i.e. the ratio of the kinematic viscosity ν to the thermal diffusivity κ of the fluid. For air, $Pr = 0.71$; it is a balance between the efficiency of two mechanisms that transport energy, namely, advection (the momentum is linked to ν) and conduction.

The *Grashof number* is defined as

$$Gr = \frac{g\beta\Delta T z^3}{\nu^3} \tag{8.54}$$

where β is the air expansion coefficient, ΔT is the temperature difference, and z is the length of the path followed by the airstream generated by ΔT. Gr is a measure of the relative importance of the effects of the buoyancy (linked to β and ΔT) and inertial effects (as the airspeed increases with the path length z as shown later) compared with the viscosity and the onset of the turbulent regime occurs when $Gr > 10^9$. For example, when a smooth wall surrounded by stagnant air is only $\Delta T = 0.1\,°\text{C}$ warmer than the air, the transition in the uprising flow occurs at $z = 4$ m from the floor, where the edge of the *IBL* is located; for $\Delta T = 1\,°\text{C}$, it occurs at $z = 2$ m. However, the surface roughness of real walls, or unavoidable atmospheric perturbations, may anticipate the transition to turbulence.

The thickness δ of the free convection *IBL* and the maximum speed w^* of the airstream (Fig. 8.11), i.e. of the free convection layer, can be calculated with the formulae (Isachenko et al., 1977; Camuffo, 1991)

$$\delta = 423\left(\frac{\mu k_a z}{\beta \rho_a^2 c_p g \Delta T}\right)^{-1/4} \tag{8.55}$$

$$w^* = 3.95\delta^2\,\beta\,\frac{\rho g}{\mu}\Delta T \tag{8.56}$$

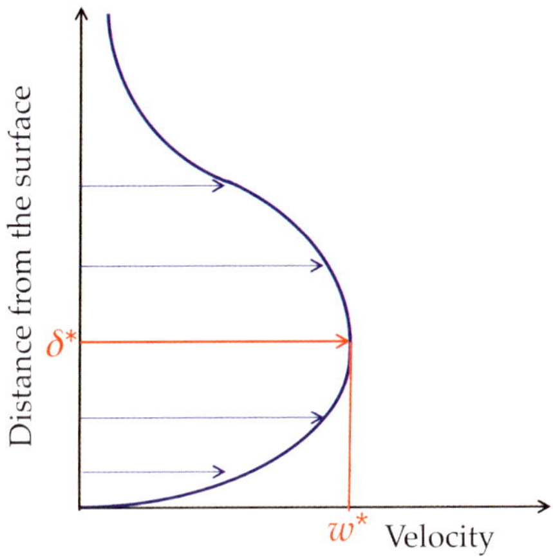

FIGURE 8.11 Velocity profile of the airstream inside the *IBL* flowing along a vertical surface with the maximum speed w^* and the distance δ^* at which the maximum speed occurs indicated.

where c_p is the isobaric specific heat of the air (see Fig. 8.12(a),(b)). The maximum speed w^* is found at the distance $\delta^* = 0.38\delta$ from the vertical surface. In order to give an example of the airspeed for $\Delta T = 0.1\ ^\circ\text{C}$, $w^* = 4.8\ \text{cm s}^{-1}$ at $z = 1$ m; for $\Delta T = 1\ ^\circ\text{C}$, $w^* = 15\ \text{cm s}^{-1}$ at the same distance from the floor. The average speed $<w>$ of the convective airstream, i.e. of the free convection layer, can be found by integrating w over the whole thickness δ of the *IBL* and is $<w> = 0.633\ w^*$.

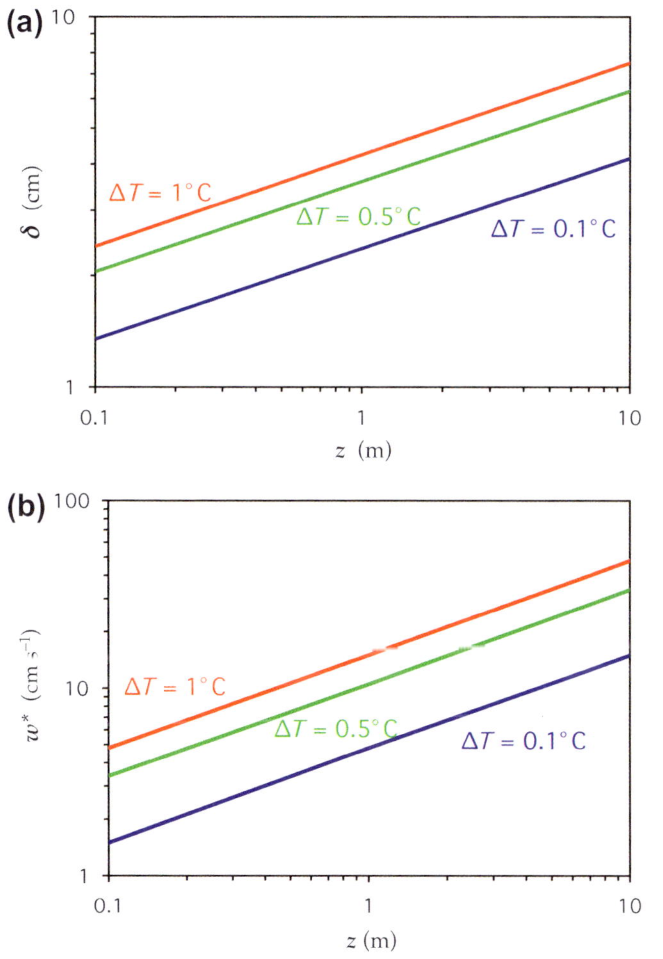

FIGURE 8.12 a) Thickness of the free convection layer δ (cm) versus the distance z (m) covered by the airstream from the edge of the layer. *Source: By kind permission of Elsevier Science. See credits.* (b) Maximum speed w^* of the airstream in the free convection layer at the distance z (m) from the edge of the layer. *Source: By kind permission of Elsevier Science. See credits.*

In the turbulent regime, the air velocity at each point of the fluid (and therefore the velocity of airborne particles) is represented by a vector that has a component perpendicular to the surface that is continually fluctuating. As a result, many particles are thrown against the surface with a large momentum even though, before colliding on it, they should first cross, by inertial motion, the very thin laminar, viscous layer that envelops the surface.

When the regime of the airflow becomes turbulent, the profile of the average airspeed in front of the surface changes: the highest speed is reduced but the speed close to the surface is increased (Fig. 8.13). In the event of turbulence, the deposition rate also increases especially for the most massive particles. The average speed of the blown air in front of the surface is not relevant; what is crucial is the intensity of turbulence, as it determines the intensity and frequency of the air movements pointing towards the surface and launching to it the most inertial particles.

In case the air flows along a rough surface, the coarseness of the surface will continuously cause local departures from a rectilinear motion so that the vector velocity will have a small fluctuating component perpendicular to the direction of the average movement and this component will alternately point towards the surface and away from it. During these sharp changes of direction, the particles with large mass will continue their inertial motion, some of them colliding with the surface. At low impact velocities the particles do not have sufficient kinetic energy to bounce and to overcome the attractive forces of the wall, so that most of the particles hitting the surface will stick on it. This especially occurs when a narrow free space exists between a painting, or a decoration, and the coarse surface of a wall. The airflow passing through the narrow space is perturbed and large airborne particles will have a high probability of being captured by the wall, with the result of a locally increased soiling. A popular example is the blackening on walls around objects hung on nails.

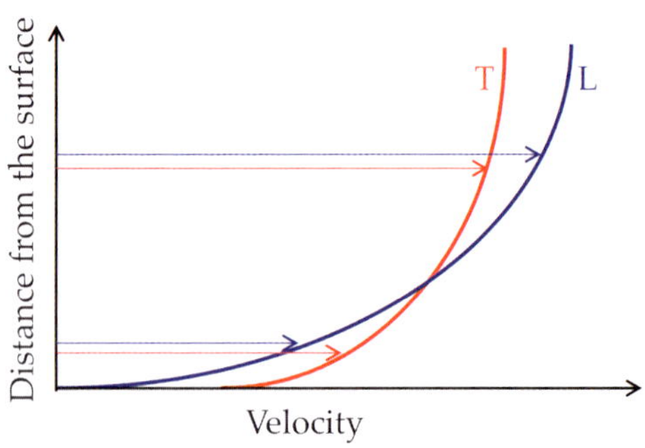

FIGURE 8.13 Change of the airflow velocity after transition from laminar (L, blue) to turbulent (T, red) regime. After transition, at some distance from the surface the velocity is reduced but in close proximity of the surface, it is increased (red arrow, compared with the blue one).

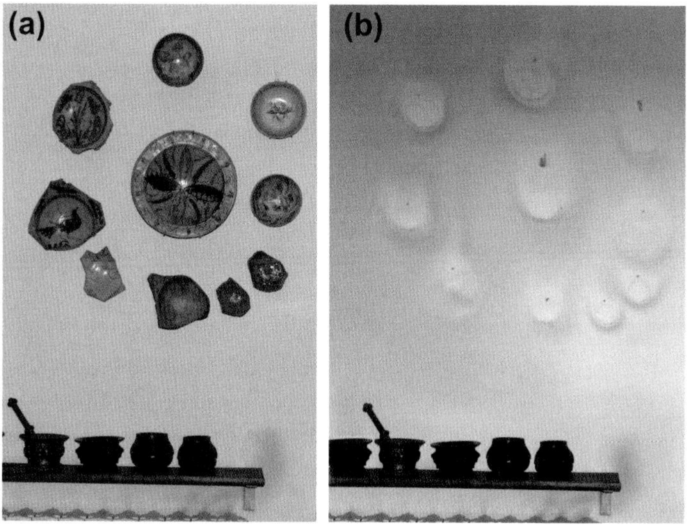

FIGURE 8.14 Wall blackening around some pottery hung on a wall. (a) The decorative pottery hung on the wall and (b) the "ghosts" left on the wall. Below the shelf, a hot water radiator is present.

In Fig. 8.14, the space where the circular base of some decorative pottery was in contact with the wall is found clear. All around, the increased airspeed and turbulence will worsen soiling leaving a "ghost" appearance. Soiling is further accelerated because the wall surface is above a hot water radiator, although in the presence of a protective shelf.

In winter, heating makes the indoor air warmer than the wall, and the contact with the cold wall will increase the density of air in a thin layer close to the wall. This small difference in density will generate a flow of air descending along the wall. This flow will develop following the general laws of an *IBL*. The opposite happens in summer. Figure 8.15(c) is a thermogram taken in winter, in an attic heated with hot water radiators. The thermogram represents the infrared (*IR*) image of some keys hung on a wall and the tilted ceiling of the attic. The edge between the tilted ceiling and the wall is colder because it is hardly reached by the descending airflow. Tips and blades, fully surrounded by ambient air, are warmer than air; bows, closer to the cold wall, are colder. The temperature difference between air and walls is a motor that drives convective motions. The airflow in close proximity of each key is locally perturbed and the microturbulence is responsible for increased deposition rate. Keys leave their soiling ghost impressed on the wall surface. Given the small temperature differences, the contribution of thermophoresis is almost irrelevant compared with inertial deposition.

If the airspeed is too fast, as corresponding to leakage for fissur es on the window frames, and the turbulence is too strong, the particles removed

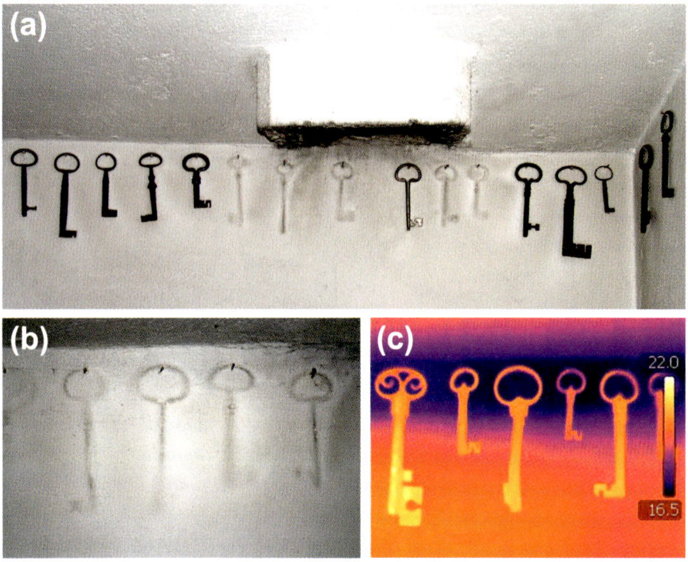

FIGURE 8.15 Wall blackening behind some keys hung on a wall. (a) Picture in visible light of a series of keys; some of them have been removed from the nail. (b) Picture in visible light of the ghosts left by the keys once they have been removed. (c) Thermogram of the keys (in winter).

from the surface become dominant in comparison with those that are deposited, inverting the net flux of deposition. Ideally, when plotting deposition versus turbulence the curve starts as an increasing function of turbulence, reaches a maximum and then decreases again.

The deposition rate depends not only on the particle mass but also on how fast the fluid runs over the surface and its roughness. This mechanism is not, however, particularly efficient in a laminar regime, especially if compared with deposition in a turbulent regime.

It should be noted, however, that in the absence of external perturbations, at any given atmospheric stability and on any given rough surface, the air velocity does, nevertheless, play a part in the development of turbulence and, for example, the *Reynolds number Re* sets it at a certain value, even though partial and conventional.

8.11. ADHESION OF PARTICLES TO PAINTINGS OR OTHER SURFACES

Once the particle hits a surface, it can bounce or stick on it. The fraction of particles that remains on the surface depends on various factors. The main ones are the nature of the adhesion forces and the physical and chemical properties of the particle and the surface. Surface roughness and particle size and shape are also important. Macroscopic roughness can hold fine particles in the surface cavities and large particles better adhere on very smooth surfaces. In this respect, wax or painting varnish have a negative influence, although they ultimately separate the underlying surface from the particle deposit (Zimon, 1982; Phenix and Burnstock, 1990).

van der Waals forces, of molecular nature, are very effective adhesive forces. A particle that is approaching a surface will experience an attractive force that is proportional to the particle diameter but inversely proportional to the distance between the separation distance between the particle and the surface. Large particles close to the surface are more easily captured. Once the particle sticks on the surface, the wider the contact area, the stronger the adhesive force. Micro-deformation at the contact zone (e.g. elastic particles or presence of wax or varnish) increases the contact area and the adhesion force. Small or flat particles have in general a larger ratio between contact area and total volume. In the range from fine to micrometric particles (i.e. $D < 2\ \mu m$), particle size is of the same order of magnitude as the sub-micrometric surface roughness and the adhesion force is inversely proportional to the particle size, being largest for the finest particles. The smaller the diameter, the closer to the surface the particle will be. The particle will immersed in the viscous layer and will be shielded by the surface roughness. In such a case it may hardly be removed by atmospheric turbulence. In addition, adhesive forces are much greater than the weight of small and fine particles. For large particles having size range $2 < D < 200\ \mu m$, the situation is very variable depending on the roughness scale of the surface where the particles are deposited. Giant particles (dust), i.e. $D > 200\ \mu m$, are mainly found on the horizontal or very rough surfaces, as their weight exceeds ordinary adhesion forces on dry surfaces.

Electrostatic charges can be found either on suspended particles or on surfaces and particularly on paint surfaces. Electric forces are exerted when two bodies have opposite charges or when the particle or the surface is electrically charged and the charged body (e.g. the particle) induces the image charge on the other body (e.g. the surface). Coulomb attraction between a charged body and the induced image charge is the most common mechanism, as only one charged body is sufficient to exert forces on all the neighbouring bodies. This attraction governs the transport of particles in the electric field, which has been established between the actual charge and the image charge, and the adhesion of the particles to the surface. This adhesive force continues until the actual charge is dissipated; however, after the charge has disappeared, the particles stick again on the surface for the continuous action of the molecular and possibly other forces. In particular, when a surface is damp, charges are rapidly dispersed and electric forces disappear. Wax-covered surfaces become electrically attractive at low *RH* levels. Electric forces are also generated when fly ash, soot, dust and other particles are charged via contact charging, such as *donor–acceptor* forces for semiconductors, as already discussed.

Liquid film adhesion (also called *capillary force*) is due to the presence of a liquid film that covers the solid surface of monuments or buildings. Sand castles built by children on the seashore are the most popular example of the great strength of this force. When a particle touches the wet surface, the liquid film adheres to it and the surface tension tends to (partially) envelop and possibly include the particle forming a meniscus all around the contact area. This phenomenon is also called '*capillarity*' and is common in all wet materials and ice, which are always covered with one monolayer or a few molecular layers of liquid water. The force exerted by liquid film adhesion is so large that a wet surface captures every particle that collides with it and for this reason, a wet surface is called a '*black surface*'.

Surfaces contaminated with deliquescent salts increase the time of wetness and worsen soiling. In principle, in humid environments, wax and some varnishes make the surface hydrophobic (i.e. the wetting angle tends to 180° and the surface is water repellent) but the advantage against wettability is lost in dry environments due to the action of two negative factors: the greater deformation of the surface in the contact zone and the smooth surface that favours adhesion of large particles, especially by the action of electrostatic forces.

Also, biofilms exert adhesive properties, which increase the dry deposition of aerosols and particles and the formation of nonbiological crusts over biological precursors (Koestler et al., 1996).

8.12. VERTICAL DISTRIBUTION OF PARTICLES IN STILL AIR AND THEIR RESUSPENSION BY TURBULENCE

In equilibrium with air, airborne particles that do not coagulate between them behave as a gas and the kinetic theory of gases can be applied to them. An example has been the computation of the thermal velocity and the Maxwell–Boltzmann velocity (Eqn (8.20) and Eqn (8.21)). We can similarly apply the barometric formula for an isothermal atmosphere, with no vertical air mixing. The *Dalton* law for gas mixtures, according to which *each gas is distributed in space with a partial pressure which is independent of the partial pressures of other gases*, leads to the consequence that each gas can be considered independently, and the vertical pressure distribution for the ith gas or type of particle can be written:

$$p_i(z) = p_{io} \exp\left(-\frac{zM_i g}{\Re T}\right) = p_{io} \exp\left(-\frac{z}{H_i}\right) \quad (8.57)$$

where p_{io} is the pressure at the level $z = 0$; M_i is the molar mass of the gas or particle and $H_i = \Re T/M_i\,g$ the scale height, i.e. the height z where the pressure reduces by a factor $1/e$. For example, for molecular hydrogen, ($M = 2$) $H(H_2) = 425$ km; for molecular nitrogen, ($M = 28$) $H(N_2) = 30$ km; and for molecular oxygen ($M = 32$) $H(O_2) = 27$ km. This differentiation between the molecular species is not found in the free atmosphere, for all the vertical motions and large-scale turbulence that favour homogenization. In the case of particles that are much heavier than gases, the scale height is accordingly reduced and a gravitational separation of different sizes is possible in a stagnant atmosphere inside a room. The above equation can be rewritten in terms of concentration

$$C_i(z) = C_{io} \exp\left(-\frac{zM_i g}{\Re T}\right) = C_{io} \exp\left(-\frac{z}{H_i}\right) \quad (8.58)$$

The barometric distribution of particles in equilibrium in an isothermal, stagnant atmosphere has been calculated for the particle density $\rho = 1$ g cm^{-3} (Fig. 8.16); in this figure, for each particle diameter, three levels z_{90}, z_{50} and z_{10} have been computed, which represent the level above the floor (Z) where the concentrations become 90%, 50% and 10% of the concentration at the soil level, respectively. From the graph, it appears that in a room the distribution of particles with $D = 0.001$ μm is practically unaffected; at $D = 0.01$ μm, $z_{90} = 8.27$ cm, $z_{50} = 54{,}6$ cm and $z_{10} = 181$ cm.

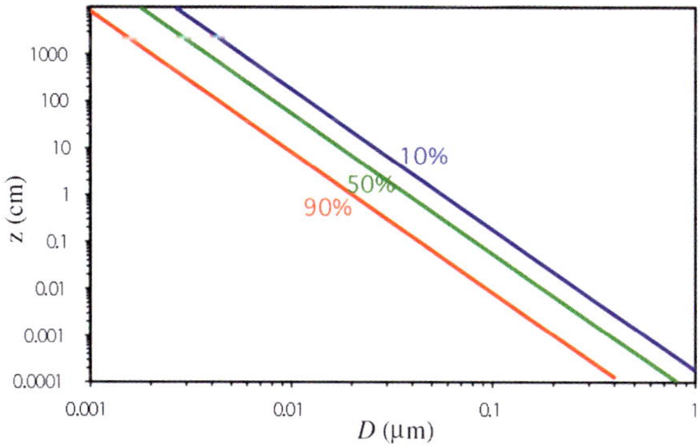

FIGURE 8.16 Barometric distribution of particles in equilibrium in an isothermal stagnant atmosphere. For each particle diameter, three levels z_{90}, z_{50} and z_{10} have been computed that represent the heights where the concentrations become 90%, 50% and 10% of the value at soil level, respectively.

At $D = 0.1$ μm, these heights are exactly reduced by three orders of magnitude and so again at $D = 1$ μm; i.e. the near totality of the particles with $D = 0.01$ μm is contained in the first 2 m, with $D = 0.1$ μm in the first 2 mm and $D = 1$ μm in the first 2 μm.

In still air, gravity and kinetic theory force the medium and large particles to follow the barometric distribution and for each diameter size, a floor-based concentration boundary layer is formed, which contains the near totality of these particles still suspended in the air. This means that every source of turbulence that causes a mixing near the floor, e.g. walking and using vacuum cleaners, is responsible for the resuspension of the particles with $D >> 0.01$ μm, whereas the particles with $D << 0.01$ μm do not stratify but are continually mixed by Brownian diffusion. *As in stagnant air large and giant particles accumulate in very thin air layers near the soil, it would be possible to activate in museums and historical buildings a nocturnal drainage with floor suction to remove most of the pollutants that lie in lowermost layer of a stable atmosphere, after the barometric equilibrium has been reached*. This theoretical finding has never been applied but seems a cheap and effective way of removing suspended particulate matter without using filters and avoiding the negative consequences that they produce.

Direct observation of the concentration profiles is difficult, as the present day particle analysers usually work in the range $0.1 \leq D \leq 10$ μm and the experimental apparatus requires a very detailed vertical resolution near the floor, which is impossible for the mixing caused by the suction. Better results will be achieved when reliable measurements will be possible at $D = 0.01$ μm. The only experimental evidence is limited to observation of the increase of the concentration of large and giant particles after having generated turbulence near the floor. Turbulence is too weak for removing particles that have been deposited and stick on floor; it only raises particles lying in the floor-based concentration layer.

A short comment about vacuum cleaners: in general, only a fraction of the removed matter remains entrapped in the bag and/or filter. The air drawn into the machine transports coarse granules, which are then broken, ground, pulverized, multiplied in number, and brought back to the ambience in the form of a cloud of fine particles that will deposit later on. Only products with the most efficient filtration should be used. The best products apply a high-efficiency particulate air (HEPA) filter suitable to retain a consistent fraction of the pulverized matter or an efficient water filtration (using water as a filter). Following EN 1822: 2009, HEPA filters are divided into eight classes based on retention of particles, making reference to the so-called most penetrating particle size, i.e. 0.3 μm in diameter. The top level is class U17 > 99.999995% total retention, the lowest level is class E10 > 85% total retention.

Central vacuum cleaners (also known as *built-in* or *ducted vacuum cleaners*) are installed into a building as a semipermanent fixture. Dirt particles are dragged through tubing installed inside walls to a collection container located in a remote room together with the power unit. Inlets are installed in walls throughout the building to attach vacuum hoses. Last but not least, the low pressure inside powerful vacuum cleaners may kill over 95% of small insects.

8.13. HOW SOILING DEVELOPS

Soiling is caused by the deposition of natural or anthropogenic particles. These particles induce different optical effects because of their various sizes, shape and nature. Their range varies from just a few hundredths of micrometres to some tens ofmicrometres. They can be spherical as in silicate particles emitted by carbon combustion or they may be articulated as in pollens, irregular like grains of sand, fibrous and prismatic as in wood combustion or in the case of vegetable or animal fibres or bunches like the soot in diesel combustion. There are semitransparent particles such as alumosilicate or quartz granules or completely opaque such as the carbonaceous ones generated by combustion of candle, oil or coal.

Once deposited, the various particles have different effects, which may be physical and optical, linked to the physiological response of the eye (Bouma, 1947; Thomson, 1986; Eastaugh, 1990). The physical effects, even though complex, can be objectively defined. In the case of completely opaque particles, the paintings gradually become darker and darker as the particles are deposited and as the 'free' clean space is reduced, which diminishes exponentially. In the case of semitransparent particles, the dimming of the colours depends on the thickness of the layer formed over the painting but again the loss of visibility of the image is exponential. Phenix and Brustock (1990) found that most of the deposits in paintings consisted of a monolayer of adhered particles. Only the fine, submicrometric particles were found as multiple layers in aggregates. These particles were grey and mostly rounded, clumped or dispersed on the painting surface.

In the case of soiling resulting from opaque particles, the blackness progresses as the cumulative section $\sum(n_i X_i)$, where n_i is the number of particles having diameter D_i and X_i is the cross-section of each individual particle that has deposited, i.e. $X_i = \Pi D_i^2/4$ for spherical particles. In this case, the cumulative section varies with the second power of the radius, exactly as the total surface $\sum(n_i S_i)$ where $S_i = \Pi D_i^2$ is the individual surface. As the total surface of the suspended particles frequently reaches a maximum in the range of the small average-sized ones (a peak around a tenth of a micrometre) e.g. soot from diesel motors, which in addition is difficult to filter, and medium-sized ones, e.g. candle smoke, these particles should be regarded with the highest attention. However, in reality, this peak is not constant and the distribution may also be bimodal.

In the case of soiling due to semitransparent particles, the thickness of the deposited translucent layer varies with the cumulative volume that has deposited, $\sum(n_i\ V_i)$. The volume V_i varies with the third power of the radius of the particle and the spectral distribution $n_i\ V_i$ increases greatly as the diameter increases, especially at the micrometric and supermicrometric sizes (Seinfeld, 1986). However, the distribution of the particles in the atmosphere depends on various factors, so that there are many deviations with respect to the above rule. In the case of closed environments, not visited by the public, the large particles sediment more quickly, while the fine particles are suspended in the air for longer; in this case, the very small number of large particles that remain in suspension may cause a decrease of the total surface and total volume for the great diameters or also bimodal distributions (Fig. 8.17). In this case, the most critical ones are the large particles, e.g. quartz granules and other alumosilicates.

The deposited layer is not homogeneous and internal diffraction and refraction tend to cause the underlying colours to be put out of focus and lighten. The scattering depends on the dimension and shape of the particles. Moreover, the particles themselves are coloured so that the original colours tend to take on false shades in addition to false colouring due to the reflection of surrounding surfaces. As a result, the painting

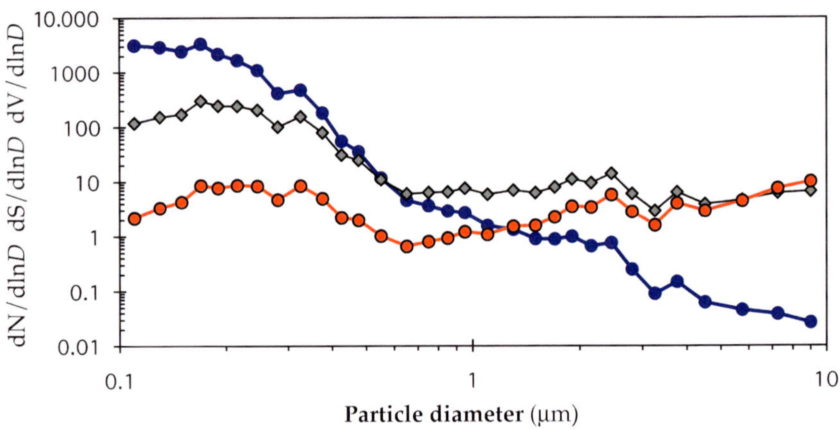

FIGURE 8.17 Spectral distribution of the number N of suspended particles, total surface S and volume V expressed in terms of dN/dln$D$ (cm^{-3}, blue), dS/dln$D$ (μm^2 cm^{-3}, grey), dV/dln$D$ (μm^3 cm^{-3}, red), where D is the particle diameter (μm). Kunsthistorisches Museum, Vienna, 26 August 1992, 3 p.m.

gradually darkens and blackens and the colours are toned down or may change.

In the case of paintings, uniquely based on the perception of colours, the dry deposit covers the surface reducing the chromatic yield and legibility, diminishing the artistic value of the work of art and violating its interpretation, as, for example, the famous case of the Michelangelo paintings in the Sistine Chapel (Fig. 8.18).

This means that the soiling should be removed using the necessary methods, even though these are, basically, undesirable, because they always expose artwork to various risks due to handling. In the case of sculptures or architectural works, the volumetric shape remains distinguishable but the enjoyment of the aesthetic beauty might be severely compromised. Famous examples are the Grand Place, Brussels, and the magnificent buildings in Venice that were originally white with decorations and then turned to black because of the carbonaceous particles deposited on them. However, while the deposited pollutants remain dry, the damage is only aesthetic (even though this may be a considerable problem), but the physical integrity of the material remains unaffected. The worse problem is when the deposit is moistened, because water

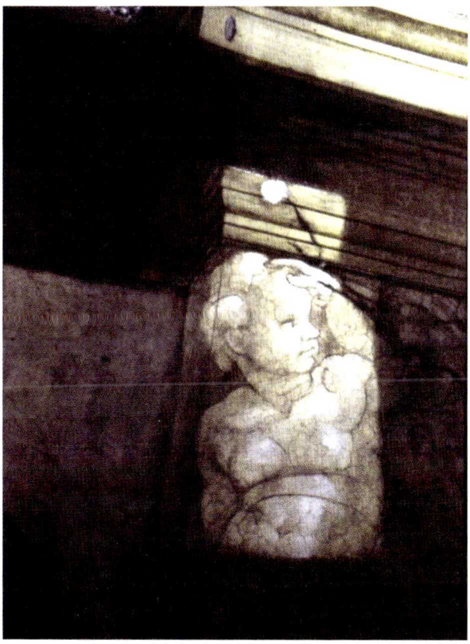

FIGURE 8.18 The Sistine Chapel, Rome, during restoration in 1986, leaded by G. Colalucci. This putto was just cleaned and makes clear the difference with the not-yet-cleaned surface, showing the soiling accumulated over five centuries. Picture taken by the author, by kind permission of Vatican Museums and Galleries.

brings into solution and activates the noxious chemical substances able to attack, dissolve and corrode materials.

A direct correlation between artwork deterioration and concentration of pollutants suspended in air (and remaining there) is of little help. As opposed, a correlation can be found between deterioration and pollutants deposited on monument surface. The number of particles with a given diameter *D* that will deposit on a surface is proportional to the concentration of airborne particles having the same diameter *D* but the deposition velocity will not be the same for particles with different diameters and not all the particles are equally dangerous. We have seen that the deposition mechanisms vary with the particle diameter, surface characteristics, microclimate and air–surface interactions. The deterioration of monuments depends, therefore, on the quality and quantity of the pollutants deposited, the presence of catalysts, and the way in which the surface is wetted (e.g. pouring rain that washes over the dirty surfaces or fine rain that activates the pollutants without removing them), the frequency and duration of wetting and other intrinsic factors (e.g. physical and chemical characteristics of the material, surface roughness and porosity).

The knowledge of the chemical and physical characteristics of individual particles, their relative abundance in the atmosphere and the deposition mechanisms that favour their transport and selective deposition should lead to a profound review of the dynamics of chemical aggression and should also clarify ideas about what can or cannot be done to reduce the negative effects of pollutants. It also helps to interpret the traditional chemical monitoring. For example, a traditional chemical investigation of suspended particles that are collected on filters would show which elements dominate and that there may be only traces of others. The spontaneous reaction would be to concentrate attention on the prevalent chemical compounds, neglecting those present in only minimum quantities. However, keeping in mind that the mass is proportional to the volume, it can easily be seen that sometimes there are a few coarse particles that characterize the total mass but are located in few points and have a small or no influence on the whole surface. On the other hand, however, there may be many hundreds of thousands of tiny aggressive particles that are spread all over the surface, but were, traditionally, considered negligible because of their small total volume.

Several case studies have shown where the microclimate plays a fundamental role in deposition processes (Camuffo, 1990, 1991; Camuffo and Bernardi, 1986, 1991, 1995). It would be possible to resolve the problem by going to the root and completely removing the particles suspended in the air by means of accurately filtering the air and leaving the microclimate free to meet its destiny. However, filtering is very costly and the complete absence of particles is practically impossible in environments with visitors. It is, therefore, necessary to find a balance: halving the number of particles is not halving the rate at which blackening occurs, under the same conditions. Deposition is not a linear mechanism. Filtered air would reduce the concentration of suspended particles but might increase the airspeed and accelerate some deposition mechanisms, especially inertial impaction and, ultimately, might increase soiling. The best results are reached only if a convenient microclimate is established, especially avoiding extreme or unnatural conditions, internal gradients and air movements.

8.14. WHAT IS THE MOST APPROPRIATE HEATING AND AIR CONDITIONING SYSTEM TO AVOID SOILING?

Bulk deposition is the result of the combination of a number of deposition mechanisms, some of them with synergistic and other with opposed contributions. Brownian deposition is always active whatever be the orientation of the surface and the same holds for aerodynamic impaction and electrophoresis. Gravitational settling acts on the horizontal surfaces facing upwards. Thermophoresis, diffusiophoresis and the Stefan flow may be either directed towards the surface (whatever its orientation) or in the opposite direction. We have already seen that diffusiophoresis and Stefan flow always combine slightly reducing the efficiency of the Stefan flow. However, the efficiency of the various deposition mechanisms is largely affected by the mass (and therefore the size) of particles (Fig. 8.19).

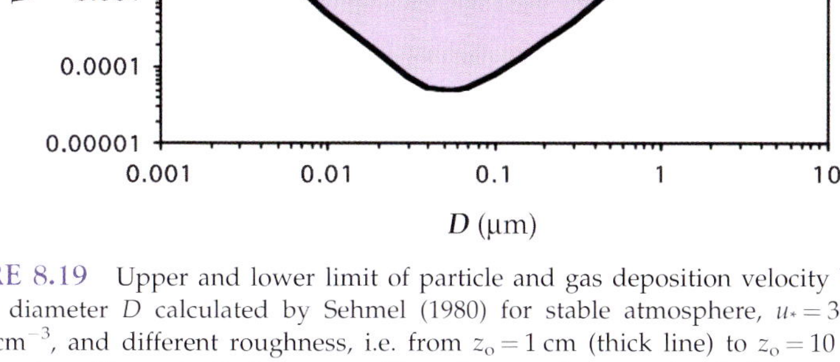

FIGURE 8.19 Upper and lower limit of particle and gas deposition velocity V versus particle diameter D calculated by Sehmel (1980) for stable atmosphere, $u_* = 30\ \mathrm{cm\ s^{-1}}$, $\rho = 1\ \mathrm{g\ cm^{-3}}$, and different roughness, i.e. from $z_o = 1$ cm (thick line) to $z_o = 10$ cm (thin line). The deposition velocity increases with surface roughness and turbulence.

Somebody imagined counteracting deposition with an appropriate thermophoretic effect (Nazaroff and Cass, 1987). However, it should be remembered that each mechanism has a different efficiency that depends on the size of the airborne particles, so that it is impossible to find a general solution for the whole size spectrum, i.e. compensating exactly all the mechanisms in a part of the spectrum generates unbalances in other parts of the spectrum. Brownian deposition is particularly active on the finest particles; thermophoresis acts prevalently on the submicrometric particles, while impaction dominates in the micrometric range. Creating a positive thermophoresis means having the surface warmer than the surrounding air, thus generating an upward convective flow and aerodynamic deposition. Positive thermophoresis can partially oppose Brownian deposition in the middle of the spectrum but it cannot prevent deposition of large and giant particles by impaction, and anyway, a warm surface will blacken. In the case of a cold surface, thermophoresis acts negatively against the surface itself, giving rise to thermophoretic deposition of the smaller particles, adding to the effects of the inertial deposition of the larger ones. The smallest damage is, therefore, obtained when the temperatures of the wall and the air facing it are exactly the same.

The efficiency of inertial impaction compared with thermophoresis can be deduced by looking at the blackening found above radiators. Blackening is not uniform as expected if only a drop of temperature between the wall and the air were involved. The grey '**V**' zones develop above the brackets that attach the radiators to the wall (Fig. 8.20(a)) and are explained in terms of the turbulent wakes generated by the brackets

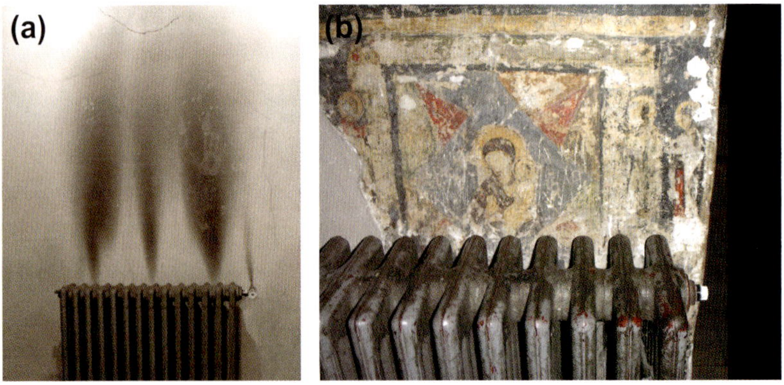

FIGURE 8.20 (a) 'V'-shaped blackening above the radiator wall-mounting brackets in a historic building. (b) It is easy to imagine what will happen to these frescoes of the sixteenth century in this monastery of the 16th century in Bucovina, Romania where radiators have been recently installed for churchgoer thermal comfort.

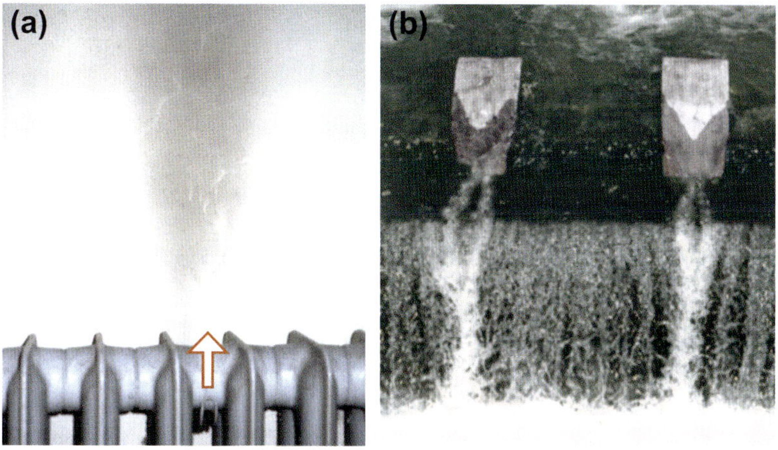

FIGURE 8.21 (a) 'V'-shaped blackening above the radiator wall-mounting brackets, due to the turbulent wakes generated by the brackets on the uprising flow. (b) The water overflowing from a dam spillway crest shows the same effect, i.e. hydrodynamic turbulence generated by baffle blocks perturb the laminar flow of water before the fall.

(Camuffo, 1991). In the space between the warm radiator and the wall, the air gains buoyancy and starts rising up, originally in laminar regime. When the air meets the wall-mounting brackets, the brackets perturb the flow and generate instability and determine the onset of turbulence, which is responsible for the intense blackening. This is the same mechanism that we see when water in laminar regime overflows the spillway crest of a dam to enclose a basin reservoir (Fig. 8.20(b)). Where the overflowing water meets the baffle blocks, these perturb the laminar flow and generate a foaming V-shaped turbulence, starting downstream of the blocks.

The same analogy is shown in Fig. 8.21, but in (a) the airflow is upward due to the buoyancy of warm air and in (b) the water flow is downward due to the high density of water compared to air. The direction is opposite because in the former, the density of the buoyant air is smaller than the density of the ambient air and in the latter, the water density is greater; however, the mechanism is the same.

Above the radiator, the area of uniform wall blackening is due to the combined effects of thermophoresis, modest aerodynamic impaction and Brownian deposition. It is, however, possible to observe the thermophoretic effect combined with the Brownian one, but separately from aerodynamic deposition, looking at some microcracks that form in the paint coating on the wall (Fig. 8.22). These microcracks locally deviate the stream of the uprising air current by a few millimetres with the consequence of forming downstream of the microcracks a narrow clear band

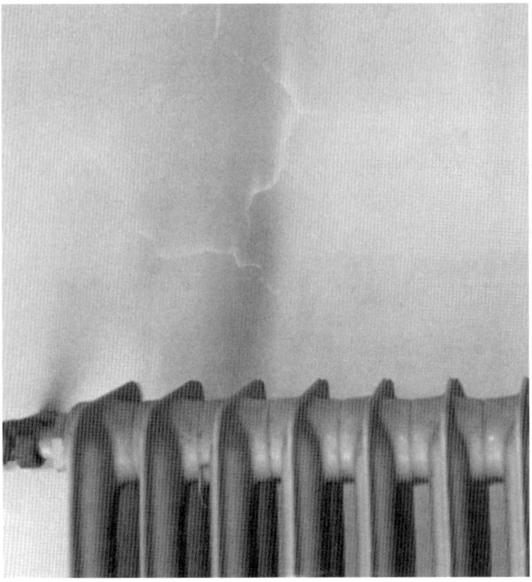

FIGURE 8.22 Narrow clear bands develop at the downstream side of fine cracks found on the wall colour coating layer.

not reached by the uprising flow and therefore protected against inertial impaction. By simply looking at a soiled wall, it is possible to recognize the dominant air movement: it develops in the direction of the narrow clear band contouring microcracks.

The marked soiling above brackets shows that inertial impaction is generally dominant over thermophoresis, but in general, it is not easy to distinguish among the different contributions because they occur simultaneously. For example, Fig. 8.23(a) shows the blackening that develops above an *IR* radiant heater with an electric resistance at red colour temperature.

The hot heater generates an upward turbulent flow, with the synergistic occurrence of inertial deposition thermophoresis and Brownian deposition for the elevated air temperature. It is possible to calculate the contribution of the various mechanisms but it is clear that hot sources should not be located in proximity of objects or walls.

Even more noxious is the *IR* radiation that directly hits paintings, furniture or other objects (Fig. 8.23(b)): *IR* is not visible to the naked eye but by the time the eye recognizes the signs of decay, it will be too late.

The conclusion is that the ideal microclimatic conditions to reduce deposition of airborne particulate matter are when the indoor air is thermally stratified in order to suppress turbulence and reduce convective motions. Atmospheric stability is reached when temperature increases with height and no heat sources or thermal imbalances that

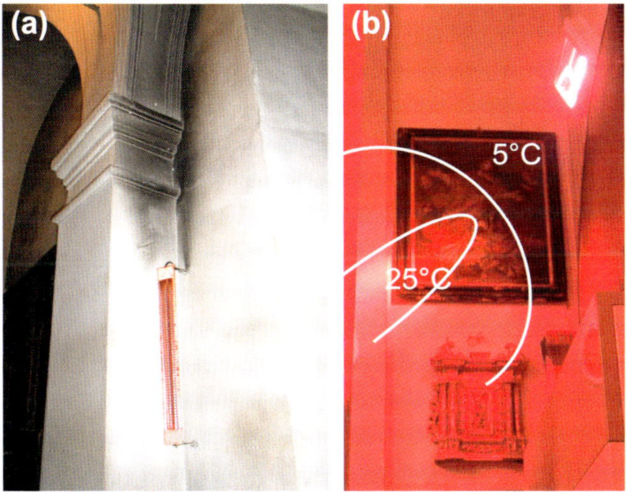

FIGURE 8.23 (a) Blackening formed above an electric incandescent radiator. (b) The painting on the side of the incandescent radiator is partially hit by *IR* radiation: the upper right corner is at 5 °C and the lower left at 25 °C. The pink colour is due to the pink brightness emitted by the incandescent heater visible on the top, to the right.

could induce convective motions are present. This means that the floor temperature should be lower than the air temperature. The lower the floor temperature, the greater the air stability. Similarly, the ceiling temperature should be higher than the air temperature. However, in tall rooms, the vertical temperature profile should be controlled, e.g. not to exceed $1\,^{\circ}\mathrm{C\,m^{-1}}$, because large gradients and internal layering might create other problems, especially in winter when people need a comfortable temperature in the first 2 m above the floor without overheating the ceiling. In addition, at all heights, walls and air should always have the same temperature level in order to avoid airflows along walls.

Heating and air conditioning systems, as well as humidifiers, are generally installed to control room temperature and *RH*, but rarely do they consider whether they will increase the deposition rate. Although they blow filtered air, the deposition mechanism is not linear and the indoor turbulence and the mixing they cause will increase the overall deposition rate and the soiling on collections. The installation of these devices should care in a global way for the control of both the thermal and dynamic state of the air and of the efficiency of the actual deposition velocity that is a consequence of the microphysical processes that have been established at the interface between the air and the objects. Both the inflow of new air and the extraction of room air should be considered from this global point of view.

Heating systems are unable to keep temperature constant in time and homogeneous in space. The distribution of heat generates preferential paths for warm air, with or without forced ventilation; in the latter case, strong gradients, turbulence and inertial deposition are generated in the rooms. In addition, a sharp rise, or a drop, in temperature occurs every time an HVAC system is turned on/off. The problem is not in designing new powerful or sophisticated machines but rather in being able to distribute blown air in a room with continuity and homogeneity. The problem of time variability might be solved using a modest but uninterrupted emission of heat that remains active for the whole season, day and night, only gently modulated with the seasonal needs, just to balance the heat loss with the exterior. This is the idea behind the so-called *conservation heating*, i.e. some kind of heating supplied to avoid frost or for the whole season in which exceeding dampness should be mitigated. The popular practice of short-time powerful heat emissions followed by cooling should be definitely abandoned. The space distribution should be made with many, extended and well-distributed sources, avoiding the use of forced ventilation, as fan coils do. However, the heating system and heat distribution are often conditioned, especially in historic buildings by several aesthetic reasons or architectonic needs and in many cases, the priority is absurdly given to the container instead of to the content. Another problem is the cultural principle that gathers together exhibits in a room with a major interest in associating objects on the ground of history of art and aesthetics, to which conservation problems are sacrificed. In the most fortunate case, gathering in the same room materials with common conservation problems is allowed only in stocks deposits. In most cases, exhibition needs and history of art are kept in highest consideration, without considering the negative impact on the expectation of life of the exhibits. This happens whenever HVAC systems are installed following the above directives by technicians not adequately expert in conservation science.

A further problem is that the climate for thermal comfort of visitors and guardians is not always the best for objects or collections. Differentiated solutions may be found; e.g. in this case, it would be advisable to separate persons from objects with a glass pane and create two independent microclimates. However, very often this is not possible and one of the two, either people or objects, should adapt: more often the result is determined by the fact that objects cannot protest as people do. It should be remembered, however, that the EU standard EN 15759: 2011 establishes that a compromise between the two needs (i.e. conservation and people's thermal comfort) should be found and in the case of conflict, conservation needs should have priority.

Heating needs and strategies depend on a number of variables, e.g. outdoor climate, building type and use; type of object; and materials involved. For instance, churches in continuous use for faith or cultural tourism purposes need continuous heating. A very moderate heating continually operated over the whole cold season might seem an excellent solution, although soiling will be very probably increased. In reality, continuous heating is possible only in the case of soft heating in mild, humid regions because the indoor *RH* may reach an acceptable level when the high moisture content of the air will balance the effect of a soft heating. However, in cold regions, the moisture content in air is small and the heating will lower too much the *RH*, reaching inappropriate levels for conservation. The psychrometric chart shows that if the external temperature is $-10\,^{\circ}\mathrm{C}$, the saturation mixing ratio (*MR*) is about $2\ \mathrm{g\ kg^{-1}}$; an indoor heating at $10\,^{\circ}\mathrm{C}$ will bring to $RH \le 30\%$ and at $20\,^{\circ}\mathrm{C}$, $RH \le 15\%$. If the external temperature is $0\,^{\circ}\mathrm{C}$, the saturation *MR* is about $2\ \mathrm{g\ kg^{-1}}$; the same indoor temperature levels will produce $RH \le 50\%$ at $10\,^{\circ}\mathrm{C}$ and $RH \le 27\%$ at $20\,^{\circ}\mathrm{C}$. If the external temperature is $10\,^{\circ}\mathrm{C}$, the saturation *MR* is about $8\ \mathrm{g\ kg^{-1}}$ and one will get $RH \le 55\%$ at $20\,^{\circ}\mathrm{C}$. This means that in severely cold climates, indoor heating will cause an exceedingly low drop in *RH*. One could consider that the use of humidifiers might solve this problem. However, historical buildings of churches have very low energy efficiency and a number of cold surfaces (e.g. stained glasses, vaults and walls) on which condensation would occur and the moisture supplied to mitigate the *RH* level will only feed mould and algal colonization.

The problem is even worse for occasional use, e.g. one or two days per week: the sharp heating of the whole church is very noxious for temperature peaks and *RH* drops, uncomfortable, expensive and often requires unaesthetic solutions. In this case, it is better to abandon the idea of a general heating and pass to localized heating, e.g. using *IR* heaters or pews might be gently heated in order to form a warmer and more comfortable area where people stay. The key problem is that not always there is awareness of the real needs and heating system convenient for large industrial rooms or other purposes are used in churches. If priority were really given to conservation, some minor cultural or practical sacrifices would be accepted by owners, conservators and visitors. After that, more appropriate heating systems will appear with new market opportunities.

Warm air heating systems sometimes have inlets from the floor, which generate strong convective movements that continually mix the air in the whole room. In addition, they uplift and diffuse the coarser suspended particulate matter that is mostly concentrated inside a floor-based boundary layer. Sometimes the warm air inlets are at a middle height on the wall and a jet of blown air crosses the room, forming two main vortexes that increase deposition of particles on the surfaces and may favour evaporation and wood shrinkage. Other times, the warm air inlets are in the ceiling. The warm air that is blown tends to stratify forming a vertical temperature gradient that gives little benefit to people who hardly

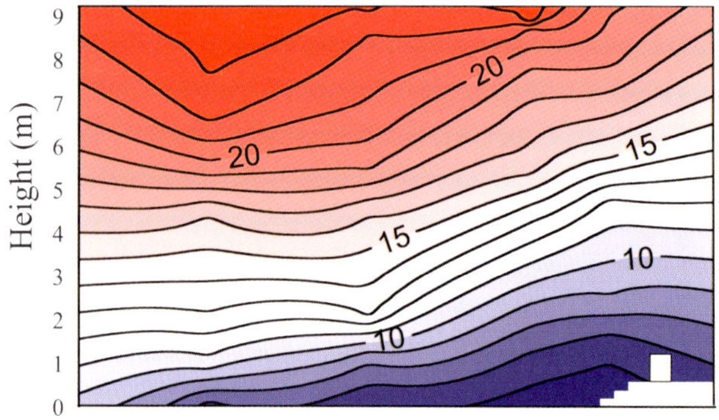

FIGURE 8.24 Thermal layering inside a church with warm air heating. *Source: Rocca Pietore, Italian Alps.*

receive the warm air (Fig. 8.24). A common strategy to reduce thermal layering is to increase the speed at which warm air is blown into the room, may be reducing its temperature. It should be noted, however, that reducing air temperature might be beneficial for thermophoresis and Brownian deposition but increasing airspeed is noxious for inertial impaction that has larger soiling efficiency compared with other deposition mechanisms.

Also, the location of the outlets for the extraction of the air is important. When they are installed close to the inlets, the air goes immediately into the outlets with small advantage for the room. When they are on the opposite side of the inlets, and there is no atmospheric stability, the warm air flows directly from the inlets to the outlets forming a dangerous path of hot air that crosses the room. If the room atmosphere is stratified, the warm air inlets are on the top and the suction is on the bottom and all the room air is progressively substituted, layer by layer, before the new air goes out. In addition, when the atmosphere is stable, the floor-based outlets suck away the concentration layer that contains most of the suspended particulate matter having large or giant size. The existence and drainage of the concentration layer near the floor have never been considered up to now.

The best solution is to avoid central heating and apply local heating, i.e. gentle heating sources located close to people or using *IR* radiation. The key point is to keep heat localized to where people stay, avoiding heat dispersion within the room, as far as possible.

A similar problem might be discussed for summer air conditioning. Cold air inlets from the ceiling form downward rivulets of cold air, which are eventually mixed with the local air; this method is very effective but generates small eddies with all the negative effects generated by turbulence. When inlets are allocated near the floor, the cold air blown into the room spreads in the bottom forming a cold layer, whose thickness grows, being topped with an inversion layer. It is sufficient to cool the layer where people stay (i.e. the first 2 m) leaving untouched the upper layers. It is important, however, that no vulnerable objects are damaged by the air conditioning, especially in the case they are located at the level reached by the top of the inversion layer. In addition, the cold air inlets should be in the lower part of the rooms but above the concentration layer. In this case, the formation of the concentration layer is not perturbed and with a moderate suction from other outlets at floor level, it is possible to drain out the main part of the large and coarse particles.

Stefan flow is often originated when the wall paintings cool during the daily temperature cycles adsorbing moisture into the micropores. In order to avoid this drawback, the room temperature should be as stationary as possible and much higher than the *DP*.

The best preservative methods are permanent, passive ones (such as, thermal insulators, screens against direct radiation and the use of materials with an elevated thermal or hygrometric capacity). These are not exposed to risk of bad functioning or going out of order as may happen to active equipments, even if technologically advanced. Usual temperature regulation systems, air conditioning, humidifiers and other active devices have two negative features: the risk of dramatic temperature–humidity shocks as a result of breakdown and the practical impossibility of distributing heat and moisture in each room without creating gradients in proximity to the points where heat or moisture is emitted or adsorbed.

8.15. INAPPROPRIATE POSITIONING OF PAINTINGS

Not every portion of wall is appropriate for hanging paintings. The problem is not only to avoid the disturbing glare of the opposed windows or lamps and the most frequent errors are the following.

1. *Painting hung over a heater* (Fig. 8.25). Every time the heater starts to operate, the painting will be immersed in a warm, dry air current, which alters the temperature and the equilibrium moisture content of the painting, shrinking and generating internal tensions in the panel, the frame, the canvas and the colour coating. New tensions and swelling occur when heating is turned off.

As the heating system is turned on/off very often, this position generates everyday many cycles and fatigue that damage paintings and cause fading. In addition to the mechanical deterioration, blackening also occurs because the hot airflow generates inertial deposition of airborne particles, thermophoresis, electrophoresis and increased Brownian deposition rate. A corbel put over the heater attenuates only a bit the dramatic effects.

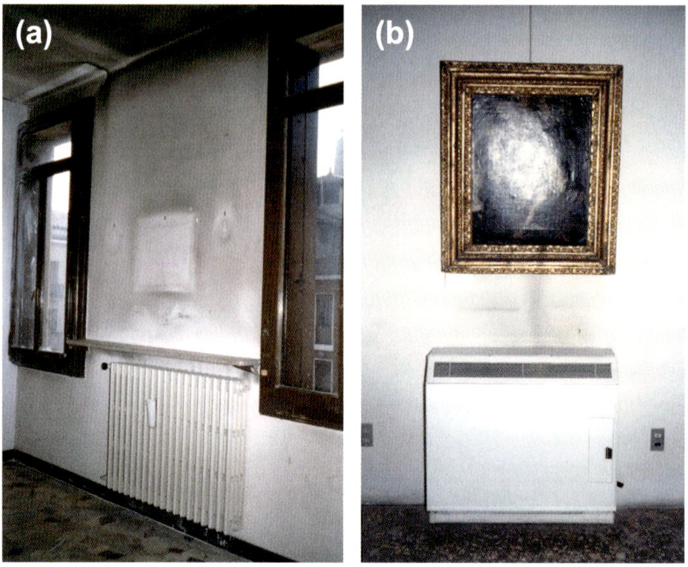

FIGURE 8.25 Incorrect position of paintings in museums. (a) The painting was hung over a heater and has been removed. The contrast between the clean wall on its back and the soiling all around is evident. (b) The painting is hung over a fan coil heater.

However, in the case of heating with blown air, e.g. fan coil, the whole area being affected by the air movement should be avoided. Often, this means that the wall in front of the blower too is not convenient.

2. *The painting is near humidifiers or dehumidifiers.* Often these devices are placed at strategic points in order to counteract the *RH* variations caused by thermal changes. Humidifiers form clouds of humid, and often also cold, air that envelope all the paintings nearby and then move through the room, transported by the air movements. The paintings are alternatively immersed in and out of these clouds, with the cycles of internal stress that have been described above. As these devices are mainly placed in zones protected in corners of rooms or near walls, the clouds that form are even more concentrated and very dangerous. All the outlets for the emission of vapour or dry, hot or cold air should be as far as possible from the works of art and distributed in a capillary way in order to avoid the formation of humidity or temperature gradients. A better alternative is to place these devices at the centre of the rooms, far from the objects, where the visitor can benefit from the air conditioning.
3. *Direct light beams from windows or lamps.* Light may favour chemical reactions (mainly oxidation), discolouring and heating, with all the negative effects that have already been discussed. Also, the position of lamps is important, especially if the lamps are of incandescent type. The heat dissipated by incandescent lamps form a convective cell that

develops growing vertically, over the lamp, as the downward development is in general impeded by the air stability that is very often present in the indoor environments. The air movement associated with the convective motion is responsible for heavy soiling of the surfaces. This means that lights should be placed as high as possible to limit the extension of the convective cells, and soiling, into spaces over the exhibits (Camuffo and Schenal, 1982; Camuffo, 1987). For this reason, it is advisable, whenever possible, to light paintings with light beams coming downwards.

4. *Doors and windows*, when they are simultaneously open for the morning cleaning, generate very dangerous air currents. All the objects affected by the current experience more intense exchanges of heat end especially moisture, with very negative effects.
5. Another common error is to hang wood panels or frames with two or more brackets *rigidly sealed* to both the wall and the wood. When the wood expands or contracts for temperature or humidity changes, the sealed brackets oppose these movements and generate tensions that may overcome the resistance of the material, especially if composed of different parts bound to each other.

It is also to be remembered that *ambient monitoring* is useful only if done in the correct way. Thermohygrographs are generally placed in a corner of room and very often on the floor. Some examples in Chapter 1 have shown that in the horizontal plane it is usual to find a difference of several tenths of degree between one part of a room and another; however, the natural air layering inside close environments makes the vertical gradient much more remarkable, reaching often a few degrees difference between ceiling and floor. In the case that the heating (or the air conditioning) is done by means of emission of hot (or cold) air, the stratification is much more conspicuous, reaching several degrees. Although air stability plays a positive role in suppressing indoor turbulence, measurements taken in only one or few points, and especially near the floor, are not representative of the actual ambient conditions in which works of art, hung at higher levels, are immersed. Attention should be paid that tall objects, e.g. wood statues and paintings, preserved in an excessively stratified air, have their base and their top immersed in air layers with very different temperature and humidity. In the long run, this generates internal tensions to the material as well as migration of moisture and heat with dramatic effects.

8.16. UPLIFTING OF GIANT PARTICLES AND WIND EROSION

In addition to the gentle deposition, the violent impact of giant particles and wind erosion must also be considered. This is often confused with other phenomena. Alveolization of stones may be due to several causes and is often incorrectly interpreted as *'wind erosion'*. Apparently, the action of the wind is very often involved: e.g. formation of severe pitting and interconnected deep holes with loss of granular material with a honeycomb-like structure is very common in carbonatic rocks with biochemical or detritic origin, having an internal nonhomogeneous texture like biocalcarenites (Fig. 8.26(a)) or marine sandstones (Fig. 8.26(b)).

Rainwater dissolves the bonds between granules; wetting/drying cycles enhanced by wind or solar radiation cause solution, mobilization and crystallization of internal salts; and thermal cycles may add internal stress. The consequence is the breaking down of the weaker parts. This weathering is intrinsic to the nature of this rock (Zezza, 1976, 1994). For example, biocalcarenites have a texture of organic origin produced by burrowing organisms during the sedimentation in the Miocene. The bioturbations, of rounded and elongated form, constitute the denser and more resistant part of the calcarenite deposit, with granules cemented better than in the rest of the sediment in which they are contained. When the less-resistant

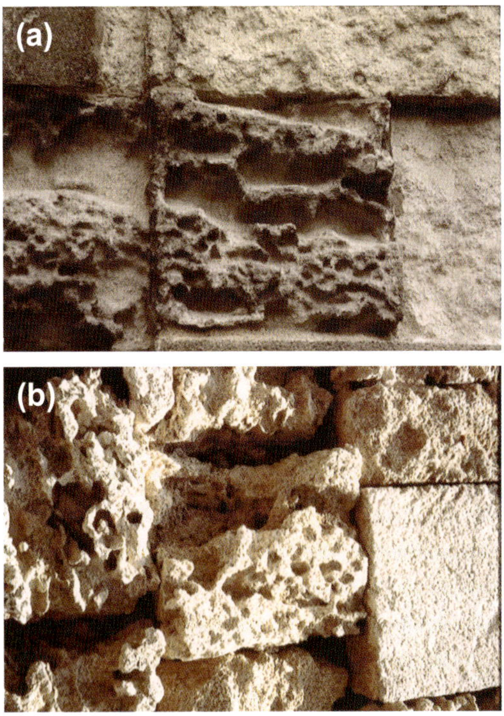

FIGURE 8.26 (a) Biocalcarenite weathering with honeycomb-like structure that is improperly attributed to wind erosion (Seville, Spain). (b) Similar weathering in marine sandstone (Rhodes, Greece).

part is weathered, any mechanical factor may be responsible for the ultimate fall of the loose granules. In particular, moisture is one of the factors involved in the adhesive forces (interparticle surface forces and electrochemical forces) acting on the independent granules (Fletcher, 1976). For this reason, the loss of material is more visible on windy days when the wind reduces the adsorbed moisture as well as the cohesion forces and the turbulence favours the fall of the disaggregated granules in an unstable equilibrium. Although the action of wind is partly involved in this form of weathering, we cannot speak of '*wind erosion*'.

Wind erosion is when wind scrapes the surface of stones, metals etc by exerting a mechanical abrasion by means of airborne abrasive particles. This mechanism is also called '*corrasion*'. The most severe erosion occurs when abrasive particles impact with great energy and for long time, causing extensive micromechanical damage. In the long run, wind erosion causes two kinds of macroscopic effects, i.e.: fringing of the softer part of stones (Fig. 8.27(a)) and formation of round holes (Fig. 8.27(b)) in

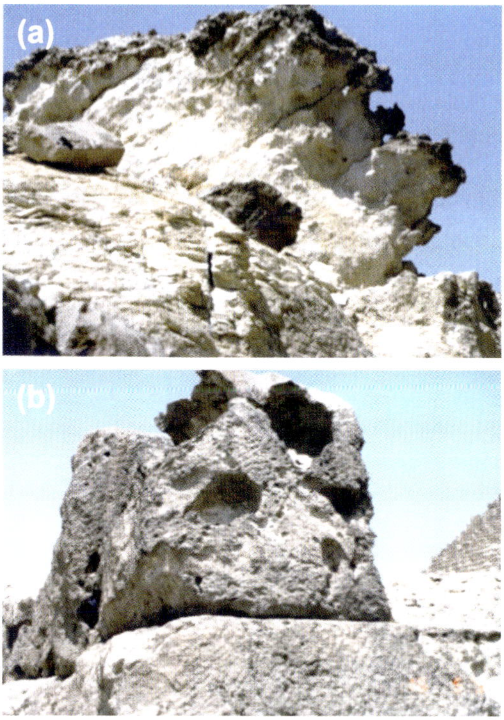

FIGURE 8.27 Two examples of corrasion: (a) fringing of the softer part of stones and (b) formation of hemispherical holes in stones that originally had some small cavities where sand granules in the presence of wind are forced to whirl eventually smoothing, eroding and enlarging the hosting cavity. Giza desert, Egypt.

stones that initially had some small cavities. With fresh wind, the hard sand granules are obliged to whirl, eventually smoothing, eroding and enlarging the host cavity. The process starts on existing cavities, e.g. micropitting, or on the so-called *tafoni*, i.e. natural rounded cavities found in granular rocks especially in coastal areas or desert regions. These cavities are mostly due to salt weathering; they may grow and reach the dimension of small caves (Smith and McAlister, 1986). The erosion rate increases with the cavity diameter and centrifugal force of sand granules accelerated by wind drag.

The dust entrainment occurs in the presence of wind and turbulence. The problem is not yet clear in all details, as the small grains are into the laminar boundary layer or shielded by coarser grains. A mechanism (Bagnold, 1941; Gillette, 1980) is that some grains are dragged by wind, start rolling and creeping on the coarse soil, bounce on the existing granules and during leaps and bounces may be uplifted by wind. When a flying granule impacts on the soil, the collision can raise other granules that are then uplifted by wind.

A more efficient uplifting mechanism is the pressure difference generated by whirlwind and dust devils (Fig. 8.28(a)) or, in the smaller scale, the drag of turbulence for wind over uneven soil or the onset of convective cells over hot sand. Devils and whirlwinds are nonstationary free vortexes (i.e. like minitornadoes) and the core is a vortex line around which the air rotates forming hyperboloids of rotation (Fig. 8.28(b)). The core of the vortex line is characterized by a very low pressure and for this reason, the vortex line should form a closed loop or should start and end at the boundary of the fluid. Devils and whirlwinds are spinning flows with the core ending and moving on the soil level and they suck in all the

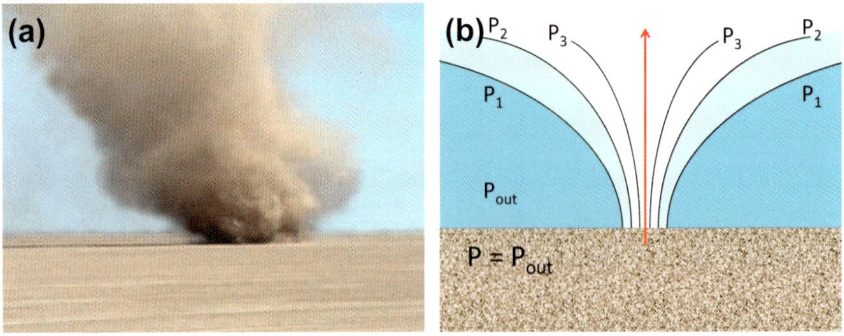

FIGURE 8.28 (a) A dust devil over a sandy soil. (b) A vortex in air has the core characterized by very low pressure and moves wandering over the soil surface. P_{out} is the external air pressure, outside the sand; inside the sand dune, air pockets have the same atmospheric pressure P_{out}. P_1, P_2, P_3 are decreasing pressure levels going towards the minimum in the core. A sand granule is ejected and uplifted as indicated by the red arrow to fill the pressure gap.

granules they meet, like a giant vacuum cleaner. The soil is composed of granules and the free spaces between them are air pockets at normal atmospheric pressure. When the whirlwind core moves wandering along the soil surface, it determines a pressure difference of about 1 atm (i.e. 1013 hPa) across the very surface layer. This enormous pressure difference causes the air pockets below the surface to explode: the granules are ejected upwards into the low-pressure core where they are entrained and uplifted by the spiralling airstreams to fill in the pressure drop.

There is a threshold velocity for wind to impress motions to granules and uplift them. The larger the granule and its weight, the higher the threshold velocity, which varies with the square of the diameter of the particles, i.e. proportional to the aerodynamic cross-section of the granule to which the wind drag applies and not to the weight, which varies with D^3. However, for very small granules, the mechanism is different. On decreasing the size of the particles, other factors arise: small granules are better shielded among the coarser ones, they are protected by the viscous layer in which they are immersed, and the adhesive forces (especially van der Waals forces) that bind the grains to the soil are stronger. The finer the granule, the higher the wind speed threshold V_t. In practice, the relation between the granule diameter and threshold velocity shows a dependence similar to a parabola (with a logarithmic distribution of D), with the minimum at $V_t = 4\ \mathrm{m\ s^{-1}}$ and $D = 100\ \mu\mathrm{m}$. These are the earliest granules to be uplifted; then, when the wind speed increases, both smaller and larger granules are uplifted. Once uplifted, the energy of a flying granule is linearly proportional to its density, to the second power of the wind speed and to the third power of its diameter.

The finest dust particles, i.e. $D < 1\ \mu\mathrm{m}$, are airborne and follow the wind stream flying around obstacles without impacting on them. Small dust particles, which have inertial force greater than the viscosity of the air, undergo inertial interception, or even also impaction, on the surface and deposit there; the main effect is to make the surface dusty, not abraded. Broadly speaking, all the dust particles with a diameter $D < 100\ \mu\mathrm{m}$ are airborne and may or may not deposit, but they are not dangerous for abrasion. In the medium size range, i.e. $100 < D < 600\ \mu\mathrm{m}$, granules are sufficiently large to be dangerous but too heavy to fly for a long time. Granules in this intermediate size interval are temporarily lifted and soon fall and are transported by means of successive leaps and bounds, so that they proceed 'saltating'. The height of the leaps depends on the roughness of the soil but on average, this phenomenon only affects the first metre above the soil. Granules larger than 600 μm cannot erode, as they cannot be lifted by the wind and move only by creeping and rolling. Grains that are coarser than 6 mm cannot be moved by the wind (Bagnold, 1941; Pye, 1987; Barndorff-Nielsen and Willets, 1991). In conclusion, at ordinary wind speed, only grains in the leaping range

$100 < D < 600$ μm are effective in terms of erosion, but only near the base of the monuments.

This weathering is important only in sandy, windy regions and has been extensively discussed in the case of the Great Sphinx (Camuffo, 1993) that is situated near the border of the Giza desert, Egypt. In this case, signs of severe corrasion actually appear at elevated levels of the monument, i.e. between the midbody and the neck. However, other parts of the monument do not suffer erosion at all. The lower part was severely weathered and restored with stone masonry. It shows exfoliation caused by the combined action of soluble salts and groundwater. The damaged stone is flaking, with thin and weak scales (Fig. 8.29). This is a clear indication that sand blasting should be excluded in the lower part of the monument.

Again, the face is still covered with a soft red coating made of animal blood (Fig. 8.30) that was attributed to the New Kingdom (1555–1075 B.C.) (Getty Conservation Institute, 1990). The presence of this coating after some millennia is a clear demonstration that the face has never suffered corrasion.

In order to explain this apparent paradox, it is necessary to consider the local aerodynamic features and the monument history. The head of the Sphinx originally was a block of rock protruding above the ground surface. The body, which is below the original ground level, was obtained by digging all around an excavation which is some 12 m deep and 10 in width; only the back of the Sphinx is at the original soil level (Fig. 8.31). In the past millennia, the excavation was filled with desert sand and the Sphinx was submerged, except the head. The dunes and the wind removal of sand changed continually the level of the soil and the eroding level of the saltating sand varied between the midbody and the neck following the changing height of the dunes (Fig. 8.32).

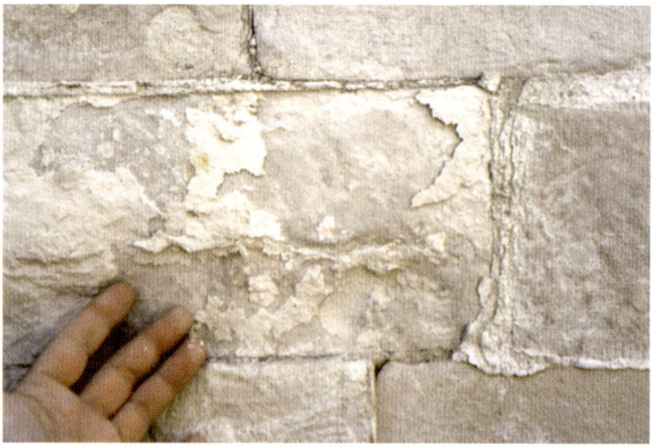

FIGURE 8.29 Exfoliation of stones on the lower part of the Sphinx. *Source: By kind permission of the International Institute for Conservation of Historic and Artistic Works (IIC). See credits.*

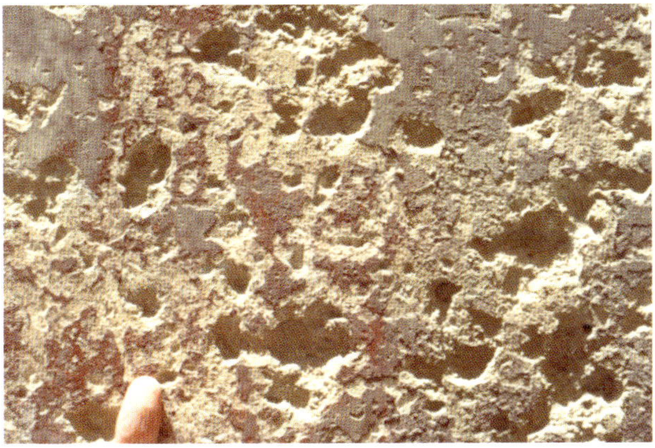

FIGURE 8.30 Fragments of red coating on the face of the Sphinx that are remains of ancient colouring made with animal blood. *Source: By kind permission of the International Institute for Conservation of Historic and Artistic Works (IIC). See credits.*

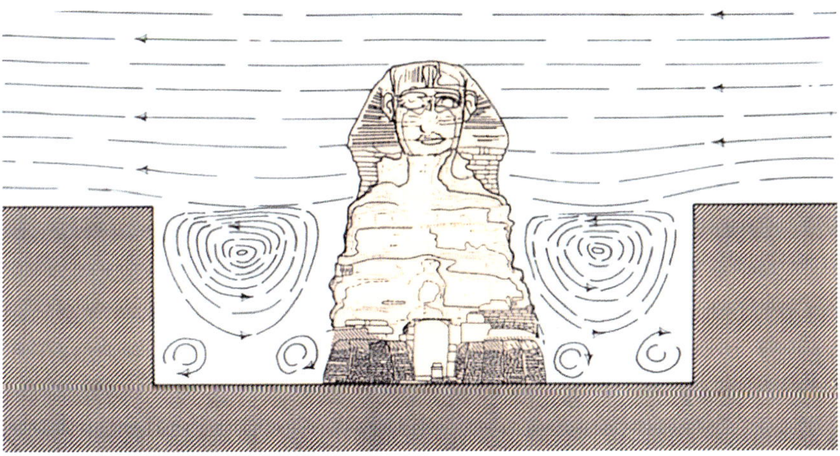

FIGURE 8.31 The body of the Sphinx was obtained by digging all around an excavation. When the wind blows, two stationary vortexes form in this excavation and they change the trajectory of coarse granules driven by the air drag, the gravity and the inertia forces. As a result, all abrasive granules fall into the excavation without reaching the monument.

Nowadays, with the excavation cleared out, the gravitational settling is sufficient to determine a parabolic trajectory to the initially horizontal inertial motion of saltating granules, which arrive at the upper edge of the excavation and fall in the bottom without impacting on the monument or hitting only the edge of the basement.

However, in the presence of cross-wind, the aerodynamic field into the excavation is far from the inertial motion: the canyon effect generates a

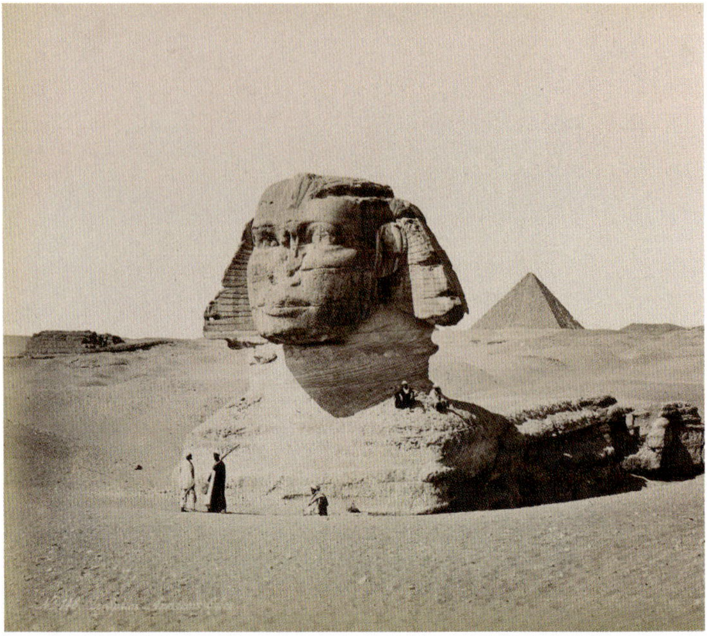

FIGURE 8.32 The Sphinx partially submersed by dune sand (photo 1860–1890). The wind erosion was due to coarse grains in the range between 200 and 600 μm blown in saltating mode. Dragged by the wind, these grains proceed by leaps and bounds and reach 1 m above the sand level. In the past millennia, the tidal motion of dunes continually varied the erosion level, which lies from midbody to the neck. *Source: By courtesy of Photographium. See credits.*

circulating region within the square cavity (Tritton, 1988), similar to a vortex with horizontal axis (as shown in Fig. 8.31), which deflects the granule trajectories, moving them away from the monument. The excavation is a sink for all the abrasive particles that arrive in the saltating mode and these cannot reach the monument. Keeping clear the excavation is the best preservation method.

Pure corrasion is extremely rare in urban areas, although wind erosion is improperly attributed to many other forms of degradation, which possibly depend on wind speed in addition to other variables. Normal weathering of biocalcarenites (mainly due to dissolution of the binding cement and salt crystallization) is a common example, as fall of disaggregate granules is visible in windy conditions, as already mentioned. In general, urban environments do not have many grains of dangerous size and buildings reduce the wind speed near the ground, except for some narrow places where wind channelling occurs; in addition, monuments are situated on a base that is usually above the dangerous height. The weathering that affects the base of urban monuments is mainly due to frequent mechanical collisions, capillary rise of rainwater, dissolved pollutants and animal urine and/or other biological attacks. In cold regions, freezing–thawing cycles are also important at the ground level.

8.17. KINETIC ENERGY AND SAND BLASTING

Besides pressurized spray water, sand blasting is also commonly used to clean monuments. This method utilizes compressed air to impress kinetic energy to abrasive granules and remove surface crusts and dirt. It can be applied in dry conditions or by adding some small jets of water in order to attenuate the impact of the granules and (partially) wash the surface. More detailed information on the method and the opportunity of using it can be found elsewhere (Lazzarini and Laurenzi-Tabasso, 1986; Ashurst and Ashurst, 1989); here, the attention is only focused on the physical principle. The formula for the kinetic energy E_c of spherical granules having density ρ, radius r and speed v is

$$E_c = \frac{2}{3}\pi\rho\, r^3 v^2 \tag{8.59}$$

In this formula, the granules of the abrasive are characterized by ρ and r, and the speed of the stream of compressed air is determined by the air pressure in the compressor, nozzle size and type, distance progressed after the nozzle and lateral distance from the axis of the stream. In

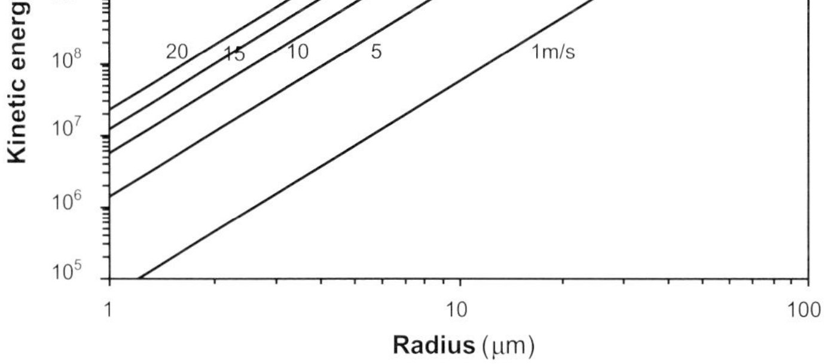

FIGURE 8.33 Kinetic energy (E_c, dyne) of grains used for sand blasting as a function of the grain radius r and for some selected airstream velocities, i.e. $v = 1, 5, 10, 15, 20\ \mathrm{m\ s^{-1}}$, and has been calculated with the density of quartz ($\rho = 2.65\ \mathrm{g\ cm^{-3}}$).

addition to the energy of the impacting granules, their hardness and shape are also important; e.g. glass spherules are less harmful than rough sand granules. However, air pressure, nozzle and abrasive characteristics are not sufficient to determine what happens at the monument surface, as the airflow speed varies with the distance from the nozzle. In addition, the same E_c can be obtained with different combinations of r and v, as shown in Fig. 8.33.

In this figure, E_c has been computed for quartz, i.e. $\rho = 2.7$ g cm^{-3}, for different granule radii and for some selected airstream velocities, i.e. 1, 5, 10, 15 and 20 m s^{-1}. The formula has three degrees of freedom, i.e. ρ, r and v, and v is not fully defined with the compressor pressure and nozzle characteristics, so that among the variables instead of finding the distance of the nozzle from the surface, it is possible to find included the 'skill of the operator' and the 'supervision of the work', which are not objective parameters, but very important factors.

This example has been chosen in order to show that not always can all the variables be controlled in a scientific way and that personal experience is also an important factor, which becomes even much more useful if it can be transmitted in strictly objective terms. Science operates to abandon personal variability and transform empiricism into general laws.

References

Theory and general applications

Bagnold, R.A., 1941. The Physics of Blown Sand and Desert Dunes, third ed. Chapman and Hall, London, 1984.

Barndorff-Nielsen, O.E., Willetts, B.B., 1991. Aeolian Grain Transport, 1: Mechanics, 2: The Erosional Environment. Acta Mechanica Supplementums 1 and 2. Springer-Verlag, Wien.

Bouma, P.J., 1947. Physical Aspects of Colour. Philips Gloeilampenfabrieken, Eindhoven, 312 pp.

Buffle, J., Van Leeuwen, H.P., 1992. Environmental Particles. Lewis, Boca Raton. Vol. I: 554 pp., Vol. II: 426 pp.

Cadle, R., 1965. Particle Size, Theory and Industrial Applications. Reinhold, New York, 158 pp.

Caporaloni, M., Tampieri, F., Trombetti, F., Vittori, O., 1975. Transfer of particles in non-isotropic air turbulence. Journal of the Atmospheric Sciences 32, 565–568.

Chandrasekhar, S., 1943. Stochastics problems in physics and astronomy. Reviews of Modern Physics 15 (1), 2–89.

Einstein, A., 1905. Über die von der molekularischen Theorie der Wärme gefordete Bewegung, von in ruhenden Flüssigkeiten suspendierten Teilchen. Annalen der Physik 17, 549–560. This and other related papers have been translated and reprinted in Einstein, A., 1956. Investigations on the Theory of the Brownian Movement, edited with notes by Fürth, R. Dover, New York, 119 pp.

Epstein, P.S., 1924. On the resistance experienced by spheres in their motion through gases. Physical Review 23, 670–733.

Fletcher, B., 1976. The incipient motion of granular materials. Journal of Physics D: Applied Physics 9, 2471–2478.

Friedlander, S.K., 1977. Smoke, Dust and Haze. Wiley, New York, 317 pp.

Gillette, D., 1980. Major contributions of natural primary continental aerosols: source mechanisms. In: Kneip, T.J., Lioy, J. (Eds.), Aerosols: Anthropogenic and Natural, Sources and Transport. New York Academy of Sciences, New York, pp. 348–358, 618 pp.

Goldsmith, P., Delafield, H.J., Cox, L.C., 1963. The role of diffusiophoresis in the scavenging of radioactive particles from the atmosphere. Quarterly Journal of the Royal Meteorological Society 89, 43.

Hicks, S.B.B., 1982. Wet and dry surface deposition of air pollutants and their modeling. In: Baer, N.S. (Ed.), Conservation of Historic Stone Buildings and Monuments. National Academy Press, Washington D.C, pp. 183–196, 365 pp.

Hidy, G.M., 1984. Aerosols, an Industrial and Environmental Science. Academic Press, San Diego, 734 pp.

Hidy, G.M., Brock, J.R., 1970. The Dynamics of Aerocolloidal Systems. Pergamon, Oxford.

Hidy, G.M., Brock, J.R., 1972. Topics in Current Aerosol Research. Pergamon, Oxford, 384 pp.

Hinds, W.C., 1982. Aerosol Technology: Properties, Behavior, and Measurement of Airborne Particles. Wiley, New York, 424 pp.

Isachenko, V.P., Osipova, V.A., Sukomel, A.S., 1977. Heat Transfer. Mir, Moscow.

Landsberg, G.S., 1979. Ottica, vol. 2. MIR, Moscow, 478 pp.

Nazaroff, W.W., Cass, G.R., 1987. Particle deposition from a natural convection flow onto a vertical isothermal plate. Journal of Aerosol Science 18, 445–455.

Phenix, A., Burnstaock, A., 1990. The deposition of dirt: a review of literature. In: Hckney, S., et al. (Eds.), Dirt and Pictures Separated. The U.K. Institute for Conservation, Belton, England, pp. 11–18, 56 pp.

Prodi, F., Tampieri, F., 1982. The removal of particulate matter from the atmosphere: the physical mechanisms. Pageoph 120, 286–325.

Pruppacher, H.R., Klett, J.D., 1980 and 2003. Microphysics of Clouds and Precipitation, first and second ed. Reidel, Dordrecht.

Pye, K., 1987. Aeolian Dust and Dust Deposits. Academic Press, London, 334 pp.

Schlichting, H., 1979. Boundary-Layer Theory. McGraw-Hill, New York, 817 pp.

Sehmel, G., 1980. Particle and gas deposition: a review. Atmospheric Environment 14, 983–1011.

Seinfeld, J.H., 1986. Atmospheric Chemistry and Physics of Air Pollution. Wiley, New York, 698 pp.

Slinn, W.G.N., Hales, J.M., 1970. Phoretic Processes in Scavenging. Atomic Energy Commission Symp. Ser. 22. U.S. Atomic Energy Committee. pp. 411–422.

Smith, B.J., McAlister, J.J., 1986. Observations on the occurrence and origins of salt weathering phenomena near lake Magadi, Southern Kenia. Zeitschrift für Geomorphologie 30 (4), 445–460.

Talbot, L., Cheng, R.K., Schefer, R.W., Willis, D.R., 1980. Thermophoresis of particles in a heated boundary layer. Journal of Fluid Mechanics 1014, 697–758.

Tritton, D.J., 1988. Physical Fluid Dynamics. Clarendon, Oxford, 519 pp.

Vittori, O., 1973. Scavenging of atmospheric particles by growing ice crystals: a contribution to a proposed mechanism. Journal of the Atmospheric Sciences 30 (2), 321–324.

Vittori, O., 1984. Transient Stefan flow and thermophoresis around an evaporating droplet. Nuovo Cimento 7C (2), 254–269.

Waldmann, L., Schmitt, K.H., 1966. Thermophoresis and diffusophoresis of aerosols. In: Davies, C.N. (Ed.), Aerosol Science. Academic, New York.

Wang, Ming Chen, Uhlenbeck, G.E., 1945. On the theory of the Brownian motion II. Reviews of Modern Physics 17 (2/3), 323–341.

Wax, N., 1954. Noise and Stochastic Processes. Dover, New York, 345 pp.

Zimon, A., 1982. Adhesion of Dust and Powder. Consultants Bureau – Plenum, New York.

Applications to conservation

Ashurst, J., Ashurst, N., 1989. Practical Building Conservation. In: Stone Masonry, vol. 1. Gower, Aldershot, U.K., 100 pp.

Camuffo, D., 1987. L'illuminazione negli ambienti di conservazione; First Part *Rassegna Beni Culturali*, 3 (4), 40–44 and second part, item 3 (6), 38–40.

Camuffo, D., 1990. Ambiente e monumenti: microclimatologia di ambienti chiusi e conservazione di opere pittoriche. *Accademia Nazionale Lincei.* Giornata Dell'Ambiente 82, 157–166.

Camuffo, D., 1991. Wall temperature and soiling of murals. Museum Management and Curatorship 10, 373–383.

Camuffo, D., Bernardi, A., 1986. Dinamica del microclima e scambi termoigrometrici tra pareti e atmosfera interna nella Cappella Sistina. Bollettino Monumenti Musei Gallerie Pontificie 6, 211–257.

Camuffo, D., 1993. Controlling the aeolian erosion of the great Sphinx. Studies in Conservation 38, 198–205.

Camuffo, D., Bernardi, A., 1991. The microclimate of Leonardo's "Last Supper"; joint edition. special issue. European Cultural Heritage Newsletter on Research, and Bollettino Geofisico 14 (3), 1–123.

Camuffo, D., Bernardi, A., 1995. The microclimate of the Sistine Chapel, joint edition. European Cultural Heritage Newsletter on Research 9, 7–32 and Bollettino Geofisico, 18(2), 7–32.

Camuffo, D., Schenal, P., 1982. Microclima all'interno della Cappella degli Scrovegni scambi termodinamici tra gli affreschi e l'ambiente. In: Ministero dei Beni Culturali ed Ambientali "Giotto a Padova", special issue of Bollettino d'Arte. Poligrafico dello Stato, Rome, pp. 107–209.

Eastaugh, N., 1990. The visual effects of dirt on paintings. In: Hackney, S., et al. (Eds.), Dirt and Pictures Separated. The U.K. Institute for Conservation, Belton, England, pp. 19–23, 56 pp.

Getty Conservation Institute, 1990. Conservation Research Proposal for the Great Sphinx Presented in Cairo. Getty Conservation Institute Newsletter, 5 (8.2), 1–2.

Koestler, R.J., Warsheid, T., Nieto, F., 1996. Biodeterioration: risk factors and their management. In: Baer, N.S., Snethlage, R. (Eds.), Saving Our Architectural Heritage: The Conservation of Historic Stone Structures. Wiley, Chichester, pp. 25–36.

Lazzarini, L., Laurenzi-Tabasso, M., 1986. Il Restauro Della Pietra. Cedam, Padova, 320 pp.

Marshall, K.C., 1984. Microbial Adhesion and Aggregation. Springer Verlag, Berlin, 424 pp.

Thomson, G., 1986. The Museum Environment. Butttherwords, London, 293 pp.

Zezza, F., 1976. Caratteristiche litogenetiche e forme della degradazione delle pietre da costruzione calcaree di origine biochimica e detritica. Rassegna Tecnica Pugliese – Continuità 10 (2), 3–28.

Zezza, F., 1994. Evaluation criteria of the effectiveness of treatments by non destructive analysis. In: Zezza, F. (Ed.), Stone Material in Monuments: Diagnosis and Conservation, 2nd Course of the Community of Mediterranean Universities. Adda, Bari, pp. 198–212.

CHAPTER

9

Consequences of the Maxwell–Boltzmann Distribution

OUTLINE

Microclimate for Cultural Heritage, Second Edition
http://dx.doi.org/10.1016/B978-0-444-63296-8.00010-X

The equations for the Maxwell distribution of molecules by speeds and the Boltzmann's one for energy levels, the thermal emission of bodies by Wien and Planck, the chemical kinetics by Arrhenius as well as the saturation pressure by Magnus-Tetens, Sonntag, Fermi and Planck have the same mathematical vest and a profound physical connection between them. This Chapter is aimed to clarify this topic and discuss further findings.

9.1. THE MAXWELL–BOLTZMANN EQUATION AND THE DISTRIBUTION OF MOLECULES BY VELOCITIES

The Maxwell–Boltzmann distribution concerns the distribution of an amount of energy between identical but distinguishable particles. It represents the probability for the distribution of the states in a system having different energies. A special case is the so-called *Maxwell distribution law of molecular velocities*. Maxwell introduced this law for the velocities of particles of a fluid (Maxwell, 1868), e.g. a gas or a liquid solution in thermal equilibrium, and gives the probability $f(v)\,\mathrm{d}v$ that the velocity v will be found in the infinitesimal range v to $v+\mathrm{d}v$ and gives the frequency of occurrence of the specific value v for the chosen velocity component v_x, v_y, v_z per each direction x, y, z.

Maxwell found that the molecules are distributed by velocities following a skewed, bell-shaped plot (Fig. 9.1(a)) given by the equation

$$f(T) = 4\pi^{-\frac{1}{2}}\left(\frac{m}{2\Re T}\right)^{\frac{3}{2}} v^2(T)\exp\left(-\frac{mv^2(T)}{2\Re T}\right). \tag{9.1}$$

where m is the mass of the molecule and T is the ambient temperature. In the case of a mixture of gases having different mass m_i, one obtains a plot for each i gaseous species. In the Maxwell distribution, the mode is the most probable velocity v_{mp}

$$v_{mp} = \sqrt{2\frac{kT}{m_i}} \tag{9.2}$$

The lower limit for the velocity is 0 m s^{-1}. In the atmosphere, the upper limit is the escape velocity v_e from the Earth, i.e. $v_e = 1.12 \times 10^4$ m s^{-1} (=28 v_{mp}). For nitrogen at $T = 300$ K, $v_{mp} = 394$ m s^{-1}, i.e. faster than the velocity of sound $v_S = 344$ m s^{-1}.

From velocities to energy levels: if, however, we pass from the velocity (Maxwell) to the energy (Boltzmann) domain, the distribution is represented in terms of kinetic energy E_c as an independent variable

$$E_c = \frac{1}{2}m_i v^2 \tag{9.3}$$

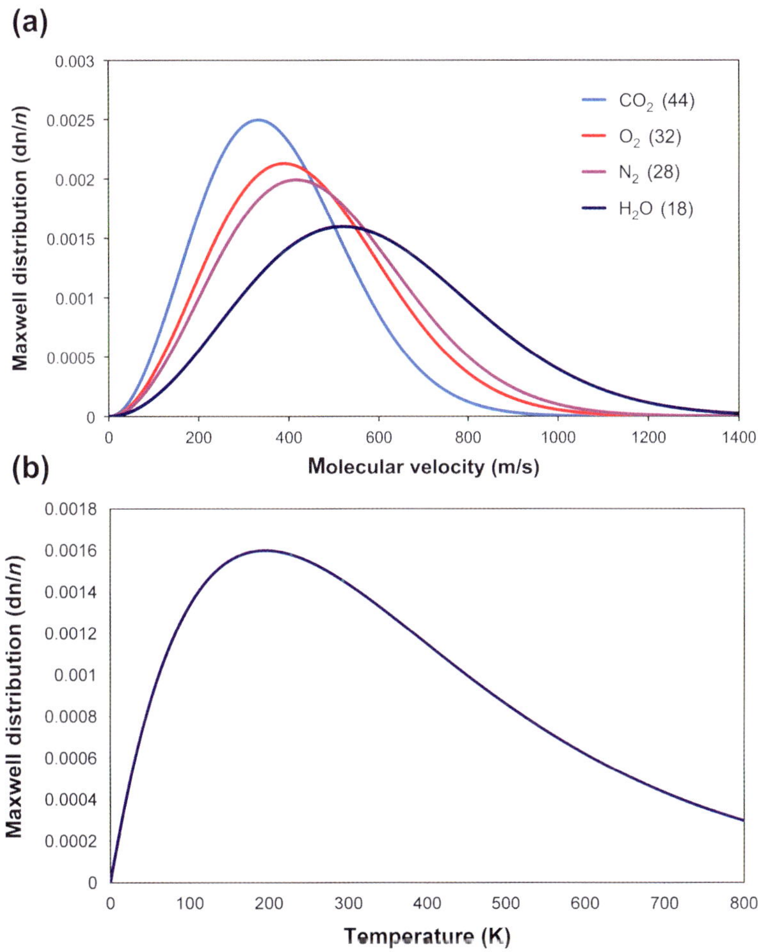

FIGURE 9.1 (a) Maxwell distribution of molecules by velocities for the most common gaseous molecules in air. Each of them has its plot depending on the molecular mass m, i.e. CO_2 ($m = 44$), O_2 ($m = 32$), N_2 ($m = 28$) and H_2O ($m = 18$). (b) The same distribution but expressed in terms of molecular temperatures. All gases have the same plot, irrespective of their molecular mass. $T_{amb} = 20\,°C$ and $p = 1013$ hPa.

and one obtains a plot of $f(E_c)$ which is irrespective of the individual molecular mass m_i of the gas. The same can be done in terms of temperature, i.e. $f(T)$, because T is proportional to the translation kinetic energy:

$$T = \frac{mv^2}{3k} \tag{9.4}$$

where m is the mass of the molecule and k is the Boltzmann constant (Fig. 9.1(b)). In thermodynamics, temperature is an average property describing a system; however, in this case, it will be useful to extend to each molecule the concept of individual temperature (e.g. cold, mild and hot) according to Eqn (9.4) relating temperature with kinetic energy.

Boltzmann followed Maxwell, but found a more general distribution, i.e. *the distribution of energy levels* (Boltzmann, 1868). If the energy level associated with some state or condition of a system is ε, then the frequency with which that state or condition occurs, or the probability of its occurrence, is proportional to

$$e^{-\varepsilon/kT}. \tag{9.5}$$

The Maxwell law of molecular velocities $f(v)$ is a particular case of the Boltzmann distribution of energy levels, if one equals to ε the kinetic energy E_c. From this point of view, the velocity distribution function $f(v)$ becomes

$$f(v) = \sqrt{\frac{m}{2\pi kT}}e^{-mv^2/2kT} \tag{9.6}$$

The advantage of $f(T)$ is that the plot becomes, per any ambient temperature, always the same, irrespective of the molecular mass m of the individual gas species (Fig. 9.1(b)).

This general distribution, known as the *Maxwell–Boltzmann distribution*, is referred to as an increasing series of small intervals composed of molecules characterized by the same kinetic energy or the same thermal level.

The Maxwell–Boltzmann distribution has been recognized to be valid for the solid state too, including the potential energy of position and the internal energy of molecules (Langmuir, 1920). More recently, it has been verified for the thermal vibrations and rotational states of molecules in solids as well as for interactions between excited electronic bands and lattice vibration processes (Vollmer and Möllman, 2011).

For any large number of molecules in thermal equilibrium at the temperature T, the probability P that a given molecule has energy lying between E and $E + dE$ is given by

$$P = Ce^{-\frac{E}{kT}} = Ce^{-\frac{mv^2}{2kT}} \tag{9.7}$$

where the coefficient $C = (m/2\pi kT)^{3/2}\,4\pi\, v^2$ varies with T but at negligible rate compared with the exponential factor $\exp(-E/kT)$. The exponential factor gives the fraction of the most energetic molecules whose energy exceeds a given level E that might be arbitrarily chosen, e.g. a specific energy threshold.

In the following, we will discuss some consequences of the Maxwell–Boltzmann distribution of the molecular energy levels, i.e. the thermal emission of bodies, the Arrhenius equation and the vapour pressure at saturation that will be related to the incoherent water in the liquid state and, finally, how the molecules of water will behave in the vapour phase.

9.2. THERMAL EMISSION OF BODIES

The Planck theory of radiation (Planck, 1901, 1913) introduced the quantum of energy $h\nu$ (where $h = 6.62606957 \times 10^{-27}$ erg s is the Planck constant and ν is the frequency characterizing the radiation) to overcome difficulties encountered with classical thermodynamics and find a general equation for the intensity $I(\nu, T)$ of blackbody emission valid for all frequencies. In terms of frequency, the empirical *Planck equation of thermal radiation* is

$$I(\nu, T) = \frac{2h\nu^3}{c^2} \frac{1}{e^{\frac{h\nu}{kT}} - 1} \tag{9.8}$$

where c is the speed of light. When for high values of ν/T the exponential factor $\exp(h\nu/kT)$ becomes much larger than 1, e.g. for $\nu > 10^{12}$ Hz at usual temperatures, the term 1 can be neglected and the Planck equation of thermal radiation becomes the *Wien displacement law* (Wien, 1886, 1898). The latter is known in the form of the power density $u(\nu,T)$ of the radiation with frequency ν emitted by unit area of a blackbody at absolute temperature T,

$$u(\nu, T) = a\nu^3 e^{-\frac{b\nu}{T}} \tag{9.9}$$

where $a = 2h/c^2$ and $b = h/k$ are constants experimentally determined from observations of blackbody radiation. This formula holds for ultraviolet (*UV*) radiation, visible radiation and infrared (*IR*) thermal emission, but fails to accurately fit the experimental data for low-frequency emission, i.e. microwaves and radio waves.

Einstein (1909, 1917) noted that the formal similarity between the Maxwell–Boltzmann distribution of molecular velocities and the Wien chromatic distribution for thermal radiation suggests that the Wien law can be derived from the Maxwell–Boltzmann distribution. The same formula means the same underlying mechanism. Einstein supposed that the internal energy distribution of molecules was related to their collisions among themselves (i.e. the Maxwell distribution) and that the collisions were related to radiative exchanges, in terms of emission or absorption of radiation. The probability that a given molecule having thermal energy E

passes to another allowed energetic level E' emitting or absorbing photons was considered physically based on 'changes of state due to irradiation'. The above consideration led to the derivation of Planck's radiation law and other advances in quantum theory.

In practice, the Wien law and the equivalent Planck law establish that thermal emission (i.e. *IR*, luminous and *UV* photons) is related to Boltzmann distribution of the most energetic levels of matter (i.e. molecules/electrons) at a given temperature T. The most unstable levels spontaneously emit the excess energy to pass to a more stable state.

9.3. THE ARRHENIUS EQUATION

The Arrhenius equation (Arrhenius, 1889; see Chapter 1) for chemical kinetics was experimentally derived for aqueous solutions and electrolytic dissociation. It was known that the temperature T influences the reaction rate, expressed in terms of the so-called equilibrium rate constant $\kappa = \kappa_1/\kappa_2$ representing the ratio between the individual rate constants κ_1 and κ_2 of the forward and reverse reactions, respectively. The heat δQ necessary for a reaction to occur was considered a form of energy collected from the liquid medium and the reaction was regulated by the general equation

$$\frac{\mathrm{d}\ln\kappa}{\mathrm{d}\frac{1}{T}} = -\frac{\delta Q}{\Re} \tag{9.10}$$

Arrhenius recognized that in the above equation it was possible to split the forward and reverse individual rate constants, i.e. κ_1 and κ_2. The process was interpreted in terms of molecular activity, the individual molecules following the Maxwell distribution of velocities and the energy needed for the reaction being cumulatively supplied by the molecular hits. In order to take an active role, the energy transfer should overcome a certain threshold called *'activation energy'* E_a, which corresponds to the energy barrier that the reactants must surmount in order to react. Chemical reactions are favoured in the case of higher temperature of the liquid because it is composed of a higher percentage of fast and energetic molecules able to transfer a consistent amount of their energy on behalf of the reaction. Briefly, Arrhenius rewrote a general equation in the form:

$$\frac{\mathrm{d}\ln\kappa}{\mathrm{d}\frac{1}{T}} = -\frac{E_a}{\Re}. \tag{9.11}$$

From the mathematical point of view, if E_a is a constant, the above equation can be integrated to yield:

$$\ln\kappa = \ln A - (E_a/\Re T) \tag{9.12}$$

where ln A is an integration constant. Since ln A has unit s^{-1}, it can be physically interpreted as a frequency factor (Lewis and Randall, 1921) that depends in some way on the frequency of molecular collisions or in terms of the entropy of the system (Kohnstamm and Scheffer, 1911). The Arrhenius equation can be rewritten in the most popular form:

$$\kappa = A \exp(-E_a/\Re T) \tag{9.13}$$

This formula was empirically derived and the meaning of its terms has been investigated over one century to build a conceptual framework to interpret kinetic data (Laidler and King, 1983). At the beginning, the key question was: if some molecules react, why do not all molecules react? Why do some molecules behave differently from others in the same group? Arrhenius concluded that a molecule should have and exceed some threshold energy E_0 before it can react, and the fraction of molecules that possess this amount of energy is given by the function $\exp(-E_0/kT)$, which is formally identical to Boltzmann equation (9.4). Chemical reaction rates increase with increasing temperature for the simple reason that a larger fraction of reactant molecules possesses energy in excess of the reaction energy barrier (Connors, 1990). This is another consequence of the Maxwell–Boltzmann distribution, very similar to the above discussion for the thermal emission of bodies.

9.4. SATURATION PRESSURE OF WATER VAPOUR IN AIR

Another mechanism related to the Maxwell–Boltzmann distribution is the saturation pressure of water vapour p_{sat}. This is quite accurately calculated with the empirical exponential formula by Magnus (1844), later improved by Tetens (1930), i.e.

$$p_{sat}(t) = p_{sat}(0) \times 10^{\frac{at}{b+t}} \tag{9.14}$$

where $p_{sat}(0) = 6.112$ hPa, $a = 7.5$, $b = 237.3$ °C with reference to water, and t is the temperature in degree Celsius. The two coefficients a and b have been more precisely evaluated by Sonntag (1990), i.e. $a = 7.65$, $b = 243.12$ °C. However, for a theoretical discussion, it is preferable to make reference to theoretically derived equations, although they are a bit less accurate than the empirical ones, for the more elegant and convenient mathematical formulation in terms of absolute temperature T expressed in Kelvin (Camuffo, 2004).

A theoretical equation was derived by Planck (1926) from the specific energy of the water vapour phase under the hypotheses of a perfect gas, under an isothermal process, negligible external work and specific volume of the liquid phase

$$p_{sat} = C_1 T^{(c_p - c_v)/\Re} \exp\left(-\frac{C_2}{T}\right) \tag{9.15}$$

where C_1 and C_2 are constants and c_p and c_v the specific heats of the vapour at constant pressure and volume, respectively. The pre-exponential factor weakly depends on T.

Fermi (1937) theoretically derived another similar equation from the Clapeyron (1834) equation concerning the coexistence curve, i.e. the line separating the gaseous from the liquid phase on the pressure–temperature diagram. Fermi derived his formula under the two assumptions that the volume of the liquid phase is negligible compared with the vapour phase and that the equations for perfect gases can be applied, i.e.

$$p_{sat} = C_3 \exp(-L_v/\Re T). \tag{9.16}$$

where the pre-exponential factor C_3 is constant, independent of T. This relationship between the saturation pressure p_{sat} and the absolute temperature quite accurately reproduces the observed data and once again, it is related to the Boltzmann equation (9.4).

9.5. RELATIVE HUMIDITY AND MUTUAL DISTANCE BETWEEN H_2O MOLECULES

In the vapour phase, a balance is established between kinetic and potential energy of H_2O molecules. Far from saturation, kinetic energy dominates, molecules are rarefied or have a high thermal level, or both. Under such circumstances, intermolecular forces are very weak and the vapour behaves like a gas. In the case of close molecular distance or low temperature, the potential establishes mutual interactions that weakly involve the molecular population as a whole and some correlation between them is established. Intermolecular forces strengthen with the degree of saturation and have an inverse dependence on the thermal level.

The potential energy of intermolecular interaction is a complicated function given by the Boltzmann distribution, related to temperature. At large intermolecular distances (IDs), forces are weak. At high temperatures, violent thermal motions disturb or overcome the mutual orientating effects of attractive forces. At low temperatures and short IDs, interactions are stronger. Near saturation, when IDs are shortest and/or kinetic energies are lowest, intermolecular forces are strongest and the molecular population as a whole is characterized by a certain internal organization until some molecules will undergo a phase transition.

At any temperature, the absolute humidity (AH) gives the number of H_2O molecules per unit volume and the inverse gives the volume

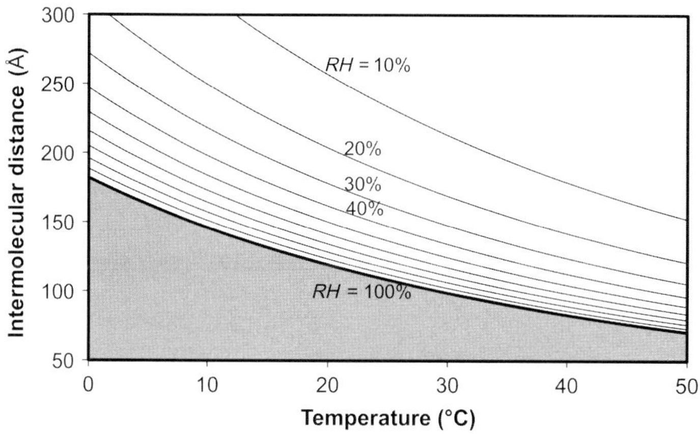

FIGURE 9.2 Intermolecular distance of water molecules at saturation (thick line) and at selected *RH* levels, i.e. *RH* = 10%, 20%, 30%, ... 1 Å = 10^{-8} cm. The grey area is not allowed for a stable equilibrium; the liquid and the gaseous phases may only coexist in metastable equilibrium. It ranges between *ID* = 60 molecular diameters at 0 °C to *ID* = 23 molecular diameters at 50 °C. *Source: After Camuffo, 2010; by kind permission of Nardini Editore, see credits.*

available per each molecule, from which the average *ID* may be calculated. At saturation, the shortest *ID* is reached, i.e. the threshold for change of phase to occur. Similarly, one obtains the *ID* per each relative humidity (*RH*) level, as summarized in Fig. 9.2

The closer the H_2O molecules, the stronger the attractive forces. Molecules cannot approach towards each other below a certain threshold, which is determined by the balance between attractive forces (i.e. the potential) that tend to establish some order, and kinetic energy (i.e. the temperature) that tends to generate chaos and destabilize order. *RH has an inverse dependence on the third power of IDs and reflects, per each temperature, to which extent the vapour is approaching the critical ID for phase transition* (Camuffo, 2010). This explains why *high RH levels favour H_2O molecular layering on surfaces and chemical reactions, e.g. oxidation, corrosion or favourable biological habitat.*

9.6. THE LIQUID STATE AND THE FREE H_2O MOLECULES IN IT

The exact structure of liquid water is still unknown. It is believed to contain many transient hydrogen-bonded networks. Most likely, water molecules are in rapid motion, constantly making and breaking hydrogen bonds (H bonds) with adjacent molecules. In the liquid state, the molecules have a reduced distance compared with the gaseous state and the forces exerted by the water molecules are responsible for a number of interactions including dynamic associations of molecular swarms. Frank and Wen (1957) developed an innovative *'flickering cluster'* model for the structure of liquid water in which small, tetrahedrally coordinated, icelike regions held together by H bonds exist for short periods (10^{-10} to 10^{-11} s), continually breaking up and re-forming. According to this dynamic model, liquid water is composed of a mixture of *'coherent'* and *'incoherent'* water (Fig. 9.3(a)).

Coherent water is composed of organized clusters, i.e. molecular aggregates composed of a variable number of molecules temporarily bound among them by intermolecular forces. *Incoherent water* is composed of single H_2O molecules, individually free. The adjectives *'coherent'* and *'incoherent'* mean that there is, or there is not, any connection, common behaviour or coordination among molecules. This was a good start for quantum chemistry. Proceeding from the flickering cluster model and using statistical thermodynamic methods, Nemethy and Scheraga (1962) developed an early model for liquid water where H-bonded icelike clusters of H_2O molecules are in equilibrium with a swarm of non-H-bonded liquid. The model was quantitatively described in terms of the molecular species participating in different numbers of H bonds in clusters. Thereafter, several models have been developed to describe the role of H bonds in forming various types of clusters ranging from small ring and cage structures up to gigantic icosahedral networks (Fig. 9.3(b)) (Ludwig, 2001; Maheshwary et al., 2001; Burnham and Xantheas, 2002a,b;

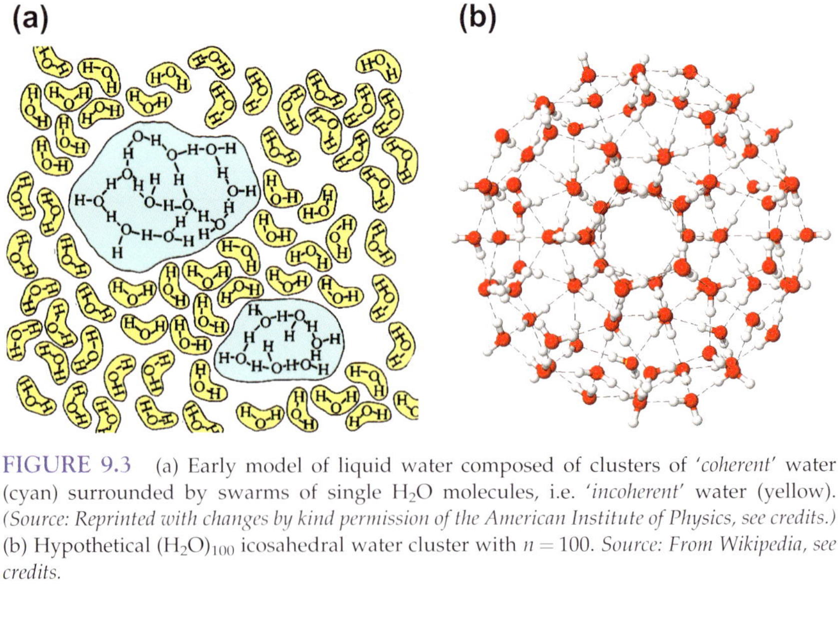

FIGURE 9.3 (a) Early model of liquid water composed of clusters of *'coherent'* water (cyan) surrounded by swarms of single H_2O molecules, i.e. *'incoherent'* water (yellow). *(Source: Reprinted with changes by kind permission of the American Institute of Physics, see credits.)* (b) Hypothetical $(H_2O)_{100}$ icosahedral water cluster with $n = 100$. *Source: From Wikipedia, see credits.*

Plumridge and Waigh, 2002; Fanourgakis et al., 2005; Smith et al., 2005; Shields and Kirschner, 2008; Quian et al., 2009; Dhir, 2011).

Mechanisms and models have been refined in the course of years but were based on the above basic ideas. Briefly, the liquid state is composed of a dynamic mixture of short-lived $(H_2O)_n$ clusters composed of a number, $n = 1, 2, 3, 4, \ldots$, of molecules, i.e. '*single*' ($n = 1$, i.e. *incoherent water*) or '*in aggregation*' ($n = 2, 3, 4, \ldots$, i.e. *coherent water*). The life of water clusters is short, being limited by unstable H bonding or by crashes due to the ceaseless thermal motion, especially of fast single H_2O molecules. The average size of clusters decreases with increasing temperature because some heat is used to separate some water molecules from each other or from the cluster (Lampert and Sommer, 2007; Pinet, 2009). In addition, increasing temperature means to increase the density of population of the swarm of freely moving incoherent water molecules and the frequency and the energy of collisions. The increased frequency and violence of molecular bombing that will follow any increase in temperture will oppose the formation of large aggregates.

At high *RH* levels, i.e. $RH > 80\%$, some coordination was observed for the vapour phase too, including the presence of small molecular clusters (e.g. water dimers or trimers, i.e. two or three water molecules loosely bound by a H bond.) (Poole and Owens, 2003; Pruppacher and Klett, 2003; Jensen, 2010). However, despite advanced modelling and sophisticated laboratory observations including Raman spectroscopy, molecular clustering in the bulk liquid and in the vapour phase remains unknown and is considered a still unsolved problem.

For the sake of simplicity, in this section we will follow a thermodynamic approach only assuming that liquid water is composed of a dynamic mixture of water molecules, some of them single and freely moving and most of them forming aggregates with various sizes. The whole molecular population will reach equilibrium with the kinetic energy distributed according to the Maxwell–Boltzmann law. As in the kinetic energy, the speed is inversely proportional to the square root of the mass; aggregates composed of a large number of molecules will have their speed slowed down. When large aggregates will impinge on a surface or another molecule different from water, the outer electronic shells will have a soft impact, i.e. no electron exchanges, and no chemical reactions will occur. As opposed to this, the single H_2O molecules will likely reach the highest velocities and when they impact on clusters, they will crash them, conferring further dynamism and instability to the whole system. When these fast molecules will violently impact on a surface or another molecule, the electronic interaction in the outer shells will be strong and chemical reactions are likely to occur. This explains what we have seen with the Arrhenius equation and the activation energy.

The swarm composed of single molecules, independent of each other, freely moving among clusters or impacting on them, can be considered an internal gas enclosed within the volume occupied by the bulk liquid. Their number will increase with the liquid temperature because they are highly energetic and require energy to be produced. These free molecules, being the most fast and energetic ones, have the highest probability of escaping from the liquid. Should this occur, they require some energy, i.e. the latent heat for evaporation, to cross the potential at the interface (i.e. the surface tension layer) and to re-establish the dynamic equilibrium with the internal clusters, which should break some H bonds to release other free molecules in substitution of the lost ones. They play a key role, e.g. making possible chemical reactions, making possible evaporation, crashing clusters and conferring internal mobility to water as well as avoiding liquid water to pass to the solid phase, as we will see later.

Within the volume occupied by the bulk liquid, the swarm of freely moving molecules exerts some pressure, as it was a real gas, or a vapour, enclosed within the same volume. On the other hand, outside the liquid, the molecules of vapour dispersed in the air also exert their partial pressure. At the interface between the air and the liquid, i.e. the free water surface, a pressure gradient will be established between the partial pressure of the gaseous water molecules dispersed in the air and those inside the bulk liquid. Should the internal vapour pressure be dominant, a flow of vapour molecules would be transferred from the bulk water to the atmosphere (i.e. evaporation), and the reverse occurs if the external water vapour pressure will exceed the internal pressure (i.e. condensation). Net flow is governed by the partial pressure gradient across the interface. At saturation, the net flow stops, the two opposed flows reaching equilibrium. As a consequence, the pressure gradient will also vanish and the two pressures are exactly the same, i.e. the saturation pressure, which is the maximum reachable vapour pressure at that temperature, either within the air or in the bulk liquid water. Therefore, the pressure exerted within the bulk liquid by the internal vapour equals the saturation pressure p_w of the water vapour measured in air.

9.7. THE RAOULT LAW FOR IDEAL SOLUTIONS

A related item is the Raoult law (Raoult, 1887) for ideal solutions. It establishes that the vapour pressure of an ideal solution directly depends on the vapour pressure of each chemical component and the mole fraction of the components present in the solution. In particular, the saturated water pressure over a solution is always lower than the pressure of pure solvent and it decreases proportionally to the mole fraction of the solute. This can be interpreted considering that the solute combines with part of the incoherent water, reducing its internal pressure and, consequently, the saturation pressure (in air) in equilibrium with the solution.

One may arrive to the above conclusions from other considerations. Ebullition occurs when the partial pressure of the water vapour equals the atmospheric pressure. At any temperature, the partial pressure of the internal vapour equals the saturation pressure. On increasing the liquid temperature, the internal vapour pressure too will increase. When the pressure of the internal vapour (that equals the saturation pressure) will reach the external atmospheric pressure, the liquid is not anymore compressed by the external pressure. The liquid becomes unstable and starts boiling to pass from the liquid to the gaseous phase.

9.8. EBULLITION AND FREEZING

The liquid state is in some way an intermediate phase between the ordered solid state, characterized by a potential energy largely dominant over the kinetic energy of the molecules, and the fully disordered gaseous state, where the molecules are linked with a very weak or even negligible potential energy compared to the very high kinetic energy. The liquid state is possible only in the presence of an external pressure that, acting in synergism with the attractive forces of the water molecules, keeps the bulk fluid constraint within a certain volume. Should the external pressure decrease, the fluid will start to boil for the excess of internal energy and the H_2O molecules will pass from the liquid to the gaseous phase. The other way to obtain ebullition is to increase the liquid temperature (and the internal kinetic energy as well) until it reaches a critical level. These two ways are apparently distinct; however, they have the common feature that ebullition occurs when the (internal) saturation vapour pressure equals the (external) atmospheric pressure, irrespective of whether the saturation pressure has been increased by increasing the liquid temperature or by decreasing the external pressure.

If one rejects the idea that the bulk liquid includes an internal vapour and assumes that the water vapour should only be considered in the air, externally to the liquid, the ebullition can hardly be explained. How might an external partial pressure influence the state of liquid water kept in a vessel, when the external partial pressure equals the air pressure? In the ebullition, vapour bubbles are directly created from the liquid itself and not from some existing air bubbles. This is because the activation energy for the nucleation of vapour bubbles (Terao et al., 1964) is twice as great as that for evaporation on a free surface, whilst the energy for the development of the vapour bubbles is equal to the heat of evaporation. What physical mechanism driven by the external pressure might generate macroscopic quantities of water vapour inside the bulk liquid?

The explanation is simple under the hypothesis of free vapour molecules dispersed inside the liquid. At boiling point, the internal and external pressures are the same and the liquid is no longer compressed and controlled from the exterior by an overwhelming atmospheric pressure. The whole of the liquid becomes unstable and tends to be transformed into vapour and be transferred to the air.

The internal vapour is essential to keep water in the liquid state because it controls the number of clusters and prevents further aggregation between them, i.e. the passage to the solid state. Per unit water volume, the number of molecules constituting the internal vapour equals the absolute humidity at saturation (AH_{sat}) (Fig.2A.4). When the population density drops below a certain threshold, the freely moving molecules are too few to crash enough clusters and confer internal mobility to water. Below this threshold, the efficiency of the attractive forces between H_2O molecules cannot be longer counteracted and the liquid passes to the solid state. The freezing point occurs when the swarm population density falls below 4.95 g m^{-3} or $4.95/N = 8.22 \times 10^{24}$ molecules per cubic metre.

9.9. AN ADDITIONAL ASPECT OF RELATIVE HUMIDITY

After the above conclusions, we can interpret in a different way the definition of RH taking into consideration the water vapour dispersed in both the gaseous (p_w) and the liquid (p_{wl}) phases, i.e.

$$RH = 100\frac{p_w}{p_{wl}} \tag{9.17}$$

i.e. RH is the ratio between the *'external'* partial pressure of the water vapour dispersed in the air (i.e. p_w) and the *'internal'* pressure of the vapour dispersed inside the liquid water (i.e. $p_{wl} = p_{sat}$) (Camuffo, 2010).

Briefly, RH indicates how strong the tendency of water (or how intense is the net flow of molecules) is to pass from the liquid to the gaseous phase (when $RH < 100\%$), to stay in equilibrium between the two phases (when $RH = 100\%$) or to pass from the gaseous to the liquid phase, i.e. when the liquid or the surface temperature is below the dew point ($RH > 100\%$ at the interface).

9.10. THE THREE CLASSES OF WATER VAPOUR

In the Maxwell–Boltzmann distribution, the molecules of water vapour are distributed by energies and for explanatory purposes, we will subdivide the whole population into three classes characterized by low,

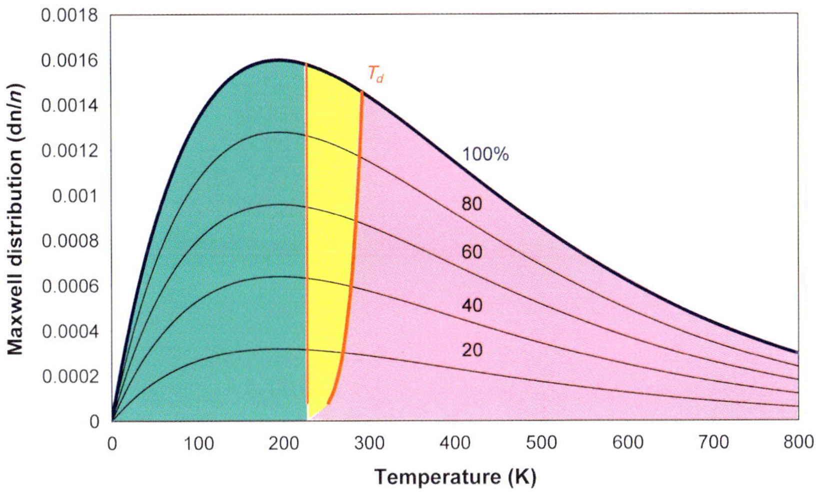

FIGURE 9.4 The three classes of water vapour, i.e. *'frozen type'* (cyan), *'wet type'* (yellow) and *'dry type'* (magenta). T_d is the dew point temperature. $T_{amb} = 20\ °C$ and $p = 1013$ hPa. *Source: After Camuffo, 2010; by kind permission of Nardini Editore, see credits.*

average and high kinetic energy (Fig. 9.4). The transition between classes is continuous and it is impossible to make a net separation; however, we are forced to make choices about borders to facilitate discussions.

In the following discussion, we will name these classes making reference to the three phases of water, i.e. frozen (low kinetic energy), liquid (average kinetic energy) and gaseous (high kinetic energy) (Camuffo and Pagan, 2006; Camuffo, 2010). The names do not reflect reality but are inspired from what would happen to a molecule impinging on a surface having the same thermal level as itself. All calculations have been made for ambient temperature $T_{amb} = 20\ °C$ and standard atmospheric pressure $p = 1013$ hPa.

Class one: the 'frozen type' (FT) covers the interval 0 K −228 K where water may exist only in the ice phase (cyan in Fig. 9.4). The upper limit has been chosen at 228 K because this is the lowest temperature to which the properties of liquid water can be extrapolated (Pruppacher and Klett, 2003), the limit of supercooled water in cloud droplets (Mason, 1971) and in porous materials (Fagerlund, 1973).

FT molecules have the lowest kinetic energy and the lowest interaction potential. They may potentially pass to the ice phase if they hit any surface at the same temperature. They will hardly interact with materials (Warnatz et al., 2001) for four reasons.

1. They have a small kinetic energy and can hardly cross the viscous sublayer around bodies that tends to slow down velocities (Kundu, 1990). Therefore, they are unlikely to reach the surface of bodies.

2. They have very low kinetic and rotational energy and generate weak electromagnetic forces; the same holds for the interactions between each other and with materials.
3. They have a small kinetic energy and the chemical reaction rate k given by the Arrhenius equation depends on energy (i.e. temperature). Chemical reactions are hardly probable to occur because the molecular energy falls below the activation threshold.
4. The molecular adsorption rate k_{ads} on surface decreases with decreasing temperature:

$$k_{ads} = S_{stk}\sqrt{\frac{\Re T}{2\pi M}} \tag{9.18}$$

where S_{stk} is the sticking coefficient and M is the mass of the sticking molecule.

FT covers 49% of the whole molecular population at standard ambient temperature and pressure, but is composed of hardly reactive molecules, which pass substantially unobserved. Of course, the actual number of molecules per cubic metre is obtained by multiplying their normalized number by AH_{sat}, and then by RH.

Class two: the 'wet type' (WT) (228 K $< T < T_d$) lies in the central part of the distribution (yellow in Fig. 9.4). The upper limit of this class has been established at the dew point T_d that depends on the amount of vapour mixed with air and constitutes the threshold for condensation.

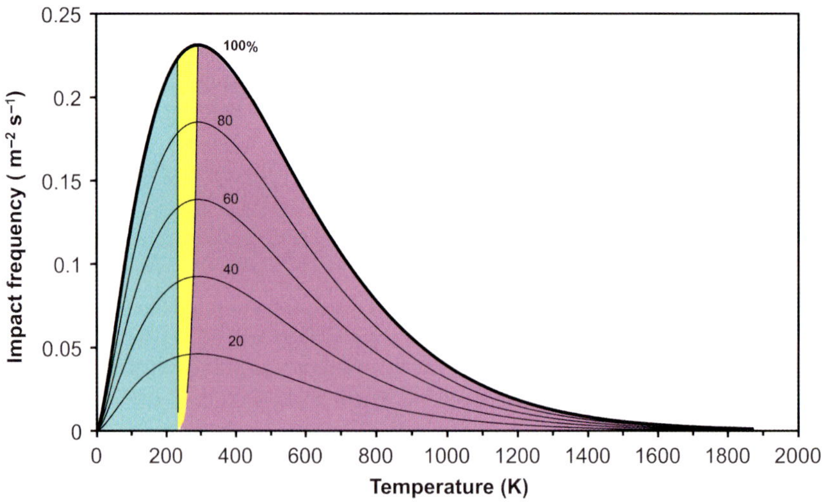

FIGURE 9.5 Impact frequency distribution of H_2O molecules impinging on a surface per unit of surface area (1 m^2) and per unit time (1 s) at selected RH levels: 20%, 40%, 60%, 80%, 100%. *'Frozen type'* (cyan), *'wet type'* (yellow) and *'dry type'* (magenta). Normalized vapour density ($N/V = 1$); $T_{amb} = 20\,°C$ and $p = 1013$ hPa.

Class WT is composed of potentially liquid, metastable molecules. When they come into contact with a surface at the same temperature, they change to the liquid phase. For this reason, and because in this class the highest frequency of molecular collision occurs on surfaces (Fig. 9.5), WT is the most relevant class for interactions with materials.

WT covers 12% of the whole molecular population at standard ambient temperature and pressure and is composed of 'active' molecules, able to reach a surface, stick to it and take part in physical mechanisms, chemical reactions or biological life.

Class three: the 'dry type' (DT) ($T > T_d$), is composed of the 'hot', fastest molecules (magenta in Fig. 9.4). DT molecules are highly energetic and Fig. 9.6 shows that they are largely responsible for most of the energy that is transferred to a surface. The same figure shows that the energy flux of the FT class is very modest, followed by that of the WT class. When DT molecules impinge on a surface, hits are violent, exceeding the adhesion force, and they bounce off. If their kinetic energy is enough, they may remove some water molecules that were sticking on the surface. This justifies why the thickness of monolayers of absorbed water on a metal surface does not increase over time, because an equilibrium is reached between the number of water molecules that will adhere on the surface and those that will be removed by violent molecular impacts.

Class DT covers 39% of the whole population at standard ambient temperature and pressure, and these molecules behave like a nonreactive gas because they have a very low probability of being captured by a surface and reacting with materials or bio-organisms.

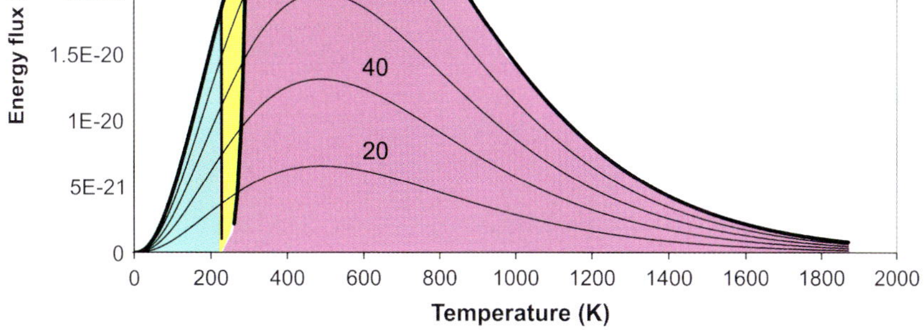

FIGURE 9.6 Total energy flux transferred by H_2O molecules impinging on a surface per unit of surface area (1 m^2) and per unit time (1 s) at selected *RH* levels: 20%, 40%, 60%, 80%, 100%. *'Frozen type'* (cyan), *'wet type'* (yellow) and *'dry type'* (magenta). Normalized vapour density ($N/V = 1$); $T_{amb} = 20\,°C$ and $p = 1013$ hPa.

9.11. CONCLUSIONS

The comparison between the equations concerning apparently different phenomena, e.g. molecular velocities, thermal emission, chemical kinetics, etc., is appropriate because all of them are directly or indirectly related to the same probability distribution of the energetic levels within a thermodynamic system. This explains the parallel between the Arrhenius equation for chemical kinetics and the thermal emission of bodies, in particular the Wien law.

A further advancement has been discussed: the quantum chemistry hypothesis that liquid water is composed of a mixture of molecular aggregates and a swarm of gas-like molecules constituting an internal vapour pressure is supported by statistical thermodynamics and the Maxwell–Boltzmann distribution. The internal vapour composed of isolated molecules (incoherent water) freely moving within the bulk liquid water structure plays a key role in controlling the liquid state because these molecules are fast and concentrate their kinetic energy in a small cross-section in comparison with large molecular clusters (coherent water). This water vapour dispersed into the bulk liquid exerts an internal pressure equal to the saturation pressure and justifies a number of physical chemistry phenomena (e.g. evaporation, condensation, boiling, freezing and depression of the saturation pressure for dilute solutions), including the Arrhenius activation energy.

The net transfer of molecules between the liquid and the gaseous phases, i.e. condensation and evaporation, is governed by the gradient in vapour pressure at the gas–liquid interface. The saturation pressure within the liquid is governed by the same equation as in the atmosphere and is a function of temperature only. At equilibrium, the same saturation pressure is found both in the gaseous phase external to the liquid and in the vapour phase inside the liquid.

Physical laws governing the internal vapour are the same as for the external one, including the saturation pressure. Both of them, i.e. the internal and the external swarm of free water molecules, follow the Maxwell distribution by velocities and the Boltzmann distribution by energy levels.

References

Arrhenius, S., 1889. Über die Reaktionsgeschwindigkeit bei der Inversion von Rohrzucker durch Säuren (On the reaction velocity of the inversion of cane sugar by acids). Zeitschrift für Physikalische Chemie, Stöchiometrie und Verwandtschaftslehre 4, 226–248.

Boltzmann, L., 1868. Studien über das Gleichgewicht der lebendigen Kraft zwischen bewegten materiellen Punkten. Wiener Berichte 58, 517–560.

Burnham, C.J., Xantheas, S.S., 2002a. Development of transferable interaction models for water. III. Reparametrization of an all-atom polarizable rigid model (TTM2-R) from first principles. Journal of Chemical Physics 116, 1500–2159.

Burnham, C.J., Xantheas, S.S., 2002b. Development of transferable interaction models for water. IV. A flexible, all-atom polarizable potential (TTM2-F) based on geometry dependent charges derived from an ab initio monomer dipole moment surface. Journal of Chemical Physics 116, 5115–5124.

Camuffo, D., 2004. Thermodynamics for cultural heritage. In: Martini, M., Milazzo, M., Piacentini, M. (Eds.), Physics Methods in Archaeometry. International School of Physics 'Enrico Fermi', Varenna, pp. 37–98 (IOS Press, Amsterdam).

Camuffo, D., 2010. The role of temperature and moisture. In: Camuffo, D., Fassina, V., Havermans, J. (Eds.), Basic Environmental Mechanisms Affecting Cultural Heritage – Understanding Deterioration Mechanisms for Conservation Purposes. COST Action D42 "Enviart". Nardini, Florence, pp. 9–30.

Camuffo, D., Pagan, E., 2006. What is behind Relative Humidity? Why it is so a relevant variable in the conservation of Cultural Heritage? In: Lefèvre, R.A. (Ed.), The Materials of Cultural Heritage in their Environment. European University Centre for Cultural Heritage, Ravello, Edipuglia, Bari, pp. 21–38.

Clapeyron, M.C., 1834. Mémoire sur la puissance motrice de la chaleur. Journal de l'École Polytechnique 23, 153–190.

Connors, K.A., 1990. Chemical Kinetics: The Study of Reaction Rates in Solution. VCH Publisher, New York, Weinheim and Cambridge.

Dhir, V., 2011. Beauty of Water: Continuum Model, Mixture Models, Flickering Cluster Model & Frank – Wen Flickering Cluster Model of Liquid Water. LAP Lambert Academic Publishing. www.lap-publishing.com.

Einstein, A., 1909. On the present status of the radiation problem. Physikalische Zeitschrift 10, 185–193.

Einstein, A., 1917. On the quantum theory of radiation. Physikalische Zeitschrift 18, 167–183.

Fagerlund, G., 1973. Determinations of pore-size distribution from freezing point depression. Materiaux et Constructions 6, 215–225.

Fanourgakis, G.S., Aprá, E., Dejong, W., Xantheas, S.S., 2005. High-level ab initio calculations for the four low-lying families of minima of $(H_2O)_{20}$. II. Spectroscopic signatures of the dodecahedron, fused cubes, face-sharing pentagonal prisms, and edge-sharing pentagonal prisms hydrogen bonding networks. Journal of Chemical Physics 122 (13), 134304.

Fermi, E., 1937. Thermodynamics. Prentice-Hall, Upper Saddle River, New Jersey, 160 pp.

Frank, H.S., Wen, W.Y., 1957. Ion-solvent interaction. Structural aspects of ion-solvent interaction in aqueous solutions: a suggested picture of water structure. Discussions of the Faraday Society 24, 133–140.

Jensen, J.H., 2010. Molecular Modelling Basics. CRC Press, Taylor and Francis, London, New York, Boca Raton (Fla), 189 pp.

Kohnstamm, P., Scheffer, F.E.C., 1911. Thermodynamic potential and velocities of reactions. Proceedings of the Koninklijke Nederlandse Akademie van Wetenschappen. Amsterdam 13, 749–800.

Kundu, P.K., 1990. Fluid Mechanics. Academic Press, San Diego, Ca, 368 pp.

Laidler, K.J., King, M.C., 1983. The development of transition-state theory. Journal Physical Chemistry 87, 2657–2664.

Lampert, W., Sommer, U., 2007. Limnoecology. Oxford University Press, Oxford.

Langmuir, I., 1920. Radiation as a factor in chemical action. Journal of American Chemical Society 42 (11), 2190–2205.

Lewis, G.N., Randall, M., 1921. Activity coefficients of strong electrolytes. Journal of American Chemical Society 43, 1112–1154.

Ludwig, R., 2001. Water: from clusters to the bulk. Angewandte Chemie International Edition 40 (10), 1808–1827.

Magnus, G., 1844. Über die Ausdehnung der Gase durch die Warme (On the expansive force of Steam). Poggendorff Annalen der Physik und Chemie LXI., 225–247. translated from German by W. Francis, pp. 218–239 in R. Taylor (Ed.), 1846. Scientific Memoirs Selected from the Transactions of Foreign Academies of Science and Learned Societies and from Foreign Journals, vol. IV. R. & J. Taylor, London.

Maheshwary, S., Patel, N., Sathyamurthy, N., Kulkarni, A.D., Gadre, S.R., 2001. Structure and stability of water clusters $(H_2O)n$, n = 8-20: an ab initio investigation. Journal Physical Chemistry 105 (46), 10525–10537.

Mason, B.J., 1971. The Physics of Clouds. Clarendon Press, Oxford, 671 pp.

Maxwell, J.C., 1868. On the dynamical theory of gases. Philosophical Magazine Ser. 4 35, pp. 129–145, 185–217.

Nemethy, A., Scheraga, H.A., 1962. Structure of water and hydrophobic bonding in proteins. A model for the thermodynamic proprieties of the liquid water. Journal of Chemical Physics 36, 3382–3400.

Pinet, P., 2009. Invitation to Oceanography. Jones and Barlett, London.

Planck, M., 1901. Über das Gesetz der Energieverteilung im Normalspektrum (On the Law of the Energy Distribution in the Normal Spectrum, Translated from German by Kuyanov). Annalen der Physik. Leipzig 4, 553–564.

Planck, M., 1913. Waermestahlung (The Theory of Heat Radiation, Translated from German by M. Masius). Blakiston's Son & Co, Philadelphia.

Planck, M., 1926. Treatise on Thermodynamics. Dover, New York.

Plumridge, T.H., Waigh, R.D., 2002. Water structure theory and some implications for drug design. Journal of Pharmacy and Pharmacology 54, 1155–1179.

Poole, C.P., Owens, F.J., 2003. Introduction to Nanotechnology. J. Wiley and Son, Hoboken, New Jersey.

Pruppacher, H.R., Klett, J., 2003. Microphysics of Clouds and Precipitation. Kluwer Academic Publishers, Dordrecht.

Qian, P., Lu, L., Song, W., Yang, Z., 2009. Study of water clusters in the n = 2–34 size regime, based on the ABEEM/MM model. Theoretical Chemistry Accounts 123 (5–6), 487–500.

Raoult, F.M., 1887. Loi générale des tensions des vapeurs des dissolvants. Comptes Rendus 104, 1430–1433.

Shields, G.C., Kirschner, K.N., 2008. The limitations of certain density functionals in modeling neutral water clusters. Synthesis and Reactivity in Inorganic, Metal-organic, and Nano-metal Chemistry 38 (1), 32–39.

Smith, J.D., Cappa, C.D., Wilson, K.R., Cohen, R.C., Geissler, P.L., Saykally, R.J., 2005. Unified description of temperature-dependent hydrogen-bond rearrangements in liquid water. USA Academy of Sciences, PNAS 102 (40), 14171–14174.

Sonntag, D., 1990. Important new values of the physical constants of 1986, vapour pressure formulations based on the ITS-90, and psychrometer formulae. Meteorologische Zeitschrift 40, 340–344.

Terao, K., Hatano, F., Sano, F., 1964. Kinetische Untersuchung über das Sieden der Flußigkeit. Japanese Journal Applied Physics 3, 688–732.

Tetens, O., 1930. Uber einige meteorologische Begriffe. Zeitschrift Geophysik 6, 297–309.

Vollmer, M., Möllman, K.P., 2011. Infrared Thermal Imaging: Fundamentals, Research and Applications. Wiley-VCH, New York.

Warnatz, J., Maas, U., Dibble, R.W. (Eds.), 2001. Combustion. Springer, Berlin, p. 299.

Wien, W., 1886. Über die Energieverteilung in Emissionspektrum eines schwarzen Körpers. Annalen der Physik 58, 662–669.

Wien, W., 1898. Über die Fragen, welche die translatorische Bewegung des Lichtäthers betreffen. Annalen der Physik 301 (3), I–XVIII.

PART II

PERFORMING MICROCLIMATE FIELD SURVEYS

CHAPTER

10

Introduction to Field Measurements

OUTLINE

10.1. WEATHER STATIONS AND OBSERVATIONS FOR MONUMENT CONSERVATION

Meteorological measurements are routinely made to describe atmospheric conditions for weather forecasting, airport assistance, agriculture, climate analysis and so on. Every station is considered as part of a global network that covers the whole earth. As all the measurements must be comparable, the type and class of instruments, exposure, site, observation time, methodology and procedure as well as data logs are regulated by a precise normative. This is given by the World Meteorological Organization (WMO) in *The Guide to Instrument and Methods of Observation* (1983;

Microclimate for Cultural Heritage, Second Edition
http://dx.doi.org/10.1016/B978-0-444-63296-8.00011-1

2006) and *Compendium of Lecture Notes on Meteorological Instruments* (1986). Other useful details or suggestions are given by National Meteorological Services (e.g. UK Meteorological Office, 1981).

In the case of specific microclimate studies, every measurement is finalized to monitor one or more microclimate variables in view of increasing knowledge about a specific problem. The purpose of the investigation is the key factor in determining the choice of instruments and operational procedures. Only when survey objectives and the accuracy needed for records have been established, is it possible to decide the type of sensors, the number and location of sampling points, the sampling frequency and duration, the appropriate period for measurements and all other details. Advanced research cannot be limited by standards or normative, because by definition, research is cutting the edge outside the borders of normality. As opposed to this, routine measurements and specialized investigation services take benefit from previous studies and apply methods and technologies that have been proved to be satisfactory. In particular, standards provide an authoritative reference and a guidance to obtain sound results. The literature provides a large number of case studies that might be considered useful examples of methodologies to be considered or rejected. For any measurement there is always a margin of doubt. However, the uncertainty should be kept within acceptable limits and errors should be totally avoided. In the following chapters, we will share the experience derived from research and long-term practice in this field to assist conservators, restorers, architects and other professionals in the difficult job of performing microclimate investigations finalized to the conservation of cultural heritage.

Making reliable measurements is not simple: the operator's capability is not less relevant than the instrument class. In fact, the most common errors are due partly to the instrument (i.e. insufficient quality of the sensor, instrument calibration, scale graduation, signal conversion, data processing and recording) and partly to the operator (i.e. influencing the variable with the operator's presence; using inappropriate operative methodology, exposing the instrument in an inappropriate way, and incomplete shielding against external disturbances; choosing measuring points scarcely representative of the site under study; choosing sampling times inappropriate to describe the natural evolution of the variable; and choosing inappropriate periods for surveys). It is evident that, when a good class instrument has been purchased, all errors should be referred to the operator.

Not always may observations needed for conservation be found in standard weather stations or aerodrome observatories, because the topography near the station may be different from the monument site. Not all the required variables are measured; the sampling intervals, the accuracy, the data analysis and how they are reported may be inappropriate for this aim. The most frequently required observations for microclimate purposes are as follows:

Surface and *air temperatures*, their *space gradients* (both on the horizontal plane and on the vertical) and *time rates*. Particularly important are gradients close to the artefact surface. Frequency distributions or average values, daily and seasonal spans and short-term variability are also important.

Dew point (*DP*) and *dew point spread* (*DPS*); *relative humidity* (*RH*) and *mixing ratio* (*MR*) (or specific humidity (*SH*)): as discussed for air temperature.

Time of wetness: distribution on the monument surface and daily and seasonal values for some points with different exposure.

Atmospheric stability: vertical gradient of the air temperature and measurement of air turbulence.

Direct (and possibly *diffuse*) *solar radiation* falling on the monument surface at different exposures and during the yearly cycle. Computer simulation is precious to improve the knowledge of the problem.

(*Micro*) *ventilation* at several points near the artefact surface: average values and fluctuations for turbulence analysis. Direct observations at the site are extremely important, as the wind flow pattern is strongly perturbed in the presence of obstacles and in the wake behind them the return flow has the opposite direction, so that records taken with standard weather stations may be completely different. However, weather data of *wind speed* and *wind direction* (dominant direction and variability) are also useful for a better interpretation of the field observations in a wider meteorological context. Another important characteristic of wind is the *gustiness*, i.e. the rapid changes of intensity about a mean level, called *gusts* (peaks) and *negative gusts* or *lulls* (valleys). Frequency and strength of gusts should be noted, when available. They constitute a basic indication of the eddy turbulence.

Precipitation frequency, duration, intensity and amount and length of dry intervals from local measurements or standard weather stations. Note that tall buildings may substantially affect the precipitation in the area nearby for two reasons: the aerodynamic perturbation to the wind field and the electric field that buildings may generate. The time interval between one precipitation and the next is another important variable, in order to know whether the monument has been dried forming a wetting–drying cycle, as well as the time available for dry deposition.

Hydrometeors roses, i.e. *direction and frequency of wind-borne droplets or ice*, either generated by precipitation or suspended in the air. The roses are obtained reporting on the compass, per each direction, how many times each type of hydrometeor was associated with wind blowing from that specific direction. These roses are very useful not only to clarify the climatic features of a region but also to put into evidence which side of a monument is exposed to frequent wetting, washout and so on. The concerned hydrometeors are: *drizzle* (very light liquid precipitation composed of small droplets, with diameter between 0.2 and 0.5 mm, which fall slowly to the ground), *rainfall* (falling drops greater than 0.5 mm in diameter), *shower* (a sudden and short-duration, violent liquid precipitation from convection clouds; drops are usually very large, up to 6 mm in diameter), *hail* (solid precipitation in the form of balls or pieces of ice, called *hailstones*, with diameter ranging from 5 to 50 mm, falling from cumulonimbus clouds), *snow* (solid precipitation that occurs in large *snowflakes* at temperatures near 0 °C, or in a variety of minute *ice crystals*, when the temperature is well below 0 °C) or *fog* (which is not a precipitation, but is composed of fine droplets more or less suspended in the air, which reduces the visibility to less than 1 km). Also, the energy supplied by hail or raindrops, when they impact on the monument, is important. The Rome case study (Fig. 10.1) shows that drizzle, rain and hail are prevalently associated with the Sirocco wind blowing from southeast, i.e. along the sea coast; showers are generated by the same wind as well as the sea breeze that blows perpendicular to the coast (i.e. from southwest); snow arrives from every inland side and never from the side facing the Tyrrhenian Sea; fog has mainly a continental origin and descends to Rome following the valley of the River Tiber (north—northeast).

With the aim of illustrating the theory presented in the previous chapters, the experience of many field surveys will be herewith summarized. The conservation scientist needs to know which parameters are available from the WMO or National Meteorological Services, as well as purposes, characteristics and limits of these observations (for this reason, some essential notes have been here reported) before deciding to take new punctual observations, with different aims. The purpose is not to summarize a treatise on microclimatological methods (e.g. Munn, 1970) or experimental techniques (these can be found in several textbooks, i.e. the above-mentioned reference reports, and in addition in Wexler, 1965; Moses, 1968; Doebelin, 1990; Linacre, 1992; Lipták, 2003), but to show the differences between standard meteorological methods and research needs, to discuss the main problems and give a number of useful suggestions to the operators working in the field of conservation. For this purpose, only the measurements of temperature, humidity, air motions and precipitation, which are the most frequently needed observations for conservation, will be illustrated.

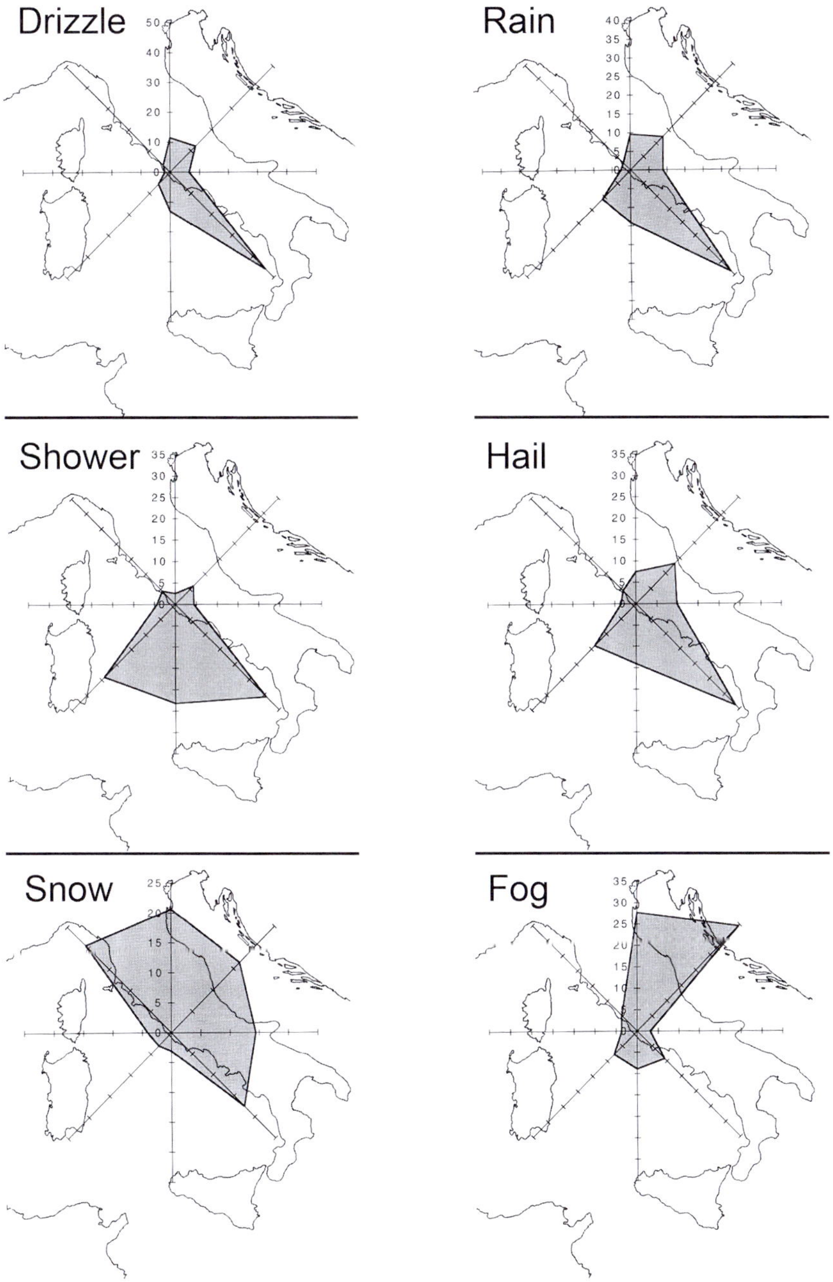

FIGURE 10.1 Precipitation roses (i.e. direction of the wind-borne drizzle, rainfall, shower, hail, snow and fog) at Rome. Drizzle, rain, shower and hail typically arrive with the Sirocco wind (from southeast); showers are mainly related to the sea breeze veering from southeast to southwest; snow with winds blowing from inland regions and fog to light breeze blowing along the Tiber River valley.

10.2. STATISTICAL REPRESENTATION OF THE DATA

Statistical representations in terms of averages are very popular but in recent times, the frequency of occurrence of specific phenomena has been preferred in climatology, as the climate of a site is better characterized by the repetition of some meteorological events. The occurrence of a rare extreme event does not change the local climate but affects the average values.

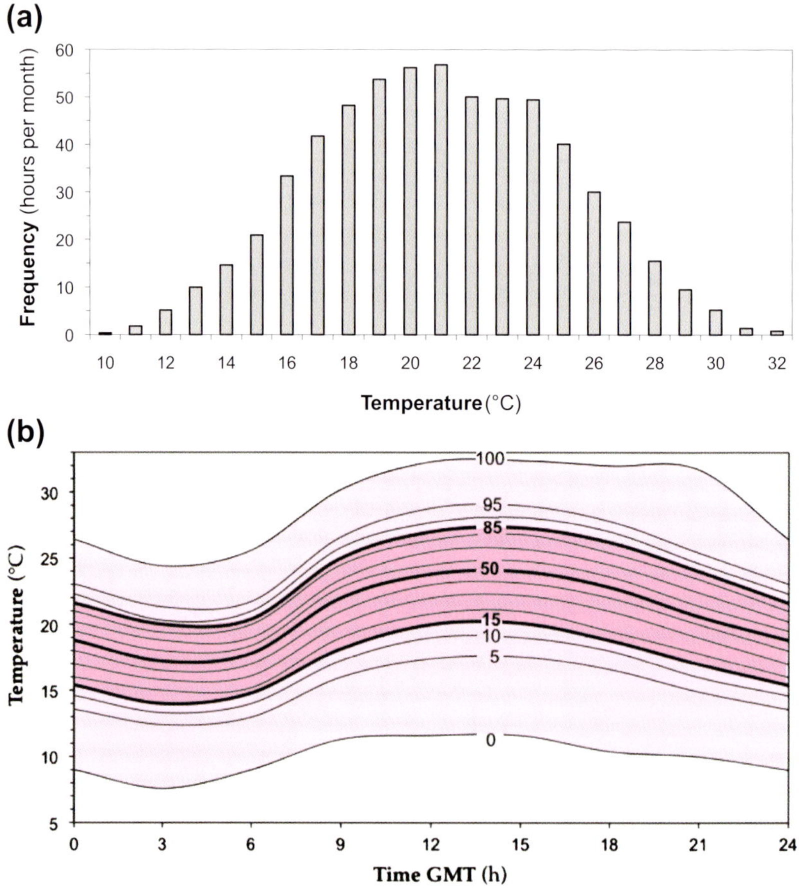

FIGURE 10.2 a,b) Hourly distribution of air temperature (in June, at Venice) in histogram and *percentile* form. Period: 1961–1994. In this example, the air temperature approaches a Gaussian distribution in a first approximation but not the relative humidity (*RH*) (Fig. 10.2(c), (d)). In the case of a Gaussian distribution, the 15th, 50th and 85th percentiles (thick lines) correspond respectively to $<y> - \sigma$, $<y>$ and $<y> + \sigma$. (c,d) As for Fig. 10.2(a), (b) but for *RH*. In the *RH* distribution, the dotted lines at 58.66%, 70.08%, 80.79%, 89.88% and 96.51% show the critical levels at which condensation occurs in micropores with radius $r \leq 0.002$ μm, 0.003 μm, 0.005 μm, 0.01 μm and 0.03 μm, respectively. The intercepts with the percentiles give the frequency of the condensation–evaporation cycles in the above-mentioned micropores.

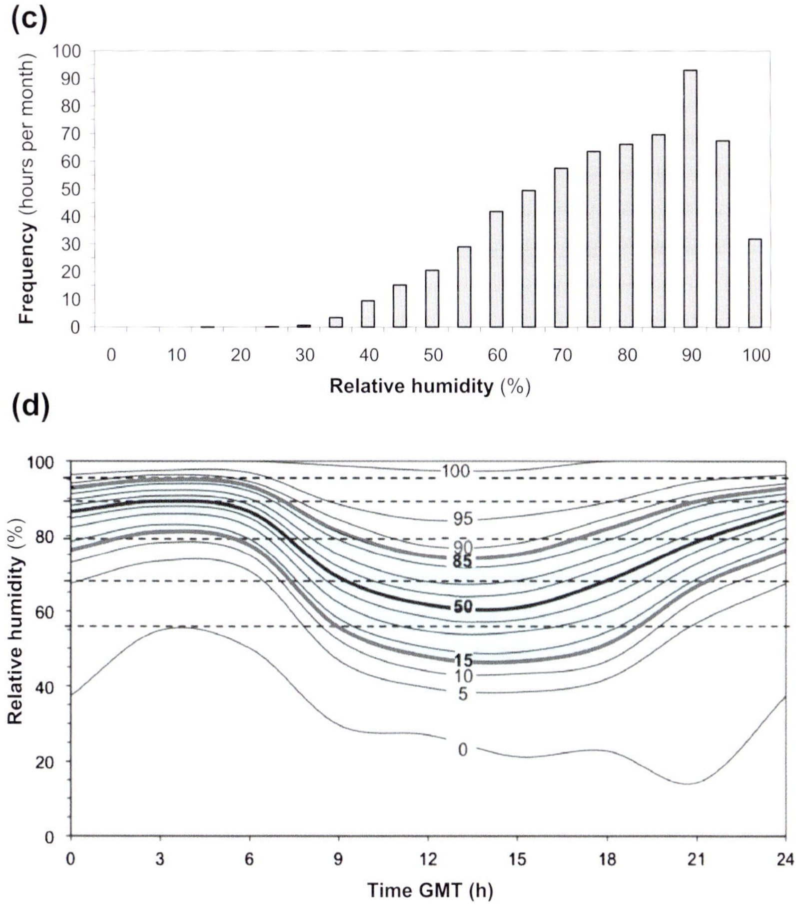

FIGURE 10.2 *Continued*

The consideration that local weather characteristics are more soundly based on the frequency of occurrence of some typical phenomena suggests disregarding anomalies and concentrating on the frequency distribution of the events. The well-known wind roses and the less familiar *hydrometeor roses*, i.e. *precipitation roses*, are two examples of frequency-oriented representation of only one (i.e. wind) or two (i.e. wind and precipitation) combined variables.

Percentiles have the further advantage of avoiding the problem of computing averages of nonlinear parameters, e.g. *RH* or pH, or of non-unimodal distributions, e.g. wind direction.

In terms of frequency, the presentation of data is usefully made by means of the *percentiles* (Fig. 10.2), (in a group of readings, the percentile of a given value is determined by the percentage of the values that are smaller than that value) and of particular interest are the 15th, 50th and 85th percentiles, which in the case of Gaussian distribution of the data y_i correspond, respectively, to $<y> - \sigma$, $<y>$ and $<y> + \sigma$. Also, the *mode*, which is the value that occurs most frequently and is representative of the so-called 'typical' climatic situation, is preferred to the *mean*, which is a mathematical abstraction, based on the actual values of the whole set of observations. Of course, in the case of unimodal symmetrical distributions, the 50th percentile, the mode and the mean are coincident. In the case of a bimodal distribution (e.g. spring and autumn precipitation and sea and land breeze), the 50th percentile and the mean are still coincident but represent anomalous events, whereas the two modes show the two typical situations.

For the above reasons, the most appropriate representation should be decided according to the specific aims of the research, e.g. in the case of *RH* at Venice, it is more useful to know the monthly values of the 50th percentile, which vary with the season, instead of the mode, which is 100% for nearly the whole year.

10.3. FREQUENCY OF OBSERVATION

The frequency of observation is important not only to arrive at a sound statistical description of the local microclimate but also to reconstruct from a limited set of data the trend and departures of a given parameter. The interval between two observations is chosen with the aim of recording the time history of the parameter with the minimum number of observations. It strictly depends on the type of the problem as well as the frequency of the oscillations of the variable and the level of smoothing that is required. Theory and problems related to sampling frequency and record duration are discussed in some useful textbooks (Lumley and Panofsky, 1964; Pasquill, 1962; Pasquill, 1974; Plate, 1982; Tennekes and Lumley, 1973; Vinnichenko et al., 1980).

If an atmospheric variable has periodic oscillations and one observation is made at sampling intervals slightly shorter or longer than the oscillation period, each sampling monitors a point of the next oscillation but in a slightly different phase lag and the plot of the observations apparently shows a new wave with a period and wavelength close to a multiple of the original oscillations (Fig. 10.3). This wave is generated by the many observations that have been necessary before reaching the same phase (or a value close to it) in the short-term oscillations. This phenomenon is named *aliasing* and, in order to elude it, Munn (1970) suggested avoiding measurements of instantaneous values at fixed intervals of time in the case of periodic oscillations in turbulence analysis.

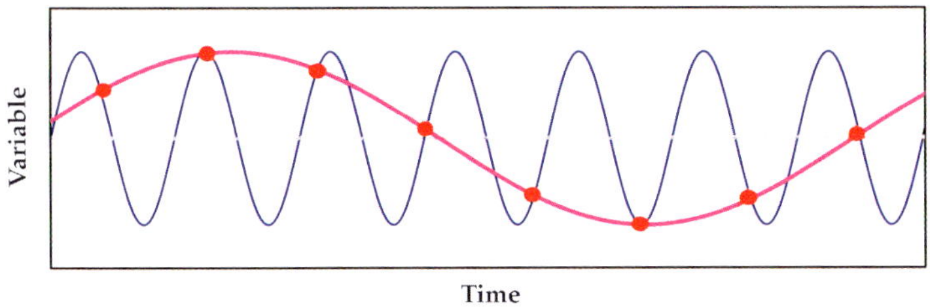

FIGURE 10.3 When a periodic variable (blue line) is observed at fixed intervals of time (red circles) close to the cycle period, the data analysis generates a new wave (pink line), with longer period and wider wavelength. This phenomenon is called *aliasing*.

As atmospheric variables continually fluctuate in a few seconds, it is fundamental to have clear ideas about what kind of information is needed, e.g. main trends or variability. From the former point of view, fluctuations constitute *noise;* from the latter, *signal*. The choice may give a very different result according to the aim: (1) if only the daily cycle of one or few parameters (e.g. the air temperature or wind speed and direction) is needed, all the short period fluctuations are smoothed out using sensors with a sufficiently long time of response, or making several sampling and numerically filtering the data before recording; in this case, a few tens of observations per day can be sufficient; (2) if the measurement is concerned with the study of the air turbulence, a high-frequency sampling and very fast sensors are necessary to monitor the short period fluctuations; in this case, the number of observations is so large that high-capacity recorders are required.

In meteorology, for synoptic purposes, all the variables are averaged over 10 min every 6 h, i.e. at 6, 12, 18, and 24 h (principal synoptic hours) or every 3 h, i.e. 3, 6, 9, 12 h,... (secondary synoptic hours), whereas for flight assistance, over 10 min every hour. The practice of measuring only over the last 10 min of each hour leads to a high probability (i.e. 50/60 = 83%) of missing important short-lived phenomena, such as passage of fronts, whirlwinds and so on. For instance, in 1994, the force of a whirlwind severely damaged roofs in the famous *Piazza dei Miracoli* in Pisa, which is in the United Nations Educational Scientific and Cultural Organization List of the World Cultural Heritage. The whirlwind passed in the gap between two sampling times of two meteorological stations located in the same site and in the nearby airport. Both stations missed recording this extreme event.

For this reason, it is necessary to measure 'continually', without gaps or with an elevated frequency. In order to reduce excessive amount of data, it is possible to compute and memorize average values of outside temperature, humidity, sunshine or rainfall every 10 min. In the case of high variability, e.g. wind or indoor temperature and humidity changes near heating, ventilating or air conditioning (HVAC) systems, the extremes

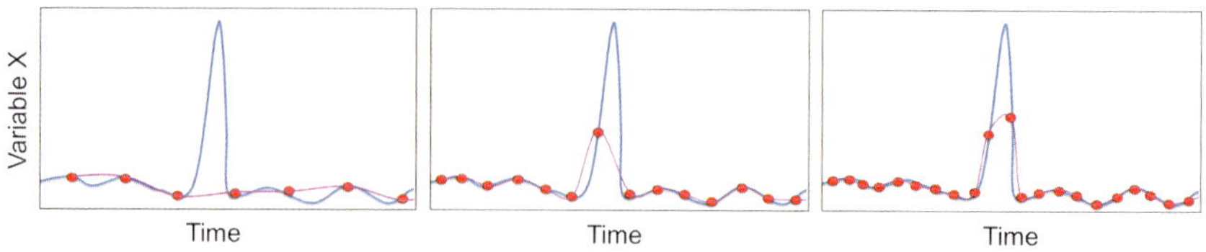

FIGURE 10.4 Effect of the sampling interval. The blue line represents the actual variable and its variability; red dots represent the sampled readings. The pink line is the output we can get connecting the readings. The sampling interval on the left is too long and the peak passes unobserved. Increasing the frequency of sampling, the peak is better documented and needs at least three ($\pi = 3.14$) samplings within its duration (Δt) to be quite accurately represented.

and possibly the variance should also be recorded continually or with very high frequency. The choice of the length of the sampling interval Δt is determined by the *Nyquist* frequency $\pi/\Delta t$. This is the highest frequency that can be used in Fourier transform and, consequently, gives the longest period that is detectable from a sampling with finite length (Wei, 1990). Too long sampling intervals compared with the duration of peaks or drops we want to observe will provide missing, distorted or incomplete information, as shown in Fig. 10.4.

However, as problems and applications are so many and different, it is not convenient to state thumb rules except for the general concept that *the sampling frequency, the interval for averaging and the length of the observation period should be decided together with the characteristics of the experimental apparatus as a function of the specific problem and the specific variable.*

10.4. LENGTH OF OBSERVATION PERIOD

Measurements should be taken over a period sufficiently long to give a complete description of the ambient variability (i.e. if there are cycles, they should be completely monitored and the most important cycles are the daily and the yearly ones) and field surveys should cover a certain number of cycles to be statistically representative. If a fraction of cycle is monitored, it will show a misleading trend and the record will be misinterpreted. If turbulence is investigated, or special transient phenomena, e.g. the perturbation caused by the turning on or off the HVAC or lighting systems, the monitoring can be only done over the relatively short period that is necessary and, if possible, should be repeated a number of times in order to increase the confidence in representativeness.

Very often the problem is to intervene as soon as possible, so that it is unbelievable to collect data for years. Therefore, it becomes customary to follow the seasonal cycle with surveys of one or a few weeks made at the most important parts of the seasonal cycle, mainly in the heart of the winter (January or February) when the climatic wave reaches extreme values, in the early spring (April) when the thick walls are still cold preserving a memory of the cold season and the air becomes warm and humid, so that surface condensation is favoured, and in midsummer (e.g. July or August) when extreme values of the warm season are reached. In winter and summer, the HVAC systems work at their maximum and it is important to know how they control or perturb the microclimate in order to reach a more homogeneous and steady distribution.

For instance, the standards EN15757: 2010 *Conservation of Cultural Property – Specifications for temperature and relative humidity to limit climate-induced mechanical damage in organic hygroscopic materials*, and EN15759-1: 2011 *Conservation of cultural property – Indoor climate – Part 1: Guidelines for heating churches, chapels and other places of worship* establish that the historic climate of vulnerable objects should be respected. However, if the historic climate is not known, it should be measured over a whole calendar year or multiples of it (EN15757: 2010). The requirement of whole multiples is obvious, being justified by the fact that it does not make sense to produce an average composed of uneven seasonal contributions. Only one calendar year might be a short period, because that particular year might include a relatively anomalous weather period. However, the acceptance of only 1 year has been determined by practical reasons, i.e. the fact that a person who is willing to host an object for a temporary exhibition, or plan or install a heating system, should start with such monitoring, and it is not reasonable to require too longer periods (i.e. 2 or 3 years) before going on with the subsequent planned activities.

When a short survey is made, it is advisable to gather the synoptic weather analysis and the regional meteorological data, in order to relate the observations taken to a wider climatic context and get a clear idea of the representativeness of the survey.

10.5. RESPONSE TIME OF A SENSOR

When average values or trends are needed, slow response sensors are preferred, as a long time constant tends to smooth out the rapid fluctuations and noise is filtered without needing an elevated number of observations. On the other hand, fast response sensors are necessary in other cases, especially when monitoring the impact of rapid or transient phenomena. Therefore, it is important to know the time response of each transducer and then select the most convenient one in every opportunity. When a sharp change of the parameter under examination occurs, e.g. a drop in temperature ΔT, the sensor gradually approaches the new equilibrium, and the time elapsed to reach a given fraction of the sharp change is called *response time*. In the case the sensor depends on only one physical variable, e.g. thermometers, the equation that describes the rate at which the sensor approaches the final temperature level T_f is

$$\frac{dT}{dt} = -\frac{T - T_f}{\tau} \tag{10.1}$$

where T is the temperature of the sensor and τ is a characteristic parameter with dimension of time, called *time constant*.

Especially in the case of air temperature measurements, it is useful to remember that the specific heat of the sensor is much greater than that of air and that the sensor temperature follows the air temperature with some delay. For very fast sensors, the time lag is one or a few seconds; for slow ones several minutes or longer. The response of a wet bulb thermometer is conditioned by both the sensible and the latent heats and, in general, it is not longer than the dry bulb. In water, τ is one order of magnitude shorter than in air and it slightly decreases when the fluid from still becomes stirred. This finding is useful not only for temperature but also for RH as well as the other parameters and the above example will merit a more precise mathematical treatment.

We will start with the transient response of first-order linear systems, i.e. depending on only one physical variable, when this variable undergoes a sharp change that may be represented in terms of the Heaviside step function. The Heaviside step function is a discontinuous function equal to 0 before a selected instant t_o and 1 after t_0. In our example, the temperature will sharply change by $\Delta T = T_0 - T_f$, where T_0 is the temperature level before the instant $t_o = 0$, and T_f the level after t_o. In such case, Eqn (10.1) can be integrated and gives

$$T_o - T(t) = \left(T_o - T_f\right)\exp\left(-\frac{t}{\tau}\right) \tag{10.2a}$$

that can be rewritten:

$$1 - \frac{T_o - T(t)}{\Delta T} = 1 - \exp\left(\frac{t}{\tau}\right) \tag{10.2b}$$

After the time $t = \tau$ is elapsed, exp $(-1) = 0.368$ and the output has reached $1 - (1/e)$, i.e. 63.2% of the sharp change ($\Delta T = T_o - T_f$). This particular time interval is defined as '*time constant*' and corresponds to the time elapsed to reach the level 0.368 in Fig. 10.5.

In the above figure, the time constant is obtained plotting the normalized difference between the output and the final equilibrium level, i.e. $1 - \exp(-t/\tau)$ (ordinate, logarithmic scale) versus the elapsed time t (abscissa, linear scale). The plot is a tilted straight line from the starting point 1 when $t = 0$ (i.e. instant of the sharp change) and continues as far as the measured value of the normalized heating/cooling becomes smaller

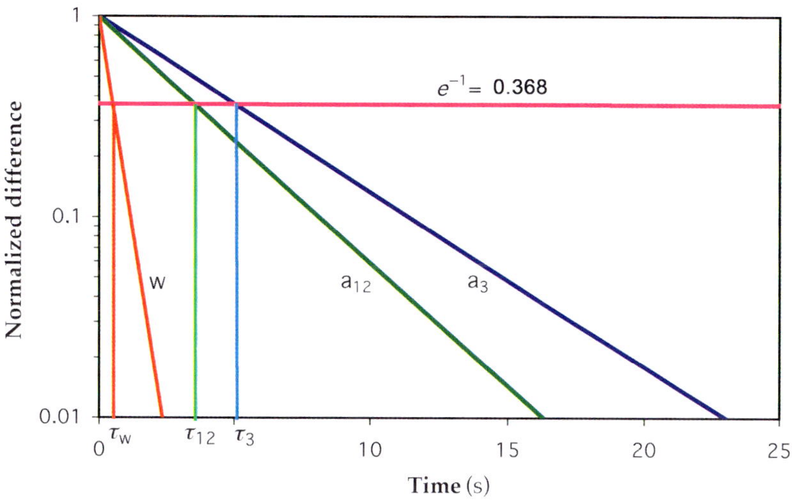

FIGURE 10.5 In a linear logarithmic paper, the plot of the normalized difference between the output and the final equilibrium level $1 - \exp(-t/\tau)$ to obtain the *time constant* is a tilted straight line, which begins at 1 when $t = 0$ (i.e. instant of the sharp change) and continues until the measured value of the normalized heating/cooling becomes smaller than instrument repeatability or ambient fluctuations, e.g. 0.01. In this graph, the three straight lines show the responses of the same fast thermistor but in three different conditions, i.e. immersed into water (red, label: w); in air at 3 m s^{-1} ventilation rate (blue, label: a_3) and at 12 m s^{-1} (green, label: a_{12}). The ordinates of the intercepts with exp (-1) level give the time constants τ_w, τ_3, and τ_{12}, respectively.

than one of these two limiting factors: instrument repeatability or ambient fluctuations. At this point, the straight line starts to show random fluctuations and the graphic representation should be interrupted.

The time constant is independent of ΔT but varies with the rate at which the sensor exchanges heat with the surrounding medium. The exchange rate is determined by several factors, i.e. heat capacity of the sensor, type of medium, specific heat, conductivity and possibly convective motions that increase the capability of exchanging heat, especially in the case of blown air or turbulence. If the medium is air, and ventilation increases, the time constant decreases, being roughly inversely proportional to the square root of the ventilation speed (Fig. 10.5).

In addition to the time constant, the above straight line methodology is useful to determine other characteristic response times, i.e. the times to respond to 90%, 95% or 99% of ΔT. These are the abscissas corresponding to the ordinates $y_{90} = 1 - 0.90 = 0.10$, $y_{95} = 1 - 0.95 = 0.05$ or $y_{99} = 1 - 0.99 = 0.01$ which are, respectively, $t_{90} = 2.3\tau$, $t_{95} = 3\tau$; or $t_{99} = 4.6\tau$.

This also means that it is useless, or even incorrect, to take a sampling time longer than 5τ e.g. waiting until the output has reached a constant value. In fact, after 5τ, the sensor has already reached equilibrium and any subsequent change will be due to the evolution of the system, e.g. changes in ambient temperature or arrival of other air masses. In case we are mapping room temperature, late reading may loose connection with the previous ones. Too long a time elapsed before taking subsequent readings may be an error, not a synonym of more accurate observation.

The air temperature continually fluctuates in the range of one or a few degrees in a few seconds. When it is relevant to monitor this variability, fast sensors are used with continuous or high-frequency sampling. This especially occurs when monitoring abrupt changes of temperature generated by turning on or off HVAC systems, when temperature fluctuations are generated by atmospheric turbulence or when quick measurements are required. Slow sensors are used to define the average meteorological conditions, as, e.g. routine aerodrome measurements. However, fast sensors are more flexible, as they can be used for either instantaneous or average values. An average can be obtained from them in different ways:

1. taking several close observations and computing the 50th percentile (i.e. frequency-oriented statistics) or the arithmetic mean (i.e. level-oriented statistics);
2. integrating a continuous measurement (e.g. a variable frequency and a voltage) with electronic devices;
3. damping the output by applying a low-pass filter;
4. increasing the time constant of the sensor, e.g. in the case of a thermometer by covering the sensor with a substance (e.g. a droplet of bee wax) that increases the heat capacity and reduces the heat exchanges, acting as a thermodynamic low-pass filter.

The first method requires a large memory capacity in the data logger or an electronic device that computes the 50th percentile or the average before recording it; the second and third methods need an integrating or damping filter before monitoring the data; and the fourth system is the simplest and cheapest one. In addition, the wax protects the sensor against water and shocks. If more than one sensor is used, attention should be paid that the mass of wax that covers each sensor is exactly the same and this can be controlled with a precision balance.

It is important to remember that *well-calibrated sensors with different time constants give exactly the same output in a calibration bath or in steady-state conditions, but a different response in dynamic conditions, as they have a different memory of the recent past values*. This is a very general statement that suggests that measurements should always be taken with the same instrument and that it might be misleading to compare measurements of the same parameter taken with a different type of instrument.

Moving a sensor before it has reached thermal equilibrium may determine important distortions to the measurement. An example is

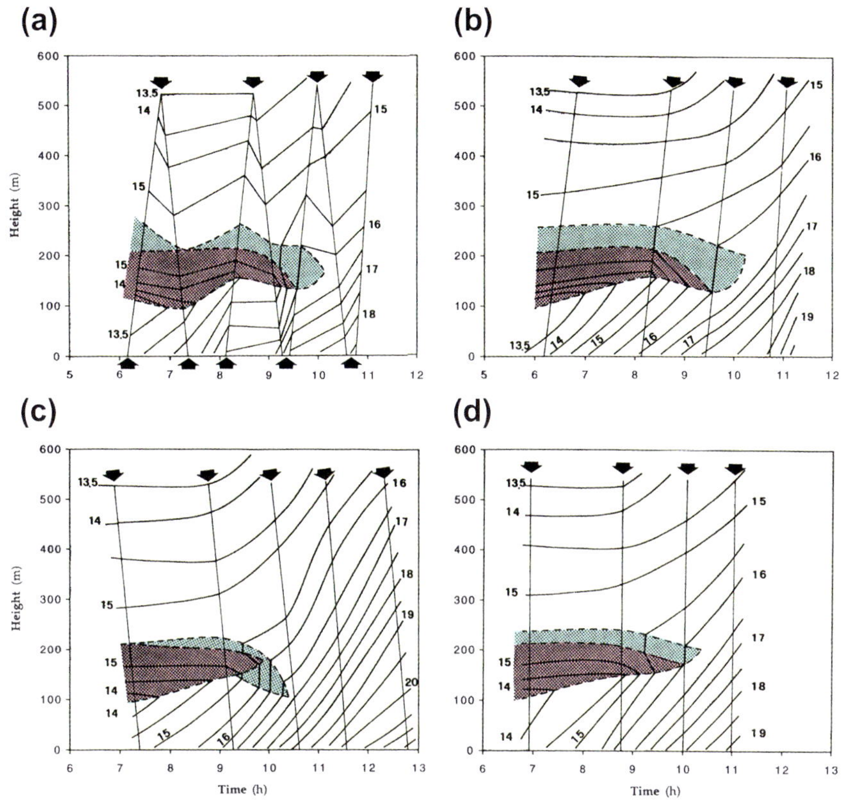

FIGURE 10.6 Vertical time cross-section of the air temperature and the final part of the erosion of a nocturnal inversion (inversion, violet; isothermal layer, cyan), measured by launching and pulling down a high-resolution radiosonde. The time and space coordinates of the radiosonde during the ascent and descent are indicated with a thin line and evidenced with an arrow. The graphs show: (a) the zigzagging path of the isotherms, reflecting the thermal inertia of the sensor, i.e. a profile from warmer to colder air, and vice versa; (b) the apparent isotherm distribution that is obtained by using only the ascent runs; (c) the same but for the descent runs; and (d) the temperature distribution obtained averaging the data transmitted during matched ascent and descent in order to combine the opposite effects.

given in Fig. 10.6, which shows a vertical time cross-section of the air temperature and the final part of the erosion of a nocturnal inversion. The measurement was performed with up and down runs obtained by launching a high-resolution radiosonde with a tethered balloon and then pulling it down. The operation was made with a motor winch and a tethering rope made of a nylon fishing line 1000 m long and 1.2 mm thick. The ascent and the descent speed were prudentially kept at 1 m s^{-1}, while the producer suggested the exceedingly high speed of 3 m s^{-1}. However, the above test shows that even 1 m s^{-1} was excessively fast. In fact, the graphs show: (1) the zigzagging path of the isotherms, reflecting the thermal inertia of the sensor, i.e. in the uprising profile it was passing from warmer to colder air, and vice versa for the descent; (2) the apparent isotherm distribution that is obtained by using only the ascent runs; (3) the same but for the descent runs; and (4) the temperature distribution obtained averaging the matched readings at the same levels, transmitted during the ascent and the descent, in order to combine the opposite effects.

The situation is even more complex in dynamic conditions when a parameter depends on a number of variables and these may interfere in a different way with their time constant. This is typical of compact systems composed of a battery and an electronic apparatus, e.g. a data logger or a transmitter, housed in a miniature plastic case together with a temperature and a humidity sensor, which communicate with the exterior through some ventilation slits. The sensors are relatively fast and data are sampled at short or programmable time intervals but nothing is specified about the heat capacity of the whole encased device and the delay introduced by the case. For example, a popular commercially available instrument, sized $7 \times 4 \times 2.5$ cm, had fast sensors but its mass and the very low ventilation through miniature slits required more than 1 h to reach equilibrium. This fact passes unobserved because the producers generally report the time constant of the individual sensors but not the effective time once they are assembled in the probe or built into the instrument. An important improvement was obtained by cutting the plastic case around the sensors in order to facilitate air exchanges. After having opened this new window, the time constant dropped to 7 min (nearly halved), i.e. 21 min to reach 95% of the change.

An illustrative experiment is shown in Fig. 10.7, where the data logger was subjected to a square wave temperature and humidity change, displacing it at the time t_0 from an initial ambient at $T_i = 12.2$ °C and $RH_i = 31\%$ to another well-ventilated ambient, warmer and dryer, at $T_f = 22.2$ °C and $RH_f = 25.8\%$. The plot shows that after t_0 the temperature rose from T_i to the final higher value T_f approaching it asymptotically, as expected. On the other hand, the fast RH sensor gave a generally unexpected result: the output immediately rose with a marked overshoot (peak value greater than 41%); afterwards, the RH decreased following the gradual heating of the case and reached equilibrium only when the case temperature had reached equilibrium.

This apparently anomalous result is explained as follows. The fast humidity sensor (a thin-film capacitive sensor) passed nearly immediately from the initial value of mixing ratio $MR = 2.7$ g kg^{-1} to the final value 4.3 g kg^{-1} and the MR values calculated from T and RH showed that this parameter was always correctly measured (graph not reported in the figure). However, being respondent in terms of RH, the transducer was

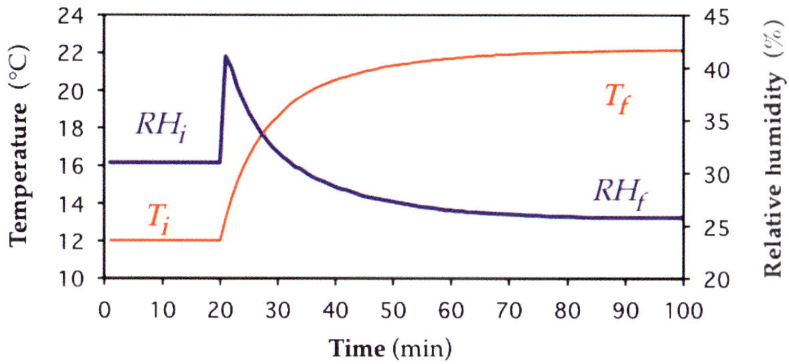

FIGURE 10.7 A miniature data logger has been moved at time $t_o = 20$ min from an initial ambient at temperature (red) $T_i = 12.2$ °C and relative humidity (blue) $RH_i = 31\%$ to another ambient at $T_f = 22.2$ and $RH_f = 25.8\%$. The plot shows that at t_o the temperature increases from T_i to the final higher value T_f, approaching it asymptotically; on the other hand, the fast *RH* sensor is conditioned by the temperature of the case: the *RH* output immediately increases and then decreases following the gradual heating of the case and reaches equilibrium only when the case temperature has reached its equilibrium.

conditioned by the case temperature. The new air, which was moister in terms of *MR*, at the beginning was cooled by the cold case and this caused the initial rise of *RH*, i.e. the overshoot; the progressive warming of the case determined the decrease of the *RH*, which stopped when the case had reached the new equilibrium.

From this example, it is evident that the temperature transducer constitutes a first-order linear system characterized by a simple time constant as in Eqn (10.2) and, after a sharp change, the output passes gradually from the initial to the final value.

On the other hand, the humidity transducer is a more complex second-order dynamic system described in terms of two independent state variables, i.e. temperature and moisture content. Second-order systems have responses that depend on two dissipative elements in the system and their synergistic combination does not follow Eqn (10.2), except in the case that one of the two variables, either the temperature or the moisture content, remains unchanged (Fig. 10.8). Under the latter hypothesis, the system has only one effective variable and returns to the first order with a gradual transition from the initial to the final level: the transition is faster if the temperature remains unchanged because the system has a very fast response to moisture; it is longer if the *MR* remains unchanged because the thermal inertia is generally large.

The various combinations of the *RH* instrumental response to sharp changes in temperature or *MR* are illustrated in Fig. 10.8 where the sharp change is applied to the instant t_0. The first two examples (a and b) with temperature (*T*) or *MR* unchanged are regular first-order transient

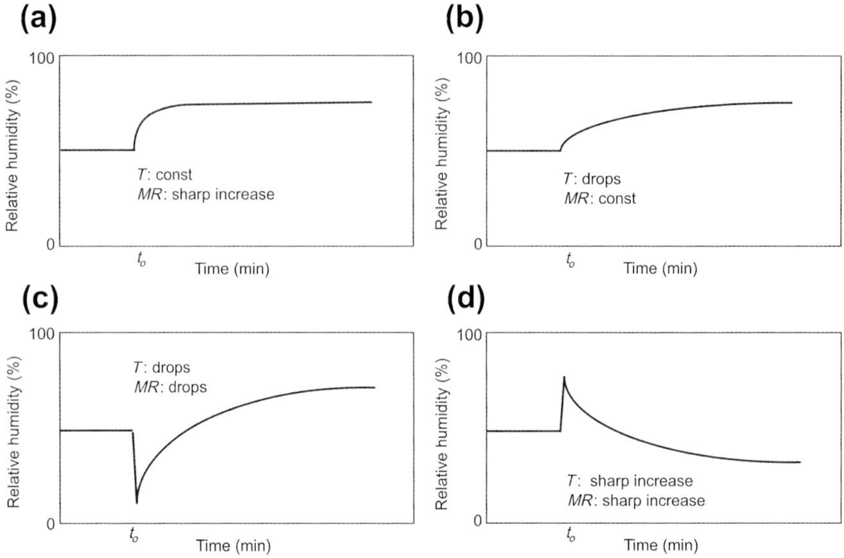

FIGURE 10.8 a) The *RH* response is gradual and fast, depending only on *MR* change. (b) The *RH* response is gradual and slow, depending only on temperature change. (c) The drop in *RH* is due to the prompt response to *MR* change while the probe is still warmer. (d) The peak in *RH* is due to the prompt response to *MR* change while the probe is still colder.

systems but with different time constant. It is evident that the actual second-order responses to a sharp temperature and *MR* change (c and d), both in the same direction, are characterized by overshoots or undershoots that depart in the opposite direction compared with the final equilibrium level. Overshoot peak and undershoot drops may go outside the span characterized by the initial and final values. Only after this critical transient period corresponding to the shorter of the two time constants the *RH* gradually reaches the final value. When temperature and *MR* have opposite departures, the system is partially compensated and the departure is attenuated.

If the change in temperature is gradual, the peaked overshoots or undershoots are smoothed out and apparently not visible, the effect is masked and the *RH* approaches gradually, with a certain delay, the new value but passing through uncertain values. This was tested with a sinusoidal temperature oscillation forced in a climatic chamber where the *MR* was kept constant, as follows. Three capacitive *RH* sensors were introduced in the chamber, i.e.: (1) the same fast sensor described above, but housed in a compact case as commercially available, with relatively high thermal capacity and low ventilation; (2) another identical transducer, i.e. housed in the same case but with the ventilation hole opened by us, i.e. the same used in the previous test; and (3) a third identical fast

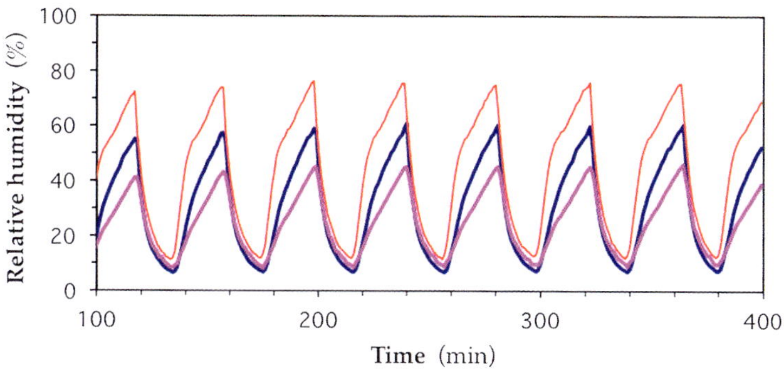

FIGURE 10.9 Response of the same capacitive relative humidity transducer, but with different probe support, to a sinusoidal temperature oscillation forced in a climatic chamber, at constant *MR*. The tested transducers are: (1) a fast sensor housed in a compact case as commercially available, with relatively high thermal capacity and low ventilation (pink line); (2) another identical transducer, i.e. housed in the same case but with the ventilation hole opened by us (blue line); and (3) the same fast sensor but externally connected to the case in order to avoid the effect of the thermal inertia (red line). The heating and cooling cycles are repeated periodically, reaching every time the dew point.

sensor but externally connected with the case in order to avoid the effect of the thermal inertia (Fig. 10.9). In the chamber, the heating and cooling cycles repeated periodically, approaching every time the dew point, so that the *RH* followed a continuous cycling.

The fast sensor was able to follow the cycle; the sensor in the case with the additional open window showed a cycle topped at lower *RH*, whereas the sensor in the original unaffected case had a cycle topped with a much lower *RH*. The outputs were the same in the dry region but different in the wet, showing an asymmetrical response.

These examples show that the response in dynamic conditions is not easily predictable and that large errors may originate when humidity transducers have different thermal capacity.

10.6. DRAWING AIR TEMPERATURE AND OTHER ISOLINES

Three-dimensional computer graphics give a very elegant visual presentation but are less useful for diagnostic purposes, as several details are masked by the perspective view (Fig. 10.10). Two-dimensional maps are much more convenient, as they show with great detail and without distortion the actual distribution and the existing gradients. Computer graphics have been developed to interpolate with linear, quadratic or more complex functions intermediate values between two sampling points.

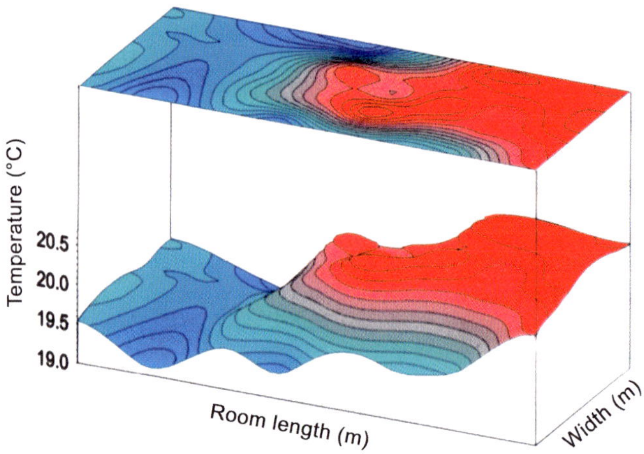

FIGURE 10.10 Three-dimensional representation of the temperature distribution at the Sainsbury Centre for Visual Arts, Norwich, UK (9 December 1996 at 16.30, shown in Fig. 1.11) and projection of the contour levels in a horizontal plane, which furnishes the two-dimensional diagram. Several details of the three-dimensional representation are masked by the perspective view.

Computer graphics is very convenient for a single room with a simple geometry, e.g. a rectangle, but becomes more complex to represent the temperature distribution in a complex array of rooms in a building. In fact, all the data are interpolated in the same way either in the free air or across walls so that the interpolation is exactly the same for areas in the same room or including walls and different rooms, without considering that the heat flow by advection through an open door is different from the conductive flow across a wall. Sometimes computer graphics hardly reach the level of sophistication of hand-drawn isolines in complex architectural systems, drawn by keeping in mind the whole dynamics of the system. The reason is that this is a problem not only of mathematical interpolation but also of a correct physical interpretation of the data. In the case of building temperature or humidity maps, the distribution is complicated by the presence of sources or sinks of heat or moisture, horizontal advection transports, vertical convection movements, irregular geometry of architectural structures and other perturbation factors. Although today most of the interpolation maps are made with the mathematical tools available for PC, drawing isolines is a good practice that helps in understanding. For this reason, some basic guidelines are summarized here.

A beginner might proceed with successive elementary steps. The first step is to write, on the side of each grid point $O_{x,y}$ the observed values $V(O_{x,y})$. Next, one starts by choosing an arbitrary point $O_{x,y}$ (it is convenient, however, to choose a maximum or a minimum) and drawing the segments that join this point with all the neighbouring points, e.g. the eight points $O_{x-1,y-1}$, $O_{x-1,y}$, $O_{x-1,y+1}$, $O_{x,y-1}$, $O_{x,y+1}$, $O_{x+1,y-1}$, $O_{x+1,y}$, and $O_{x+1,y+1}$ in the case of a regular square or rectangular grid. Once the map resolution is chosen, i.e. the unit step from isoline to isoline (e.g. 0.1 °C for air temperature *T* or DPS, 1% for *RH*, 0.1 g kg^{-1} for *SH* or *MR*, 0.1 g m^{-3} for absolute humidity), all these segments are divided in equal parts, which are determined by the number of times the unit step is included in the difference between the numeric values of the correspondent grid points $V(O_{x,y}) - V(O_{x-1,y-1})$, $V(O_{x,y}) - V(O_{x-1,y+1})$, and so on. For example, if $V(O_{x,y}) = 18.3$ and $V(O_{x-1,y-1}) = 18.7$, $\Delta V = 0.4$ and the segment $O_{x,y}\ O_{x-1,y-1}$ is divided into four subsegments, which are divided by the equally spaced points 18.4, 18.5 and 18.6. In this way, it is possible to note at the extremes of each subsegment all the values that have been obtained by interpolation. Continuing with this method, it is possible to interpolate the space between all the pairs of points in the grid. Joining by a line all the points having the same value, observed or interpolated, a map is obtained that corresponds to a linear interpolation. All the isolines are closed or end on the walls. This is done in a very short time by several computer programs.

However, hand-drawn isolines may take into account several other factors. It is possible to distribute more gradually the gradients from flat zones to more perturbed ones instead of using a rude linear interpolation. It is also possible to take into account the presence and location of heat or moisture sources or sinks (e.g. diffusion intakes or suction outlets for warm, cold, humid or dry air) and improve the detail. Even more important is to consider the dynamics of the room, with advection transport from doors or windows, and vertical convection. The dynamics is well represented with analogy to the weather maps and the representation of fronts. In the weather maps, a continuity is assumed not only on the values of the parameter but also in its first- and second-space derivative. This means that all the isolines and their curvatures (i.e. concavity and convexity) present a progressive and coherent increase or decrease and that all the points of greatest curvatures are distributed along a line, which in a weather map may represent a front and has a physical meaning also in a thermal or hygrometric map. In the actual case, when external air masses with different characteristics from the inside air can penetrate through a door and flow horizontally for a certain path, several successive isolines start from the door and protrude inside having the maximum curvature lying on an imaginary line, which is the core of the air stream.

Two methods are possible: the *line-oriented* point of view, i.e. connecting with an isoline all points having the same value, which is common in topography, and the *area-oriented* criterion, i.e. including between two isolines the belt that has the same values $V(O_{x,y})$ or values that fall within the upper and the lower limit represented by the two contiguous isolines. The field variability obliges to prefer the first or second criterion: *line oriented* is necessary with a wide range of parameter variability, e.g. climatic maps, whereas the *area oriented* is necessary in the case of a little variability. For instance, if in a room, the observed values span a few tenths of degree, or 1 °C, the *area-oriented* method is appropriate with resolution of 0.1 °C; if the span is several degrees, it is practically impossible to keep the same resolution and draw so many lines and it is convenient to choose a larger resolution, e.g. 0.5° or 1 °C. In this case, a large amount of data fall within each class and it is preferable to use the *line-oriented* method. The *area-oriented* method is necessary when there are several scattered points having the same value and when determining homogeneous areas. For instance, in the case of two contiguous zones characterized by the values 18.3 and 18.4, the band 18.3 includes all the points from 18.30 to 18.39 but they are not distinguishable because of the truncation at the first decimal. The separation line ideally joins the points between 18.39 and 18.40, which of course do not appear in the graph. Suppose that we have a very accurate instrument, e.g. a psychrometer with high resolution sensor (e.g. 0.01 °C), but with a low resolution display, e.g. with numbers truncated after the first decimal digit. In such a case all the values between 18.30° and 18.39 °C will be read 18.3 °C. This shows that all data having the same figures on the display are only apparently identical, being truncated and directly classified within a class of values.

In any case, when a map is drawn, all the space is subdivided with vertical steps and horizontal belts. The *line-oriented* method joins with the vertical steps the few identical values that can determine a line and leaves on the belt the many values between an isoline and the next; *the area-oriented* method groups on the same belt all the many identical values that are scattered over a wider area, where the values are apparently identical, being the result of a truncation, or only slightly different, but falling within the values of two contiguous isolines. The two methods are conceptually the same, if one considers that in the *area criterion* an isoline instead of separating two areas ideally connects all the nonexisting points having a value between the two contiguous classes, as in the example of the truncation.

A very practical method for beginners is to facilitate drawing with colours that evidence the maxima and the minima in the map. For example, all the maxima can be evidenced with red and the minima with blue. All around the maxima, the red colour is attenuated (i.e. reduced colour saturation) as the observed values $V(O_{x,y})$ decrease, forming concentric coloured areas with the same pencil but less and less marked.

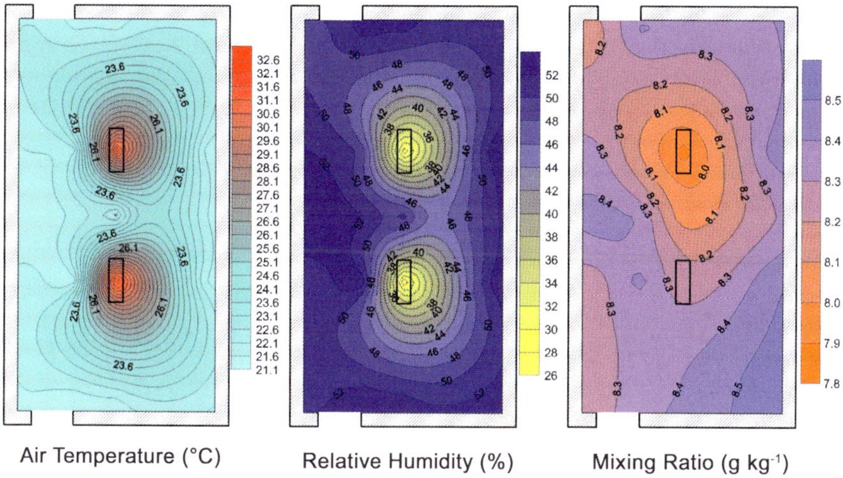

FIGURE 10.11 **Koninklijk Museum voor Schone Kunsten, Antwerp, Room T - 13 Feb. 1999 Local time: 10:30.** The indoor climate is controlled by two units located in the middle of the room to avoid microclimate disturbances to paintings on walls. The map of temperature has two hot spots because both units are operating and supply heat. The map of relative humidity has minima corresponding to the hot spots. The map of mixing ratio indicates that only one of the two units is absorbing some moisture (i.e. a local minimum); the other neither absorbs nor emits moisture.

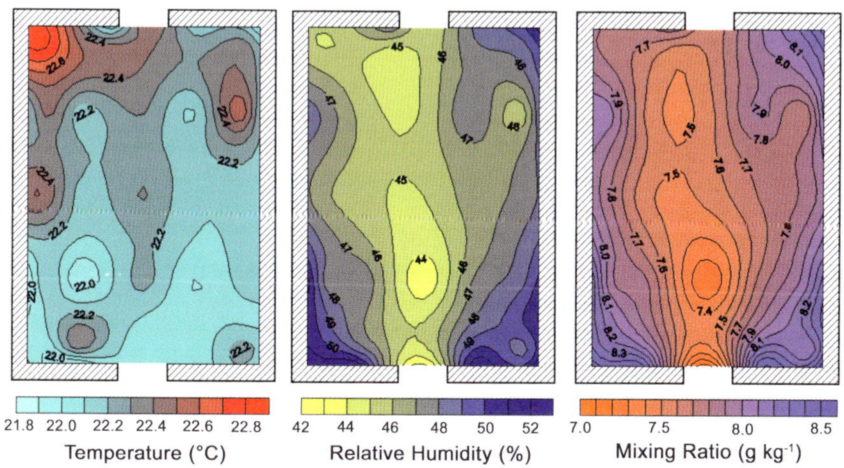

FIGURE 10.12 **Koninklijk Museum voor Schone Kunsten, Antwerp, Room K - 10 Feb. 1999 Local time: 16.00.** The map of temperature shows a homogeneous situation, except for a modest heating on the upper corner on the left, where a monitor of the microclimate recording unit was operating. The map of relative humidity shows a less homogeneous situation, characterized by the pattern of the mixing ratio. The map of mixing ratio points out that some dryer air is entering through the door in the bottom, and this air spreads through the room, and enters the next room.

The same can be made with blue, starting from the minima and leaving untouched the intermediate values. Drawing isolines in a coloured map becomes much easier. After having sketched the first coloured draft, a better and more detailed map can be drawn with a pencil on a new paper sheet.

In particular, when all isolines have been drawn, a good practice is to colour again the belts, using two different colours for each parameter, one

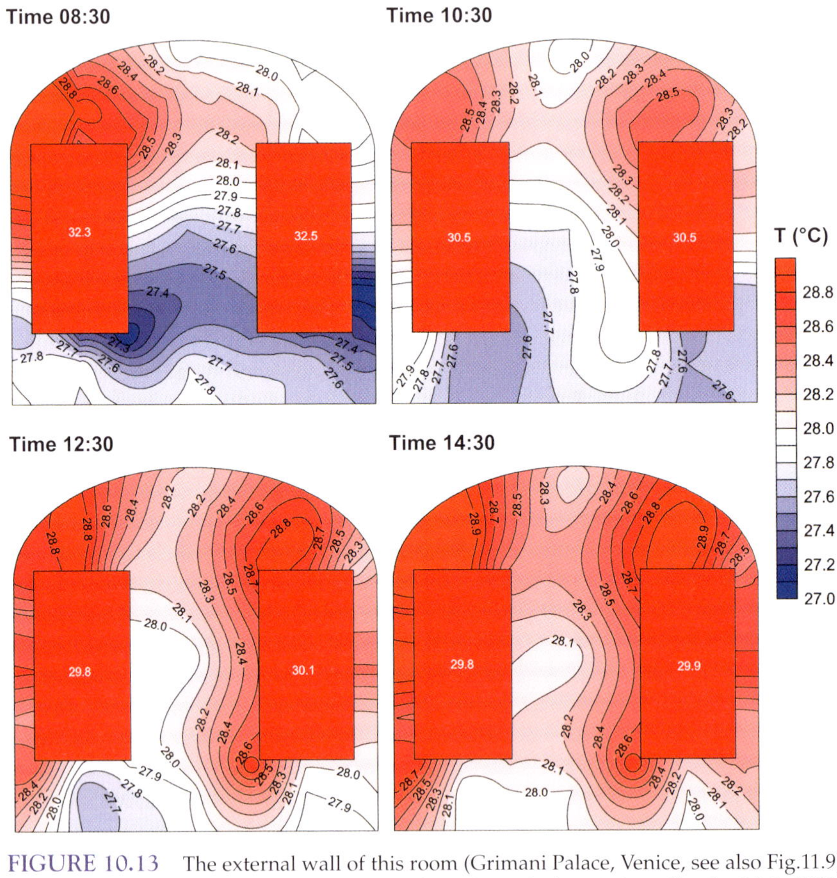

FIGURE 10.13 The external wall of this room (Grimani Palace, Venice, see also Fig.11.9) has two windows and some rain water percolates through the windowsills, the 3 July 2008. Initially the area affected by percolation is colder, for two reasons: (i) the external air temperature is lower and dampness forms a thermal bridge with the exterior with internal heat leakage (most relevant), (ii) some internal evaporation (less relevant). After a few hours the external temperature is higher and the high thermal conductivity of the damp areas show hot spots in correspondence of dampness at windowsills level. The contribution of the modest evaporation is no detectable. The solar radiation on the vertical sides of the windows, and the overheating of glass are responsible for the overheating on the left side of the windows and above them. Repeated monitoring points out the dynamics of this mechanism. Only one record, considered alone, might be misleading.

for the high levels and one for the low ones, leaving white the middle of the range (i.e. intermediate levels). The effect is improved with colour saturation decreasing from the extreme to the intermediate levels. This representation makes better the understanding of the microclimate, the forcing factors and the dynamics of the air masses.

Finally, it might be useful to compare the practice of *rounding off* of a number with that of *truncating* it, i.e. 16.6 becomes 17 in the first case and 16 in the second one. It is merely illusory to think that the former is more precise, as the number 17 is in reality 17 ± 0.5 and may indicate any number from 16.5 to 17.4, so that the span of uncertainty is 1. This is exactly the same in the case of truncation, as 16 represents any value from 16.0 to 16.9. The main consequence is that the average of a large population of random numbers is different if the numbers are truncated or rounded off, being 0.5 lower for truncated numbers, i.e. truncation makes levels to the lowest value of the last digit and rounding off makes levels to the middle value of it. In the case of a map of isolines drawn following the former or the latter criterion, the distribution is absolutely equivalent, although not identical, but with values (and isolines) displaced by 0.5 of the last digit. A few examples of microclimate maps and their interpretation are reported in Figs. 10.11, 10.12 and 10.13.

References

General theory on sampling and turbulence

Lumley, J.L., Panofsky, H.A., 1964. The Structure of Atmospheric Turbulence. Wiley, New York, 239 pp.
Munn, R.E., 1970. Biometeorological Methods. Academic Press, New York, 336 pp.
Pasquill, F., 1962. Atmospheric Diffusion, first ed. Van Nostrand, London, 297 pp.
Pasquill, F., 1974. Atmospheric Diffusion, second ed. Wiley, New York, 429 pp.
Plate, E.J., 1982. Engineering Meteorology. Elsevier, Amsterdam, 700 pp.
Tennekes, H., Lumley, J.L., 1973. A First Course in Turbulence. MIT Press, Cambridge, Mass, 298 pp.
Vinnichenko, N.K., Pinus, N.Z., Shmeter, S.M., Shur, G.N., 1980. Turbulence in the Free Atmosphere. Consultants Bureau, New York, 310 pp.
Wei, W.W., 1990. Time Series Analysis. Addison Wesley, Redwood City, Ca, 478 pp.

Instruments and measuring techniques

Doebelin, E.O., 1990. Measurement Systems – Application and Design. McGraw Hill, New York, 960 pp.
Linacre, E., 1992. Climate Data and Resources. Routledge, London, 366 pp.
Lipták, B.G., 2003. Instrument Engineers' Handbook: Process Measurement and Analysis. In: The Instrumentation System and Analysis (ISA), vol. 1. CRC Press, Boca Raton, Florida.

Moses, H., 1968. Meteorological instruments for use in atomic energy industry, 257–300. In: Slade, D.H. (Ed.), Meteorology and Atomic Energy 1968. U.S. Atomic Energy Commission, Div. Tech. Info.

Wexler, A., 1990. Humidity and Moisture Vol. 1: Principles and Methods of Measuring Humidity in Gases. Rehinold, New York, 647 pp.

World Meteorological Organisation, 1983. The Guide to Instrument and Methods of Observation. WMO Technical Publication No 8, Geneva.

World Meteorological Organisation, 1986. Compendium of Lecture Notes on Meteorological Instruments for Training Class III and Class IV Meteorological Personnel. WMO Technical Publication No 622, Geneva.

World Meteorological Organisation, 2006. Guide to Meteorological Instruments and Methods of Observation. WMO Technical Publication No 8, Geneva.

CHAPTER

11

Measuring Temperature

OUTLINE

Microclimate for Cultural Heritage, Second Edition
http://dx.doi.org/10.1016/B978-0-444-63296-8.00012-3

11.1. MEASURING AIR TEMPERATURE

11.1.1. Choice of the Sensor

Several kinds of good sensors are commercially available, not all equally suitable for microclimate study. The principal types are liquid-in-glass thermometers (i.e. mercury or spirit thermometers), bimetallic thermometers; electrical resistance thermometers (metal resistance sensors and thermistors) and thermocouples. The choice is determined by several factors, i.e. compatibility with the ambient conditions, measuring range, resolution, accuracy, response time, drift, compatibility with the recording instrument and cost.

The outdoor measurements generally require a much greater temperature range than the indoor ones. If the sensor has a sufficiently elevated resolution, by limiting the range and expressing data as a percentage of the range, more accurate measurements can be obtained. In this sense, indoor measurements can be more precise than outdoor ones.

11.1.2. Thermometers

A short description is reported here of the most commonly used instruments, discussing their advantages and limits, especially in view of their application in the field of microclimate analysis and diagnostics.

11.1.2.1. Mercury-in-Glass Thermometers

Mercury-in-glass thermometers are, or at least were, the most popularly used instrument for routine measurements and a secondary laboratory standard for comparing other kinds of sensors that undergo rapid ageing or drift. Meteorological station mercury thermometers are some 40 cm long and cover the range $-30\,°C$ $-60\,°C$ with scale division $0.2\,°C$. The advantages of mercury as thermometric liquid are small thermal capacity, high thermal conductivity, small deviation from linearity, high boiling point and freezing point at $-38\,°C$. This is a problem only for mountain sites or polar regions, but with the addition of thallium the freezing point is lowered to $-58\,°C$. The change of output, expressed as change of the mercury column height Δh due to the change in temperature ΔT, is given by the equation:

$$\Delta h = \Delta T \frac{V_o(\beta_1 - \beta_2)}{A} \tag{11.1}$$

where V_o is the volume of mercury inside the bulb at standard temperature, β_1 and β_2 are the cubic coefficients of thermal expansion of mercury and thermometric glass, i.e. $\beta_1 = 1.81 \times 10^{-4}\,°C^{-1}$, $\beta_2 = 2.53 \times 10^{-5}\,°C^{-1}$ respectively and A is the cross-sectional area of the capillary. The sensitivity of the mercury thermometer is given by the ratio $\Delta h/\Delta T$ and the larger the reservoir V_o and the smaller the section A of the capillary, the higher is the sensitivity. This can be represented by the equation

$$\frac{\Delta h}{\Delta T} = \frac{\Delta V}{\pi r^2 \Delta T} = \frac{V_o(\beta_1 - \beta_2)\Delta T}{\pi r^2 \Delta T} = \frac{V_o(\beta_1 - \beta_2)}{\pi r^2} \tag{11.2}$$

where r is the capillary radius. This equation shows that the sensitivity increases proportionally to the bulb volume and with the inverse of the square of the capillary radius. However, the larger the volume V_o, the greater the heat capacity and the longer the time constant of the thermometer, so that sensitivity and time of response have opposite requirements.

The main errors of this kind of thermometer are:

1. *Thermometer heating:* when the observer goes too close to the thermometer for reading the scale, and remains for too long a time, influencing its temperature or when the thermometer is not adequately shielded against infrared (*IR*) radiation.
2. *Parallax*: when the reading is made with the eye placed not at the same height as the top of the mercury column.
3. *Emergent stem*: when the temperature of the bulb is not the same as that of the surrounding medium, i.e. when the thermometer is not completely surrounded by only one medium having the same temperature. This error may be important for measurements of solid bodies.
4. *Drift*: when the characteristics of the thermometer change slowly with time, e.g. the bulb of the thermometer tends to contract slowly over a period of years, thus increasing the temperatures.
5. *Departure from linearity* and inequality in the expansion of the liquid and glass.
6. *Capillarity:* this may influence the height of the mercury in the capillary tube.
7. *Elastic errors:* occurs due to exposure to a large range of temperature in a short time or to large changes in external pressure.
8. *Scale division and calibration.*

Spirit or other thermometric liquids are also used for their greater sensibility (thanks to a much larger expansion coefficient) but they have some problems: adhesion to the glass, stronger deviation from linearity, drift due to polymerization and slow changes of the liquid properties as well as breaking of the liquid column. Their use is not recommended.

11.1.2.2. Liquid-in-Metal Thermometers

Liquid-in-metal or pressure thermometers consist of a sensitive bulb, an interconnected capillary tube and a pressure measuring device such as a Bourdon tube. They follow an equation similar to the Eqn (11.1), which has been discussed for liquid-in-glass thermometers, i.e.

$$\Delta V = V_o(\beta_1 - 3\alpha)\Delta T \tag{11.3}$$

but with the appropriate coefficients and β_2 substituted with 3α, where α is the coefficient of linear thermal expansion of the bulb material. In meteorology, this sensor is used to drive the arm of the recording pen of thermographs and in the United States, this is accomplished according to the Weather Bureau specification No. 450.1201. The time constant is some minutes, so that all the short-period temperature fluctuations are smoothed out. This kind of transducer is used in industry and sometimes in meteorology but is not relevant in microclimate studies.

11.1.2.3. Bimetallic Thermographs

The bimetallic thermograph is commonly used both in standard meteorological stations and in museums. It is mainly associated with a hair hygrometer and the resulting instrument is the well-known thermohygrograph. The bimetallic sensor is composed of two thin metal strips having different coefficients of thermal expansion roll welded together along one of its flat sides. It provides a mechanical output, i.e. the displacement of the free end of the strip. This end is usually connected with a pen, whose movement is amplified and used to trace the temperature records in clock thermographs. It is useful to remark the main characteristics of this instrument:

1. The thermograph is a *linear* transducer, as the displacement of the free end of the strip is a linear function of the temperature change.
2. The *sensitivity* of the sensor is directly proportional to the square of the length of the strip and proportional to the difference in the coefficients of thermal expansion of the two metals, but is inversely proportional to the thickness of the strip.
3. The *time constant* is a few minutes, so that this instrument smoothes out all the short-period temperature fluctuations.
4. The *resolution* in the strip chart recorder is generally from 1 to 1.5 mm per °C and the time scale division is 15 min for daily clocks and 2 h for weekly clocks. The resolution in reading the strip chart is some 1 °C.
5. The *accuracy* is no better than ±1% of the range, which generally varies between 45 °C and 90 °C.
6. The *friction* between pen and chart is excessive compared with the force supplied by the bimetallic strip, so that the graph is smoothed out, with underestimation of the maxima and overestimation of the minima.

These instruments are not accurate but are very popularly used in museums. The most common sources of error are:

1. Lack of maintenance and periodic calibration.
2. Exposure in a position not representative of the parts of the room where artwork are located, e.g. near room corners where air is stagnant and on the floor, at a height different from that of the exhibits.
3. Dirt, dust and pollution, which increase the friction of the instrument.
4. Excessive friction between pen and strip chart.
5. Corrosion or mechanical damage to the bimetallic strip.

To avoid the above causes of error, the instrument requires a frequent cleaning, and particularly, of the bimetallic strip and the pen, which should be carefully cleaned with alcohol. After cleaning, the instrument calibration should be controlled as well as the fine regulation of the three adjusting screws (i.e. bimetallic element, magnification and pen pressure). Monthly control and maintenance is recommended but in practice, this necessary operation is usually neglected.

11.1.2.4. *Platinum Resistance Sensors*

Although iron, copper, nickel and other metals have a temperature coefficient of electrical resistance greater than platinum, platinum resistance sensors are preferred for their stability in time and noncorrodibility, which are important characteristics especially in humid or polluted environments. The platinum resistance sensor is an electrical resistance made of very pure platinum wire, 0.1 mm in diameter or less, used sometimes as a wire and more usually wound on a glass or ceramic rod. However, thin deposited platinum films are also common, as well as several type of probes with special windings for different purposes.

For a metal, the electrical resistance $R(T)$ is described by a MacLaurin expansion of the successive powers of the increase of temperature,

$$R(T) = R(T_o)\left(1 + a\Delta T + b\Delta T^2 + \cdots\right) \tag{11.4}$$

where $R(T_o)$ is the resistance of the sensor at the reference temperature $T_0 = 0\,°C$; ΔT is the temperature change; and a, b,... are constants characteristic of the metal used. The platinum resistance is characterized by the following values: $a = 3.968 \times 10^{-3}\,°C^{-1}$, $b = -5.847 \times 10^{-7}\,°C^{-2}$, $c = -4.22 \times 10^{-12}\,°C^{-3}$, but within the interval of meteorological observations is in a very good approximation described by the first two linear terms of the series and is linear within $\pm 0.3\%$ of the whole range. As it presents an excellent stability, the measurements are reproducible within 0.01 °C. Another advantage of these sensors is that they are cheap. Metal resistance thermometers should be compared with a standard every year.

11.1.2.5. *Thermistors*

Thermistors are essentially semiconductors that behave as resistors with a high temperature coefficient of resistance. The electrical resistance varies with the temperature and is usually negative, i.e. decreases with increasing temperature unlike metals. The response function is an exponential one of the type:

$$R(T) = R_o \exp\left(\frac{A}{T}\right) \tag{11.5}$$

where all the coefficients are constant and depend on the material used. By differentiating the above equation, the temperature coefficient B is obtained

$$B = \frac{\mathrm{d}\ln R}{\mathrm{d}T} = -\frac{A}{T^2} \tag{11.6}$$

which has a negative parabolic dependence on temperature. The sensitivity of thermistors varies with temperature but at ordinary values, it is one order of magnitude higher than that of platinum resistance sensors.

It is important to note that thermistors are heated by the current load and that the supply should be very well calibrated in view of the limited natural heat dissipation. Thermistor thermometers should be compared with a standard thermometer every year if they are of good quality or every month if they are of low quality, and recalibrated because the characteristics of the sensors are not very stable and suffer ageing.

The main disadvantage of thermistors is the functional dependence that is characterized by a nonlinear resistance versus temperature. Linearization is mainly obtained using analogue circuit techniques, but sensor linearization is also possible and has been accomplished producing linear output thermistor assemblies. These consist of two or three thermistors assembled as a single thermistor and of an additional resistor. With an appropriate choice of the elements, these packages are interchangeable within a stated tolerance. Although standard thermistors are generally inexpensive, the cost of linear, interchangeable low drift thermistors is elevated. Linear response transducers have the advantage of a simpler electronic circuit or data processing and have a homogeneous accuracy on the whole range. Linear thermistors in a selected range of temperatures are obtained with a suitable combination of two subcomponents: a thermistor composite and a resistor set composed of a compensating network of two or three precision metal film resistors. Further details on these and other temperature transducers can be found in Schooley (1986), Doebelin (1990), Michalski et al. (1991), and Nicholas and White (1994).

A common method to obtain linearization of standard thermistors is to use a Wheatstone bridge circuit (Fig. 11.1) with the resistance sensor $R_T(T)$ that constitutes one arm of the bridge and measuring in the deflection mode (i.e. with the bridge out of balance) the output voltage. This is, in general, a linear function of the bridge excitation E (i.e. the bridge battery

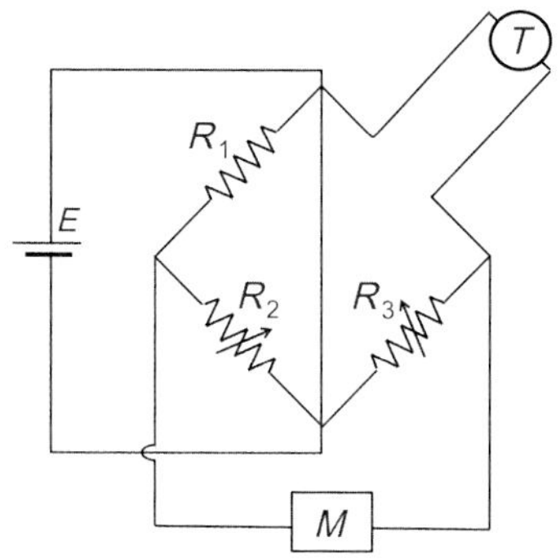

FIGURE 11.1 The basic Wheatstone bridge in deflection mode for the case of a thermistor T. The two resistors R_2 and R_3 are adjustable to set up the bridge to make linear the response of T, as seen by the meter M. E is the stabilized excitation voltage.

voltage after stabilization) but a nonlinear function of the resistance of the elements R_1, R_2, R_3 and $R_T(T)$ of the four arms. When the sensor calibration curve $R_T(T)$ is known and expressed with a polynomial regression, the appropriate resistance of R_1, R_2 and R_3 of the other three arms can be properly calculated and adjusted, so that the exponential curvature of the sensor is compensated by the nonlinearity of the bridge circuit, which has been appropriately unbalanced.

The meter M, which measures the output voltage across the terminals, instantly follows the variations of the sensor, and with an appropriate choice of R_1, R_2 and R_3, the output voltage directly equals the value of the thermistor temperature. Output accuracy and departure from linearity can be better than 0.1 °C. Main sources of errors are instability of the bridge excitation; condensation or rainfall, which forms a shunt of liquid water between the thermistor leads if these are not well insulated; and drift of the thermistor or other resistors of the bridge.

For psychrometric measurements, the assembly of two basic Wheatstone bridges for two thermometric sensors T_1 and T_2 is necessary. It is possible to change from the reading of T_1 to T_2, or vice versa, operating on a commutator. Two adjustable resistors are used to set up the bridge for the thermistor, during the calibration of the device. If the meter has elevated impedance (as most of the electronic meters have), the current across it vanishes and the meter measures the output voltage $e(T)$:

$$e(T) = \left(\frac{R_1}{R_1 + R_2} - \frac{R(T_1)}{R(T_1) + R_3} \right) E. \tag{11.7}$$

Both platinum resistance sensors and linear thermistor transducers are very small (Fig. 11.2), accurate (better than 0.1 °C), repeatable and reliable; linear or transformable into linear; interchangeable within 0.1 °C; and

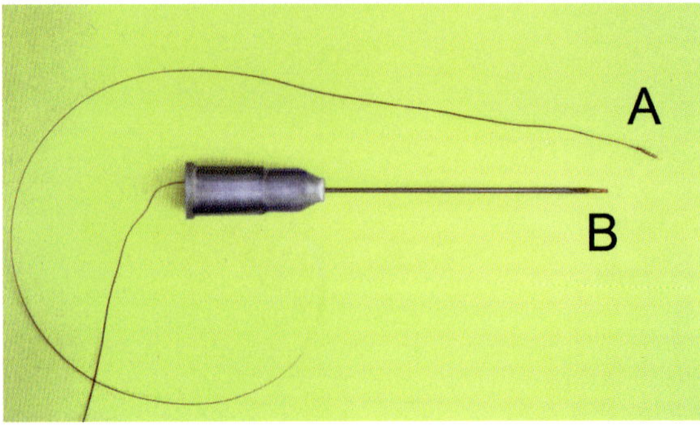

FIGURE 11.2 Small thermistors used for fast response in psychrometers: one is free (label A) and one is inserted into a hypodermic needle (label B), forming a fast response probe.

have fast response, e.g. time constants <1 s can be found. However, it is advisable to buy one or some dozen(s) of sensors and calibrate all together in a calorimetric bath. It is then possible to individuate those that give a closer response, so that they can be substituted for each other or match two or more of them with a very similar response. In practice, the precision of 0.1 °C is satisfactory in most cases and matching of the order of ±0.01 °C can be done without difficulty. Linear thermistors are not only much more sensitive than platinum resistance transducers but also far more expensive.

11.1.2.6. Thermocouples

Thermocouples are based on the *Seebeck effect*, i.e. a small thermoelectric current is generated when two different metal wires are put into contact at both ends with their junctions having a different temperature. If one junction is open, a contact electromotive force is generated. The current, or the electromotive force, is to a first approximation proportional to the temperature difference ΔT between the two junctions. A better approximation is obtained with a MacLaurin expansion with the second power of ΔT. The electromotive force of some of the most common junctions is reported in the junction table:

Junction	EM Force	Junction	EM Force
Iron–constantan	52 μV/°C	Copper–constantan	43 μV/°C
Copper–manganin	41 μV/°C	Manganin–constantan	41 μV/°C
Chromel–alumel	41 μV/°C	Platinum–constantan	34 μV/°C
Platinum–rhodium	6.4 μV/°C		

The strongest electromotive forces are obtained with the less-expensive junctions not with metals that are generally used to resist to high temperatures. Metals used for thermocouples should be carefully aged by electrical annealing because the electromotive force is sensitive to mechanical deformations.

The main advantages they offer are that they have a very low sensitivity to *IR* radiation and a short time constant, of the order of 1 s. If more than one set of thermocouple junctions is used in series, the electromotive force is increased proportionally to the number of junctions.

As thermocouples respond directly to differences in temperature, they are convenient to measure temperature gradients. However, the wire is thermally conductive and in the case of gradients, or different thermal levels between a surface and the air, it may transport heat and form a thermal bridge, attenuating the temperature difference. Thermal bridges can be avoided if they are long enough to minimize the heat transport from one joint to another. With this precaution, they are convenient to measure the wet bulb depression in psychrometers, as this instrument requires better accuracy in detecting the temperature difference between two sensors than in knowing the exact value of the dry bulb thermometer. However, the electromotive force generated by the wet bulb depression is very weak. The sensitivity of a thermocouple to temperature changes is much lower than that of a platinum resistance sensor and especially to that of a thermistor.

Thermocouples need a cold reference junction that is usually immersed in melting ice. This fact limits the practical use of these sensors. Another alternative is to compensate the cold junction temperature by using a resistance bridge compensation circuit (WMO No.622, 1986) or other methodologies. The change of resistance of the thermocouple with changing ambient temperature creates an out-of-balance bridge potential, compensating for the missing cold junction. A further possibility is to measure with an independent thermometer the reference temperature at one junction of the thermocouple and then measure with the other junction the differences in temperature.

11.1.2.7. Quartz Thermometer

The sensor is a quartz piezoelectric resonator that is connected to an electronic solid-state oscillator. The latter supplies a small amount of power to the resonator that acts as a highly selective filter that holds the oscillation frequency very close to the natural frequency of the resonator. The resonant frequency of the quartz crystal sensor undergoes a change in frequency that depends on the change in temperature, following a third-order polynomial equation of T. By a proper choice of the cutting planes of the crystal, the coefficients of the second- and third-order powers of T can be nullified and in this case, the resonant frequency becomes a linear function of the temperature (Michalski et al., 1991).

Once the probe has reached equilibrium with the surrounding fluid, the quartz thermometer readings are very accurate and may constitute a good reference standard for calibrating other thermometers or may be used in precision calorimeters. The response time of a probe is a few seconds in stirred water and is quite long in still air. However, by counting all the oscillation pulses, it is possible to integrate with a great precision over a time interval: the longer the sampling time, the higher the resolution. For this reason, commercially available instruments have options for different sampling intervals and corresponding resolutions, e.g. on increasing by 10 the sampling period, the last digit represents a resolution 10 times greater, so that incredibly high resolutions, as far as 10^{-5} °C or 10^{-6} °C, are attained. It should be remembered, however, that these readings are accurate only in the case the probe has really reached equilibrium with the fluid and the temperature is stationary; otherwise the precision is merely illusory and misleading, being only a precise average of the sensor temperature during the sampling time interval.

11.1.3. Blackbody Globe Thermometer

It is important to avoid the simultaneous use of different kinds of thermometric transducers. Any thermometer furnishes the temperature of the sensor, which is not exactly the same as that of the air, as each sensor responds in a different way to the *IR* radiometric forcing as well as to conductivity and convection. When different kinds of well-calibrated sensors are immersed in a water or oil bath, they give the same output because their temperature is conditioned only by the bath temperature. In fact, the external *IR* radiation is shielded by the bath vessel or is completely adsorbed after the radiation has penetrated a few micrometres into the liquid. The specific heat of water is three orders of magnitude greater than that of air. Sensors responding in the same way to the air temperature, but not to the radiometric forcing, may provide different readings and generate confusion. Screens may attenuate this problem but not solve it.

The rise of temperature of a thermometer exposed to environmental *IR* radiation is the principle on which the *blackbody globe thermometer* (also called *black globe thermometer* or, simply, *globe thermometer*) is based. It consists of a temperature sensor placed at the centre of a hollow 'black' sphere that absorbs, and emits, *IR* from surrounding surfaces. The equilibrium is reached after the contributions of conductivity (determined by ambient air temperature), *IR* balance (radiation exchanges with the ambient and hot sources) and convection (ventilation and convective motions). Frequently, these globes are made of metal (e.g. copper, aluminium) and the sphere is painted with black varnish. There is no doubt that the sphere absorbs the visible radiation and appears black to the eye but the metal below the coating may reflect *IR* radiation, as it often occurs if the coating is not adequately chosen. Emissivity of thick oil paints generally ranges from 0.90 to 0.97 with mode around 0.94. For this reason, the coating should be made of pigments absorbing in the *IR* region, which do not necessarily appear black to the eye. After the sphere has reached equilibrium with air temperature, air movements and *IR* income, the temperature measured at the centre of the sphere gives an approximate measure of the effective temperature experienced by people or objects in that room. The overheating, i.e. the globe thermometer temperature minus the air temperature, gives a crude indication of the contribution of the radiant heat. This method is used in health physics to study the heat stress in workshops.

This principle, which furnishes the above useful application, is also a drawback in that *IR* radiation affects any thermometric measurement and is negligible only for very thin metallic sensors that reflect *IR* radiation, e.g. thermocouples and platinum resistance transducers.

Abundant literature exists on this subject and special reference has been made to UK Meteorological Office (1981), Benedict (1984); Schooley (1986), WMO (1986; 2008), Michalski et al. (1991), Wylie and Lalas (1992), Nicholas and White (1994), Bentley (1998), Michalski (2000), Liptàk (2003) and Camuffo and Fernicola (2010).

11.1.4. Screen

The result of a temperature measurement is a signal provided by the transducer as a function of the sensor temperature. However, the temperature of the sensor does not necessarily coincide with that of the air or the surface. The thermal equilibrium of the sensor includes not only the conductive heat exchange with the air that comes into contact with the sensor but also the energy obtained from visible or *IR* radiation from remote bodies. In fact, the air is transparent to the visible and to the main part of the *IR* spectrum. The incoming radiation is an important source of error and may cause departures from some tenths to several degrees. Under extreme conditions the difference may reach 25 °C. A screen is necessary to reflect this income. However, it is never possible to reflect completely the energy income or avoid the nocturnal radiative loss. The double louvred screen of standard weather stations may introduce an error of +2.5 °C during strong sunshine and calm of wind and of −0.5 °C on clear, calm nights (WMO, 2008). Another source of error is rainfall, when the wet screen evaporates and approaches the wet bulb tem-perature.

For outdoor measurements of temperature and humidity, the instruments should be installed inside a shelter to minimize the effects of sunshine, rainfall, and other adverse weather situations. In the case of standard meteorological observations, the Stevenson screen, i.e. a box made of a low conductivity material (generally wood or plastic), painted white inside and outside to reflect radiation, having double louvred walls that enable good ventilation while minimizing the effects of radiant heat is popularly used. The roof is made of two layers of wood with an air layer between them and the floor is slatted to permit free air circulation. The door is on the northern side and enables the personnel to observe or maintain the instruments. These are placed between 1.25 and 2 m above the ground. Meteorological shelters need to be cleaned frequently and repainted every year but this recommendation is often neglected.

Several kinds of screen have been designed to house electrical-type temperature sensors. They are characterized by a single or double, highly reflective shield (mirror or white paint) with natural or forced ventilation. The shield may consist of one or two concentric tubes and a single or double dome or multiple plates, made of metal, plastic or mirrored glass. The multiple plate nonventilated type is an extension of a model with acrylic plastic dishes developed by Hadlock et al. (1972). In windy, rainy regions, the radiation error is negligible, ventilation is not necessary and the shield must protect the sensor against rainfall arriving from any angle. In sunny, low wind regions, a motor blower to force ventilation (e.g. 3 m s^{-1}) in the area of the sensor is applied to reduce stagnant air and overheating. For nonventilated shields, the overheating error varies with the wind speed, e.g. in a sunny day, the overheating of a white thermoplastic multiplate radiation shield is +1.5 °C at 1.5 m s^{-1} wind speed, +0.7 °C at 2 m s^{-1} and +0.4 at 3 m s^{-1}.

The best results are obtained with a double-walled glass tube, silvered on all interior surfaces, then evacuated and sealed, like heavy-duty vacuum bottle with open bottom. The cylinder is kept vertical, with the open bottom facing downwards and with a blower located at the upper top that continually forces the ventilation inside. In this case, the error is limited within ±0.05 °C.

Also, indoor measurements need a screen, but when direct solar beams or incandescent lamp light do not reach the sensor, the screen might be less sophisticated. Good results are obtained with two methods commonly used for radiosondes: (1) a tube is made of white polystyrene foam that is reflective, is a bad conductor and has a very low thermal capacity and (2) a tube is built with a very thin foil of aluminium, perfectly reflecting on the outer side and blackened on the internal one. In the case of metal shield, a lower disturbance is obtained with a double concentric shield and this caution can be used in the absence of forced ventilation. A common size is a cylinder with some 10 cm diameter and 20 cm height. The sensor is placed in the middle of the tube and radiosonde ascent provides ventilation; for use in a fixed location, a fan can be included on the bottom. When the sensor is located in sites where all the surfaces have a nearly homogeneous temperature and there are no important sources of radiative perturbation, there is no need for shields.

11.1.5. Instrument Location

The site of a standard weather station should be meteorologically representative of the area in which it is located and free from local perturbations generated by trees, buildings, water bodies and air pollution. A plot of level ground, in size 6 by 9 m and covered with short grass is generally used for the installation of meteorological instruments.

Meteorological phenomena take place in time and space scales completely different from those of monuments, which are governed by the local microclimate. In contrast to standard meteorological or airport observations, it is not possible to state a similar guideline for conservation. In the case of microclimate measurements for the cultural heritage, observations must be made when and where necessary, according to the aim of the survey.

However, it might be useful to remember that the air temperature is not the same at all heights and that the choice of the height is very important. Outdoor, the air temperature is essentially governed by the soil temperature: during night-time, the soil is coldest and the temperature increases with height; during daytime, the soil is warmer and the temperature decreases with the height. In summertime, in the central part of the day, bare sand may reach 70 °C and the air temperature rapidly decreases by 30 °C–40 °C in the first 2 m of atmosphere. This shows how the 'air temperature' may apparently change by locating the sensor a bit up or down. The essential point is that there does not exist a generic 'air temperature' T, but T is a punctual and instantaneous value of a function that is variable in space and time coordinates. Also in closed rooms the temperature is not the same at all heights but tends to stratify in horizontal layers, the warmer air being trapped below the ceiling and the colder and denser air being closer to the floor, with a difference that is of the order of 1 °C or 2 °C in usual conditions but that may reach 20 °C or more. Measurements made to test the horizontal homogeneity of the temperature in one room or on one floor must all be exactly made at the same horizontal level; in the case of churches with warm air heating, where a vertical gradient of some 7 °C m^{-1} was found (Chapter 1, Fig. 1.15), a vertical displacement of 10 cm of the sensor will introduce an apparent change of temperature of 0.7 °C.

When ambient monitoring can be performed using only one instrument (e.g. a thermohygrograph), it is extremely important to know the representativeness of the instrument location, with reference to the whole room or the site. In fact, temperature and humidity vary either temporally or with space, e.g. responding to the solar radiation through glass panes, the opening of doors and windows or the switching on or off of light or other heating, ventilating and air conditioning (HVAC) systems. Measurements made at points particularly shielded or exposed to air currents, in proximity to HVAC, or perturbed by the presence of hot or cold water pipes are nonrepresentative of the real situation and their interpretation may induce wrong management of the ambient conditions. A micromapping survey should be made to know the temperature distribution inside a room in order to choose the most representative point for the location of the instrument and especially of the sensors for the control of the temperature and humidity in the room.

11.1.6. Measuring Vertical Profiles of Air Temperature and Room Atmospheric Stability

Vertical profiles of air temperature are important because they constitute a measurement of the atmospheric stability that is responsible for the dispersion of airborne pollutants and their deposition. Outside, profiles are measured with tall towers or mini radiosondes carried aloft by small spherical balloons (in urban areas, a small cluster of balloons with parachute is preferred for safety reasons, Fig. 11.3).

In the case of heavy radiosondes or moderate to strong wind, the preferred carriers are tethered balloons, e.g. *kytoons* (Fig. 11.4), which substantially are moored balloons, sized 40 or 80 m^3. The primary advantage of a kytoon is that it remains up and at a reasonably stable position above the tether point, irrespective of the wind, except for very strong winds. Stability is provided by some soft wings on the back. The soft wings are normally floppy and have some inflation openings at the front. In the case of wind, air blows through the openings, inflates wings and stabilizes the kytoon by damping the oscillations caused by wind gusts and turbulence.

The gas used to inflate balloons is hydrogen but this gas should be handled only by well-trained personnel because it is very dangerous and there are risks of explosion and burning especially in proximity to electrically charged bodies, e.g. the operator when the air is dry. Helium is safe but much more expensive and provides slightly less buoyancy (the mass of helium is twice the mass of hydrogen molecule). The suggested practice is to fill balloons with a mixture of hydrogen and helium popularly used to inflate balloons for children because it is relatively light and nonexplosive. In urban areas, for the onset of wind, the balloon may reach soil far from the operator and this constitutes a potential risk.

Meteorological radiosondes and low troposphere radiosondes (resolution 0.1 °C, 1% *RH* and 0.5 hPa atmospheric pressure) with lattice or neoprene balloons are made to be launched only once and be lost.

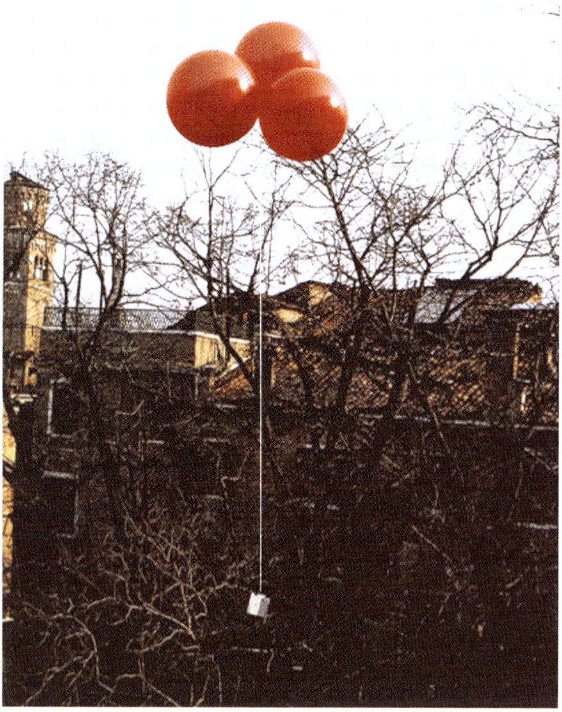

FIGURE 11.3 A mini radiosonde carried by a cluster of small lattice balloons launched from the garden of Papadopoli Palace, Venice.

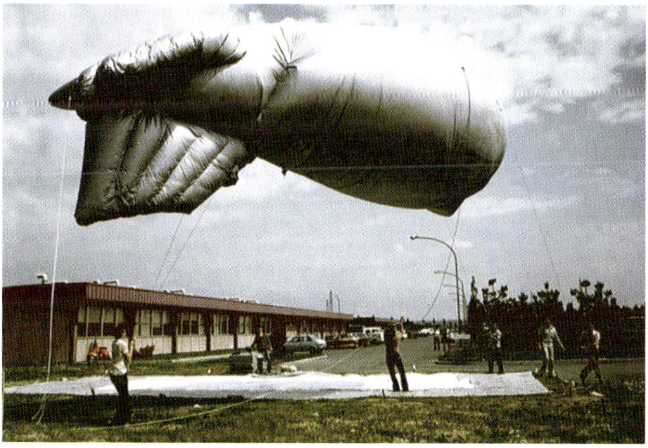

FIGURE 11.4 A tethered kytoon used as radiosonde carrier. The kytoon is shaped like a traditional aerostat but with floppy wings at the back. In the case of wind, air blows into the wings through some inflation holes; wings are self-blown up and the kytoon is stabilized. Launch from the CNR campus, Padua.

However, if the study is limited to the lower part of the planetary boundary layer (PBL), e.g. the nocturnal inversion layer and first kilometre of the diurnal mixed layer, it is also possible to use the same radiosonde several times. The method consists in tethering the balloon, which can be launched and recovered with a nylon fishing line 0.7–1.2 mm diameter (Camuffo, 1980, 1982). During the ascent, the buoyancy force of the balloon turns freely the wire roll; the descent is made by winding the fishing line on the roll by means of an electric motor. This operation is possible only in case of calm or low wind speed. In the case of no wind, the forces are only two, opposed along the vertical: F_b, the upward buoyancy and F_{wt}, the downward traction of the wire. The operation is easily made by controlling the downward force. In the presence of wind, the balloon is subjected to three forces (Fig. 11.5): F_{wd} horizontal, moderate to strong, due to the wind drag; F_{wt} along the wire due to the tension that becomes dominant during the recovering operation and forms an obtuse angle with F_{wd}; and finally, the weak vertical force due to the buoyancy F_b. The resultant of these three forces will make the balloon to descend slantwise and the wire curvature risks to approach, or even touch the ground. In urban areas, this can constitute a strong attraction for children.

Carrier balloons are carefully inflated to such a size that they will provide the proper lift. In urban environments, miniaturized systems are often preferred but small balloons risk exploding for the latex tension when they are recovered. If two or three balloons are used to diminish the risk of explosion and fall, the turbulence generated by tethered balloon clusters forces violent vibrations and increases the risk of bursting. An effective remedy has been found by enveloping the cluster, or only one balloon, with a light hunting or fishing net and attaching the tethered fishing line to the net envelope. The surface stress of the latex is therefore

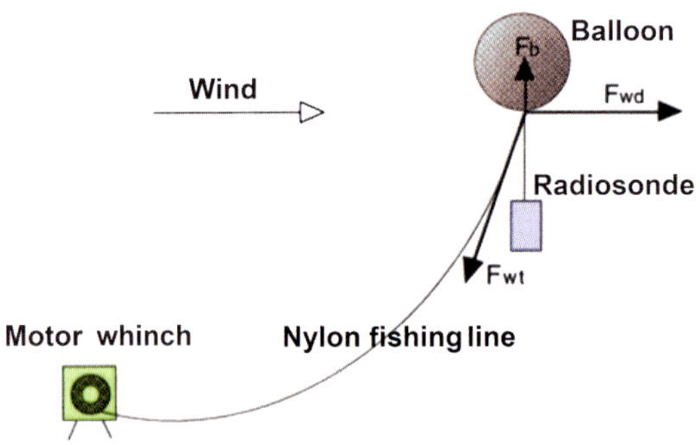

FIGURE 11.5 Forces acting on a tethered balloon: F_{wd} horizontal, very strong, due to the wind drag and F_{wt} (tilted) along the nylon fishing line (or a rope in case large forces are involved) due to the tension that becomes dominant during the recovering operation. F_{wt} forms an obtuse angle with F_{wd}. The third force is the weak vertical force due to buoyancy F_b.

better distributed and the breakdown risk reduced. Minisondes carried by a balloon (size: 1 m^3) or a balloon cluster make possible vertical runs also in city centres, operating in very small gardens or small roof terraces (Fig. 11.3). Using tethered balloons means using a number of times the same radiosonde, thus abating costs. Using the same radiosonde a number of times, the data of each run are taken with the same instrument and are absolutely comparable without the problem of intercalibration errors. Regarding the carrier for ascent, the use of minisondes borne by small balloon clusters is much cheaper than the use of 80 m^3 tethered balloons (e.g. kytoons) with heavy envelope. By the way, the latter need a much larger operating space (Fig. 11.4), hence they can be used only in rural environments. By reporting the results of each run in a diagram and drawing the isolines, vertical time cross-sections of the thermodynamic properties of the PBL are obtained.

Fixed towers or portable masts can be used for profiles on a smaller vertical extent. Some 10 m folding masts with a tripod basement are commercially available. Extensible masts, with the antenna composed of an assemblage of five concentric sections with clamp collars to which instrument can be hung, are also easy to use. Masts are extended by pumping air with a hand pump. On opening the valve and loosening the clamp collars the mast retracts.

Inside buildings, under unperturbed conditions, the air is often stratified in horizontal layers, whose density decreases with height. The layers are not horizontal, or the stratification may even disappear, when there are sources of momentum (air currents) or heat (formation of convective cells). In addition, in museums, vertical profiles show the difference of temperature of the air masses that come into contact with the lower part and the top of a painting or a statue. In museums, small portable masts with sensors placed at the floor level and at 1, 2 and 3 m are in general sufficient, as this height is representative of the air layer where most exhibits are exposed.

When rooms or churches with large domes are too tall to be controlled with a portable mast, if they have small openings on the ceiling, e.g. for raising/lowering chandeliers, it is possible to let down a rope with sensors and record the temperature and humidity profile on that vertical. This method was used several times, e.g. for the Giotto's Chapel, Padua (Camuffo and Schenal, 1982), with thermistors or high-resolution radiosondes for the lower troposphere. An easier system, not conditioned by the presence of openings, is to fill with a hydrogen and helium mixture a black tethered meteorological balloon and control its vertical ascent. The balloon envelope made of thin latex reaches thermal equilibrium quickly with the surrounding air. With a laser pointer radiometer, we can read the latex temperature that equals the air temperature at selected heights, determined from the length of the tethering wire. The balloon should not be inflated too much in order to preserve elasticity and reduce vulnerability in the case it accidentally comes into contact with a rough surface; dark colour is preferable to avoid reflection of visible radiation. In order to also furnish a well-visible target in tall rooms, the balloon may be 2 m in diameter and 300-g latex balloons are needed to reach this size. For rooms that are not exceptionally tall, 100-g latex balloons may be enough. The tethered balloon methodology is more attractive than easy.

A common sampling time for indoor temperature and humidity is 10 min or less, in order to monitor transient phenomena such as room cleaning, windows and doors opening for room ventilation and indoor air exchange. Longer intervals (e.g. one acquisition per hour) might not monitor important perturbations, e.g. room cleaning with open windows.

In Chapter 7, several criteria have been presented for the measurement of atmospheric stability. Most of them consider that the atmospheric turbulence is determined by two main factors:

1. vertical temperature gradients generated by the contrast of temperature between air and soil, which in the case of warm soil causes convective mixing and instability; in the opposite case of colder soil, thermal layering and stability;
2. eddies generated by the wind and expressed in terms of wind gustiness, shear and speed.

It is evident that outside the wind plays an important role but inside it is practically absent or is substituted by weak ventilation or local convective motions. For this reason, inside the thermal factor becomes largely dominant and, in general, it is sufficient to measure the vertical profile of the air temperature with a chain of sensors or to measure the floor and ceiling temperatures with two radiometers, as it has been previously discussed.

However, dealing with a closed environment, several causes of stability and instability must be considered in addition to the floor and ceiling temperatures. Stable conditions are generated when HVAC systems introduce in the room new air with density very different from the ambient air. In the case of warm air, it rises due to its buoyancy till it is trapped by the ceiling and diverges and part of it mixes with the ambient air. The warm air cushion aloft, the cold ambient air at the bottom and the mixed air at the intermediate levels determine some thermal layering and a stable atmosphere. Similarly, when cold air is introduced, it sinks generating again a thermal layering. In the absence of HVAC, thermal stability is generated only when the floor is colder, the ceiling is warmer and the walls have intermediate temperatures. A cold ceiling generates instability like a warm floor for the symmetric reason; warm walls induce uprising air currents and cold walls induce descending ones; in general, this perturbation is appreciable only close to the walls, but these are often painted or have paintings or tapestries. As floor, ceiling and wall temperature anomalies cannot be excluded, the indoor stability is better monitored by an integrated system composed of a vertical chain of air temperature sensors and some fixed radiometers pointing at the floor, the ceiling and the walls.

If some ventilation (either natural or forced) exists, a measurement of the air speed should be included. If only one air speed sensor is used, e.g. a hot wire anemometer, the stability can be expressed in terms of the Högström ratio, which considers the ratio between the vertical gradient of temperature and the destabilization effect of the wind kinetic energy, measured at only one point, intermediate between the two points at which the temperature sensors are placed.

Vertical temperature gradients might be measured with thermocouples. This kind of sensor simplifies the problem of the perfect intercalibration of sensors and is theoretically convenient when differential measurements are needed, because the electromotive force that is generated is a function of the temperature difference between the two junctions. However, in order to overcome the problem of a weak signal generated by a small temperature difference, well-calibrated and matched platinum resistance sensors or thermistors are generally preferred.

11.2. MEASURING ARTWORK SURFACE TEMPERATURE ACCORDING TO EN 15758: 2010

Measuring the temperature of cultural heritage objects is a very particular task, because the surface is precious, vulnerable and always exposed to risk when handled or investigated. Although the sensors may be the same as are used in other fields (e.g. meteorology, industry and agriculture), use and limitations are mostly different. A number of standards exist on temperature measurements, but the European standard EN 15758:2010 is the only one specifically tailored for cultural heritage. For this reason it is useful to comment it. In practice, three methodologies can be used to this aim:

- contact sensors placed in physical contact with the sensed object;
- quasi-contact sensors placed in close proximity to the surface but without any physical contact with it;
- remote sensing, with the instrument located at safe distance from the surface.

In theory, it would be preferable to avoid or limit direct contact with cultural heritage objects but this is not always possible. The first methodology is based on the exchange of heat between two bodies (i.e. the sensor and the sensed object) entering into physical contact. The other two are based on the thermal *IR* emission of bodies and are conditioned by the surface emissivity of the material under investigation.

For ethical reasons, contact sensors cannot be used when the physical contact is potentially noxious for the object. For practical reasons, they cannot be used when the physical contact is difficult or the object is remote, in motion, is very thin or with extremely low thermal capacity (e.g. lace, tapestry, a paper sheet). Contact sensors may be used with stone, glass, metals or wood. On the other hand, they constitute the only possible method to measure the temperature of polished metals or other surfaces with low *IR* emissivity.

Quasi-contact sensors constitute an excellent alternative to contact sensors, except in the case of polished metals. Although quasi-contact sensors collect the total radiation (i.e. both direct and diffuse radiation) emitted by a surface and are therefore independent of the actual value of the specific emissivity ε of the surface, the surface should have a reasonably high emissivity (e.g. $\varepsilon > 0.7$), otherwise readings are largely uncertain and contact sensors are preferable. There is no problem in using quasi-conatct sensors for thin objects or low-thermal-capacity materials.

Remote sensing is the most difficult technique because it requires high precision and very expensive instruments and the readings are affected by both the surface emissivity and the *IR* emitted by third bodies and reflected by the sensing surface. It has the great advantage of measuring remote surfaces (e.g. a ceiling without needing scaffolding) or mobile objects. The surface should have reasonable emissivity (e.g. not possible for polished metals) and should be opaque to the specific wavelengths of the thermal radiation monitored by the instrument (e.g. not convenient for glass).

Measurements of cultural heritage objects are regulated by the European standard EN 15758: 2010 '*Conservation of Cultural Property – Indoor Climate – Procedures and instruments for measuring temperatures of the air and of the surfaces of objects*' that recommends the most convenient methodologies for the measurement of the temperature in air and at the surface of the cultural heritage materials. The measurements in air should be of good quality and be representative of the process that we investigate. In addition, the measurements of the surface temperature require a special care for vulnerable surfaces and should consider the special characteristics of some materials. For instance, polished metals are resistant and have low *IR* emission. For them *IR* monitoring is impossible and contact sensors are needed. Stones and masonry are in general resistant to gentle use of contact sensors but their surface emissivity is high enough and *IR* measurements are possible and often preferable. In the case of tapestry, paintings on canvas or panels, books, mummies, wooden objects, murals, heavily oxidized metals or other very vulnerable surfaces with medium to high emissivity it is preferable to avoid any risk of contact with sensors. For this reason, quasi-contact devices based on total *IR* emission (see later) are preferable. By the way, this technique is the only one technically possible for surfaces like paper sheets, paintings on canvas or tapestry, if we exclude remote sensing. Remote observations of high emission bodies made with *IR*-based instruments (e.g. pyrometer, thermocamera) are possible and sometimes particularly useful, especially in the absence of scaffolding, but it should be considered that the accuracy is generally low, except for high-quality, very expensive instruments (Camuffo and Fernicola, 2010; Camuffo, 2010). In the following discussion, we will return to characteristics, advantages and problems of the various methodologies and instrument type according to EN 15758: 2010.

11.2.1. Contact Sensors

Contact sensors have a flat surface that is put into contact with the object to investigate. The equilibrium is reached when, at the interface between the sensor and the measurand surface, the exchange of heat stops; however, a bad contact may also led to the same result, leaving the sensor and surface at different temperatures. This particularly happens when the surface is rough so that only a few points of the artefact are in contact with the sensor, because the small air pocket between the artefact surface and the sensor acts as a good insulator. In order to improve the thermal contact, grease or gelatine can be smeared on the flat side of the sensor. A commonly used material in industry is silicone grease with some metal powder included to improve the heat transmission. This medium, however, can not be applied on cultural heritage objects, as it stains the surface, and other noninvasive, clean methodologies should be used.

Another problem is that the outer side of a contact sensor is exposed to the atmospheric agents, e.g. different air temperature and solar radiation. To minimize these perturbing factors, this side is isolated but if it is exposed to direct solar radiation, it needs an extra shield

A further problem is that the presence of the sensor perturbs the surface temperature, especially in the case of a surface that is warmed by solar radiation. In fact, the skin of the surface that is hit by solar beams is warmed but the small area to which the sensor is attached remains shielded and colder. The sensor exchanges heat with the deeper layers below it, in the colder area, and mainly comes in equilibrium with them, although it receives some heat that converges laterally from the warm lighted surface. Consequently, the surface sensor instead of measuring the skin temperature measures the temperature of a colder, subsurface layer. This is a problem of the measuring approach used, not an instrumental error.

The opposite occurs during the night-time, when the monument looses heat by *IR* radiation, except the small area that is in contact with the sensor. Similarly, when the monument is wet and its surface temperature drops to the wet bulb temperature, the area in contact with the sensor is not affected by forced evaporation and a different temperature is measured. As always, the variable that is monitored is the sensor temperature and the actual surface temperature remains unknown, except for the contact sensor approximation.

In steady-state conditions, it is assumed that a body is in equilibrium with its surrounding atmosphere. When a sensor comes in contact with the surface, it alters the radiative balance and the conductive and convective exchanges, so that the thermal distribution of the body surface is locally altered. The sensor generates a new heat flow from (or to) the surface, and where the thermal contact is not ideal, a thermal contact resistance is introduced. Therefore, it can be concluded that a sensor will always interact with the surface under investigation, so that measurements will be perturbed in any case.

When passing from the steady-state to the dynamic conditions, the difference between the actual temperature and the measured value will increase. This is apparently expected when the thermal response of the sensor is slower than the response of the body surface but in practice, this always happens because the presence of the sensor perturbs the heat exchange between the surface and the environment.

Only when the radiative balance does not affect too much the surface temperature and the ambient conditions remain stationary for a sufficiently long time, does the measurement become representative of the surface temperature. When the radiative gain or loss makes the skin temperature of the body different from the subsurface layer, a possible way to monitor the real temperature is to leave free the surface and then touch it with the sensor for a very short time. However, the heat capacity of the sensor affects the temperature of the few protruding points of the surface in contact with the sensor. For this reason, it is necessary to proceed with further steps. The method consists in touching the monument surface with the sensor at different points from that under observation in order to bring the sensor to a closer and closer temperature so that the observed point will not be affected by the sensor's influence, as follows. When the sensor touches a surface, in a short time it reaches a new equilibrium and its temperature goes closer to the unknown value under examination. Repeating a number of times this operation, a better approximation is obtained with the sensor approaching more and more the undisturbed skin temperature. After some of these manual operations, when the output readings remain unchanged, the asymptotic value is reached and it is possible to read the temperature of the point under observation minimizing the influence of the sensor. This method has been successfully applied several times, e.g. the Aurelian and Trajan columns, Rome (Camuffo and Bernardi, 1993), or Pisa Tower; the drawback is that it needs an operator.

11.2.2. Quasi-Contact Sensors

Quasi-contact sensors are placed at short distance from the surface but not in physical contact with it and are also called *total radiation sensors* because they receive both the *direct* and the *diffuse IR* radiation emitted by the surface. The sensor is located on the focus of a parabolic, mirroring cavity made of stainless steel, which has a low emissivity $\varepsilon = 0.07$ (Fig. 11.6). In addition to the direct *IR* emitted by the surface towards the sensor, the internal mirroring parabolic cavity conveys on the sensor all diffuse *IR* emitted by the surface. In practice, the sensor is virtually enveloped by the mirror image of the surface and in this way, it is only hit by the radiation emitted by the object surface, if we disregard the small contribution (i.e. 7‰) of the mirror. The external mirroring surface shields the sensor from all the external *IR* emitted from third bodies. In practice, the sensor shielded and encased in the mirroring probe is only (or almost only) hit by the *IR* emitted by the object surface and this apparatus transforms the efficiency of the emitting body surface, as it had emissivity $\varepsilon = 1$ or, more exactly, $\varepsilon = 0.993$ if we consider the influence of the stainless-steel mirror. In conclusion, quasi-contact sensors can be used with any type of material with high or average emissivity. The external shield removes all reflections and the parabolic mirror multiplies the radiation emitted from the target. The mirrored image behaves as the original target surface. The instrument virtually transforms a high or average emissivity into an almost perfect blackbody. This instrument cannot be used with materials having very low emissivity, e.g. polished metals, because the low *IR* emission is insufficient and comparable to noise. A polished metal target is indistinguishable from the metal mirroring cavity.

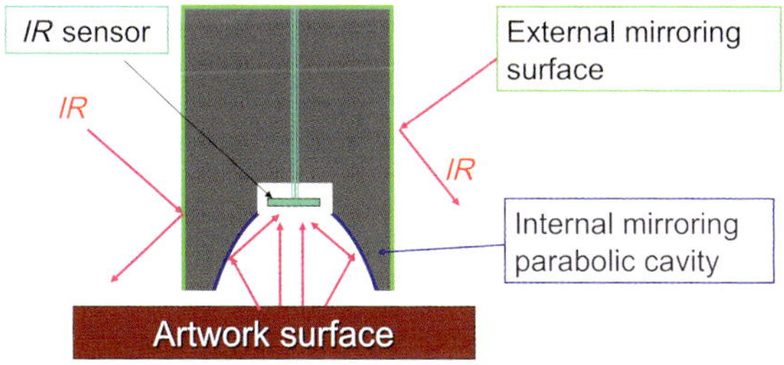

FIGURE 11.6 Operation principle of a quasi-contact sensor. The sensor is positioned on the focus of a mirroring parabolic cavity and receives the total radiation (i.e. direct and diffuse) emitted by the object surface. The external metal case reflects all *IR* radiation emitted by third bodies. In practice, the sensor receives only the radiation emitted by the surface under investigation. *Source: After Camuffo (2010); by kind permission of Nardini Editore, see credits.*

11.2.3. Radiometers and Remote Sensing

Radiometers provide indirect measurements of small- or large-area temperatures, based on the principle of measuring the *IR* emission of bodies. In order to reduce the effects of atmospheric absorption of some *IR* bands due to the presence of water vapour (which is the most important greenhouse contributor), CO_2, O_3 and other greenhouse gases, the 8- to 14-μm window is generally used. Radiometers are essentially of two types: those that reproduce the thermal image of the objects and the nonimaging transducers that measure the total power of the *IR* radiation that reaches the sensor. These measurements have the advantage that they do not physically perturb the surface under investigation and do not cause damage to the works of art to which contact sensor cannot be stuck. They constitute a non-contact and nondestructive method that is particularly appreciated in the field of works of art.

Another important aspect is that they are remote sensing devices and make possible measurements on ceiling or other surfaces that are reached with difficulty, or on moving surfaces. Finally, their output is representative of a more or less large area, being less conditioned by local departures or fluctuations. On the other hand, measurements are affected by: (1) the presence of extraneous reflected *IR* radiation that constitutes a serious problem for outside measurements during the daytime, (2) the emissivity of the surface and its Lambertian nature and (3) suspended aerosols or moisture that adsorb the *IR* signal. For further details, see Wolfe and Zissis (1989), Kondratyev et al. (1992).

Imaging instruments are thermocameras, i.e. thermal imaging cameras, similar to a normal camera but reproducing images in false colours and sensitive to some wavelengths in the near *IR* band, and thermovision, based on television sensing systems using electron beam scanning. Such instruments are expensive but a thermal image is of immediate understanding for specific problems, showing gradients, warm or cold spots, energy loss and leakages, and cold zones that might possibly be generated by water percolation or other unknown reasons. They are useful in building diagnostics, especially when rapid temperature changes and uneven heat transmission pinpoint to surface temperature anomalies due to subsurface inhomogeneity. To this aim, the methodology is used either in active or in passive mode. The active mode is when the surface under investigation is deliberately heated to monitor temperature pattern emerging during the quick heating or cooling phase. Subsurface discontinuities may appear for the different heat transmission. The passive mode is when the surface is observed during its natural conditions, either stationary or variable. It is also possible to recognize internal discontinuities especially when the object is naturally affected by temperature changes, either for the daily cycle or for the alternation of sunshine and shade.

The false colour images, called thermograms, are constituted of pixels that in practice indicate temperature levels distributed over a bidimensional matrix x–y, i.e. the geometric coordinates of each pixel on the surface. This map is useful in calculating the distribution of the *RH* in the viscous layer at the interface between the surface and the air. If the mixing ratio (*MR*) inside the room is measured, and it is homogeneously distributed, one can calculate the *RH* at the interface from the *MR* in air and the surface temperature, repeating the operation pixel by pixel in order to obtain a visual image of *RH* in false colours. This practice is very useful in environmental diagnostics, e.g. to see whether some heating caused by exhibition lamps or by sunbeams passing through windows may be noxious to paintings, furniture or other objects, as discussed in Chapter 4 (Fig. 4.4 and Fig. 4.10).

Nonimaging instruments (*IR* thermometers) are commonly employed in field surveys for measuring the temperature of inaccessible surfaces. They measure the heating of a sensor placed at the point of convergence of the cone of the flux of *IR* radiation, whose optical angle is controlled by a diaphragm. The diaphragm is fixed or adjustable; adjustable diaphragms allow the use of the instrument for close or remote monitoring or for averaging the temperature over a small or wide area. Close position and/or small angles give high resolution and point measurements; long range and/or wide angles give the temperature averaged over a wide area. A cause of error is the diffraction of the collecting radiation at the limiting aperture, which depends on the *IR* wavelength and therefore the object temperature.

A real surface both *emits* and *reflects* radiation with a relative intensity that varies from one material to another; the measurement is based on the emitted component but the radiometer (as well as the operator) is unable to distinguish it from the reflected one. The mechanism is also complicated because the reflected radiation from liquid or solid bodies consists of two components: the well-known *surface reflection* and the *bulk reflection*, i.e. the portion of the radiation that was transmitted into the material and was reflected by internal backscatter; the latter is independent of the surface condition. The problem of taking reliable measurements especially arises when there are other important sources of *IR* radiation or the surface under investigation has a low emissivity and a high reflectivity. In this case, the measurement is not representative of the body temperature. When a metal or another reflecting surface is investigated, special care should be taken in order to avoid the *IR* radiation emitted by the operator, or other bodies, in particular the sun and clouds. If the surface is Lambertian (this approximation is particularly good for a rough surface), the operator influence can be avoided by observing the surface at slanted angles. However, for most surfaces, the emissivity depends on the angle of view and drops near grazing incidence. In this case a nearly normal incidence is preferred.

Radiometric measurements are based on the knowledge of the surface emissivity ε and the spectral radiance in the band used by the instrument. The simplest case (blackbody) is $\varepsilon = 1$ and corresponds to a known, continuous spectrum; however, in general the uncertainty of these two characteristics is a source of error. Bodies with the same temperature but different emissivity generate different radiometer outputs. An adjustable emissivity control in the instrument permits to correct the output, adapting the transducer to the characteristics of each surface, in order to obtain accurate temperature measurements. However, the emissivity of the object is unknown. One method used is to cover the surface target with a coating (e.g. soot) or a film with known emissivity in the *IR* band used for the radiometric measurement. However, this may alter the body–atmosphere interactions and the surface temperature. The actual value of ε can be found empirically, adjusting ε on the radiometer setting until the true surface temperature (known by means of a surface thermometer) is indicated.

It is necessary to consider that the emissivity of a surface may change with time from a value typical of the dry material to a value close to 1 (i.e. surface wet by rainfall or covered with a film of water when the surface temperature is below the dew point), and this happens frequently with outdoor monuments. The emissivity changes during the day with the water content of the surface layer, in the presence of drying–wetting cycles, and with the surface temperature or wind speed. However, surface soiling, particle deposits, efflorescence, hydrophilic salts, biopatinas and other factors may substantially alter the surface emissivity. The same occurs for the surface reflectivity and this also changes with the angle of incidence. For instance, the reflectance R from a water surface varies from 2.5% at normal incidence (i.e. at 0° angle incidence from the normal) and becomes $R = 8\%$ at 60°, 35% at 80° and 97.5% at 90° i.e. at grazing incidence. As the emissivity is $\varepsilon = 1 - R$, this factor is very important, as a wet surface observed at small incidence (grazing) angles becomes a pure reflector and the emissivity of the water film that envelops the body vanishes. In such a condition, the only *IR* radiation that reaches the observer is caused by third bodies.

Another important factor is that the emissivity is a function of the spectral wavelength λ, i.e. $\varepsilon = \varepsilon(\lambda)$ and materials have a more or less deep valley in the spectral emissivity, which can fall in the instrument bandwidth, thus changing the apparent radiometric temperature. Most organic materials, e.g. wood and parchment, have a low emissivity in the visible part of the spectrum, and high emissivity in the *IR* region. On the other hand, metals covered with a thin layer of oxide seem very dark but the oxide becomes transparent at longer wavelengths, so that in the *IR* range the surface becomes reflecting and with low emissivity, typical of a pure metal (Nicholas and White, 1994).

Of course, it is convenient to compare radiometric observations with other homogeneous (i.e. radiometric) measurements, and for this reason, a complete set of measurements (e.g. ceiling, floor, walls and murals) should be made with the same type of transducer. The instrument calibration should be verified before and after use, by pointing the radiometer at a reference surface, which behaves as a perfect blackbody. A used reference is the free surface of water in a bucket as in the normal direction $\varepsilon = 0.96$ i.e. close to 1. However, care must be taken that the water tends to stratify in layers with different density and the surface layer is affected by evaporation so that its temperature tends to drop to the wet bulb temperature. As the *IR* radiation is absorbed in a few micrometres of water, the radiation emitted originates in the very surface layer that has a temperature different from the bulk water, where the bulb of the truth thermometer is immersed.

The best reference is with the radiometer pointing at a blackbody cavity, which has an emissivity that approaches unity. This can be obtained with a can blackened inside having a very large internal size and a very small aperture, so that the internal reflections are so many that there is equilibrium inside and an external radiation that penetrates is practically extinguished. If the internal surface of the cavity has gradients in temperature, the blackbody emission is a combination of spectra and introduces an error in the calibration. In order to be isothermal, the can should be immersed in a mixed liquid with known temperature. Mixing should be made by generating turbulence with up and down movements, not with rotation that does not destroy density layering.

When the reflectivity of an external radiation is zero, the emissivity also equals unity. A practical formula that determines the effective value of the emissivity ε_{eff} of a cylindrical cavity with a small opening is

$$\varepsilon_{\text{eff}} = 1 - (1 - \varepsilon_{\text{c}})\frac{r_{\text{a}}^2}{r_{\text{c}}^2} \tag{11.8}$$

where ε_{c} is the emissivity of the internal surface of the cavity and r_{a} and r_{c} are the radii of the aperture and the cavity, respectively (Nicholas and White, 1994).

Real materials emit radiant energy at a fraction of the perfect blackbody reference. It is useful to know the emissivity of some materials. Table 11.1 lists the emissivity of some selected materials used for cultural heritage or taken as an example. The table suggests that stones or heavily oxidized metals can be measured with a radiometer but this method is not recommended for polished or slightly oxidized metals or other *IR*-reflecting surfaces.

As an example of the characteristics of commercially available instruments, the main technical specification of the *IR* transducers most popularly used for microclimate measurements are: spectral band pass ranging from 8 to 14 μm, field of view from 4° to 20°, scale range from −30 to 100 °C, accuracy of ±0.5 °C, repeatability of ±0.1 °C, resolution of

TABLE 11.1 Emissivity ε of Selected Materials

Material	Emissivity	Material	Emissivity
Water	0.96	Plastics	0.91
Grass	0.90–0.98	Asphalt	0.96
Snow (old-fresh)	0.82–0.99	Basalt	0.90–0.92
Clay (dry-wet)	0.95–0.97	Dolomite	0.96
Sand (dry-wet)	0.84–0.96	Limestone	0.90–0.96
Soil (dry-wet)	0.90–0.98	Dunite	0.89
Lacquer (white-dark)	0.92–0.97	Feldspar	0.87
Oil paint	0.87–0.98	Gypsum	0.85–0.93
Paper (white)	0.93	Granite	0.81–0.93
Wood (oak)	0.90	Quartz (agate)	0.71
Hardwood across grain	0.82	Silicon sandstone	0.91
Hardwood along grain	0.68–0.73	Brass (polished-oxidized)	0.03–0.61
Glass	0.91–0.94	Bronze (polished)	0.1
Porcelain, glazed	0.92	Copper (polished-oxidized)	0.05–0.78
Brick (glazed-red)	0.75–0.93	Iron (polished-oxidized)	0.21–0.78
Mortar	0.87–0.94	Iron rust	0.75
Plaster (rough coat)	0.91	Iron heavily rusted	0.91–096
Concrete	0.92–0.95	Lead (polished-oxidized)	0.05–0.63
Soot on a solid surface	0.91–0.94	Steel (polished-oxidized)	0.07–0.79

Source: Platridge and Platt (1976), Oke (1978), Green and Maloney (1984), Wolfe and Zissis (1989), Lide (1990).

±0.1 °C, noise effective temperature less than 0.05 °C, response time less than 1 s, operating environment ranging from −10 to 50 °C and $RH \leq 90\%$. More sophisticated instruments are also available, with accuracy, repeatability and resolution improved at least by an order of magnitude, but they are far more expensive. They are particularly useful in detecting anomalous areas during transient conditions. However, the better the instrument characteristics, the more elevated the cost and the thinner the surface layer that takes advantage of the finer investigation, as very small, and short period, temperature changes (i.e. $\Delta T \approx \pm 0.01$ °C) are smoothed out in a very short depth so that only skin, or just subskin, disturbances can be detected.

Comparing a radiometer having accuracy $\pm 0.5\,^\circ$C with a contact thermometer having the same class, one should expect observations with $\pm 1\,^\circ$C maximum departure. However, this is not always true, even if we exclude errors due to *IR* reflection from third bodies. The reason is that a radiometer and a thermometer are based on different physical principles and measure temperature at different depths below the surface. The radiometer measures the *effective radiation temperature,* which is representative of a temperature below the surface at a depth of $1/A(\lambda)$, where $A(\lambda)$ is the spectral absorption coefficient of the material. In absorbing materials, this layer is the skin. On the other hand, to reach equilibrium, a contact thermometer exchanges heat for a deeper layer, whose extent depends on the heat capacity of the sensor and the initial difference of temperature, so that the measurement is more properly representative of a deeper subsurface layer. This problem arises when there are subsurface gradients, as it always occurs in dynamic conditions, or when the body is irradiated by direct or diffuse solar radiation, is cooling via *IR* emission or is utilizing latent heats of evaporation or condensation. Historic buildings or stone monuments may sustain very large temperature gradients, as their heat transfer to the atmosphere is very low. This makes difficult a comparison between radiometric and thermometric observations.

11.2.4. Blackbody Strip

The effective temperature resulting from a thermal balance between air temperature, *IR* radiation and air convection is determined with a fast responding method, i.e. the blackbody strip.

The blackbody strip was first used in some field surveys to solve some problems concerning fast response and high space resolution. It was introduced to assess the level of the thermal comfort in churches among the pews during the continually variable temperature when a warm air heating system is turned on oroff or when *IR* heaters are used. This was critical because heating should be reduced to a minimum acceptable level and/or to restricted areas. At the same time, it was useful to detect the areas hit by *IR* and to evaluate the impact that the *IR* radiation may have on paintings, tapestry, furniture etc. or to verify what may happen in the case that some *IR* heaters will be installed or displaced from one position to another (Camuffo et al., 2007, 2010a,b).

In principle, it is similar to the globe thermometer but is fast responding and more convenient for cultural heritage purposes. The blackbody strip is a low thermal inertia, blackbody target, like a strip of black textile having high emissivity (Fig. 11.7). It soon reaches equilibrium with the incoming radiation, ambient air temperature and ventilation. The equilibrium temperature of the blackbody strip should be accurately measured using the procedures for the measurement of surface

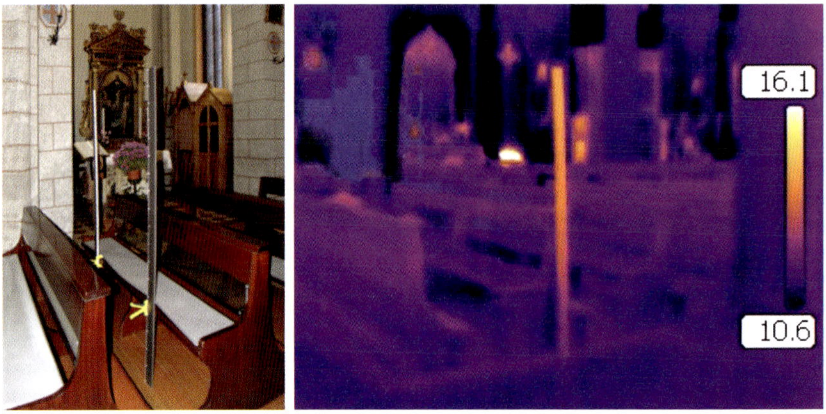

FIGURE 11.7 Blackbody strip to monitor the efficiency of heating systems, either based in *IR* emission or warm air, or both. The blackbody strip temperature is sampled with a radiometer at regular height levels (e.g. every 10 cm) to obtain a vertical profile or is recorded with a thermocamera. In this example, blackbody strips are used to evaluate the thermal comfort for standing persons inside the pew area. One can observe that the head is warmer and feet remain cold, which is not the best for churchgoer comfort. *Source: After Camuffo (2010); by kind permission of Nardini Editore, see credits.*

temperature described for a quasi-contact thermometer, using a radiometer, or using a thermocamera.

In general, a blackbody strip is useful to measure spot temperature levels at selected positions and space gradients and to determine vertical or horizontal profiles, with high spatial resolution. After the above tests, it was included in the EU standard EN 15758: 2010 '*Conservation of Cultural Property – Procedures and instruments for measuring temperatures of the air and the surfaces of objects*'.

11.2.5. Good Practices and Misleading Interpretations

Generally speaking, environmental diagnostics based on the results of only one methodology might be misleading and a good norm is to cross-compare findings derived from independent methodologies. The problem is that a number of different mechanisms may produce the same thermal image and a thermogram may be useful, but if it is considered alone, it is neither necessary nor sufficient to identify which of the various mechasnisms is going on. While temperature measurements constitute a useful investigation tool, thermograms are not per se sufficient to draw conclusions without the support of other specific investigations that are always necessary for confirmation. In the following discussion, some useful examples will be reported in order to familiarize with this methodology.

Critical factors are cold spots, capillary rise and water percolation. Most handbooks suggest that dampness in masonry, e.g. capillary rise

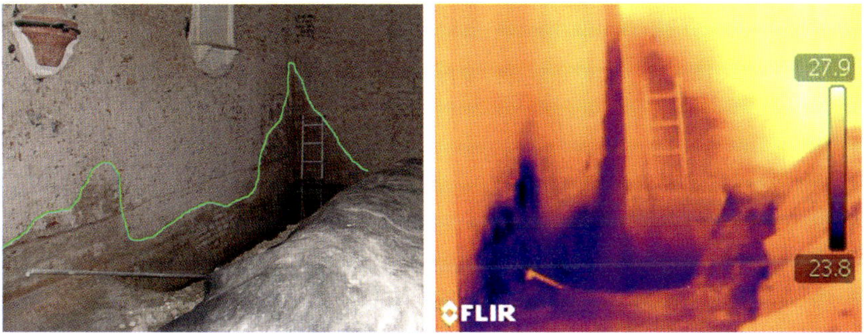

FIGURE 11.8 Visible and *IR* image of dampness for capillary rise and gutter percolation on the corner. The damp area is contoured with the green line in visible picture and appears as the colder part in the thermogram.

and water percolation are easily recognized from the cold spot generated by evaporation from the damp surface. The reason is that the latent heat of evaporation is supplied partly by the air and partly by the surface that is consequently cooled. This is generally true, as shown in Fig. 11.8.

However, several examples have been found where the coldest part of the thermogram is not damp, or even where dampness is found in the warmest part of it, as shown in Fig. 11.9. In this historic palace, unsealed windowsills allowed rainwater to percolate into the wall on both sides forming damp spots. The spots were cold in the morning as generally expected but warm in the afternoon. The reason is that dampness in the wall constitutes a thermal bridge, connecting the inside with the outside

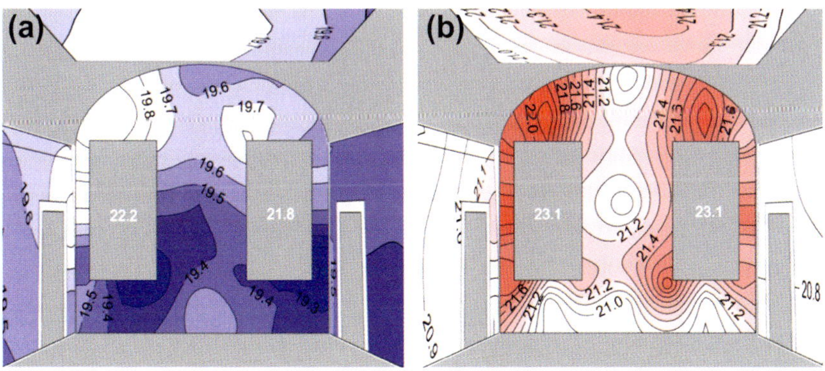

FIGURE 11.9 Temperature (°C) mapping of a room with damp spots on a wall. Damp spots of percolated water lie on both sides of the windowsills, on 14 May 2008. At 9:00 (a), the damp spots constitute the coldest areas; at 15:00 (b), they are warm for the thermal bridge established with the exterior. Other warm spots correspond to heating from window posts and glass panes. Thermogram executed sampling with a nonimaging radiometer and plotting with computer mapping. *Source: Grimani Palace, Venice, after Camuffo et al. (2011); by kind permission of Nardini Editore, see credits.*

surface. In the early morning, the damp spots were the coldest areas not only because of the small evaporation but also because of the heat transported by conductivity from inside to outside. In the afternoon, the heat flow was reversed: the external surface, facing east, was overheated during the whole morning and the external heat entering the room formed the hot spots. Dampness increases thermal conductivity and heat flow, and this effect may largely dominate over a weak evaporation cooling. Also, the window posts facing the solar beams and the glass panes were overheated and formed secondary warm bands inside.

In the next thermogram (Fig. 11.10), windows of a historic palace are contoured by cold bands. The season is winter. Possible hypotheses are: (1) the wooden frame is affected by leakage and cold air blows in, cooling the nearby masonry; (2) some water percolated into the wall from the window and the evaporation causes the cooling in the damp area; and (3) the outdoor temperature is lower, and heat naturally passes from higher to lower temperature levels crossing the wall and the window pane, which is colder. However, some heat crosses diagonally the wall via a shorter path going out through the windowsill and the two lateral posts, forming a cold band all around the window. In this case, the vertical cold bands on both

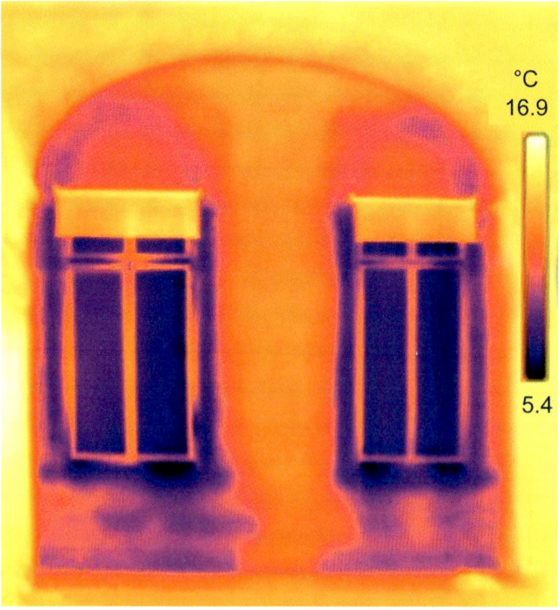

FIGURE 11.10 These windows are surrounded by cold bands. Cold bands might be caused by air leakage through the frame, water infiltration and shorter path for heat transfer. See text for explanation. *Source: Grimani Palace, Venice. After Camuffo et al. (2011); by kind permission of Nardini Editore, see credits.*

sides were due to heat transferred across the shorter diagonal path; the cold area under the window, where individual bricks are distinguishable, is due to water percolation that locally increases the heat conductivity across the masonry, forming a thermal bridge with different efficiency corresponding to bricks or mortar. The problem with thermograms is that different mechanisms may have the same appearance.

Other two examples of warmer evaporating surfaces: in winter, the water flowing in the canals at Venice is mild being continually exchanged with the sea that acts as a huge thermal buffer. In contrast, the temperature of the air in this region is cold or very cold. Palaces built on the side of the border of canals have their basement immersed in water and continually receive heat by conductivity. In the first example (Fig. 11.11), the basement is damp and characterized by a green–brown belt of algal infestation. Although this band is damp and evaporates, it is warmer than the upper part of the building, which is dry, but is in contact with the cold air without benefitting from the heat supplied from the seawater.

In the second example, a funeral monument in a church has the basement that is in contact with the soil, which is mostly damp due to a variable water table fed by the tide level (Fig. 11.12). The underground water migrates transporting heat. The damp funeral basement has modest indoor evaporation, but especially benefits from the heat supply. Once again, this evaporating surface is warmer than the rest of the church.

We have seen some examples of thermograms where the evaporation was occurring in the warmest part of the image. Now we will consider another potentially misleading example: a cold area, but not related to

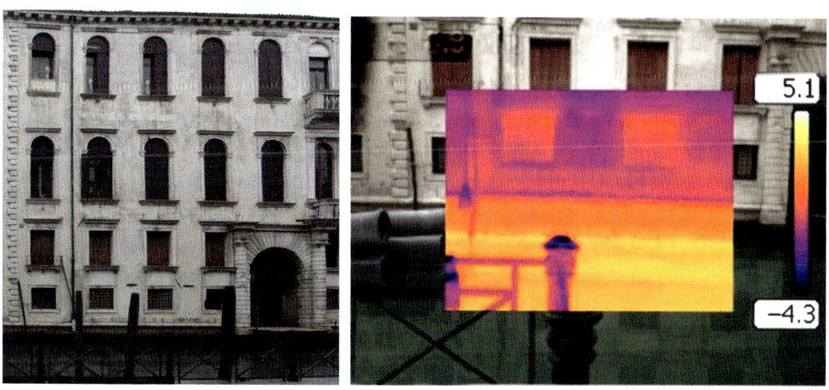

FIGURE 11.11 Grimani palace in Venice, visible and *IR* mixed picture. In winter, the water temperature in the canal is at 4 °C (yellow–orange) and the damp band infested by algae on the building (orange) has almost the same level. The upper part of the wall, dry, is colder (magenta). The heat supplied by the canal is dominant over the heat lost due to evaporation. The evaporating band in the basement is warmer.

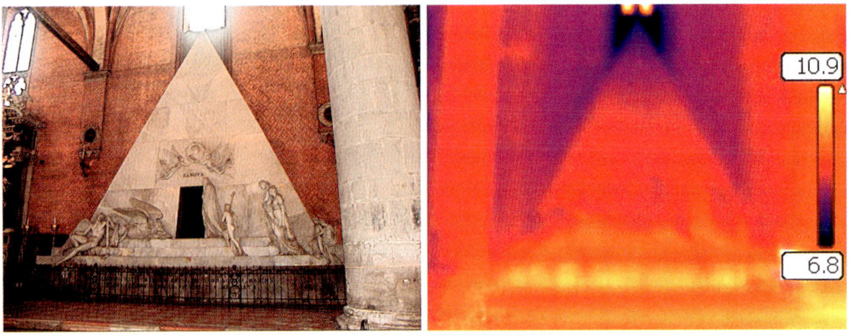

FIGURE 11.12 The funeral monument of Canova, inside the Basilica dei Frari, Venice, is affected by capillary rise of water originating from the canal. However, the indoor evaporation occurs at a very slow rate and the heat transported by the mild canal water is dominant over the loss of latent heat due to evaporation.

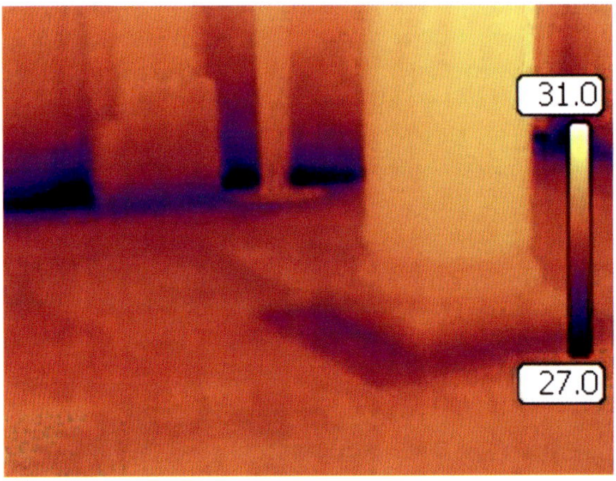

FIGURE 11.13 A ventilated crypt in the warm season. The external air warms pillars and walls but hardly reaches corners and edges, which remain colder. The floor is in brick and the column is based on a squared stone block, with higher conductivity. For this reason, the floor at the base of the column is colder. This is not capillary rise but higher conductivity compared with bricks (by the way, bricks are more porous and favour capillary suction).

dampness. In the warm season, in this crypt (Fig. 11.13), warm air is penetrating from outside, heating pillars, ceiling and walls. However, corners and the edge between floor and walls are hardly reached by the warm air entering the crypt.

The result is that all corners and edges are colder and sometimes this cold may cause elevated moisture levels and a favourable habitat for mould colonization. In this case, the cold areas are not due to capillary rise and evaporation but only due to reduced ventilation; for this reason, they might be affected by condensation (that is the opposite of evaporation).

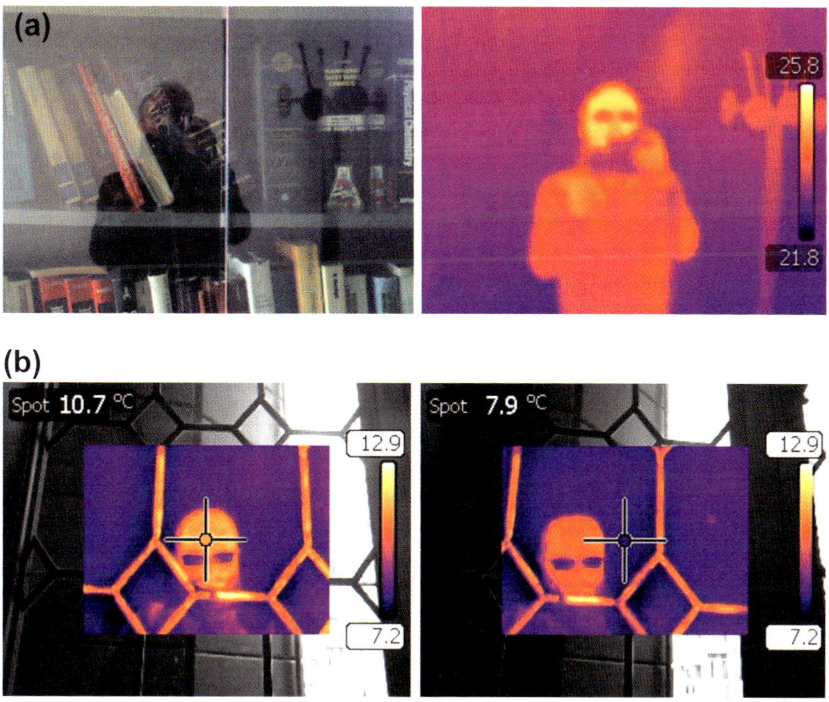

FIGURE 11.14 a) Visible light and *IR* reflection on the glass panes of a glass door bookcase. The operator reflection apparently increases by some 4 °C the glass temperature. (b) *IR* reflection on stained glass panes. The temperature difference pointing just at the forehead or outside it is 2.8 °C. The metal frames appear warm because they reflect the *IR* emitted by the operator. *Source: After Camuffo (2010); by kind permission of Nardini Editore, see credits.*

Glass is partially transparent, partially reflecting and partially absorbing either in the visible or in the *IR* region. For this reason, radiometric reading is affected by both what is behind the glass (transmission) and what is in front of it (reflection), e.g. the operator (Fig. 11.14).

Metals are good reflectors and behave as mirroring surfaces. For this reason, radiometers are not suitable to measure glass or metal temperature. A trick solution is to stick on the glass or the metal surface a leaf of paper (paper is opaque to *IR* and has a quite high emissivity, i.e. $\varepsilon = 0.93$) that will adopt the temperature of the glass or metal, and then to measure the temperature of the paper target.

A good use of paper is to materialize the air temperature. Air is transparent to *IR*, but a sheet of paper takes on the air temperature and a radiometer, or a thermocamera, may measure it. This practice has the advantage of taking readings of both the surface and the air

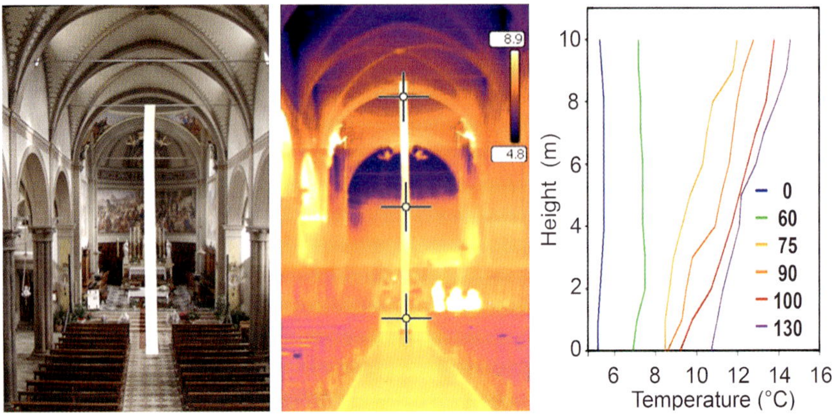

FIGURE 11.15 Plaster, wood and paper have almost the same emissivity. With a vertical strip of paper, it is possible to monitor both the surface and the air temperature with the same thermogram. The vertical temperature profile is obtained from the pixels sampled on the paper strip. On the right, the vertical temperature profiles made with repeated thermograms starting when the warm air heating was turned on, i.e. at time 0 min (blue line), and after 60, 75, 90, 100 and 130 min (colour codes in the legend). After 2 h, the top level was heated by 10 °C, churchgoers half this value. *Source: Church in Agordo, Italian Dolomites, after Camuffo (2010), Camuffo et al. (2010b), by kind permission of Nardini Editore, see credits.*

temperature using the same instrument thus avoiding errors due to the particular response (or calibration) of different instruments and/or methodologies.

A very advantageous practice is to use a roll of paper to obtain vertical temperature profiles, as in Fig. 11.15, during a field survey in a church to control efficiency and risks of a warm air heating system. Once the paper roll is raised vertically, it is sufficient to take repeated thermograms at regular time intervals to monitor how heat is distributed and thermal layering.

The same thermograms will show you, in addition to the thermal profile in air, the direct impact of heating on objects, decorations and structures that will respond in a different way depending on their heat conductivity. This method is particularly useful because the image provides at the same time forcing factor (i.e. warm air) and effect (i.e. response of all objects). The thermogram is accurate because paper and the materials most commonly found inside historic buildings, i.e. mortar, bricks, plaster, frescoes, tapestry and wood, have all almost the same emissivity, i.e. $0.90 < \varepsilon < 0.94$. This methodology is useful for environmental diagnostics and preventive conservation purposes, as well as for tuning the heating system, e.g. lowering the warm air temperature or increasing the air blowing rate to reduce internal layering and overheating in the upper part of the building.

References

Benedict, R., 1984. Fundamentals of Temperature, Pressure, and Flow Measurements. Wiley, New York, 532 pp.

Bentley, R.E., 1998. Handbook of Temperature Measurement: Temperature and Humidity Measurement. Springer Verlag, Berlin.

Camuffo, D., 1980. Fog and related diffusion potential at Venice: two case studies. Boundary Layer Meteorology 18, 453–471.

Camuffo, D., 1982. The nocturnal IBL over an hilly island with reference to the diffusion of radioactive nuclei. Boundary Layer Meteorology 22, 233–240.

Camuffo, D., 2010. The role of temperature and moisture. In: Camuffo, D., Fassina, V., Havermans, J. (Eds.), Basic Environmental Mechanisms Affecting Cultural Heritage – Understanding Deterioration Mechanisms for Conservation Purposes, pp. 9–30. COST Action D42 "Enviart", Nardini, Florence.

Camuffo, D., Bernardi, A., 1993. Microclimatic factors affecting the Trajan Column. Science Total Environment 128, 227–255.

Camuffo, D., Fernicola, V., 2010. How to measure temperature and relative humidity. Instruments and instrumental problems. In: Camuffo, D., Fassina, V., Havermans, J. (Eds.), Basic Environmental Mechanisms Affecting Cultural Heritage – Understanding Deterioration Mechanisms for Conservation Purposes, pp. 31–41. COST Action D42 "Enviart", Nardini, Florence.

Camuffo, D., Schenal, P., 1982. Microclima all'interno della Cappella degli Scrovegni: scambi termodinamici tra gli affreschi e l'ambiente. In: Ministero dei Beni Culturali ed Ambientali: Giotto a Padova, special issue of Bollettino d'Arte. Poligrafico dello Stato, Rome, pp. 107–209.

Camuffo, D., Pagan, E., Schellen, H., Limpens-Neilen, D., Kozlowski, R., Bratasz, L., Rissanen, S., Van Grieken, R., Spolnik, Z., Bencs, L., Zajaczkowska-Kloda, J., Kloda, P., Kozarzewski, M., Mons. Santi, G., Chmielewski, K., Jütte, T., Haugen, A., Olstad, T., Mohanu, D., Skingley, B., Sáiz-Jiménez, C., Bergsten, C.J., della Valle, A., Don Russo, S., Bon Valsassina, C., Accardo, G., Cacace, C., Giani, E., Giovagnoli, A., Nugari, M.P., Pandolci, A.M., Rinaldi, R., Acidini, C., Danti, C., Aldrovandi, A., Boddi, R., Fassina, V., Dal Prà, L., Raffaelli, F., Bertoncello, R., Romagnoni, P., Camuffo, M., Troi, A., 2007. Church Heating and Preservation of the Cultural Heritage: A Practical Guide to the Pros and Cons of Various Heating Systems – Il riscaldamento nelle chiese e la conservazione dei beni culturali: guida all'analisi dei pro e dei contro dei vari sistemi di riscaldamento. Electa Mondadori, Milano, 240 pp.

Camuffo, D., Bertolin, C., Fassina, V., 2010a. Microclimate monitoring in a church. In: Camuffo, D., Fassina, V., Havermans, J. (Eds.), Basic Environmental Mechanisms Affecting Cultural Heritage – Understanding Deterioration Mechanisms for Conservation Purposes, pp. 43–47. COST Action D42 "Enviart", Nardini, Florence.

Camuffo, D., Pagan, E., Rissanen, S., Bratasz, Ł., Kozłowski, R., Camuffo, M., della Valle, A., 2010a. An advanced church heating system favourable to artworks: a contribution to European standardisation. Journal of Cultural Heritage 11, 205–219.

Camuffo, D., della Valle, A., Bertolin, C., Leorato, C., Bristot, A., 2011. Umidità e Diagnostica Ambientale in Palazzo Grimani, Venezia. Proceedings of the Symposium: Edifici storici e destinazione museale: conservazione degli edifici e delle opere d'arte progetti per il restauro e integrazione di impianti esistenti. Milan, 1–2 April 2010.

Doebelin, E.O., 1990. Measurement Systems - Application and Design. McGraw Hill, New York, 960 pp.

Green, W., Maloney, G.O. (Eds.), 1984. Perry's Chemical, Engineers Handbook, sixth ed. McGraw-Hill, Singapore.

Hadlock, R., Seguin, W.R., Garstang, M., 1972. A radiation shield for thermistor development in the atmospheric boundary layer. Journal of Applied Meteorology 11, 393–399.

Kondratyev, K.Y., Kozoderov, V.V., Smokty, O.I., 1992. Remote Sensing of the Earth from Space: Atmospheric Correction. Springer-Verlag, Berlin, 478 pp.
Lide, D.R. (Ed.), 1990. CRC Handbook of Chemistry and Physics, sixty seventh ed. CRC Press, Boca Raton, Fla.
Lipták, B.G., 2003. Instrument Engineers' Handbook: Process Measurement and Analysis. In: The Instrumentation System and Analysis (ISA), vol. 1. CRC Press, Boca Raton, Florida.
Michalski, S., 2000. Guidelines for Humidity and Temperature in Canadian Archives. Canadian Conservation Institute, Ottawa.
Michalski, L., Eckersdorf, K., McGee, J., 1991. Temperature Measurement. Wiley, New York, 514 pp.
Nicholas, J.V., White, D.R., 1994. Traceable Temperatures – an Introduction to Temperature Measurement and Calibration. Wiley, New York, 358 pp.
Oke, T.R., 1978. Boundary Layer Climates. Methuen, London, 372 pp.
Platridge, G.W., Platt, C.M.R., 1976. Radiative Processes in Meteorology and Climatology. Elsevier, Amsterdam, 318 pp.
Schooley, J.F., 1986. Thermometry. CRC Press, Boca Raton, Fla, 245 pp.
UK Meteorological Office, 1981. Handbook of Meteorological Instruments. In: Measurement of Temperature, vol. 2. Her Majesty's Stationary Office, London.
Wolfe, W.L., Zissis, G.J., 1989. The Infrared Handbook. Environmental Research Institute of Michigan.
WMO, 2008. Guide to Meteorological Instruments and Methods of Observation. WMO Technical Publication No 8, World Meteorological Organisation, Geneva.
WMO, 1986. Compendium of Lecture Notes on Meteorological Instruments for Training Class III and Class IV Meteorological Personnel. WMO Technical Publication No 622, World Meteorological Organization, Geneva.
Wylie, R.G., Lalas, T., 1992. Measurement of Temperature and Humidity. WMO 759. Secretariat of the World Meteorological Organization, Geneva.

CHAPTER

12

Measuring Humidity

OUTLINE

http://dx.doi.org/10.1016/B978-0-444-63296-8.00013-5

12.1. MEASURING AIR HUMIDITY ACCORDING TO EN16242: 2012

Several instrument types, based on various principles, have been devised to measure the amount of vapour in air. In the 1970s, with the advent of electronics and the progress in material science, new sensors were developed and the measurement of relative humidity (*RH*) was radically changed, except for those who were not updated and kept the unjustified tradition of hair hygrometers and thermohygrographs.

From the user point of view, hygrometers can be grouped into four categories:

1. Dew point meter, an optical device based on accurate temperature measurements, useful for calibration purposes. It is a high-quality, primary instrument serving to calibrate secondary instruments.
2. Electronic fan psychrometers, also based on accurate temperature measurements, are useful for field measurements (when possible). It is a high-quality instrument; if pressure too is considered in the formula, it is useful to make spot checks to control whether secondary instruments need to be substituted or calibrated.
3. Thin-film capacitive or resistive hygrometers useful for field measurements. These are secondary instruments, of average quality, but able to provide reasonably accurate readings if periodically checked and calibrated.
4. The old hair hygrometer that has been the best methodology in the past, i.e. before the 1970s, but today has the lowest quality and unsuitable as a standard instrument for measuring humidity; i.e. it's use is no longer justified.
5. A variety of instruments based on different principles, with variable reliability, not recommendable for standardization purposes.

The novel EU standard EN16242: 2012 '*Conservation of cultural property – Procedures and instruments for measuring humidity in the air and moisture exchanges between air and cultural property*' recommends instruments of types 1 to 3. The use of the hair hygrometer is not recommended and is only accepted in exceptional circumstances for visual inspection, in the absence of vulnerable objects that need a precise microclimate control. More details and comments on the use of this standard are reported elsewhere (Camuffo and Fernicola, 2010).

Looking at the physical operational principle, the following classification is possible:

1. Hygrometers based on *condensation*, e.g. dew point (*DP*) or frost point hygrometers and water equilibrium hygrometers for particular saturated salt solutions.

2. Hygrometers depending on the *addition or removal of water vapour*, e.g. psychrometer and diffusion hygrometer and gravimetric, volumetric and pressure methods. The psychrometer should be kept in consideration; other methods are of less relevance.
3. Hygrometers based on *sorption methods*, e.g. mechanical hygrometers with human hair or parchment sensors; electric hygrometers with thin-film capacitive sensor; with resistance polymer sensor; with aluminium oxide, polyelectrolyte or carbon; and piezoelectric sensors. The measuring principle may be a mechanical displacement; a change of the electric resistance, capacity or inductance; or a change of vibration frequency. These sensors, and in particular, thin-film capacitors or resistors are largely used in commercial instruments because of their low cost, high resolution and fast response. The negative aspect is drift, especially after contamination.

None of them is fully reliable at extreme (i.e. very low and very high) *RH* levels. All of them present important drifts or departures after contamination and/or ageing. For this reason, careful periodic maintenance and frequent calibration are needed.

An analysis of all these transducers is too long and of limited interest, as only few types can be recommended for this field. In the following discussion, only the most important types will be commented and further details are given by Wexler (1965), Meteorological Office (1981), WMO (1986; 2008), Carr-Brion (1986), Doebelin (1990), Harriman (1990), Bentley (1998), Wylie and Lalas (1992) and Lipták (2003). The hair hygrometer will be considered in detail, in the hope of updating those who are still using it. Some specific papers have addressed this topic for conservation purposes (Macleod, 1983; Michalski, 2000).

12.2. HYGROMETERS

12.2.1. Dew Point Meter

This instrument is based on the detection of the temperature of a cooled mirror at the point at which condensation forms, i.e. the *DP*. Being only based on the accurate temperature measurement of the chilled mirror, it provides accurate readings and EN16242: 2012 recommends the dew point meter as a primary instrument for calibration. The mirror is electrically cooled with a Peltier plate and the temperature level is controlled by an electronic feedback to keep dynamic equilibrium between evaporation and condensation, thus closely following the dew point temperature changes. An infrared (*IR*) beam impinges on the chilled mirror and the intensity of the reflected beam is continually monitored. When the mirror starts to mist over by initial condensation, the reflection drops. The instrument provides a reading of the critical temperature at which evanescent microdroplets start to form or cease to leave. For this reason, it is known as *chilled mirror dew point meter*.

If the chilled mirror can reach low temperatures, it might be used to detect the frost point too.

The chilled mirror dew point meter provides the most accurate humidity measurements and is recommended as a reference instrument for calibration of other hygrometers. It operates in the full *RH* range, from about 0% to 100%. To guarantee high precision, the instrument requires periodic control and cleaning of the optical components.

Other instruments exist in the market under the improper name '*dew point meter*' because they are normal hygrometers but with a chip able to calculate the *DP*. They are based on a combination of temperature and *RH* sensors (generally capacitive or resistive *RH* sensors) and from their outputs they calculate the dew point value. The output of hygrometric sensors is not as accurate as temperature readings, which lowers the instrument quality. EN16242: 2012 warns against using such instruments as primary instruments.

12.2.2. Psychrometer

In 1888–1892, Adolf Richard Assmann developed an instrument for accurate measurement of atmospheric humidity and temperature, based on two thermometers shielded from solar radiation, one of them having a wet bulb. The hygrometric variables were obtained from the two readings with the help of a diagram or tables. The instrument, called psychrometer, was soon produced by Rudolf Fuess' factory in Germany. The historic Assmann-type psychrometer was composed of two identical mercury-in-glass thermometers, ventilated by a mechanical fan and at that time, it provided the most accurate humidity measurements. However, continuous *RH* monitoring was not possible and any reading of the dry and wet bulb temperatures required the help of tables or diagrams to be interpreted. For this reason, the mechanical psychrometer was rarely used for conservation purposes and was confined to laboratory calibration of lower accuracy instruments, i.e. the hair hygrometer.

Today, the historic Assmann psychrometer has been abandoned and has been substituted with automatically operated electric fans and more advanced temperature sensors, e.g. platinum resistance, thermocouples and thermistors. Electronic psychrometers are recommended by EN16242: 2012 to periodically control if other humidity sensors (e.g. thin-film capacitive or resistive sensors) provide accurate readings or need calibration. However, it can be also connected to a data logger and with a water reservoir that allows medium-term autonomy, it may provide routine monitoring at selected sampling intervals.

The psychrometer is an instrument based on the readings of a matched pair of sensitive thermometers (i.e. they must read alike at any given temperature), joined side by side: one being normal, called the 'dry bulb' to be distinguished from the other that has its bulb covered with a wet cotton cloth wick, called 'wet bulb'. At equilibrium, i.e. when the heat Q_v lost by evaporation from the wet bulb equals the sensible heat Q_s transferred from the ambient air to the colder wet bulb, the psychrometric formula is obtained as discussed in Chapter 1, i.e.

$$e = e_w - Ap(T - T_w) \tag{12.1}$$

where A is the psychrometric coefficient, which is independent of the evaporating surface area, i.e. the size of the wet bulb wick covering, but is slightly dependent on ventilation rate and, to an even minor degree, on psychrometer design, size and dimension of the dry and wet bulb, ambient temperature and RH. However, when ventilation is in excess of 2.5 m s^{-1} (and particularly in the range between 3 and 5 m s^{-1}), A is nearly constant and suggested values for the classical Assman psychrometer with mercury-in-glass thermometers are $A = 6.2 \times 10^{-4}$ K^{-1} (WMO No.8, 1986, 2008) and $A = 6.667 \times 10^{-4}$ K^{-1} (UK Meteorological Office, 1981). Some values of A versus ventilation rate v are (WMO No.622, 1986):

v (m s^{-1})	0.12	0.50	1.0	2.0	4.0
A (K^{-1})	13.0×10^{-4}	9.0×10^{-4}	7.8×10^{-4}	7.1×10^{-4}	6.7×10^{-4}

In order to get an idea of the error derived by an incorrect value of A, at $t = 20$ °C, when A changes by 10%, the error that is generated in RH, i.e. ΔRH, varies with the actual RH level and is (UK Meteorological Office, 1981):

RH (%)	0	20	40	60	80	100
ΔRH (%)	4.0	3.0	2.2	1.4	0.7	0

At temperature and humidity levels usually found in museums and galleries, the error is of the order of ±2%; it decreases at higher temperature and increases at lower temperature.

Both dry and wet bulb thermometers are ventilated with a forced air flow with speed generally ranging from 3 to 5 m s^{-1}, where the above proportionality coefficients become more constant and less dependent on the ventilation rate. Please note that WMO No.8 (1986, 2008) suggests a wider interval from 2.5 to 10 m s^{-1}, but 2.5 m s^{-1} is close to the limit of error (2 m s^{-1}) and every slow down of the fan, e.g. for a not well-charged battery, may cause a super evaluation of the wet bulb temperature and, consequently, of the measured *RH*. On the other hand, a faster ventilation rate causes unnecessary power consumption. For elevated *RH* values, slower ventilation rates are sufficient but when the *RH* drops, fast ventilation is necessary to cool the wet bulb to its full depression. The fan speed is a limiting factor in dry environments.

Accurate measurements taken with a psychrometer can be considered as a home standard, i.e. as a useful reference to compare and check calibration of other instruments. In particular, when *RH* is very high or approaches saturation, the psychrometer is superior to all other sensor types (Wiederhold, 1975, 1997). However, it has two important limits. First, it cannot be conveniently used when the wet bulb is below the freezing point. This means that readings may be impossible in dry air at temperature below 10 °C. Although reference may be made to the latent heat and saturation pressure for ice, instead of water, the measurement becomes less accurate; the water supply is interrupted and the water reservoir may be damaged. Second, as the *RH* drops below 20%, it becomes difficult to cool the wet bulb to the equilibrium depression, even when the sensors are aspirated at an airstream rate of 10 m s^{-1}. The reason for this is that the water of the wet bulb evaporates before the wet bulb reaches its maximum depression temperature (Fisher et al., 1981). For this reason, the use of this instrument is suggested in the *RH* range from 20% to 100%, which fortunately covers the main parts of the practical cases. No other sensors provide a wider range of reliability.

The most common causes of error as follows:

1. Breathing in proximity of the sensor or keeping the instrument close to the observer's body will cause an exceedingly high value of moisture. Another frequent misuse is to handle a psychrometer with raised arm, as the air heated by the body follows the upward chimney path along the arm and is then aspirated by the psychrometer fan. The correct position is to handle the instrument with a lowered arm or to hang the instrument from an extension pole.
2. Reading the thermometers (in particular the wet bulb one) before it has reached equilibrium: it might be useful to remember that the fan does not affect the temperature of the dry bulb but lowers that of the wet bulb. Although covering the bulb with a thin tubular cotton wick reduces the thermometer time constant, when the fan is switched on, the dry sensor remains in the previous condition of equilibrium but the wet sensor begins to lower its temperature and needs time to reach equilibrium. This error is frequent with sensors having a long time of response.

3. Errors in the thermometer reading: this error may be important with mercury-in-glass thermometers, as the operator must stay with his face close to the thermometers for too long a time to read the small scale divisions of both of them and the face *IR* emission and the breath may warm the sensors. This problem has been eliminated in electronic psychrometers where the display is well visible and placed far from the sensors. Some of them have an automatic recording and there is no need to read the display, except for controlling if the equilibrium has been reached.
4. Insufficient ventilation of the wet bulb: this error is frequent and becomes important for ventilation below 2 m s^{-1}. Controlling the airspeed of ordinary psychrometers, it is very frequent to find insufficient ventilation speed. Often it is possible to find a solution by diminishing the screen section to increase the airstream speed. In clockwork aspirated psychrometers, observations made too early (i.e. with reference to the instrument time response) are affected by error, as well as those made too late, when the spring is loosing its energy and the fan is slowing down. Electrical fans should be preferred to spring-driven mechanical fans that might have uneven ventilation speed.
5. Tubular cotton wick covering of the wet bulb is not completely wet.
6. Contaminated wick covering or use of nonpure distilled water: after some time has elapsed, the forced ventilation will cause contamination of the covering. Monitoring in coastal area (i.e. marine aerosols) or polluted environments requires frequent changes of the wick covering. When a new cotton wick is used, it must be previously boiled in distilled water to remove extraneous substances that may alter the surface tension of the absorbed water.
7. Too thick a wick covering of the wet bulb (or even frost) may increase the time constant.
8. Temperature below 0 freezes the wick, stopping the water supply, and the equilibrium is reached with ice instead of water.

All the hygrometric parameters can be obtained after the dry bulb temperature and wet bulb depression with the help of tables, diagrams or formulae, as discussed in Chapter 2A, but the same can also be obtained from the air temperature and *RH*.

The psychrometer gives very accurate measurements only when it is correctly operated. The effect of an error in the wet bulb reading varies with the temperature and humidity level. An example of the propagation of errors is shown in Fig. 12.1 by supposing that $RH = 50\%$ and that T_w is read with an error of $-0.1\ ^\circ C$. In the most frequent range of indoor climate, i.e. $10 \leq T \leq 30\ ^\circ C$, the error is relatively small, i.e. less than 1% for *RH*, between one to a few tenths for specific humidity (*SH*), absolute

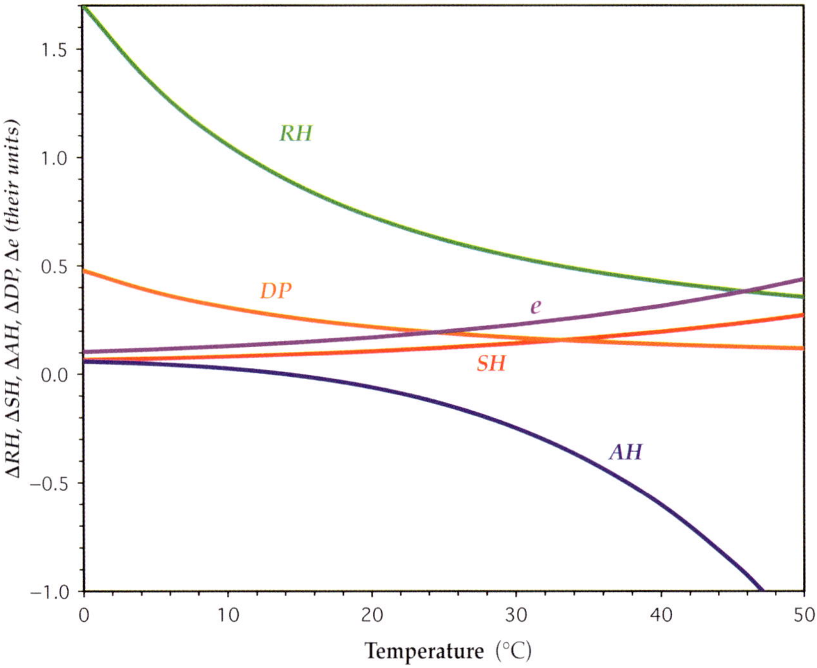

FIGURE 12.1 Error generated in computing relative humidity (*RH*, %), specific humidity (*SH*, g kg^{-1}), absolute humidity (*AH*, g m^{-3}), dew point (*DP*, °C) and vapour pressure (*e*, hPa), when the wet bulb temperature T_w is affected by −0.1 °C error, at ambient *RH* = 50%.

humidity (*AH*), *DP* and *e* (vapour pressure). In the meteorological range of variability, the error of RH increases exponentially with decreasing temperature below 0 °C but in this span, psychrometric measurements are impossible due to the formation of ice. For elevated temperatures, the error of *AH* becomes important. However, it should be noted that this propagation of errors affects the accuracy of all the measurements, but systematic errors affect very little the differences from point to point or from time to time. It is very important to check before use if the two sensors, both dry, have exactly the same temperature readings. Even in the case that they measure the actual temperature with a small error (although identical for both sensors), the error is systematic, affects in the same way all readings, and is therefore negligible in the calculation of the variability of the hygrometric parameters in terms of gradients over space or trends over time. This is especially true when plotting the distributions of the above variables in horizontal maps, as differential values are involved, and the departures from the average of the local minima and maxima remain practically unchanged.

In order to monitor and map the distribution of the temperature and the humidity over a horizontal cross-section of a room, a number of samplings with dry and wet bulb readings should be made in a short time. This means that the instrument should have quick response and reading repeatability. Precision electronic psychrometers have been designed and built in our laboratory, with a better accuracy than 0.1 °C and fast response. Although sensors with a time constant better than 1 s were used, the overall time constant was 5 s, which is inclusive of the thermal inertia of the screen, mechanical parts and electronics. The combination of different time constants reflects in the fact that on plotting the normalized temperature change for the step ambient variation versus t in a logarithmic paper, the plot departs from a straight line. The resulting time constant limits the number of observations per run. The critical factor is that total time needed to perform a whole run should not exceed a certain duration over which the ambient conditions may be considered stationary. For instance, in the case that the acceptable total duration is 10 min and that the time needed to move from one point to another and to take readings is 30 s, only 20 sampling points are possible, i.e. 30 s $\times$ 20 $=$ 600 s. Runs should be made with the fan uninterruptedly operating even when moving from one sampling point to the next one, in order to reduce the time to reach equilibrium. In fact, during the move, when the operator approaches the next position, the sensor too was at the same time passing to levels closer and closer to the final equilibrium point.

All components (i.e. sensors, screen and electronic circuits) of the psychrometers built to this aim should be tested separately and then in conjunction to optimize the overall time constant and accuracy. Either linear or nonlinear thermistors, or platinum resistance sensors, are appropriate fast sensors. Linear outputs avoid unnecessary electronic transformations of the signal but miniaturized thermistors can equally be linearized and have a shorter time constant, which is a very important feature, especially when several measurements are made in each run. Two types of screen are suggested: (1) a white polystyrene foam that is reflective, has a very low thermal capacity and is a good thermal insulator; its response time, measured with a radiometer, is less than 2 s and (2) a thin aluminium foil reflecting outside and blackened inside.

Low-power-consumption fans are utilized in order to reduce weight, volume and cost of rechargeable batteries. The velocity of blown air at the matched sensors is regulated by varying the internal section of the tube corresponding to the sensors or operating on the fan speed. The dry and wet bulb sensors are placed side by side or the wet bulb is midway between the dry one and the aspirating fan.

The output should be clearly visible in a display; it may be printed on paper or saved to a computer.

12.2.3. Thin-Film Capacitive Hygrometer

This sensor is composed of a thin film of hydrophilic material and its moisture content (*MC*) quickly reaches equilibrium with the ambient *RH*. The thin film is used to vary the capacity in an electronic circuit and its routine use for measuring or controlling *RH* levels is recommended by EN16242: 2012. A dielectric film (e.g. acetal polymer or cellulose acetate butyrate) absorbs water vapour from the air establishing a state of humidity in equilibrium with it. The capacitance changes proportionally to the equilibrium *MC* of the film, which in turn is proportional to the *RH* of the air, so that the output is linear and independent of the actual value of the air temperature. The change in dielectric constant is measured with a capacitance bridge. This kind of sensor is very fast, i.e. response time varies from 1 or a few seconds to a few minutes; it has a low hysteresis, i.e. ±1% (or less) from 5% to 95% *RH*; it is relatively accurate, i.e. ±1% to ±2%, accuracy; and it has high resolution, which may vary from 0.1% to 1%.

Thin-film capacitive hygrometers are recommended by EN16242: 2012 for routine indoors/outdoors measurements because they constitute the best method when the temperature drops below 0 or when measurements should continue unattended for a relatively long time.

The main problem is the drift, especially in polluted environments. For this reason, the probe is often inserted into a filter that protects the sensor against air pollution, dirt or splashing droplets; also, oil and grease vapours may form a film that prevents the absorption of water molecules. However, the presence of a filter slows down the free exchange with the atmosphere and increases the actual time of response. This is not a problem for several applications (e.g. long-term ambient monitoring) but is an important limit when fast response measurements are needed.

They need calibration every few months, or every year in the most optimistic case. In the best instruments, in order to avoid the problem of drift, the calibration runs automatically after setup, activating two calibration points, at low and high *RH*, obtained by means of reference saturated salt solutions.

12.2.4. Thin-Film Resistive Hygrometer

This sensor is made of hydrophilic materials and its *MC* quickly reaches equilibrium with the ambient *RH*. Its electrical resistance changes with the *MC*, which makes possible a calibration of resistance versus *RH* at equilibrium. Its performance is similar to that of thin-film capacitive sensor and is equally recommended by EN16242: 2012. The resistive sensor is generally stable and accurate and may provide long-term monitoring on-site either indoors or outdoors, except in the case of extreme environmental conditions.

In case the sensitive element is an electrolyte, it may be damaged after a long period at very high humidity, in wetting conditions or in the presence of contaminants.

12.2.5. Electrolyte Hygrometer

This sensor measures the impedance of a hygroscopic substance, an electrolyte, which absorbs water molecules to produce a solution. The hygroscopic electrolyte tends to establish equilibrium with the surrounding atmosphere, without drift, and has a very fast response. Commercially available sensors are declared having a time constant less than 10 s, a repeatability of 0.1% *RH* and a range unusually extended towards dryness, i.e. from 0% to 95%. However, although this sensor has a good performance, it is fast only after it has established thermal equilibrium with ambient air and this stabilization requires 2–3 min. This time is the limiting factor and should be considered as a more realistic time constant in the field, when both temperature and humidity are variable. The manufacturer suggests calibration with adjustment every 6–12 months and more care is suggested in the case of operation in a polluted environment.

12.2.6. Hair Hygrometer

The most popular instrument is the traditional hair hygrometer that, for almost two centuries, has served a huge variety of users and is especially appreciated for its apparently easy use, if one disregards periodic maintenance and calibration. The hair hygrometer was invented in 1783 by Horace Bénédict de Saussure. This was the best hygrometer developed in the eighteenth century, able to provide reasonably correct, direct reading on an analogue scale. Combined with a spring clock motor and a rotating drum, continuous records on a strip chart were possible. Records were made by means of an ink pen fixed to a mobile arm connected to the hair whose elongation was proportional to the equilibrium *RH*. Not only in weather stations but also in other application fields, the thermohygrograph, i.e. a combination of a bimetal thermometer and a hair hygrometer, became very popular. Air temperature and *RH* were plotted with two ink pens on weekly or monthly charts fixed to a rotating drum. The most famous producers in the nineteenth and early twentieth centuries were Richard Frères in France. As a matter of fact, although the hair hygrometer produced less-accurate measurements compared with the Assmann psychrometer, it constituted the most practical or even the only possible methodology up to the 1970s, when electronic sensors were produced. The thermohygrograph become very popular in museums, galleries, archives and libraries At first sight, the routine use of such instrument is very simple, as it is sufficient to replace the strip chart at fixed intervals and the ink pens when necessary. However, this instrument requires periodic maintenance and calibration. If such care is omitted, readings may depart very much from the truth and misleading readings might cause dangerous consequences when controlling the *RH* in archives, museums or galleries.

The user needs to be informed about errors (typically between ±5% and ±30%), advantages and limitations of this dated methodology (Davey, 1965). The response of the hair sensor is only slightly affected by temperature but is changeable with hair hydration (i.e. memory of the past humidity levels), hair tension (i.e. mechanical load), contamination and ageing. These factors combine to make hair hygrometers or hygrographs unsuitable as a standard instrument to measure humidity (UK Meteorological Office, 1981).

The length of human hairs, when the natural grease has been thoroughly removed, increases by a value ranging from 1.7% to 2.5% when the RH rises from 0% to 100%, and the elongation $\Delta L/L_o$, where L_o is the dry air length, is approximately proportional to the logarithm of *RH*, at least for not too dry environments, i.e. for $RH > 20\%$, and is usually calculated by means of the equation

$$\frac{\Delta L}{L_o} \approx k \ \ln(RH) \tag{12.2}$$

where k is a coefficient of proportionality. However, better approximations on the whole range $0 \le RH \le 100\%$ for increasing values of *RH* and hydrated hair are given by the functions:

$$\frac{\Delta L}{L_o} \approx k_1 \sqrt{RH} \tag{12.3}$$

$$\frac{\Delta L}{L_o} \approx k_2 \ln^2(1 + RH) \tag{12.4}$$

$$\frac{\Delta L}{L_o} \approx k_3 \ln^2\left(1 + \sqrt{RH}\right) \tag{12.5}$$

$$\frac{\Delta L}{L_o} \approx k_4 \ln(1 + RH)\ln\left(1 + \sqrt{RH}\right) \tag{12.6}$$

$$\frac{\Delta L}{L_o} \approx k_5 \ RH^2 + k_6 \ RH + k_7 \tag{12.7}$$

which give similar results, and the constants depend on the final elongation at saturation; in general, $17 \times 10^{-4} \le k_1 \le 25 \times 10^{-4}$, $8 \times 10^{-4} \le k_2 \le 12 \times 10^{-4}$, $30 \times 10^{-4} \le k_3 \le 44 \times 10^{-4}$, $15 \times 10^{-4} \le k_4 \le 22 \times 10^{-4}$,

$k_5 \approx -2 \times 10^{-6}$, $k_6 \approx 4 \times 10^{-4}$ and $k_7 \approx 1 \times 10^{-3}$. The *'dry'*, i.e. the lower transfer function, due to either hysteresis for decreasing values of RH, or dehydration of the hair is not so well defined and a crude calculation can be done with the equation

$$\frac{\Delta L}{L_o} \approx k_8 \sqrt{RH} \ln(1 + RH) \tag{12.8}$$

where $37 \times 10^{-5} \leq k_8 \leq 54 \times 10^{-5}$. All these formulae are plotted in Fig. 12.2.

From the graphs, it is evident that approaching saturation, the hair looses sensitivity: it has the same elongation at $RH = 100\%$ and when it is immersed into water. For this reason, immersion in distilled water is an easy method to calibrate the transducer at the upper limit of its range.

The response time of the hair is not constant: it depends on temperature, stress on the hair, RH value and direction of RH change. It increases when the air temperature decreases. At room temperature, the time constant varies from 0.5 to 3 min. Measurements below freezing temperature need more and more minutes to reach equilibrium and at $-40\,^{\circ}$C the hair becomes insensitive. The response is faster for high humidity levels as well as with increasing humidity levels, following the natural hair hygroscopicity. It is also faster for hair under stress when charged with external load (e.g. the tension due to the series of linkage to the recording pen arm); however, tension generates long-term drift.

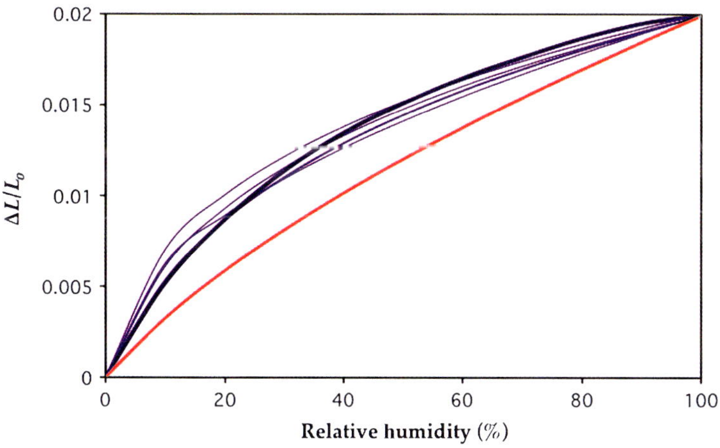

FIGURE 12.2 Elongation of a hydrated hair (thick black line) for increasing RH and elongation computed with the formulae in Eqns (12.2)–(12.7) (thin violet lines). The best approximation, i.e. $(\Delta L/L_o) = 9.3 \times 10^{-4} \ln^2(1 + RH)$ is shown with an intermediate thickness line. Tensile load: 1 g. The red line shows the Eqn (12.8) for not well-hydrated hair or hysteresis for decreasing RH.

The sensor is affected by hysteresis, so that the same output can be observed for different values of *RH* depending on the past environmental values to which the sensor has (temporarily) adapted. The largest errors especially occur in dry conditions and the permanence in dry environments decreases the hair sensitivity. The hair is composed of different layers that respond in different ways. When the sensor is well hydrated, it presents the so-called '*wet*' calibration curve to which all instruments refer; however, after some days in dry condition, it reduces the hydration level of the outer layers and follows another '*dry*' curve. To minimize the uncertainty due to hysteresis, both the calibration curves (i.e. for the '*wet*' and '*dry*' hair) are always referred to increasing humidity. The transition from '*wet*' to '*dry*' occurs only after continual permanence at low *RH* and this is the case of most heated rooms and museums in winter. For this reason, hairs should be regenerated very frequently, immersing them into distilled water. The frequency of this operation is determined by the ambient *RH* and the span of daily cycles; the dryer the ambience, more frequently should the regeneration be repeated. For instance, in an ordinary room, the regeneration should be made at least every week by keeping for a whole night the hairs immersed in water. In very dry environments, this operation should be made more frequently, at least daily. This imposes an unacceptable limitation.

Hairs are affected by chemical contaminants and it is necessary to clean them with ether every time hairs have been exposed to air pollution, contaminated with soot deposits or natural grease or touched with bare fingers. Ideally, the transducer calibration should be made weekly, when the chart strip is changed. If the reading is not precise, the calibration should be made to evaluate the drift and correct the most recent readings and to tone the instrument before the use, after having cleaned and regenerated the hairs. Without all these operations, it is not clear which ambient *RH* corresponds to the instrument output.

A clear indication that the instrument being use is out of calibration is given when the sensor reaches saturation and the pen remains for several hours very stable in the same position of maximum at elevated values of *RH*, different from 100%. Such maxima may be found at any value between 75% and 120% and show how much the upper limit of the span has changed.

The response time can be reduced by flattening the hair. Hair rolled flat is much faster than ordinary hair: at room temperature, it is 3–5 times faster; below zero, one order of magnitude or even more. The speed may also be further increased by applying a greater rolling pressure. The faster response also involves negative aspects: for rolled hairs, the '*wet*' to '*dry*' transition occurs after a few hours of dry environment. Another drawback is that rolled hairs have their strength decreased and, consequently, their drift increased.

EN16242: 2012 points out that this instrument has low accuracy level and considers its use acceptable in exceptional circumstances for visual inspection only but not for spot or routine measurements, for controlling indoor climate, or for data collection for statistical analyses for conservation purposes.

12.3. CALIBRATING HYGROMETERS

All hygrometers need periodic calibration made by an accredited laboratory in compliance with EN ISO/IEC 17025 '*General requirements for the competence of testing and calibration laboratories*'.

Instruments should be calibrated periodically in an accredited laboratory but some basic maintenance and preliminary checks should be made at home using a reference instrument. Such an instrument will be used for checking the calibration of other secondary instruments by comparison for an early discovery of departures.

Dew point meters and psychrometers can be used as a reference. Being based on temperature measurements, they are more accurate and reliable than humidity meters.

Calibration of hygrometers should be made in a closed box calibration chamber, in equilibrium, under controlled temperature and humidity levels. Two methodologies are possible.

The first methodology is based on the comparison between the primary and one or more secondary instruments at selected *RH* levels. Both the primary and the secondary instruments are kept in a calibration chamber until equilibrium is reached at a number of selected *RH* levels and at a given temperature. It is not necessary that the selected levels be exactly reproduced but that the equilibrium is reached. At this moment, one exactly knows the actual *RH* and temperature level inside the chamber (i.e. the dew point meter readings) and the individual readings of the secondary instruments that need calibration. Repeating this operation, it is possible to plot the readings of each instrument versus the reference readings and obtain the individual calibration plots. This is the most accurate method but needs a chamber and especially a good reference instrument.

The second methodology is based on the possibility of producing a number of accurate *RH* levels inside the chamber and on the assumption that the instrument can be tuned to them. The method is less expensive because no primary instruments are needed, the chamber requires a less-rigorous control and the reference *RH* levels are created with salt solutions in saturation equilibrium. A supersaturated salt solution is spread in an area as large as possible and air mixing provides homogeneity inside the box. An excess of the solid salt is suggested, in order to ensure saturation even in the case that a huge amount of water vapour is absorbed by the solution with two possible negative consequences: (1) the excess of solid salt disappears and the solution is not anymore saturated and (2) a density stratification forms, with the upper layer enriched with absorbed water, not saturated and in equilibrium with a higher *RH*.

Here, a list is reported of substances (in addition to vacuum) that might be used to this aim (Table 12.1). Please note that vacuum is the most easiest way to reach *RH*=0%, but might damage some sensors.

However, it is better to consider this method as a crude approximation (i.e. within a few percent of *RH*) but an effective way to obtain a steady-state condition (i.e. constant *RH*) around an approximately known value, and pass to the second-order approximation comparing the instrument with a reference hygrometer also located inside the same box. In this case, there is no need for temperature regulation and this simplifies procedures, reducing time and costs. The most convenient home reference is a psychrometer, but when this is placed inside the box, the moisture continually evaporating from the wet bulb causes a perturbation, generates an excess of vapour pressure and might be absorbed by the saturated solution. In order to avoid this problem, the psychrometer should initially have the cotton cloth wick dry and inserted into a capillary tube made of glass or polytetrafluoroethylene (brand name '*teflon*') empty and connected with the box exterior, for the whole time needed to reach equilibrium, and only at the moment of the calibration test, the cotton cloth wick should receive and suck the water through the tube. In this way, the perturbation is very short and negligible, the whole volume in the box having already attained equilibrium, the water being supplied at the very last minute and the free surface of the cotton cloth wick being extremely small compared with the free surface of the saturated solution. This

TABLE 12.1 Obtaining Constant Relative Humidity within a Calibration Chamber. When the Supersaturated Solution Method is used, an Excess of the Listed Substances Should Remain in Solid Phase in Saturated Aqueous Solution at 20 °C.

Saturated Solution or Vacuum	***RH***	**Saturated Solution**	**RH**	**Saturated Solution**	***RH***
Air pump vacuum	0%	LiCl*	15%	$MgCl_2 \cdot 6H_2O$*	32%
$CaCl_2 \cdot 6H_2O$	32%	$Zn(NO_3)_2 \cdot 6H_2O$	42%	$K_2CO_3 \cdot 2H_2O$	44%
$Mg(NO_3)_2 \cdot 6H_2O$*	56%	$NaBr \cdot 2H_2O$	58%	NaCl*	76%
KBr	84%	$ZnS0_4 \cdot 7H_2O$	90%	$Na_2CO_3 \cdot 10H_2O$	92%
$Na_2SO_4 \cdot 10H_2O$	93%	KNO_3*	95%	H_2O	100%

* *The substances suggested by the Smithsonian Institute.*
Source: *List (1971), Weast (1977/78), Perry et al. (1984).*

practice is far more precise and cheaper, as pure (and expensive) chemical products are not necessary. In fact, the principle is to know (and adjust) the instrument output at exactly known humidity levels, chosen at four or five different points in the humidity scale and the knowledge of the humidity level is based on the reading of a precise instrument, not on the reproduction of standard points by means of the use of chemical products at a precisely controlled temperature.

It is recommended to avoid the erroneous practice of cross-comparing two instruments of the same type, because both may be affected by the same drift and uncertainties. Finding that both exactly provide the same readings generally means that both are affected by the same types of errors, only exceptionally that both are accurate.

12.4. MEASURING HEAT AND MOISTURE EXCHANGES BETWEEN AIR AND MONUMENTS

A fast psychrometer or a combination of a fast thermometer with a fast hygrometer is used to monitor exchanges of heat and vapour between the air and surfaces, e.g. between air and frescoed walls. In analogy with molecular diffusion, fluxes are proportional to the negative gradient of the diffused property. The same theory is applied as in Chapter 7 for the diffusivity of heat, i.e. the flux of heat $\mathscr{H}$ is

$$\mathscr{H} = -K_H \frac{\partial T}{\partial n} \tag{12.9}$$

where n is the normal to the surface and K_H is the coefficient of heat exchange in the direction of n and depends on the physical mechanism responsible for the heat transport.

In the case of still air in a closed environment, K_H is determined by the thermal conductivity of air, i.e. $K_H = k_T$; when an internal dynamic boundary layer develops along the surface, the turbulent transport becomes dominant over the conductivity and when turbulence is well developed, as in outdoor environments, K_H is given by the eddy diffusivity K_E. Of course, K_E is a function of the Richardson number, increasing when Ri decreases (for experimental values see Plate, 1982). The eddy diffusion coefficient applies when the eddy motions are rapid enough and are the dominant factor in the heat transfer. However, if the wall is directly hit by solar radiation, or the temperature gradient is measured outside, in the convective boundary layer that forms over the ground when the solar radiation overheats it, the mechanism of transfer changes, being characterized by the buoyancy and not by the momentum transfer. For this reason, the subscript H is a reminder that the eddy diffusion coefficient for heat exchanges may differ from the eddy diffusion coefficient for momentum exchanges.

When mechanical transfer dominates, eddies preserve their properties over an average displacement l, called *mixing length*, and then mix with the surrounding environment. From the definition of *eddy flux*, i.e. the capability of transporting along the vertical (or the normal) mechanical mixing, the coefficient of eddy diffusivity K_E is given by

$$K_E \approx \langle w'l \rangle \tag{12.10}$$

For a neutral shear flow, the physical factors governing the mean flow are the surface stress, the distance from the surface and the fluid density. Dimensional considerations lead to find the following equation in terms of the shearing stress τ, which is linked with the dynamic viscosity and the wind shear

$$K_E = C_E \sqrt{\frac{\tau}{\rho}} \tag{12.11}$$

where C_E is a constant (Goody, 1995).

In conclusion, when an observation of the temperature gradient near a surface is made, a quantitative evaluation of the heat exchange needs a careful examination of the dynamic state of the air motion, as the exchange coefficient varies with it. Hence, it might be better to limit the analysis to pinpoint the presence, the sign, the time of appearance, and the place where temperature gradients appear and avoid the risk of macroscopic errors. The only reliable quantitative case is found in closed rooms, with calm, stable stratified air. Under these circumstances, the quantitative value of the molecular heat exchange is obtained by substituting in Eqn (12.9) the actual value of the thermal conductivity $k_T = 6.26 \times 10^{-5}$ cal cm^{-1} s^{-1} K^{-1} (or 26.2 mW m^{-1} K^{-1}) at $T = 300$ K, i.e.

$$\mathscr{H} = -6.26 \times 10^{-5} \frac{\partial T}{\partial n} \quad \left(\text{cal cm}^{-2}\ \text{s}^{-1}\right) \tag{12.12}$$

where the coefficient -6.26×10^{-5} gives the flux per unit gradient of T. Similarly, the flux of moisture M_v is given by

$$M_v = -\rho K_w \frac{\partial MR}{\partial n} \tag{12.13}$$

where ρ is the air density and K_w is the diffusivity for water vapour. As in the previous case, K_w is not constant but varies between $D_{aw} \leq K_w \leq K_E$ where the extremes are the molecular diffusivity D_{aw} of vapour in still air in closed indoor environments and the eddy diffusivity K_E outdoors, where turbulence and mixing dominate.

In order to get an idea of the numerical value of the moisture flux in a closed room with stable air, it is possible to substitute in the above equation the actual values of $\rho = 1.16 \times 10^{-3}$ g cm^{-3} and $D_{aw} = 0.24$ cm^2 s^{-1}, i.e.

$$M_v = -2.77 \times 10^{-4} \frac{\partial MR}{\partial n} \quad (\mathrm{g\ cm^{-2}\ s^{-1}}) \tag{12.14}$$

where the coefficient -2.77×10^{-4} gives the flux per unit gradient of MR.

A flux of moisture of this order of magnitude was measured with a *Piché evaporimeter* as potential evaporation rate in the Sistine Chapel at $RH = 50\%$ (Camuffo and Bernardi, 1986). The Piché evaporimeter is composed of an inverted graduated cylinder of water with a porous ceramic disk or a filter paper seal at the mouth, which constitutes or forms the base of this instrument. Evaporation takes place from the wet base and thus depletes the water in the cylinder, so that evaporation can be read directly from the graduations sampling at regular time intervals, e.g. one or more hours. The evaporation rate is obtained by dividing the amount of evaporated water by the time elapsed between the two readings. This value is an index of the evaporation that might occur if a spot of dampness were made on the wall.

Exchanges of heat or moisture are always associated with gradients of T or mixing ratio (MR) but in many cases, exchanges are small and the gradients are difficult to measure as they are weak, close to the surface, easily dissipated by air motions and turbulence, and are easily perturbed by the experimental apparatus. For this reason, it is not only very difficult to arrive at a precise quantitative evaluation of H and M_v, but also difficult to individuate the occurrence of these fluxes. Therefore, it is practically convenient to measure in finite terms ΔT and ΔMR close to the surface and far from it; although this measurement does not give the value of the actual gradient at the interface, it in any case offers an objective observation in finite terms.

However, on positioning the aspirated psychrometer near the surface, the local air is perturbed and the excess (or defect) of heat and/or moisture is immediately aspirated by the fan. Keeping in the same position the instrument for all the time necessary for the measurement, the fan suction brings new air that quickly passes near to the surface without having time to reach equilibrium with the moisture exchanges between the air and the surface. This problem can be partially overcome by inserting on the psychrometer intake a thin insulating disk, with three soft spacers, some 3 mm thick, glued to the disk. Once the spacers are put into contact with the surface, e.g. a fresco, the air is obliged to flow between the target surface and the topping disk until it arrives at the centre of the disk where the sensors and the fan are located (Fig. 12.3). The air remains longer in contact with the wall surface exchanging heat and moisture and reaching a better (even if partial) equilibrium before reaching the sensors. In order

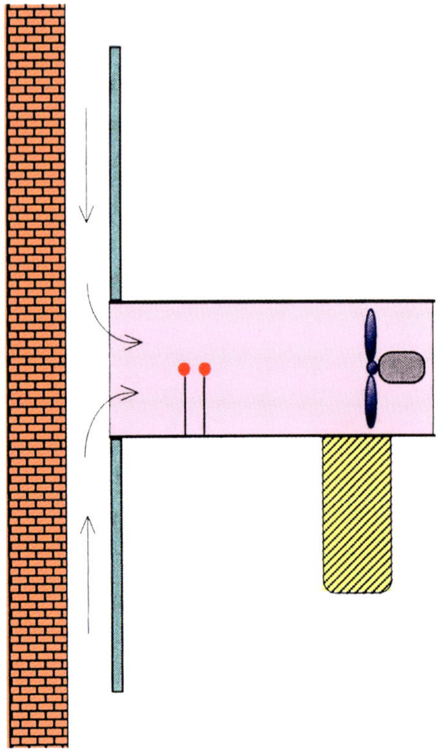

FIGURE 12.3 An insulating polystyrene disk (cyan) can be inserted on the psychrometer intake to make the air to stay longer in contact with the wall surface while passing through the thin layer between the disk and the surface (in this drawing, a brick wall is sketched on the left); in this way, the wall–air interactions are better evidenced. Red dots, wet and dry bulb sensors; pink, psychrometer tube; blue, fan blades; and yellow, handle with electronic circuits and wick water reservoir.

to avoid, or to minimize, heat exchanges between the disk and the air, the disk should be made of a material both insulating and with low density, e.g. polystyrene foam. We have used polystyrene foam disks with 30 cm diameter, to which three small spacers are attached (e.g. 3 to 4 mm) made of felt or soft silicone for two reasons: (1) to ensure a regular distance between the disk and the wall and (2) to avoid damage when the disk touches the wall.

Another, but less-reliable, method is sometimes applied. The claimed principle is that by sticking a sealed box on a wall, an air pocket remains entrapped inside the box, reaches equilibrium with the wall, and can be monitored over time. It is claimed that the air has more opportunity to interact with the wall, being not continually removed or perturbed by external factors. Just for this reason, the entrapped air pocket is unable to reproduce the exchanges that actually occur at the free interface: after some time has elapsed, the air pocket reaches a new equilibrium and any change will stop or will occur in very different conditions. On the other hand, at the free interface, where air is continually renewed by air movements and diffusion, exchanges of heat and moisture occur at a rate that is determined by the turnover of the external air and *IR* radiation, in addition to the transport and diffusion inside the wall.

This method might be compared with evaporimeters. An *evaporimeter* consists of a pan filled of water where the quantity of liquid that has evaporated in a given time is measured. The evaporation rate is governed by the difference of vapour pressure in the free air and over the pan surface as well as by the local ventilation. Suppose that an evaporimeter is exposed in summertime after a rainfall, followed by some clear days. In the first day, when the terrain is wet, the evaporation rate is all the same on the pan and the terrain. After a few days, the soil has lost most of the moisture in the surface layer and the evaporation is reduced to a minimum, or is completely stopped; on the other hand, the evaporimeter that is in a very dry environment shows an increased evaporation rate, which is nonrepresentative of the real situation at that site. Therefore, evaporimeter readings may be representative of the evaporation from a lake, not from the soil or masonry. Another quaint error, typical of this instrument, is an apparent fast evaporation near sunset, when birds drink some water in the pane.

Returning to the sealed box method, it might be supposed that it represents the equilibrium situation inside a pore in the inner part of a wall. In reality, the entrapped air pocket is external to the wall where both the *MR* and the temperature and, consequently, the *RH*, reach values different from the wall interior. In fact, the temperature of the air pocket entrapped between the wall surface and the box is affected by both these surfaces and in addition, the box has a wider area. For instance, if the box is a cube with edge length L, the ratio of the box surface immersed in air (i.e. $5\ L^2$) to the active wall section (i.e. $1\ L^2$) is 5; if the box is a hemisphere with radius r, the ratio of the cup to the wall surface is $2\pi r^2/\pi r^2 = 2$. Both are independent of the size, but the uncertainty decreases with the above ratio and theoretically the hemisphere is the best approximation. The box temperature is determined by the exchanges that occur with the wall and the air; the portion of wall covered with the box has a temperature that is governed by both the inner wall and the interactions between the wall and the box. This condition may differ from a free wall surface that interacts in another way with the room air. Finally, inside the wall pores, the saturation is governed by the Kelvin equation and the presence of hygroscopic salts: a situation that the box is unable to reproduce. Briefly, the sealed box method is useful and can be used, but keeping in mind that it is only partially representative of the wall interior.

12.5. MEASURING MOISTURE CONTENT

A number of EU and US standards exist to measure *MC* in materials, establishing either primary or secondary methodologies. The only absolute and most accurate methodology is gravimetry and is recommended in EN 322: 1993, EN 13183-1: 2002, ASTM D4442 – 07 and ASTM D2216 – 10. However, the above standards are suitable for commercial products, which we could call '*regular*' The definition of '*regular*' may be applied to a material when the main physical properties (e.g. thermal and electric conductivity, density, specific heat and thickness), the chemical composition and the molecular structure are known and homogeneous within the whole sample and are the same in all samples. This definition can hardly be applied to cultural heritage materials because these have variable density, sometimes unknown composition or transformations for past treatments; may have suffered weathering, deterioration or degradation; and, in addition, we cannot take samples of them. For the above reasons and to assist cultural heritage professionals engaged with this difficult problem, CEN TC346 is drafting a guideline entitled '*Conservation of cultural property – Guide to the measurements of moisture content in materials constituting movable and immovable cultural heritage*'.

The most popular methodologies have been described in the literature for regular materials (Dill, 2000; Healy, 2003; Lipták, 2003; Larsen, 2004; Kumaran et al., 2006; Beall, 2007; Saïd, 2007; NWFA, 2008; Palaia et al., 2008; Weritz et al., 2009; Mitchell, 2010) and will be commented in the following discussion for cultural heritage. However, not all of these methodologies are convenient for cultural heritage, some of them being invasive or destructive. Some others are neither destructive nor invasive but can hardly be applied to cultural heritage materials for two reasons: (1) the instrumental readings are in arbitrary units (AU) and (2) need interpretation and calibration to be related to *MC* (Milota, 1994; Kumaran et al., 2006; Palaia et al., 2008; Capitani et al., 2009; Mortl et al., 2011). All of them provide a useful tool for diagnostic purposes but none of them (except for calcium carbide and nuclear magnetic resonance (*NMR*)) is able to provide an absolute measurement of the *MC*, for the special reason that cultural heritage materials (except for stone) are far from being homogeneous and standardized and cannot be calibrated with gravimetry (Camuffo and Bertolin, 2012).

12.5.1. Gravimetry

MC is determined via oven drying and gravimetry and is expressed in percentage (%), with the formula:

$$MC = 100 \times \frac{m_H - m_o}{m_o} \tag{12.15}$$

where m_H is the initial mass of the test piece and m_o is the mass of the specimen after ventilated oven drying at a temperature of (103 ± 2) °C until constant mass has been reached. The constant mass at oven equilibrium m_o is by definition the *'dry'* mass md and the difference $m_H - m_o = m_w$ is the *MC* extracted from the specimen.

However, at above boiling temperature or/and at low *RH*, unstable hydrated minerals may slowly dehydrate and organic materials may suffer decomposition. For unstable materials, ASTM D2216–10 establishes lower oven temperature (e.g. 60 °C) not harmful to vulnerable materials or recommends the use of desiccators at room temperature.

Gravimetry cannot be used to obtain a continuous record of *MC* because of the necessity of removing samples from the object for laboratory testing. In particular, in the specific field of cultural heritage, removing samples is impossible for ethical reasons.

Another alternative definition exists, in terms of volume, where the volume of extracted water is also determined by gravimetry, but the volume of the dry sample should be measured by volumetric determination. The reading in percent of the wet material is slightly different from the percent of the dry material of oven-dry gravimetry. It is, however, possible to convert the water content (m_{ww}) obtained by the wet mass basis (i.e. by volumetry, without oven drying) to the above water content (m_w) determined with the dry mass basis (i.e. after oven drying) with the equation:

$$m_w = 100 \times \frac{m_{ww}}{100 - m_{ww}} \tag{12.16}$$

and return to the basic definition in (12.15) substituting $m_w = m_H - m_o$.

The volumetric definition may be equivalent to the previous one for regular materials with homogeneous spatial density constant in time but differs in an uncontrolled way for aged materials, e.g. wood affected by insect tunnelling and mould rotting. Internal cavities affect the internal volume but not the external one. Therefore, they do not change the gravimetric ratio but change the volumetric one.

As a consequence, for cultural heritage materials, it is not possible to compare gravimetric with volumetric methodologies, and in particular, to calibrate a volumetric instrument with gravimetry.

This makes it impossible to use absolute units and only arbitrary units (AU) can be considered. It may be useful to remember that AU of a given instrument are not comparable with AU of other instruments.

12.5.2. Calcium Carbide

The calcium carbide test (also known as the CM test or calcium carbide bomb) is a gas pressure test in which the moisture present in the crushed specimen of masonry, stone or soil reacts with calcium carbide to create acetylene gas and the gas pressure produced is measured to provide an *MC* reading, expressed as a percentage of moisture (Zazueta and Xin, 1994; Poděbradská et al., 2000).

Pressure readings are absolute and can be transformed into percent gravimetric mode, but the methodology is not accurate, requires a number of specimens (e.g. drilling the surface) and is destructive.

12.5.3. Conductive/Resistive Instruments

Conductive instruments (also called *resistive*) measure the electric current passing across two contact electrodes (representative of the *MC* on the surface) or two nail pins implanted in the subsurface layer, generally from 1 mm to 10 cm. In regular materials, electrical conductivity varies with the *MC* and its gradients but the accuracy decreases at elevated *MC* (Carll and TenWolde, 1996; Wilson, 1999; Dai and Ahmet, 2001; Goodhew et al., 2004; Brischke et al., 2008; Mitchell, 2010). In practice, in the presence of bound water, wood is highly resistive to the passage of electric currents, and the resistivity decreases with increasing bound water content. At the fibre saturation point (*FSP*), the process changes and the conductivity is favoured very much by the increasing presence of free water. In addition, wood resistivity decreases at increasing water temperature (Fig. 12.4).

The instrument output is affected by the type, density and temperature of the material. Instruments generally have a temperature sensor. However, only for homogeneous materials with standard physical properties,

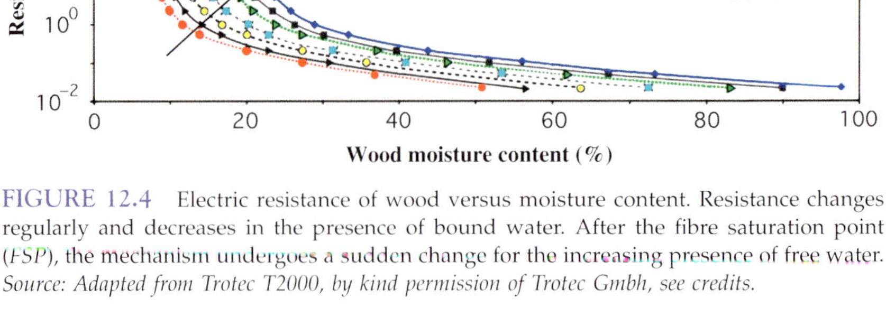

FIGURE 12.4 Electric resistance of wood versus moisture content. Resistance changes regularly and decreases in the presence of bound water. After the fibre saturation point (*FSP*), the mechanism undergoes a sudden change for the increasing presence of free water. *Source: Adapted from Trotec T2000, by kind permission of Trotec Gmbh, see credits.*

a correction can be automatically made after instrument calibration is made by the manufacturer.

Conductance in wood strongly depends on the grain orientation and past treatments with oil or wax that might have altered conductivity. Instruments are tuned for bound water conductivity and cannot be used in the presence of free water. However, in the case of cultural heritage wood, the presence of cracks, insect tunnelling and mould rotting decrease conductivity as if wood were drier. Hence, readings may be misleading.

In stones and masonry, conductance increases with *MC* and the presence of soluble salt crystals for the onset of electrolytic conductivity. Readings are severely affected by efflorescence and/or subflorescence or the presence of soluble salts for water transport due to percolation or capillary rise. Conductivity dramatically changes above or below the threshold for deliquescence of soluble salts. The nature and concentration of hygroscopic salts changes for several reasons, responding not only to external air pollution (mainly NO_x) but also to soil pollution and underground waters.

12.5.4. Capacitive/Dielectric Instruments

Capacitive instruments (also called *dielectric*) measure the change in capacity generated by the presence of water (and other polar molecules or radicals) found at short distance (e.g. a few centimetres) from the electrodes, i.e. in the subsurface layer. Therefore, it strongly depends on the material type, density and temperature; however, a calibration may be made for regular materials (Milota, 1994; Wilson, 1999; Titta and Olkkonen, 2002; Shimamura et al., 2004; Sundara-Rajan et al., 2004; Mitchell, 2010; Larsen, 2012).

In the most popular solutions, both electrodes lie on the same plane to put the sensor in contact with the object. Another popular solution uses a small spherical electrode (the other electrode is virtual, at infinity) to ensure better contact with nonplanar surfaces.

Often, electrodes are fed with high-frequency alternating voltage and emit a high-frequency electromagnetic field similar to microwave (see later), forming a hybrid system. In masonry, for a wide range of frequencies, the capacitance is almost independent of the presence of soluble salts, except in the case of deliquescence. For this reason, a dry efflorescence weakly affects the reading. In contrast, a thin film of water formed by condensation, or adsorbed in an efflorescence above the deliquescence level, will behave as a mirror of the electromagnetic field and readings go out of scale. In masonry, the methodology is also sensitive to changes in density, presence of cavities, internal discontinuities or metals. If a plaster, or a fresco, is detached from the wall, less H_2O molecules are encountered and the output might decrease as for a drier layer; however, if the internal surface is reflecting the signal, the output will be increased as for a damp layer. In practice, the interpretation might be obscure, especially in the case of beam reflections for the presence of internal discontinuities.

The instrument responds to surface and subsurface moisture and it is sensitive to the distance at which polar molecules are found, e.g. in the presence of moisture gradients. The output is not linear with the *MC* and the accuracy decreases at high *MC*. For this reason, instruments for wood are tuned to measure bound water below the *FSP*. However, in aged wood, the internal cavities due to insect tunnelling or mould rotting are largely dominant over the signal determined by water. The output will be reduced in proportion with the volume of cavities falling within the electromagnetic field. In wood, the contact electrode can be placed with any orientation on the side grain of wood (e.g. parallel to grains or across grains) with little effect on readings, except for the end grain. The output depends on the wood density, e.g. wooden species, and each species should have a specific calibration or correction factor. This is realistic only for commercial timber with regular characteristics. Local density changes are influential, i.e. a knot is interpreted as a local peak in *MC* and the presence of subsurface worm tunnelling is interpreted as a local decrease in *MC*, i.e. dryer wood.

12.5.5. Microwave Instruments

An antennae assembly radiates into the material an alternating electromagnetic field in the order of gigahertz, i.e. microwaves. When the microwave beam encounters polar molecules (e.g. H_2O and, to a lesser extent, cellulose chain or other molecules with electric charges), it interacts with them, looses energy decreasing in power intensity and has a phase shift (King and Basuel, 1993; Schajer and Orhan, 2005, 2006; Camuffo and Bertolin, 2012).

The intensity $I(x)$ of the microwave beam passing through regular matter is reduced with increasing travelling depth x, in a first approximation, following an exponential law, i.e.

$$I(x) = I(0)\exp(-\gamma x) \tag{12.17}$$

where $I(0)$ is the incident intensity when $x = 0$ and $\gamma = \alpha + i\beta$ is the attenuation coefficient determining how the beam intensity decreases, i.e. the fraction of energy removed from the beam per unit depth traversed. With increasing material density and *MC*, the attenuation coefficient also increases. The real part α indicates the attenuation in power at increasing depth x and the imaginary part β is related to the phase shift caused by the reduction in propagation speed through the medium. In real materials, for a number of disturbing factors, it might happen that γ is not constant but varies from point to point, and with time too, i.e. $\gamma = \gamma(x, t)$.

In addition to absorbing or changing phase, polar molecules reflect back a fraction of energy from the incoming beam. As the interaction with the water molecules is stronger than the interaction with the other molecules of the material, the back radiation is related to the *MC* of regular materials, with a specific relationship for each material. However, things are different for cultural heritage materials. If the material is not homogeneous, but has internal discontinuities, these may reflect back part of the power, causing beating interference, i.e. the interference between the incoming and the reflecting waves. If the beam is not fully extinguished within the material, when it reaches the ending surface, it has a back reflection that alters the reading. The same happens if the probe is not in strict contact with the target surface, i.e. the surface will cause back reflections and signal beatings, which depend on the distance. Therefore, measurements should only be made in strict contact with homogeneous materials that should have thickness larger than the complete extinction of the beam.

12.5.6. Evanescent-Field Dielectrometry (*EFD*)

Evanescent-field dielectrometry works at microwave frequencies (1–1.5 GHz), can be considered a particular development of microwave, and focuses on distinguishing the part of the signal due to H_2O from the part due to the bulk material (mortar, plaster and brick) with much lower dielectric capacity (Olmi et al., 2006). The signal is composed of a real and an imaginary part. The real part, measured through a frequency shift, depends on the *MC*, whereas it is nearly independent of the salt content. The presence of soluble salts is detected and quantified by looking at the resonance line width or the quality factor of the resonant probe.

12.5.7. Nuclear Magnetic Resonance (*NMR*)

When a magnetic field pulsed at the radio wave frequency meets hydrogen (H) atoms within a material, the H atoms are excited and subsequently relax back to their normal state, emitting a characteristic signal. This is the *NMR* technique (Hartley et al., 1994; Capitani et al., 1999; Casieri et al., 2004; Le Feunteun et al., 2011). Measurement of the relaxation signal allows the evaluation of the number of H atoms present and an absolute measure of the *MC*. To this aim, an appropriate calibration for each specific material is needed to distinguish the contribution of H from H_2O or other molecules including H (e.g. cellulose). With *NMR*, the H_2O molecules contained in a material are subjected to static and oscillating magnetic fields at right angles to each other. The molecular mobility of free water is appreciably higher than that of bound water and so *NMR* can be used to give information about the relative abundance of these two forms of internal water. This method is accurate, can be compared to gravimetry for calibration and may be applied in the field of cultural heritage.

12.5.8. Ultrasonic Pulses

The speed of propagation of ultrasonic pulses (typical frequency: 50 kHz) and also sonic pulses (typical frequency: 5 kHz) increases when water fills pores and capillaries of regular materials (Mamillan, 1992; Murphy et al., 1996; Olivera et al., 2005; Vun et al., 2006; Calegari et al., 2011). Measurements of speed of ultrasonic pulses can be taken in the transmission, reflection and surface modes. In the first method, the receiver is in the same straight line as the emitted beam, on the other side of the object; in the last two methods, the receiver is not aligned and is displaced in several positions at regular increasing distances. This allows controlling if the delay in receiving the pulses is proportional to the increased distance. In the path, attenuations, dispersions and reflections of the signal occur, especially in the presence of discontinuities, defects and cracks.

12.5.9. X- and γ Rays

A beam of X- and γ rays passing through matter is reduced in intensity according to the general exponential Eqn (12.17) where now $I(0)$ is the incident intensity of the radiation beam, $I(x)$ is the reduced intensity of the beam after having crossed the path length x inside the material, and ε is the extinction coefficient that depends on the physical characteristics of the material, its *MC* and the wavelength of the X- or γ rays (Takashi Tanaka et al., 2009; Bucurescu and Bucurescu, 2011). The measurement is made in transmission, with the sample between the emitter and the receiver.

After calibration, the analysis of these measurements gives the amount of moisture (and the cavities as well) along the path length through the material. By scanning across a surface, a measure of the moisture distribution (and the density distribution as well) can be assessed, and by scanning in three dimensions, by, for example, rotating the sample whilst scanning in two dimensions, it is possible to use tomography to build up a picture of the three-dimensional *MC*. This can produce high-resolution results of the combined effect of moisture and density distributions.

12.5.10. Neutrons

An important nondestructive method, originally devised for the soil moisture measurement, is based on the principle that high-energy *neutrons* emitted from a radioactive source are slowed down to thermal energy by elastic collisions with other nuclei (Nappi and Côte, 1997; WMO (1986; 2008); Larsen, 2004). The loss of energy depends on the mass of the colliding nuclei and increases with decreasing atomic number. The highest efficiency occurs with protons, i.e. hydrogen nuclei, which have the same mass as the impinging neutrons. A water molecule has two hydrogen atoms and these atoms are practically absent in traditional building materials such as brick, stone, and mortar, except for some water that may be adsorbed, in the liquid state, or in hydrated crystals. However, nonwater hydrogen is present in organic materials, and crystalline structures, clay minerals, hydroxides and other substances also have important neutron capture properties that add uncertainty to these measurements. After collision, neutrons are slowed down and scattered in all directions.

The flux density of transmitted (or diffused) slow neutrons is nearly proportional to the water content over an approximately spherical volume with a radius of 15–35 cm. Thermal neutrons are counted with either scintillation crystal detectors or gas-filled detectors and the standard error is proportional to the square root of the count. As a consequence, the error is larger at high water contents. In addition, changes in the air humidity and temperature cause drift to the equipment. The neutron interaction occurs across the whole thickness of the wall and the method cannot distinguish between surface and deep layers.

This method may give another information: part of the slow neutrons are captured by heavier nuclei with a gamma ray emission having an energy that is determined by the mass of the absorbing nuclei and ranges between 3.3 and 12.6 MeV. The analysis of the gamma ray spectrum detects the chemical elements. Dealing with the risk connected with radioactive materials, this method can be only applied with special precautions by authorized, well-trained personnel.

12.5.11. Thermography

Thermograms are thermal images made with a camera sensitive to *IR* radiation. They only provide the temperature of a surface but, if adequately interpreted, this may be a useful tool when it is necessary to localize damp zones. The camera is not directly sensitive to the water content but to temperature changes, i.e. space gradients or time rates, that might be locally affected by moisture. The thermocamera, or "thermal camera", may be used either in *passive mode,* i.e. taking pictures without changing the target surface temperature, or in *active mode*, i.e. with deliberate heating of the target surface to investigate anomalies when heating or cooling (Rosina and Robison, 2002; Rosina and Spodek, 2003; Palaia et al., 2008; Grinzato et al., 2011). Because different physical processes may interact with each other, it is sometimes misleading to consider damp the coldest areas. Dampness increases conductivity and may form a thermal bridge. Inside, a damp area is perceived warmer when outside the wall is hit by solar radiation or the temperature is higher. Outside, after sunset, when temperature is dropping, damp areas appear warmer due to the higher thermal capacity. For this or other examples, see Chapter 10.

12.5.12. Equilibrium *RH* (Proxy)

The *MC* is in equilibrium with ambient *RH*, and vice versa, the feedback mechanism being driven by the system with the largest water capacity, i.e. the air or the material. In the case of a hole drilled in a wall, or a small external sealed chamber fixed on the surface of the target material to simulate a cavity, the *RH* in the small air pocket will be driven by the *MC* in the material. A capacitive, or resistive, sensor enclosed in this air pocket can record how *RH* changes over time. The external sealed chamber is used to avoid drilling and damaging.

A *'proxy sensor'* is one that is used to indirectly measure a directly unobservable quantity of interest, as in the above example. Although a proxy is unable to provide a direct measure of the desired quantity, it is strongly related to it and may represent, or even substitute, the desired variable. When it is impossible to take samples, a *'proxy sample'* substitutes a real sample, from which the variable of interest can be obtained. The *proxy* methodology is applied for measuring the water absorbed on a traditional sensor or on a hydrophilic sample, e.g. chalk and terracotta, after the equilibrium has been reached in the sampling cavity or in an external sealed box. The *RH* sensor, or the hydrophilic sample, acts as a *proxy*. The proxy sample is then analysed with dry oven gravimetry and readings may be repeated over time.

The equilibrium *RH* responds to the variability of the water vapour pressure before saturation is reached at a given temperature. The water vapour reaches saturation (i.e. $RH = 100\%$) when a film of liquid water is forming inside the pores, the pores being still empty. It will be unable to provide any information when the pores will be progressively filled of liquid water and the material becomes heavier and heavier, being increasingly soaked with water. Therefore, the range of *MC* detected by the equilibrium *RH* methodology is only a small portion of the wider range covered by gravimetry. However, this methodology is convenient to follow the *MC* over time, e.g. how a wall will dry after having been soaked with floodwater.

Inside masonry, or a porous stone, the existing moisture, or the solution that will form in the presence of soluble salts, exerts a partial vapour pressure that is a particular fraction of the maximum level determined, for each temperature, by the saturation pressure of pure water. This pressure is commonly expressed in terms of '*free water*' or '*water activity*' after it has been normalized in terms of pure water saturation pressure and is expressed in percentage, similarly to the definition of *RH* but making reference to the situation in the microair pockets inside the pores of the material. This variable is representative of the water content exchangeable with the surrounding environment until the equilibrium will be reached. In a larger air pocket of an artificial cavity obtained by drilling inside a material, or in an insulated box sealed to the material surface, the free water pressure reaches equilibrium with the measurable *RH*, defining the percentage of existing free water.

An advanced passive sensor, i.e. the '*free-water sensor*' (FWS) has been developed in association with the Salvatore Maugeri Foundation, Padua, under the EU-funded Climate for Culture Project. This sensor follows a dynamic equilibrium with the free water and undergoes irreversible transformations at some selected *RH* levels. The sensor is composed of a series of soluble salts that have affinity for moisture and will absorb some water from the air when their deliquescence thresholds are reached.

The sensor is shaped as a small stick, as shown in Fig. 12.5. The envelope has a transparent band on the upper side and is permeable to water vapour on the bottom in order to reach equilibrium with the free water inside the masonry. It is divided into a number of specific sectors, all of them covered with a white paper disk. Each sector contains a tablet composed of dyes and salts that will become deliquescent at some selected *RH* levels (i.e. RH_1, RH_2, RH_3,...). When one of the above deliquescence thresholds has been reached or exceeded, the absorbed moisture dissolves the salt, the solution activates a dye and the white paper disk becomes and remains permanently coloured indicating that this specific *RH* level has been reached. Each disk and each colour is related to a specific *RH* level that can be read on the front side. The selected levels

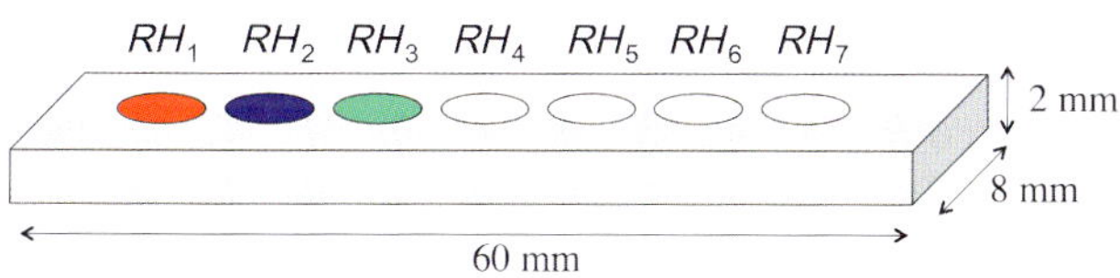

FIGURE 12.5 Scheme of a free-water sensor. In this example, the top side contains seven passive sensors and the highest *RH* level reached over the whole sampling period has exceeded the third level, but not the fourth one.

can be changed on request, e.g. multipurpose with broad intervals over the whole *RH* range or specific for dampness or dryness with higher resolution in the preferred span.

The FWS is supplied in a sealed, dry bag and becomes active once it has been removed from the dry bag. The FWS is a disposable irreversible sensor that provides useful information about the highest humidity level reached over the whole sampling period, e.g. a week, a month or a season. It is cheap in order to be single-use, or utilized in arrays for multiple sampling locations. Specific advantages are that the sensors are miniaturized, can stay in unmanned areas without power supply and are useful for preliminary field investigations to assess whether a building or an archaeological remain is at risk of moulds, has seasonal cycles in dampness, or is affected by gradients in moisture that may be an index of capillary rise, water percolation, etc. The freezing points of the selected saturated solutions are very low, e.g. from −20 °C to −60 °C.

Finally, the FWS is built with materials permeable to the vapour but not to the liquid phase. As a consequence, it can be used inside books, placed between leaves, or can stay in contact with other vulnerable surfaces without staining.

12.6. MEASURING TIME OF WETNESS

It is well known that when the junctions of a thermocouple are at different temperatures, an electromotive force, or a current if the circuit is closed, is generated. The inverse phenomenon is to cool junctions of a thermocouple array with a driving current (*Peltier effect*) and this technology is often used to cool the reflective target surface in chilled mirror dew point meters. The same optical technology used to detect the onset of droplets in a chilled mirror can be applied to detect the same phenomenon but on mirrors having the same temperature of an artefact. In this case, a *dew point signaller* is obtained, useful for monitoring the *time of wetness* (*TOW*) or for switching on a heating system when droplets begin to form on cold windows, or dewing may affect statues or frescoes. In this case, it is necessary to warm directly the dewing surface by means of *IR* radiation or electrical wire resistors; it is not convenient blowing warm air because it has less efficiency, may increase particle deposition and might generate efflorescence.

The *TOW* is an important parameter in conservation science, as several physical, chemical and biological deterioration mechanisms are possible only in the presence of liquid water. However, it is always necessary to keep in mind that the *TOW* is a complex atmospheric variable, determined by the *MR* of the water vapour in air, the object surface temperature T_S and the surface contamination. Only when T_S drops below the *DP*, the surface becomes wet and the *TOW* begins, independent of the ambient *RH*. In reality, the *TOW* does not exactly coincide with the total time during which the surface temperature remains below the *DP*, but is (slightly) longer, as when T_S rises above *DP*, some time is still needed to evaporate the water from the surface. However, for nonadsorbing materials, the difference is small and the two variables are often assumed to be coincident, at least in a first approximation.

Some dew point detectors can be found that are based on the conductivity (or dielectric capacity) of an adsorbing medium. The main problem is that they easily reach equilibrium during the wetting phase but they retain water and remain damp for some time after the surface dries. They present hysteresis in drying and might be representative of some porous adsorbing materials but not of glass or polished metals.

A dew point detector that responds better to the behaviour of nonabsorbing materials is based on the extinction (or attenuation) of a reflected *IR* beam when dew droplets are forming on the surface of the body under examination. In fact, the *IR* radiation is absorbed and extinguished in a few micrometres of liquid water, so that when the first droplets begin to form and grow on the surface, the reflected beam is immediately attenuated and its intensity progressively diminishes until it is completely extinguished; this happens for a completely wet surface. However, as only metals and few other smooth surfaces are good reflectors, a tiny mirror is introduced as reference surface, which has the same temperature as the body. The mirror is put into thermal equilibrium with the artefact but is part of the measuring device (Fig. 12.6), being mechanically sealed

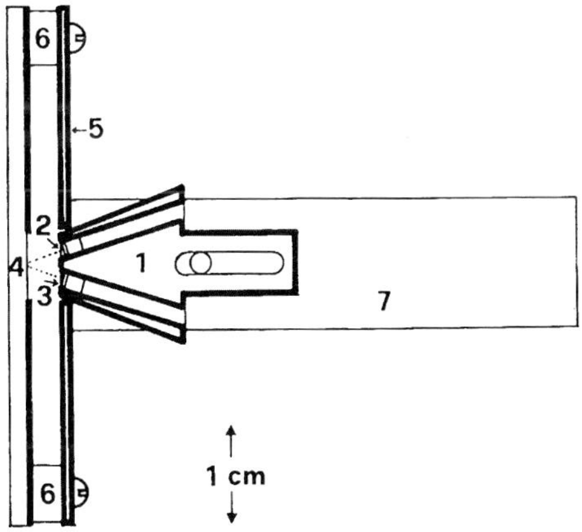

FIGURE 12.6 Dew point detector. The optical transducer (1) is composed of an *IR*-emitting diode (2), a photodarlington (3), a blackened metal board with mirror (4), a shield (5) and four spacers (6). It may be attached to the surface or hung with a holder (7). *Source: After Camuffo and Valcher (1986); by kind permission of Kluwer Academic Publishers. See credits.*

at the appropriate distance to receive the *IR* beam from a light-emitting diode (i.e. an *IR*-emitting diode) and to reflect it into the sensor (i.e. a photodarlington). The mirror is at the centre of a small blackened metal board over which a black plastic sheet is placed at short distance to allow free air circulation but shields the mirror against external radiation and avoids false reflections. The emitter is activated by a pulsed current in order to sensibly reduce the power supplied by the *IR* beam used in the detector and therefore the heating of the mirror. The output of the photodarlington is amplified, transformed into a continuous level by a peak detector and then compared with a reference voltage that can be adjusted to calibrate the threshold at a selected dimension of microdroplets (Camuffo and Valcher, 1986).

References

ASTM D2216 – 10, 2010. Test Method for Laboratory Determination of Water (Moisture) Content of Soil and Rock. American Society for Testing and Materials, West Conshohocken, Pennsylvania.

ASTM D4442 – 07, 2007. Standard Test Methods for Direct Moisture Content Measurement of Wood and Wood-Base Materials. American Society for Testing and Materials, West Conshohocken, Pennsylvania.

Beall, F.C., 2007. Industrial applications and opportunities for non-destructive evaluation of structural wood members. Maderas. Ciencia y Tecnología 9 (2), 127–134.

Bentley, R.E., 1998. Handbook of Temperature Measurement: Temperature and Humidity Measurement. Springer Verlag, Berlin.

Brischke, C., Rapp, A.O., Bayerbach, R., 2008. Measurement system for long-term recording of wood moisture content with internal conductively glued electrodes. Building and Environment 43, 1566–1574.

Bucurescu, D., Bucurescu, I., 2011. Non-destructive measurement of moisture in building materials by Compton scattering of gamma rays. Romanian Reports in Physics 63 (1), 61–75.

Calegari, L., Gatto, D.A., Martins Stangerlin, D., 2011. Influence of moisture content, specific gravity and specimen geometry on the ultrasonic pulse velocity in *Eucalyptus grandis* hill ex Maiden wood. Ciência da Madeira (Brazilian Journal of Wood Science), Pelotas 2 (2), 64–74.

Camuffo, D., Bernardi, A., 1986. Dinamica del microclima e scambi termoigrometrici tra pareti e atmosfera interna nella Cappella Sistina. Bollettino dei Monumenti, Musei e Gallerie Pontificie 6, 211–257.

Camuffo, D., Bertolin, C., 2012. Towards standardisation of moisture content measurement in cultural heritage materials. E-Preservation Science 9, 23–35.

Camuffo, D., Fernicola, V., 2010. How to measure temperature and relative humidity. Instruments and instrumental problems. In: Camuffo, D., Fassina, V., Havermans, J. (Eds.), Basic Environmental Mechanisms Affecting Cultural Property – Understanding Deterioration Mechanisms for Preventive Conservation Purposes. COST Action D42 "Enviart". Nardini, Florence.

Camuffo, D., Valcher, S., 1986. A dew point signaller for conservation of works of art. Environmental Monitoring and Assessment 6, 165–170.

Capitani, D., Emanuele, M.C., Bella, L., Segre, A.L., Attanasio, D., Focher, B., Capretti, G., 1999. 1H NMR relaxation study of cellulose and water interaction in paper. Tappi Journal 82 (9), 117–124.

Capitani, D., Proietti, N., Gobbino, M., Soroldoni, L., Casellato, U., Valentini, M., Rosina, E., 2009. An integrated study for mapping the moisture distribution in an ancient damaged wall painting. Analytical and Bioanalytical Chemistry 395, 2245–2253.

Carll, C., TenWolde, A., 1996. Accuracy of wood resistance sensors for measurement of humidity. Journal of Testing and Evaluation, ASTM 24 (3), 154–160.

Carr-Brion, K., 1986. Moisture Sensors in Process Control. Elsevier Applied Science, Amsterdam.

Casieri, C., Senni, L., Romagnoli, M., Santamaria, U., De Luca, F., 2004. Determination of moisture fraction in wood by mobile NMR device. Journal of Magnetic Resonance 171, 364–372.

Dai, G., Ahmet, K., 2001. Long-term monitoring of timber moisture content below the fiber saturation point using wood resistance sensors. Forest Products Journal 51 (5), 52–58.

Davey, F.K., 1965. Hair humidity elements. In: Wexler, A. (Ed.), Humidity and Moisture, Principles and Methods of Measuring Humidity in Gases, vol. 1. Rehinold, New York, pp. 571–573, 647 pp.

Dill, M.J., 2000. A Review of Testing for Moisture in Building Elements. CIRIA Report No. CIRIA-C538. London.

Doebelin, E.O., 1990. Measurement Systems - Application and Design. McGraw Hill, New York.

EN 13183-1, 2002. Moisture Content of a Piece of Sawn Timber – Part 1: Determination by Oven Dry Method. European Committee for Standardization (CEN TC 346), Brussels.

EN 322, 1993. Wood-based Panels – Determination of Moisture Content. European Committee for Standardization (CEN TC 346), Brussels.

Fisher, P.D., Lillevik, S.L., Jones, A.L., 1981. Microprocessors simplify humidity measurements. Transactions on Instrumentation and Measurements IM 30 (1), 57–63.

Goodhew, S., Griffiths, R., Woolley, T., 2004. An investigation of moisture content in the walls of a straw-bale building. Building and Environment Journal 39, 1443–1451.

Goody, R., 1995. Principles of Atmospheric Physics and Chemistry. Oxford University, New York.

Grinzato, E., Ludwig, N., Cadelano, G., Bertucci, M., Garfano, M., Bison, P., 2011. Infrared thermography for moisture detection: a laboratory study and in-situ test. Material Evaluation 65 (1), 97–110.

Harriman, L.G., 1990. The Dehumidification Handbook. Munters. Cargocaire, 186 pp.

Hartley, I.D., Kamke, F.A., Peemoeller, H., 1994. Absolute moisture content determination of Aspen wood below the fiber saturation point using pulsed NMR. Holzforschung 48, 474–479.

Healy, W.M., 2003. Moisture sensor technology – a summary of techniques for measuring moisture levels in building envelopes. ASHRAE Transactions 109 (Pt. 1), 232–242.

King, R.J., Basuel, J.G., 1993. Measurement of basis weight and moisture content of composite boards using microwaves. Forest Products Journal 43 (9), 15–22.

Kumaran, M.K., Mukhopadhyaya, P., Normandin, N., 2006. Determination of equilibrium moisture content of building materials: some practical difficulties. Journal of ASTM International 3 (10), 1–9.

Larsen, P.K., 2004. Moisture measurement in Tirsted Church. Journal of Architectural Conservation 10 (1), 22–35.

Larsen, P.K., 2012. Determination of water content in brick masonry walls using a dielectric probe. Journal of Architectural Conservation 18 (1), 47–62.

Le Feunteun, S., Diat, O., Guillermo, A., Poulesquen, A., Podor, R., 2011. NMR 1D-imaging of water infiltration into mesoporous matrices. Magnetic Resonance Imaging 29 (3), 443–455.

Lipták, B.G., 2003. Instrument Engineers' Handbook: Process Measurement and Analysis. In: The Instrumentation System and Analysis (ISA), vol. 1. CRC Press, Boca Raton, Florida.

List, R.J., 1971. Smithsonian Meteorological Tables. Smithsonian Institution, Washington D.C, 527 pp.

Macleod, K.J., 1983. Relative Humidity: Its Importance, Measurement and Control in Museums. Canadian Conservation Institute, Ottawa.

Mamillan, M., 1992. Méthodes d'évaluation des dégradations des monuments en pierre. In: Zezza, F. (Ed.), Weathering and Air Pollution. Adda, Bari, pp. 175–181.

Michalski, S., 2000. Guidelines for Humidity and Temperature in Canadian Archives. Canadian Conservation Institute, Ottawa.

Milota, M.R., 1994. Specific gravity as a predictor of species correction factors for a capacitance-type moisture meter. Forest Products Journal 44 (3), 63–68.

Mitchell, P., 2010. Methods of Moisture Content Measurement in the Lumber and Furniture Industries. Wood Product Notes, Department of wood and paper science, Raleigh, North Carolina, 5 pp.

Mortl, A., Muñoz-Carpena, R., Kaplan, D., Li, Y., 2011. Calibration of a combined dielectric probe for soil moisture and porewater salinity measurement in organic and mineral coastal wetland soils. Geoderma 161, 50–62.

Murphy, W., Smith, J.D., Inkpen, R.J., 1996. Errors associated with determining P and S acoustic wave velocities for stone weathering studies. In: Smith, B.J., Warke, P.A. (Eds.), Processes of Urban Stone Decay. Donhead, London, pp. 228–244.

Nappi, A., Côte, P., 1997. Non-destructive methods applicable to historic stone structures. In: Baer, N.S., Snethlage, R. (Eds.), Saving Our Architectural Heritage: the Conservation of Historic Stone Structures. Wiley, Chichester, pp. 151–166.

NWFA, 2008. Appendix C – Moisture Guidelines & Moisture Testing. National Wood Flooring Association. www.nwfa.org/member/.

Oliveira, F.G.R., Candian, M., Lucchette, F.F., Salgon, J.L., Sales, A., 2005. Moisture content effect on ultrasonic velocity in *Goupia glabra*. Materials Research 8 (1), 11–14.

Olmi, M., Bini, M., Ignesti, A., Priori, S., Riminesi, C., Felici, A., 2006. Diagnostics and monitoring of frescoes using evanescent-field dielectrometry. Measurement Science and Technology 17 (8), 1623–1629.

Palaia, L., Sánchez, R., López, V., Gil, L., Monfort, J., Tormo, S., Navarro, P., Álvarez, M.Á., 2008. Procedure for NDT and Traditional Methods of Ancient Building Diagnosis by Using Thermograph, Digital Images and Other Instrument Data Analysis. 17th World Conference on Nondestructive Testing, Shanghai.

Perry, R.H., Green, D.W., Maloney, J.O., 1984. Chemical Engineers' Handbook, Volume 6. McGraw-Hill, New York.

Plate, E.J., 1982. Engineering Meteorology. Elsevier, Amsterdam.

Poděbradská, J., Maděra, J., Tydlitát, V., Rovnaníková, P., Černý, R., 2000. Determination of moisture content in hydrating cement paste using the calcium carbide method. Ceramics – Silikáty 44 (1), 35–38.

Rosina, E., Robison, E.C., 2002. Applying infrared thermography to historic wood-framed buildings in North America. APT Bulletin XXXIII (4), 37–44.

Rosina, E., Spodek, J., 2003. Using infrared thermography to detect moisture in historic masonry: a case study in Indiana. APT Bulletin XXXIV (1), 11–16.

Saïd, M.N., 2007. Measurement methods of moisture in building envelopes – a literature review. International Journal of Architectural Heritage 1 (3), 293–310.

Schajer, G.S., Orhan, F.B., 2005. Microwave non-destructive testing of wood and similar orthotropic materials. Sensing and Imaging: An International Journal 6 (4), 293–313.

Schajer, G.S., Orhan, F.B., 2006. Measurement of wood grain angle, moisture content and density using microwaves. Holz als Roh- und Werkstoff 64, 483–490.

Shimamura, Y., Urabe, T., Todoroki, A., Kobayashi, H., 2004. Electrical impedance change method for moisture absorption monitoring of CFRP. Advanced Composite Materials 13 (3–4), 297–310.

Sundara-Rajan, K., Byrd, L., Mamishev, A.V., 2004. Moisture content estimation in paper pulp using fringing field impedance spectroscopy. IEEE Sensors Journal 4 (3), 378–383.

Tanaka, Takashi, Avramidis, Stavros, Shida, Satoshi, 2009. Evaluation of moisture content distribution in wood by soft X-ray imaging. Journal of Wood Science 55 (1), 69–73.

Titta, M., Olkkonen, H., 2002. Electrical impedance spectroscopy device for measurement of moisture gradients in wood. Review of Scientific Instruments 73 (8), 3093–3100.

UK Meteorological Office, 1981. Handbook of Meteorological Instruments. In: Measurement of Humidity, vol. 3. Her Majesty's Stationary Office, London, 43 pp.

Vun, R.Y, Bhardwaj, M.C., Hoover, K., Janowiak, J., Kimmel, J., Worley, S., 2006. Development of non-contact ultrasound as a sensor for Wood Moisture Content. ECNDT 2006 – Tu.4.2.3, 1–7.

Weast, R.C., 1977/78. CRC Handbook of Chemistry and Physics. CRC Press, West Palm Beach, Fla.

Weritz, F., Kruschwitz, S., Maierhofer, C., 2009. Assessment of moisture and salt contents in brick masonry with microwave transmission, spectral-induced polarization, and laser-induced breakdown spectroscopy. International Journal of Architectural Heritage: Conservation, Analysis, and Restoration 3 (2), 126–144.

Wexler, A., 1965. Humidity and Moisture. In: Principles and Methods of Measuring Humidity in Gases, vol. 1. Rehinold, New York, 647 pp.

Wiederhold, P., 1975. Humidity measurements part I: psychrometers and percent RH sensors. Instrumentation Science and Technology 22, 31–37.

Wiederhold, P., 1997. Water Vapour Measurement: Methods and Instrumentation. Marcel Dekker Inc, New York.

Wilson, P.J., 1999. Accuracy of a capacitance-type and three resistance-type pin meters for measuring wood moisture content. Forest Products Journal 49 (9), 29–32.

WMO, 1986. Compendium of Lecture Notes on Meteorological Instruments for Training Class III and Class IV. Meteorological Personnel, WMO Technical Publication No. 622. World Meteorological Organisation, Geneva.

WMO, 2008. Guide to Meteorological Instruments and Methods of Observation, WMO Technical Publication No 8. World Meteorological Organisation, Geneva.

Wylie, R.G., Lalas, T., 1992. Measurement of Temperature and Humidity. WMO 759. Secretariat of the World Meteorological Organization, Geneva.

Zazueta, F.S., Xin, J., 1994. Soil Moisture Sensors. Bulletin 292. Florida Cooperative Extension Service, Institute of Food and Agricultural Sciences, University of Florida.

CHAPTER

13

Measuring Wind and Indoor Air Motions

OUTLINE

13.1. MEASURING WIND SPEED AND DIRECTION

13.1.1. Anemometry

For meteorological purposes, the wind speed and direction are measured with anemometers and wind vanes, respectively. The two transducers

Microclimate for Cultural Heritage, Second Edition
http://dx.doi.org/10.1016/B978-0-444-63296-8.00014-7

may be separated or combined in only one instrument. For air pollution studies, the transducers may be three, e.g. three orthogonal propellers for the three wind vectors. For special studies, other instruments are used, e.g. hot wire, sonic, laser, and Doppler (Lipták, 2003; WMO, 2006; Tropea et al., 2007; Burt, 2012).

Anemometers are based on one of the following wind properties.

13.1.1.1. Wind Kinetic Energy

Wind kinetic energy causes rotation of cup wheel or propeller anemometers. The cup arms are supported by a vertical shaft and are insensitive to the wind direction; the propeller is continually oriented by a vane. The rotation is transformed into electrical signal by means of photo choppers, magnet system output signal, contact switches, direct current generators (i.e. tachometers), or other systems. The threshold varies with the friction and the energy subtracted by the transducer. Photo choppers do not subtract energy, reed switches subtract a very small amount and tachometers subtract a considerable amount.

This anemometer type has low sensibility but is resistant to unfavourable environments and is reliable; for this reason, it is typically used by military and weather services, especially in remote stations. Some miniature types may be used in micrometeorological studies for the dispersion of air pollutants.

Cup anemometers (*Robinson* type) are composed of three or four conical or hemispherical cups revolving about a vertical shaft. Originally, a four-cup mounting was used but the dynamic response of three cups is better and the majority of instruments follow this finding. The maximum torque produced by a single cup does not occur when the wind blows directly into the concavity of the cup but when it forms an angle of about 45°. The output is essentially linear with constant wind speed up to about 30 m s^{-1} when the cup anemometer is calibrated in a wind tunnel but, in the field, with gusts and lulls, it overestimates the speed as the rotor is more sensitive to increasing than decreasing wind speed, so that it is practically accelerated during fluctuating wind speeds (Moses, 1968; Ramachanddran, 1970; Hyson, 1972).

Propeller anemometers (either simple combinations of blades and tail, or combined aerodynamic structures like *aerovanes* or *windmills*) are composed of a propeller and a tail on the same axis. The propeller revolves about a pivoted horizontal shaft that is oriented along the direction from which the wind is blowing. The orientation is governed by the pressure/turbulent drag exerted by wind on the vane tail that is incorporated on the backside of the instrument. The rate of rotation of the propeller is linearly proportional to the wind speed from 1 m s^{-1} up to 45 ms^{-1}. The propeller responds in a different way as a function of the *angle of attack*, i.e. the angle between the axis of the anemometer and the wind direction, and indicates a lower wind speed when the angle of attack increases. As propellers are oriented by a vane; fluctuating wind directions lead to an underestimation of the wind speed because the inertia of the system forces the vane to smooth out and follow with delay the changes in direction and the angle of attack is generally different from 0. The inertia of the system is given not only by the friction and the distribution of the mass in the system, which dominate at low wind speed, but also by the moment of inertia of the blades under rotation, which dominates at high wind speed.

Propellers should respond, to a good approximation, with a cosine law to the angle of attack. Only in this case can they be used to measure the orthogonal components of the wind vector (Camp et al., 1970; Horst, 1973).

It should be noted that both cups and propeller blades are not symmetrical, when observed downwards, on the same vertical. This means that rainfall impresses kinetic energy on initially motionless anemometers. On the contrary, rainfall slows down the rotation at high wind speed, especially when large drops impact on cups having motion opposite to the wind field or when drops impact on uprising blades.

Cups and *propellers* are generally made of metal in order to be rugged and to resist bad weather conditions. However, very sensible sensors are built of thin aluminium or plastic to reduce inertia and thresholds. For low wind speed, polycarbonate cups or polystyrene foam propellers are used, with a threshold around 0.1 m s^{-1}. Mechanical sensors cannot be employed for very low wind speed, i.e. below 0.1 m s^{-1}.

Vanes may be either flat or splayed, i.e. vanes composed of two divergent flat plates. Flat vanes offer minimum resistance to the wind and for this reason they may swing or be unstable. In contrast, splayed vanes increase the resistance to wind and generate turbulence and this helps to keep the vane oriented with the wind. Therefore, splayed vanes follow even small changes in wind direction, better than flat plates. However, the heavy mass of these devices increases the inertia of the system, which is unable to follow high-frequency fluctuations in wind direction. When fast response is needed, e.g. for air pollution and dispersion purposes, low-inertia vanes should be used. Flat vanes are preferred for micrometeorological and air pollution studies and splayed vanes are preferred for meteorological applications where the average wind direction is needed (UK Meteorological Office, 1981).

Aerovanes and/or *windmills* have an aerodynamic-shaped structure like an aircraft without wings. They are very resistant but have a very high inertia due to both the distribution of mass (i.e. some 15% torque more than a flat vane with similar physical dimensions) and the moment of inertia of the rotating blades. They operate as a low-pass filter acting on the wind fluctuations and their use is recommended when average values are needed (UK Meteorological Office, 1981).

13.1.1.2. Wind Pressure

Wind pressure is the physical principle behind the pressure tube (Pitot) or pressure-plate anemometer. The *Pitot tube* is based on the Bernoulli law for moving airstreams and consists of a manometer that measures the difference between the static and the dynamic pressures sampled at two orifices made on the probe tube, made, respectively, in the stagnation point and where the wind field is unaffected. It is especially used in chimneys or aeroplanes. The wind pressure is also used to orient wind vanes and measure the wind direction.

The pressure plate measures the deflection of a flat plate that is continually oriented along the wind direction by a vane. It was invented in the fifteenth century by Leonardo da Vinci and was used only in the past (Wildt anemometer).

Wind turbulence, Reynolds stress and drag coefficient have been measured using a table tennis ball as a sensor, pierced with a vertical metal axle whose tilting is measured with two orthogonal transducers (Smith, 1970).

13.1.1.3. Wind Cooling Power

Wind cooling power is used in hot wire or hot film anemometry. The principle is to measure the current required keeping constant the temperature of an overheated wire or the change of temperature of a wire heated at constant current. As the wire is very thin, the probe is very small and sensible, shows fast response, and is particularly suitable for microclimatic studies and to monitor low-speed air movements as well as short-term fluctuations. The sensor is very fragile and must be used with great care and is more suitable for indoor studies.

13.1.1.4. Wind Property of Transmitting Sounds

The continuous-phase-type sonic anemometer transmits across the path of interest a continuous beam of sonic energy; the phase of the received signal is compared with a fixed or reference phase. The pulse-type sonic anemometer transmits sonic energy in bursts over the path of interest. Sonic pulses are transmitted in opposite directions over the same path on each axis of measurement. Pulses travelling with the wind arrive sooner than those travelling against the wind. The main advantages of this instrument are that it is linear, has a very fast response (i.e. a few milliseconds, so that it can respond to high-frequency wind fluctuations), and has a very low threshold. This anemometer is particularly used in microphysical studies of the planetary boundary layer, e.g. to measure the exchanges of heat or momentum.

13.1.1.5. Wind Property of Transmitting Light

The *Doppler effect* is the change in the observed frequency of an (electromagnetic) wave due to relative motion of the source and observer.

Laser Doppler anemometry has developed different configurations (Albrecht et al., 2003). One of the most important is the differential Doppler or *'fringe'* mode, which concerns the interference fringe of two laser beams; suspended particles, transported by the air movement across the dark and light fringe pattern, determine the frequency of electric signals that are then processed to compute the airspeed. The velocity of the suspended particles causes a Doppler shift in the frequency of the light that is scattered and this is measured with a photodetector. In a relatively clean atmosphere, instruments need the introduction of tracer particles in the airstream. The main advantages are direct method of measuring velocity leaving the flow undisturbed, very small sensing volume (e.g. a cube of 0.2 mm on a side), very high frequency (e.g. MHz) response, and use for extremely low airspeed. This system is expensive and is more suitable for scientific research than routine measurements.

13.1.2. Installing Anemometers

There are several sources of error in wind speed and direction measurements but the largest errors are due to the presence of the transducer, which perturbs the wind field, or generates turbulence, or the presence of eddies generated by buildings, trees, or other obstacles; also, the presence of the tower or mast to which the instrument is attached cannot be neglected.

Meteorological measurements of wind speed and direction are regulated by a normative that is extensively reported in many handbooks and that can be summarized as follows. The anemometer should be placed on a mast at 10 m above the ground, over open, level terrain with no obstructions within 300 m. Obstacles, if any, should be at a distance not less than 10 times their height if they are upwind or 5 times if they are downwind. The aim is to observe the wind 'undisturbed', except for the action of the soil friction, i.e. the wind that an aeroplane experiences when it is preparing to land or take off. The same normative has been then applied for use in pollutant transport and dispersion.

For pollution studies, the *wind shear*, i.e. the vertical change of wind speed and direction, is an important parameter. This vertical profile is obtained by installing some anemometers on a tower or tall mast, at heights increasing logarithmically from the soil. Also, vertical profiles of the wind field are obtained launching pilot balloons and measuring their position with two theodolites at regular time intervals (e.g. every 15 s). The wind field is then calculated with trigonometry.

13.1.3. Measuring Wind Vector Components

First of all, *an anemometer measures the total amount of air masses that in unit time is passed through the sensible part of the instrument*, supplying part of its kinetic energy to the cups or propellers that are induced to rotate. The wind gustiness and the smaller scale turbulence oblige every air parcel to fly not in a straight line but with many vicious loops around the main direction. As a consequence, the space length computed by means of the integral $\int u(t)\,dt$, where $u(t)$ is the instantaneous wind speed, is much greater than the real distance travelled by the air parcel from the sensor and is represented by the average wind vector. If this vector is needed to model the transport of pollutants, only three methods are appropriate, as follows.

1. To use a *two-component anemometer*, which measures the two orthogonal vectors of the wind speed. The anemometer is composed of two tachometric generator transducers, mounted at right angle on a common mast. Each propeller measures the component of the wind that is parallel to its axis of rotation. Propellers have been studied in order to have a response to the wind angle that approximates the cosine law. When wind and propeller rotation reverse, signal polarity also reverses. The same principle is adopted for the so-called *UVW* anemometer, which measures the three orthogonal wind vectors U and V on the horizontal plane and W along the vertical. The lateral and vertical fluctuations of the wind are important in micrometeorological investigations of the dispersion of airborne pollutants.
2. To use a *wind vane with a sine–cosine transducer* supplied with a voltage proportional to $u(t)$. This method is much more convenient but needs a special anemometer (Camuffo, 1976). This is substantially composed of a generating anemometer, e.g. a dynamo connected with the propellers (or the cups), that supplies an output proportional to $u(t)$ and this output is used to feed the terminals of the sine–cosine potentiometer of the vane that measures the wind direction. This potentiometer has a continuous-rotation resistance element that varies with the sine function of the shaft rotation and two independent brushes, spaced 90° between them: consequently, one monitors $\sin\alpha(t)$ and the other $\sin(\alpha(t)+90°)=\cos\alpha(t)$, where $\alpha(t)$ is the instantaneous wind direction. When the potentiometer is supplied with the tension $u(t)$, the two potentiometer outputs give, respectively, $u(t)\sin\alpha(t)$,$u(t)\cos\alpha(t)$. Continuous resistance elements, e.g. cermet, bifilm or conductive plastic should be preferred to wire-wound ones, as the noise is reduced when the brushes oscillate following the wind fluctuations. Generating aerovanes are preferred, as they furnish only one compact and strong instrument. The same result can also be obtained with ordinary anemometers by substituting the linear potentiometer with a sine–cosine one and using the output of the airspeed transducer to supply the power to the potentiometer.

3. To carry out a very frequent sampling of $u(t)$ and $\alpha(t)$ and then *compute* for each instant the vector components $u(t)$ sin $\alpha(t)$,$u(t)$ cos $\alpha(t)$. This method is less convenient, for it requires a huge amount of data to be monitored and processed, but can be made with ordinary anemometers.

13.1.4. Averaging Wind Direction

Application of statistical formulae in meteorology needs particular care to avoid inappropriate calculations and misleading conclusions. For instance, during the first part of the summer nights in the plain of northern Italy, cold air flows from the mountain valleys on the southern lee of the Alps and enters the plain, which is characterized by warmer and moister air masses. These cold tongues force the moist air of the plain to rise up generating local violent thunderstorms very similar to the thermoconvective ones that often occur in the early afternoon. By beginning at midnight the meteorological observations of the day, some of these thunderstorms are recorded in 2 days and attributed half to the previous and half to the next day, thus misleadingly increasing the frequency of occurrence.

The above example helps to understand the behaviour of two air masses when they meet, and vector algebra has to be applied in a correct way to interpret it. When two air masses meet, their dynamic state is represented by two wind vectors but, in general, the resultant is not the vector obtained as the algebraic sum of the two original vectors. If the density of the two air masses is not the same, a frontal situation is generated and the denser one continues nearly unaffected its motion, whereas the lighter one slips above.

It is useful to remember that a wind vane will always point a direction, with or without wind. When the wind drops, the vane remains oriented with the last direction. When the average wind direction is computed, all the directions recorded in the absence of wind speed must be rejected.

Another problem occurs when the mast or the vane shaft is not perfectly vertical. The distribution of the weight in a vane is not always balanced with the rotation axis passing across the centre of gravity of the vane. Out of balance, it shows a bias towards the tilting direction, especially in presence of low wind speed or during calm periods.

In climatology, the frequency distribution makes more sense than the arithmetic average represented by the *mean*. Consider, e.g. the case of a coastal site, with the sea breeze from east (90°) and the land breeze from west (270°). The distribution of the wind direction is bimodal with two peaks at 90° and 270°, whereas the mean gives the dominance of a non-existing wind from south (180°). However, although in the long term frequency distributions are necessarily used, in the short term of a sampling interval (e.g. 10 min), the wind is considered to oscillate around a main direction and an arithmetic average is used to compute the mean value. A common problem, not statistical but technological, is averaging the wind direction with fluctuations in the northern sector. In fact, ordinary potentiometers present a discontinuity between 0° and 360°, so that in the case the wind fluctuates symmetrically around north, averaging the data one obtains 180°, i.e. exactly the opposite direction. If recording is being performed in a strip chart recorder, the trace is a line going back and forth across the whole chart width. Several solutions have been devised to overcome this drawback. Here only a few of them will be presented for their practical interest. Other original devices have been described elsewhere (Camuffo, 1979).

The first system is composed of two identical potentiometers connected with the same vane, the first having the range 0–360° and the second 180°–540° being assembled rotated at 180°, i.e. when the brush of the first one reaches the discontinuity 0–360°, the brush of the second is positioned in the middle of the second potentiometer, on the opposite side of the discontinuity. When this happens, the output is switched from the first potentiometer to the second one, so that continuity is ensured for 1½ revolution. The range extension can also be performed with only one potentiometer and an electronic translator, in order to reduce friction and wind threshold. The range extension 0–540° ensures continuity for most of the cases, except when the upper limit 540° is reached and the output is switched to 180°, and as well as the lower limit 0° is reached and the output is switched to 360°.

The second system (Camuffo and Denegri, 1976) can be applied to standard instruments, i.e. a vane coupled to an ordinary potentiometer and only one sliding contact. The principle is as follows. Suppose that the compass is divided into 100 parts, so that each one is represented by a two-digit number from 00 to 99. By summing 1000 of these numbers, a five-digit number is obtained and the first two figures represent the mean value within the above resolution. If three-digit numbers are used, and the first figure is considered as an index of periodicity, then the two figures following the first one in the sum of 1000 numbers represent the mean value related to the compass card. The index of periodicity disappears if the first figure is neglected, even if it has been used to compute the sum. In this example, an index of periodicity from 0 to 2 can be used, which is sufficient to allow complete rotations of the wind vane during the averaging period; in practice, only one extension is necessary to avoid the discontinuity around north. The method consists in adding or subtracting only one unit to the index, depending on whether the wind vane arrives to the discontinuity rotating clockwise or counterclockwise, i.e. from west or from east. Actually, at the beginning, the index is put equal to 1 and becomes 0 or 2 during the addition. Therefore, when the wind is rotating from the direction labelled

99-00, the corresponding numbers are 199-200, without discontinuity. In our example, making 1000 samplings equally distributed around the former potentiometer discontinuity, i.e. 199-200, the total 199,500 is obtained. Disregarding the first figure and considering the last three as decimal values, the mean direction 99.5 is obtained. The resolution may be increased by dividing the compass into 1000 parts.

This principle can be applied to ordinary potentiometric wind vanes in two ways. The first is to supply the potentiometer with a fixed voltage and convert the brush output into digital numbers. These numbers, with the proper index, can be summed to obtain the mean value as explained. The second method is obtained with the help of a tension to frequency converter and by counting the pulses during the sampling interval. In this case, before conversion, the voltage applied to the terminals of the potentiometer must be added to the signal measured at the brush if this has moved to the lower value across the discontinuity, or subtracted for the opposite rotation.

Another method is based on the principle that in climatology, frequency statistics are better than performing averages. On this ground, the instantaneous direction is monitored at regular time intervals and after the sampling time, the prevalent direction is found as the middle of the sector with the largest population, i.e. where the mode is located; sometimes the mode is found by looking at the frequency distribution.

13.1.5. Measuring Wind Variance

The wind variance is a key variable in computing the dispersion of airborne pollutants and we have seen that the Brookhaven and the Pasquill stability classes were focused on estimating this parameter in the absence of direct measurements. In this case, we are dealing with the actual measurements and how to transform anemometer readings into variance. Following the definition, the standard deviation σ_θ of the wind direction can be measured as

$$\sigma_\theta = \sqrt{\frac{\sum \theta'^2}{n-1}} \tag{13.1}$$

where θ' is the angular wind fluctuations determined after recording n measurements of $\theta(t)$, taken with very short sampling time, for a certain time interval. Determination of the variance, made by means of the above mathematical definition of σ_θ, has a negative aspect: it needs huge amounts of data, which should be recorded on a data logger having a very large capacity and, therefore, is expensive. For this reason, other methods may be preferred and one of them is here described (Camuffo, 1976). In general, fluctuations are symmetrically distributed around the mean wind direction $<\theta>$ and the distribution function is fairly well approximated by a Gaussian. As a consequence, the third spectral moment of the fluctuations, called *skewness*, is 0.

As $\theta = <\theta> + \theta'$ and recalling the formula for the sum of angles in trigonometric functions,

$$\begin{aligned} < \sin\theta > &= < \sin(<\theta> + \theta') > \; = < \sin<\theta> \cos\theta' + \cos<\theta> \sin\theta' > \\ &= \sin<\theta> <\cos\theta'> + \cos<\theta> <\sin\theta'> \\ &= \sin<\theta> <\cos\theta'> = \sin<\theta> \end{aligned} \tag{13.2}$$

$$\begin{aligned} < \cos\theta > &= < \cos(<\theta> + \theta') > = < \cos<\theta> \cos\theta' - \sin<\theta> \sin\theta' > \\ &= \cos<\theta> <\cos\theta'> - \sin<\theta> <\sin\theta'> \\ &= \cos<\theta> <\cos\theta'> = \cos<\theta> \end{aligned} \tag{13.3}$$

as $\sin<\theta'> = 0$ in that the sine is an odd function and the fluctuations are randomly distributed around 0, and $<\cos\theta> = 1$ for the symmetrical reason. Therefore,

$$\frac{<\sin\theta>}{<\cos\theta>} = \tan<\theta> \tag{13.4}$$

where the mean direction is related to the average values of the trigonometric functions sine and cosine. Also in the case of nonsymmetrical distribution of the fluctuations, i.e. near an obstacle or other disturbing factors, by expanding the sine function in MacLaurin series, i.e.

$$\sin<\theta'> = <\theta'> - \frac{<\theta'^3>}{3!} + \frac{<\theta'^5>}{5!} - \frac{<\theta'^7>}{7!} + \cdots \tag{13.5}$$

it can be recognized that the error is very small, i.e. less than 1%.

When the fluctuations are randomly distributed, then the central moments are

$$<\theta'^{2n}> = 1 \times 3 \times 5 \times \ldots \times (2n-1)\sigma_\theta^{2n} \tag{13.6}$$

Substituting this equation in the MacLaurin expansion of $\cos\theta'$ in Eqn (13.2) and Eqn (13.3),

$$\frac{<\sin\theta>}{\sin<\theta'>} = \frac{<\cos\theta>}{\cos<\theta'>} = \exp\left(-\frac{\sigma_\theta^2}{2}\right) \tag{13.7}$$

the variance of the wind fluctuations is obtained:

$$\sigma_\theta^2 = -2\ln\frac{<\sin\theta>}{\sin<\theta'>} = -2\ln\frac{<\cos\theta>}{\cos<\theta'>} \tag{13.8}$$

The dispersion coefficient is simply obtained from average values measured with a sine–cosine transducer or by computing averages of these trigonometric functions.

13.2. MEASURING INDOOR AIR MOTIONS

Microclimate for conservation and weather analysis for general purposes have different aims, problems and observational methodologies. In weather analysis, we should know the unperturbed wind in open countryside far from obstacles; in conservation, the interest is focused on local effects and the interaction with the monument. On an open site, the wind is represented by the speed that is only a function of time i.e. $v(t)$, and the same for its direction α on the horizontal plane, i.e. $\alpha(t)$. In proximity of a monument, wind changes both in time and space, point by point, for the aerodynamic interactions, i.e. $v(x, y, z; t)$. In addition, it departs from the horizontal plane so that the angle of attack θ too should be considered, i.e. $\alpha(\theta;t)$. Briefly, it becomes a function of five variables x, y, z, θ, t instead of only one.

This means that standards concerning the best practice to make wind measurements in meteorology are of little help when considering architectural heritage. Field observations should be only made after a preliminary analysis of the problem, the site topography and monument features, in order to determine *why*, *what*, *where*, *when* and *how* to measure.

13.2.1. Hot Wire and Hot Film Anemometers

Observations of low airspeed, turbulence, thin airstream on a developing surface boundary layer, or thin airstream passing below a closed door cannot be made with ordinary mechanical anemometers (e.g. cup or propeller types), due to their elevated inertia and threshold or the unbalanced effect on the opposed cups or the inhomogeneous pressure distribution on blades. Miniature *hot wire* (or *hot film*) anemometry responds (although not completely) to this demand and is extensively used for its simple use and low cost (Fig. 13.1). The size of the sensor is of the order of one or few millimetres in length and the diameter is of the order of 5 μm. The hot wire measures airspeeds above 10 cm s^{-1} and the time constant is of the order of 0.001 s. A lower threshold, i.e. 5 cm s^{-1}, is obtained with a nickel thin film deposited by sputtering on a spherical glass sensor, with a diameter of 3 mm (Dantec, 1996). The relatively larger mass increases the time constant to 0.08 s and the overheating generates a convective motion that interferes with the air movement at low airspeeds. This interference determines the lower limit of reliable measurements, which is around 3 cm s^{-1}

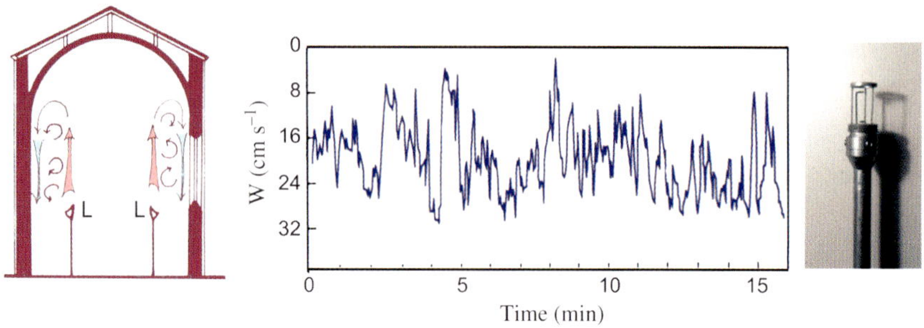

FIGURE 13.1 A hot wire anemometer (to the right) was used inside the Giotto Chapel, Padua, to monitor the vertical flow (W) of warm air (red arrows) ascending over the incandescent lamps (L) and the cold return flow (cyan arrows) along the frescoed walls (to the left). The observed flow along the frescoes has speed fluctuating between 0 and 32 cm s^{-1} (see credits).

As a single wire responds to the velocity component perpendicular to it, a variety of probes exist, mounted either single or coupled orthogonally in a plane or three-dimensionally, suspended between the tips of a fork-like support, for detecting one, two or three components of the airstream. Some probes are inserted into a cylindrical shield (a tube) in order to measure the stream component along the cylinder axis. However, the edges of the tube disturb the flow field and generate turbulence. It is convenient to remove this shield and insert the bare probe into the airstream, with the wire normal to the flow direction.

The physical principle (DISA, 1976; Doebelin, 1990; Tropea et al., 2007) is the thermal loss of an overheated wire operating as resistance sensor. The heat loss not only depends on the airspeed but also on a number of parameters such as air temperature and pressure. If only the airspeed changes, or the influence of the other parameters is compensated through the use of other sensors and suitable electronic units, the output gives the airspeed. The characteristic transfer function is in first approximation composed of an exponential and a square root function, but the signal can be linearized, so that the processed output is simply proportional to the airspeed.

Two different circuits are available for these kinds of sensors: the *constant-temperature* and the *constant-current* anemometer. The constant-temperature type consists of a Wheatstone bridge and a servo amplifier and the sensor acts as the active arm of the bridge. The current through the wire is adjusted to keep the wire temperature constant and is a measure of the flow velocity. The constant-current type has the sensor powered by a constant current supplied by a generator having high internal resistance in order to be independent of any resistance changes in the bridge. The wire reaches an equilibrium temperature that is determined by the heat exchanges with the airstream. The heat generated is the product of the electrical resistance, the square of the current intensity and the wire temperature. Briefly, the airspeed is measured in terms of the electrical resistance. In practice, constant-temperature anemometers are preferable and are effectively popular for their easy use, fast response and low cost.

13.2.2. Sonic Anemometers

The velocity of a sonic wave in a medium is known and depends on the elastic properties of the medium. When a sonic wave is superimposed on an airstream, its transmission speed equals the sum of the velocity of the sound with respect to the medium plus the velocity of the medium. Sonic pulses are transmitted in opposite directions over the same path for each axis of measurement. Pulses are exchanged between two miniature piezoelectric transducers that are used to transmit and then to receive and the sonic anemometer measures the average value of the speed of propagation of these pulses. The measurement is representative of the average airflow that crosses a cylinder, i.e. the sonic beam having the cross-section determined by the transducer size and the path length L equal to the transducers spacing. The space resolution is determined by the transducer size (typically of the order of 1 cm diameter) and the value of L, which varies with the model, e.g. 15 cm, 40 cm. However, the higher the space resolution, the greater the perturbation caused by the transducers to the fluid motion.

The operating principle (Beaubien and Bisberg, 1968; Tropea et al., 2007) is that the sound wave transmitted in still, or moving air, introduces a time lag that depends on the airspeed and direction. The first equation that governs the operating principle concerns the velocity C of sound in still air, i.e.

$$C = \sqrt{\frac{\gamma \Re T}{M}} \tag{13.9}$$

where γ is the ratio between the specific heats at constant pressure and volume and M is the molecular mass of the gas, e.g. 28.97 g mol^{-1} for dry air. At $T = 273$ K, the speed in the air is 33,145 cm s^{-1}.

Therefore, a sonic anemometer is always associated with a precise thermometer whose measurements are necessary to enter the formulae and compute the sonic velocity. However, the elevated presence of moisture may also cause departures from the measurements increasing the sonic speed. As water vapour has a mass (18.016 g mol^{-1}) noticeably different from the other gases constituting the atmosphere, the following empirical equation holds for humid air

$$C = 2006.7\sqrt{T\left(1 + 0.3192\frac{e}{p}\right)} \quad (\text{cm s}^{-1}) \tag{13.10}$$

where e is the water vapour pressure and p the atmospheric pressure. The variable vapour pressure is a source of error and the maximum error

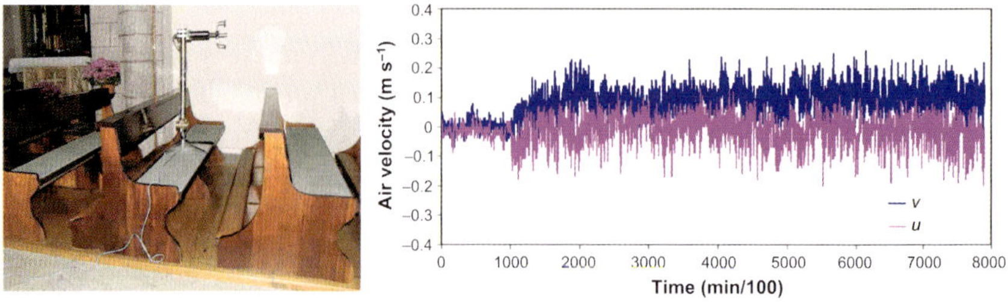

FIGURE 13.2 A bidimensional sonic anemometer positioned above pews to monitor air draught in the church when pew heating is operated. The air is initially still but when the heating is turned on, the air movements are evidenced by fluctuations from -20 to $+20$ cm s^{-1}.

is found in summer, e.g. at $T = 303$ K and relative humidity $= 70\%$, $e = 30$ hPa, and this variation is equivalent to 0.3 °C shift in air temperature. The error is smaller in the cold season.

When the airstream is moving at the speed u and with an angle θ with reference to the transducer alignment, the sonic anemometer measures the speed component $u \cos \theta$ by means of the sonic transit time Δt, which is given by

$$\Delta t = \frac{2L\, u \cos \theta}{C^2\left(1 - \frac{u^2}{C^2}\right)} \tag{13.11}$$

which is the basic equation for sonic anemometry.

The main interest for the sonic anemometer is that it has no threshold and is a totally passive instrument that does not interfere with the fluid motion, except for the presence of the transducers that may generate turbulence.

Flux measurements are accurate with a good cosine response of the sensors (Gash and Dolman, 2003; van der Molen et al., 2004). The experimental array can be composed of one, two or three axes, each having a pair of aligned transducers, depending on the number of dimensions that should be taken into account (Fig. 13.2). For the above-mentioned reasons, the measurements are not as punctual as with a hot wire and cannot go close to a surface such as a hot wire but these goals are much better attained with a laser Doppler anemometer.

Sonic anemometers are useful to measure either fast or slow air motions in the free air or in proximity of objects. This helps in evaluating deposition rates and soiling or other variables useful in preventive conservation.

13.2.3. Laser Doppler Anemometers

Laser Doppler anemometry (Albrecht et al., 2003; Strunck et al., 2004) is based on the well-known principle that a moving source emitting or reflecting a wave generates a frequency shift. A number of different configurations exist but the most used is the differential Doppler, also called *fringe mode*. The air is transparent to laser light but a number of reflecting particles introduced in the airflow may diffuse light, introducing a Doppler shift generated by their movement. By crossing two coherent light beams having plane wavefronts (i.e. two laser beams generated by the same source), an interference fringe is generated in the crossing area. The fringe spacing is proportional to the wavelength of the light λ and inversely proportional to the angle 2θ between the two beams. A particle moving in the intersection of the two beams will scatter light whose intensity will vary according to the intensity pattern of the light as determined by the brightness of the interference fringe. The frequency of the light scattered by the particles transported by the airflow is characterized by a Doppler shift generated by the velocity of the particles. A photomultiplier detects these variations and the frequency of the resulting signal is determined from the Doppler analysis (Durst et al., 1981). The frequency f of the electric signal generated by a particle moving across the fringe volume with a velocity component u normal to the fringes is

$$f = 2\,\frac{u \sin \theta}{\lambda} \tag{13.12}$$

so that for a typical wavelength and $\theta = 30°$, f is of the order of 10^5 Hz (Doebelin, 1990). The method is more sensitive when the fringe pattern is perpendicular to the airflow and hence the fringe can be rotated to obtain the highest sensitivity (and find the flow direction) or to obtain the two components of the velocity vector in the plane parallel to the surface.

This method is very accurate as the interference fringe area can be very small (i.e. with size of the order of a tenth of millimetre) and can be sited very close to the surface, or in contact with the surface, i.e. within the internal boundary layer that develops on the surface. This is the only instrument able to measure air motions very close to a surface and is potentially very useful for studying air–surface interactions and aerodynamic deposition. Another important advantage is that the measure is direct, the flow remaining undisturbed by measurement, without needing the introduction of solid probes or mobile items into the flow. The response is immediate.

The negative aspects are substantially three. The first is that the airflow must be seeded with tracers, i.e. particles that may deposit on the surface, soiling and damaging it. The second is that the introduction of these tracers near the surface perturbs the natural dynamic equilibrium under investigation. Last but not least, these devices are very expensive.

13.2.4. A Simple Analysis of Atmospheric Turbulence

Several approaches exist to study the atmospheric turbulence, and several books have been written on this subject, e.g. Sutton (1960),

Pasquill (1962, 1974), Lumley and Panofsky (1964), Tennekes and Lumley (1973), Csanady (1980), Vinnichenko et al. (1980), Newstadt et van Dop (1984), Landahl and Mollo-Christensen (1986), and Clifford et al. (1993) and many others exist on the statistical analysis of time series. However, it may be useful to report some notes on a statistical method that was originally introduced by Rice (1944, 1945) for the telephone random noise and then adapted by some oceanographers (Cartwright and Longuett-Higgins, 1956; Longuett-Higgins, 1957, 1962; Kinsman, 1965) to the analysis of the sea waves.

The methodology can be applied to the atmospheric turbulence, as the instrumental records of the instantaneous sea level and the wind speed are very similar between them: the average sea level is substituted by the mean wind speed and the fluctuating waves by eddies.

Although the theory is quite complex, the application is very simple and needs only counting the number of times the signal crosses the mean level and the total number of fluctuations. It gives the mean airspeed, average period of eddies, the modes of the gusts and the lulls, the spectral width parameter, and the first three even spectral moments.

The zero-crossing period is defined as the average period $\tau(0)$ for which a sensor placed at the average sea level (which is assumed as zero level) is alternatively submersed by waves and then emerges and is expressed as $\tau(0) = T_j/N(0)$ where T_i is the observing time interval and $N(0)$ represents the number of times that the waves have exceeded the calm sea level.

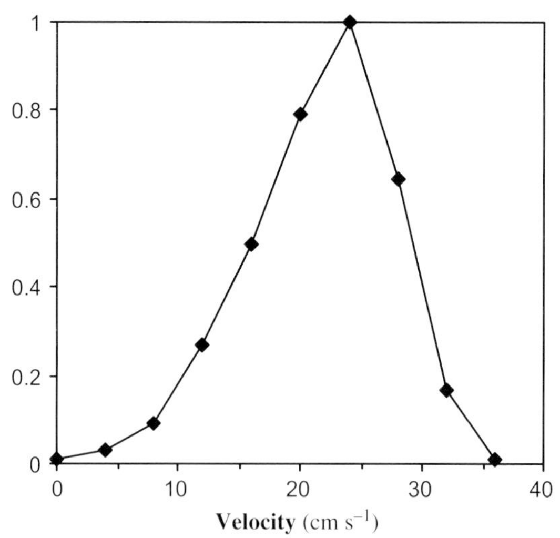

FIGURE 13.3 Normalized crossing frequency of the velocity levels 0, 2, 4, 6,... cm s^{-1} in an internal boundary layer along a wall. The distribution is skew, because the friction due to the wall roughness increases the frequency of the slow motions.

In the same way, the crossing period $\tau(\eta)$ is defined as well as the number of crossings $N(\eta)$ for every arbitrary level η, i.e.:

$$\tau(\eta) = \frac{T_i}{N(\eta)} = \tau(0)\,\frac{N(0)}{N(\eta)} \tag{13.13}$$

and the mean zero-crossing frequency $F(0)$ and the η-level frequency $F(\eta)$ are obviously defined as the inverse of $\tau(0)$ and $\tau(\eta)$, respectively. To apply this statistical method to our case, the following obvious substitutions are required: the eddy turbulence for the waves, the airspeed or wind direction for the sea level, the mode of the instantaneous air velocity or direction for the average sea level.

If m_o is the zero-order moment, i.e. the standard deviation of the wave height, the second-order moment of the spectrum is $m_2 = m_o(2\pi f(0))^2$. The plot of the distribution $N(\eta)$ versus η (Fig. 13.3) is bell shaped with the maximum at $N(0)$ and the standard deviation can be graphically obtained by measuring the half of the segment that intercepts the plot of $N(\eta)$ at the frequency level $\sqrt{e}\,N(0) = 0.607\,N(0)$. Please note that the maximum at $N(0)$, i.e. the mode, is determined as a first approximation, being conditioned by the choice of the crossing levels, the resolution being the step between two levels. The distribution $N(\eta)$ lies between two limit distributions, i.e. the symmetric Gaussian one

$$N(\eta) = N(0)\exp\left(-\frac{\eta^2}{2m_o}\right) \tag{13.14}$$

for a wide-band spectrum, i.e. random components, and the asymmetric Rayleigh distribution

$$N(\eta) = N(0)\,\frac{\eta}{m_o}\exp\left(-\frac{\eta^2}{2m_o}\right) \tag{13.15}$$

for a narrow-band spectrum, i.e. when the spectrum is sharply peaked around a definite frequency. Cartwright and Longuett-Higgins (1956) found a general analytical expression for the distribution of any shape of spectrum, given by the probability distribution $Pr(\zeta)$ of the maxima between the levels η and $\eta + \delta\eta$

$$Pr(\zeta) = \frac{1}{\sqrt{2\pi}}\left[\exp\left(-\frac{\zeta^2}{2\varepsilon^2}\right) + \zeta\sqrt{1-\varepsilon^2}\exp\left(-\frac{\zeta^2}{2}\right)\int_{-\infty}^{\omega}\exp\left(-\frac{x^2}{2}\right)dx\right] \tag{13.16}$$

where $\omega = (\zeta/\varepsilon)\sqrt{1-\varepsilon^2}$. The transition from the two limit distributions (Gaussian and Rayleigh) is determined by the value of the spectral width parameter ε, which ranges from 0 to 1 and is defined by the zeroth-, second- and fourth-order moments,

$$\varepsilon = \frac{m_o m_4 - m_2^2}{m_o m_4} \tag{13.17}$$

The Gaussian distribution is obtained with $\varepsilon = 1$ and the Rayleigh one is obtained with $\varepsilon = 0$. The above authors demonstrated that the value of ε can be simply determined by counting the number of the zero crossings and the number of maxima N_m that have occurred in the same interval, i.e.

$$\varepsilon = \sqrt{1 - \frac{N(0)^2}{N_m^2}} \tag{13.18}$$

and in addition, the mean frequency of the zero crossings $F(0)$ and the mean frequency of the maxima $F(max)$, are linked with the even moments as follows:

$$F(0) = \frac{1}{\sqrt{2\pi}}\sqrt{\frac{m_2}{m_o}}; \quad F(max) = \frac{N_m}{T_i} = \frac{1}{\sqrt{2\pi}}\sqrt{\frac{m_4}{m_2}} \tag{13.19}$$

The graphs of the incremental values $N(\eta_i) - N(\eta_{i-1})$, i.e. the number of times the fluctuations (originally: the waves) have exceeded the level η_{i-1} but not the level η_i, plotted versus η are more or less symmetrical with reference to the origin and are characterized by two peaks, one positive and one negative (Fig. 13.4). The intercept between the plot of observations and the line 0 crossings (which corresponds to the average level that is assumed as "zero" reference level, and that is characterized by the maximum of the crossing frequency) gives a better approximation of the mode. The peaks provide useful information about the distribution of the fluctuations that recur most frequently, i.e. the mode of the gusts and the lulls.

The method simply requires counting the number of times some arbitrarily chosen levels of speed (or some directions) have been crossed by the instantaneous wind speed (or direction) and the number of maxima of this independent variable during the same observing interval. There are two operational procedures: (1) to measure continually or to sample with a high frequency the variable, and the arbitrary crossing levels can be chosen after the measurement, by subdividing into equal intervals the range of variability of the measured variable; (2) to select the crossing levels before the measurement and simply record the number of crossings and the number of maxima. The former method requires storing a huge amount of data and the memory of the recording instrument should have an elevated capacity. The latter requires a special device, or a data acquisition system, programmed to this aim; the advantage is that a much smaller memory capacity is sufficient and the data processing is much simpler.

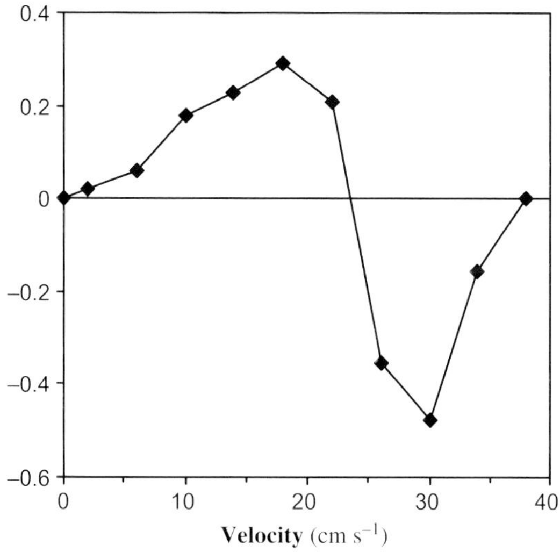

FIGURE 13.4 Plot of the incremental values $N(\eta_i) - N(\eta_{i-1})$, i.e. the number of times the speed has exceeded the velocity level η_{i-1} but not the level η_i, plotted versus the airspeed.

This approach can also be used to measure the time distribution of the wind direction, based on the principle of dividing the compass into a number of equally spaced directions and counting the number of times these directions have been crossed, which avoids the problem of averaging around the discontinuity 0–360°.

References

Anemometry

Albrecht, H.E., Damaschke, N., Borys, M., Tropea, C., 2003. Laser Doppler and Phase Doppler Measurement Techniques. Springer Verlag, Berlin.

Beaubien, D.J., Bisberg, A., 1968. The Sonic Anemometer. Cambridge System, Newton, Mass, 5 pp.

Burt, S., 2012. The Weather Observer's Handbook. Cambridge University Press, Cambridge, 456 pp.

Camp, D.W., Turner, R.E., Glichrist, L.P., 1970. Response tests of cup, vane and propeller wind sensors. Journal of Geophysical Research 75, 5265–5270.

Camuffo, D., 1976. How to obtain mean value and variance of wind direction by using a sine-cosine transducer. Atmospheric Environment 10, 167–168.

Camuffo, D., 1979. Graphic recording and averaging the wind direction. Il Nuovo Cimento 2C, 607–618.

Camuffo, D., Denegri, A., 1976. A method for measurement of mean wind direction with the use of standard potentiometric transducers. Atmospheric Environment 10, 415.

Dantec, 1996. 54N50 Low Velocity Flow Analyzer Mark II. Dantec Electronics, Skovlunde, Denmark, 8 pp.

DISA, 1976. Description of the DISA Constant Temperature Anemometry System. DISA Electronics, Skovlunde, Denmark, 56 pp.

Doebelin, E.O., 1990. Measurement Systems – Application and Design. McGraw Hill, New York, 960 pp.

Durst, F., Melling, A., Whitelaw, J.H., 1981. Principles and Practice of Laser-Doppler Anemometry. Academic Press, London, 437 pp.

Gash, J.H.C., Dolman, A.J., 2003. Sonic anemometer (co)sine response and flux measurement I. The potential for (co)sine error to affect sonic anemometer-based flux measurements. Agricultural and Forest Meteorology 119 (3–4), 195–207.

Horst, T.W., 1973. Corrections for response errors in a three-component propeller anemometer. Journal of Applied Meteorology 12, 676–725.

Hyson, P., 1972. Cup anemometer response to fluctuating wind speeds. Journal of Applied Meteorology 11, 843–848.

Lipták, B.G., 2003. Instrument Engineers' Handbook: Process Measurement and Analysis. In: The Instrumentation System and Analysis (ISA), vol. 1. CRC Press, Boca Raton, Florida.

Lumley, J.L., Panofsky, H.A., 1964. The Structure of Atmospheric Turbulence. Interscience, New York, 239 pp.

Moses, H., 1968. Meteorological instruments for use in atomic energy industry. In: Slade, D.H. (Ed.), Meteorology and Atomic Energy 1968. U.S. Atomic Energy Commission, Div. Tech. Info, pp. 257–300.

Newstadt, F.T.M., Van Dop, N., 1984. Atmospheric Turbulence and Air Pollution Modelling. Reidel, Dordrecht, 358 pp.

Pasquill, F., 1962. Atmospheric Diffusion, first ed. Van Nostrand, London. 297 pp.

Pasquill, F., 1974. Atmospheric Diffusion, second ed. Wiley, New York. 429 pp.

Ramachandran, S., 1970. A theoretical study of cup and vane anemometers. Quarterly Journal of the Royal Meteorological Society 96, 115–123.

Smith, S., 1970. Thrust-anemometer measurements of wind turbulence, Reynold stress, and drag coefficient over the sea. Journal of Geophysical Research 71, 6758–6770.

Strunck, V., Sodomann, T., Muller, H., Dopheide, D., 2004. How to get spatial resolution inside probe volumes of commercial 3D LDA systems. Experiments in Fluids 36, 141–145.

Tennekes, H., Lumley, J.L., 1973. A First Course in Turbulence. MIT Press, Cambridge, Mass, 298 pp.

Tropea, C., Yarin, A.L., Foss, J.F., 2007. Springer Handbook of Experimental Fluid Mechanics, vol. 1. Springer Handbooks, Berlin, 1557 p.

UK Meteorological Office, 1981. Handbook of Meteorological Instruments, vol. 4. Measurement of Surface Wind. Her Majesty's Stationary Office, London, 43 pp.

van der Molen, M.K., Gash, J.H.C., Elbers, J.A., 2004. Sonic anemometer (co)sine response and flux measurement – II. The effect of introducing an angle of attack dependent calibration. Agricultural and Forest Meteorology 122 (1–2), 95–109.

Vinnichenko, N.K., Pinus, N.Z., Shmeter, S.M., Shur, G.N., 1980. Turbulence in the Free Atmosphere. Consultants Bureau, New York, 310 pp.

General theory on turbulence

Cartwright, D.E., Longuett-Higgins, M.S., 1956. The statistical distribution of the maxima of a random function. Philosophical Transactions of the Royal Society, Series A 237, 212–232.

Clifford, N.J., French, J.R., Hardisty, J., 1993. Turbulence. Wiley, Chichester, 360 pp.

Csanady, G.T., 1980. Turbulent Diffusion in the Environment. Reidel, Dordrecht, 248 pp.

Kinsman, B., 1965. Wind Waves. Prentice Hall, Englewood Cliffs, N.J.

Landahl, M.T., Mollo-Christensen, E., 1986. Turbulence and Random Processes in Fluid Mechanics Cambridge. Cambridge, 154 pp.

Longuett-Higgins, M.S., 1957. The statistical analysis of a random, moving surface. Philosophical Transactions of the Royal Society, Series A 249, 321–387.

Longuett-Higgins, M.S., 1962. The distribution of intervals between zeros of a stationary random function. Philosophical Transactions of the Royal Society, Series A 254, 557–599.

Rice, S.O., 1944. Mathematical analysis of random noise. Bell System Technical Journal 23, 282–332.

Rice, S.O., 1945. Mathematical analysis of random noise. Bell System Technical Journal 24, 46–156.

Sutton, O.G., 1960. Atmospheric Turbulence. Methuen, London, 111 pp.

WMO, 2006. Guide to Meteorological Instruments and Methods of Observation. WMO Technical Publication No 8, World Meteorological Organization, Geneva.

CHAPTER

14

Measuring Rainfall and Wind-Borne Droplets

OUTLINE

Precipitation is defined as a *hydrometeor made up of an aggregate of aqueous particles, liquid or solid and crystallized or amorphous, which fall from a cloud or a group of clouds and reach the ground* (WMO, 1966). This definition, which is the most accredited one, includes drizzle, rain, shower, snow, sleet and hail but does not include dew, rime, hoar frost and mist because they are formed by direct condensation or sublimation, are not associated with a cloud and, finally, do not fall. However, such hydrometeors are sometimes included in the list, as e.g. WMO, 1988a, 2004, or are called *occult precipitation*.

It may be useful to remember that precipitation is not homogeneously distributed in time and space but is a variable amount of water or ice that falls as a consequence of the type of cloud, its past and present history (which determines the droplet or ice crystal size), the vertical cloud growth, cloud speed and path. Cumulus clouds develop vertically, with strong convective currents inside, which make faster the growth of droplets or hail. In this case, the precipitation occurs abruptly and violently, forming patches that move following the cloud and wind

Microclimate for Cultural Heritage, Second Edition
http://dx.doi.org/10.1016/B978-0-444-63296-8.00015-9

transport. On the other hand, stratus clouds that layer horizontally are rather uniform and supply droplets more regularly, with homogeneous distribution in time and space. A rain gauge located just below the trajectory of a cumulus cloud, and another collecting rainfall at a certain distance (e.g. 1 km), will measure different precipitation amounts. The same occurs at the passage of fronts. With stratus clouds, the departures are smaller and measurements are representative of a wider scale.

14.1. PRECIPITATION MEASUREMENTS

Although the standard weather records are not directly useful to evaluate the total amount of water that has fallen on monuments, they are in any case representative of the climate in which monuments live and constitute environmental information useful to understand the deterioration mechanisms in their natural context.

In meteorology, precipitation gauges (UK Meteorological Office, 1981; WMO, 1984, 1986, 2008a, 2008b; Michaelides, 2008; Srivastava, 2008; Burt, 2012) are used to monitor the total precipitation amount over a given time or to monitor the intensity, i.e. the instantaneous precipitation rate. Various instruments exist and their response is not always the same. For this reason, from October 2007 to April 2009, the World Meteorological Organization has organized in Vigna di Valle, Italy, an international intercomparison where 30 precipitation gauges and additional sensors of various types and from different manufacturers have been exposed side by side on a homogeneous terrain, i.e. flat green grass area, to detect the specific departure, behaviour and performance of each instrument at various conditions of rain intensity and wind driving force. The dispersion of data was high (i.e. reaching ±60% of the signal) for moderate rain intensity (i.e. below 30 mm h^{-1}); for average rain intensity (i.e. below 90 mm h^{-1}), the relative uncertainty was higher than the 5% recommended by the WMO Guide; and only for extreme intensities (i.e. above 90 mm h^{-1}), was the relative uncertainty of the reference found to be below 5% (Vuerich et al., 2009). This shows how difficult it is to obtain precise precipitation records and the difficulty increases very much in nonhomogeneous terrains and especially in urban environments where buildings and turbulence cause severe perturbations to the wind field and to airborne droplets.

Two main categories of instruments can be found: catching and noncatching instruments (Lanza et al., 2006). Catching instruments collect precipitated water and measure it. Noncatching instruments make an evaluation of the precipitation intensity by calculating it from some indirect, related signals but without collecting water.

14.1.1. Catching Instruments

Catching instruments are composed of a collector in the form of a funnel, with a horizontal circular aperture of given size, and the amount of water that has been collected over a certain time is then measured in one of the following ways.

1. The *tipping bucket* rain recorder is based on a bucket divided into two equal compartments, mounted on a spindle in the middle (Fig. 14.1).

The two compartments are symmetrically disposed so that the bucket is balanced in unstable equilibrium about a horizontal axis. In its normal position, it rests against one of the two stops, which prevent it from tipping over completely. Rain flowing out of the funnel falls into the uppermost compartment and when it has filled it, the centre of gravity is displaced and the bucket overbalances, tips and empties the precipitation. This pivoting movement displaces the second compartment in the uppermost position in which it receives the rainwater until is filled and tips discharging the water in it. Counting with a magnetic reed switch the number of tips, it is possible to know the amount of water that has been collected and discharged.

Of course, this method does not supply a continuous record but only proceeds by counting increments of precipitation with resolution of 0.2 mm. An error is due to the rainfall lost during the time employed in the tipping motion, which may be appreciable during heavy precipitation. The time of the very beginning and ending of drizzle or very light rain cannot be accurately measured. Notwithstanding these limits, this kind of

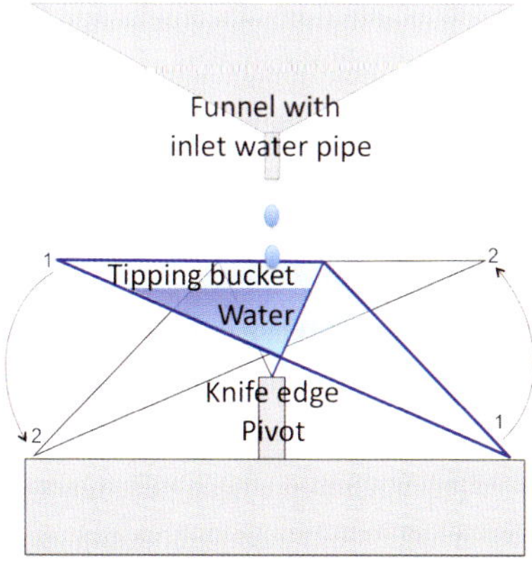

FIGURE 14.1 Mechanism of the tipping bucket rain gauge transducer with the bucket shown in the two rest positions 1 and 2 (blue and grey cross-sections, respectively). The motion is determined by the weight imbalance caused by the collected rainwater.

gauge is the most convenient and popularly used instrument because it is the most suitable for recording rainfall automatically with an electronic datalogger.

2. The *float type*, with automatic siphoning arrangements, records the movement of a light float in a float chamber that receives the rain collected by the funnel. The float is connected with a pen that registers on a drum or a strip chart the level, i.e. the rainfall amount. When the chamber is full of water, a natural or a tilting siphon system makes water to flow out of the float chamber until the level falls to the zero level; the siphoning action ceases, and the cycle starts again. Some rainfall is lost during the siphoning action. This type is used in the British Isles.
3. The *weighing type* operates by recording the total weight of the water collected in a can. The weight of the collecting can is recorded continuously, either by means of a spring mechanism (e.g. the can descends against the compression of a spring) or by the displacement of a weight. This system is accurate and continuous but normally it has no provisions for emptying itself. This is used in cold climates where it is desired to record snow, hail or rainfall and the solid precipitation should be kept frozen to be measured. In mild or warm climates, evaporation losses are reduced by adding oil to form a film 1 mm thick over the free water surface in the container.

In regions where snow or sleet are common, automatic measurements are possible only with gauges provided with a heater to melt ice and measure it as liquid water. If the funnel is not heated, surface condensation such as dew, hoar frost, rime and also fog can provide some water, which is neither rainfall nor strictly precipitation, but the gauge interprets and measures it in the same way. In the case of hail, hailstones bouncing out of the funnel constitute a relevant loss.

A number of other causes of error can be found in measuring precipitation. In northern and humid, mild climates evaporation may contribute some −1% of the total but it is much more important in warm or arid regions. Minor errors are due to droplet adhesion, i.e. −0.5%; gauge colour, −0.5%; funnel inclination, −0.5%; and large droplets splashing, +1%. However, the main departures are due to wind and exposure and this error cannot be exactly evaluated. The World Meteorological Organization suggests that the amount of precipitation collected by a rain gauge may be 3–30% less than the actual precipitation reaching the ground (WMO, 1984, 2008) and the UK Meteorological Office suggests a much wider range, i.e. 5 to 80% less (UK Meteorological Office, 1981). Large errors especially occur with strong winds. In such conditions, airborne droplets tend to be transported parallel to the horizontal plane of the funnel mouth and the turbulence generated near the funnel may disperse the droplets loosing them and decreasing the collection efficiency. This effect becomes especially important with snowflakes that are more affected by wind transport, wind gusts and local turbulence.

In order to minimize the disturbance on the wind field, the site should be possibly chosen on open flat terrain, which should be actually representative of the area under consideration. To this aim, the gauge should be free of overhanging obstructions. If the area has vegetation, this should be uniformly distributed and the funnel should be at the average height of the vegetation. Some appropriate fence structures could be installed to homogenize the area characteristics. Dynamic wind shields are suggested around the gauge orifice to shelter it from high wind speed and turbulence. These shields are built with metal strips forming a truncated cone with the vertex pointing to the base of the gauge but leaving some free space between strip and strip. The gauge should be at a sufficient distance from obstacles to avoid local eddies. In order to avoid interferences with the wind field, the minimum distance allowable for obstructions should be twice their height. Sometimes a greater distance is suggested, e.g. the WMO report No 266 (1984) suggests four times their height.

14.1.2. Noncatching Instruments

This family of instruments is based on indirect evaluation of precipitated water and the main types are the disdrometer and the radar (Lanza et al., 2006). These instruments are not yet frequently used for conservation purposes, although they have an interesting potential.

1. *Impact disdrometers*: a membrane of plastic or metal is used as the measurement surface to sense the impact of single precipitation particles. The impacts are transformed into an electrical signal either by direct evaluation or after an analysis of the frequency spectrum.
2. *Optical disdrometers*: one or two thin laser beams are used to detect particles crossing this optical system, which is able to evaluate the falling velocity and the size of drops.
3. *Radar*: small X-band radars are used to determine the spectrum of the signal backscattered by falling drops. The intensity of the backscattered signal is related to the number of falling drops and/or their water content. The spectrum is related to falling drop velocity.

Observations of precipitation with radar or satellite remote sensing are used in meteorology or for general weather purposes but are of less interest for monument preservation, except to forecast and alert for large and important storm systems. In such cases, the attention is mainly focused on the civil protection and less commonly on monument safety.

14.2. PRECIPITATION ON MONUMENTS

Standard weather measurements are taken on unobstructed open country and the precipitation is not the same as in towns. Towns are generally warmer and the heat island will generate convective motions that favour the formation of clouds. The different roughness and the building turbulence will exchange momentum and generate mixing with the effect of slowing down the wind field, making colder the lower layers and transporting heat aloft. Traffic, industry and building heating are sources of pollutants that act as condensation nuclei. All these factors contribute to increase the frequency of precipitation, particularly with showers and thunderstorms especially over large towns. It has been demonstrated that the precipitation over towns is greater than over the surrounding country and that the average frequency of precipitation in working days is higher than during Saturday and Sunday (Landsberg, 1981).

Weather measurements poorly represent the rainfall that fall on monuments located in the centre of a town where the situation is complicated by building shielding, wind channelling along the streets (i.e. Venturi effect), recirculation in the perpendicular to the streets, or other aerodynamic disturbances. In addition, metal domes or even wetted buildings may generate strong anomalies in the electric field in the presence of charged clouds. As the water molecule is a dipole, droplets may be charged by induction and attracted. In addition, all the droplets and aerosols formed by raindrop splashing are charged and their behaviour depends on the wind drag, gravity force and electric field.

Weather measurements are representative of the undisturbed rain that precipitates over an open countryside. On monuments, the rainfall intensity is relevant because intense rainfall is more effective in washing the surface and in causing erosion. This situation totally changes in windward and leeward surfaces. The airflow and the electric field may reduce the precipitation in one place and increase it in another.

Weather observations measure the water fallen into a horizontal circular aperture, whereas monuments mainly have a vertical extent and the number of raindrops intercepted by the monument varies with the wind speed, direction and local aerodynamic disturbance. For this reason, *vertical collectors* are sometimes used in close proximity of a monument surface. A vertical collector is a vertical surface on which airborne droplets may impact, with some borders to retain lateral splashes. The collected water runs to the collector bottom where it flows through a pipe into one of the above catching instruments where it is measured. Vertical collectors catch the rainfall blown by wind or local air currents having direction perpendicular to the monument surface. The collected amount is null in the absence of wind; it increases with the wind intensity, but when the intensity becomes too strong, the wind passing on the two sides of the obstacle (i.e. the monument with the collecting surface) drains away the raindrops too. Lateral dragging increases with increasing wind intensity and decreasing droplet size. This is what in reality happens on the monument surface where the instrument is located and it should be kept in mind that the impacting water changes with height from the ground and from point to point. Disdrometers may be used instead but they are not as popular as vertical collectors. No normative or standards exists for these methods.

Another possibility is to collect in some way the water that has impacted on the monument and flows down along the monument surface (*run-off water*). This methodology works except when the monument is made of a porous material that absorbs water. Often the run-off is not only composed of mere rainwater but also is '*washout water*', i.e. the solution that includes some dissolved stone, crusts and deposited pollutants removed by the running water, with eroded material. Hence the run-off is sometimes collected in order to perform chemical analyses on leached ions and dissolved stone. On historic buildings, run-off has been measured in several occasions and in a number of ways, either by collecting directly running water into a funnel placed under an edge or by building simple devices to this aim. For instance, Leysen et al. (1989) placed longitudinally against the wall of the Mechelen Cathedral a plastic cylinder with a longitudinal slit 3 cm wide, with the lower edge firmly fixed against the wall. The running water was collected into the cylinder and flowed through a hole in the bottom and a tube, arriving into a plastic bottle to be taken away and analysed in the laboratory. The same authors also used gutters pressed against the wall to collect the water running off the cathedral walls. In this way, it was possible to arrive at a rough estimate of the annual material loss and surface recession rate.

A third method, based on standard instruments, is to collect precipitation on the horizontal plane with a recording rain gauge and associate the measurement with the wind speed and direction. As the trajectory of a falling raindrop is determined by the combined action of the falling velocity and wind drag, the impact on a vertical surface can be calculated once the drop diameter and the wind speed are known. The falling velocity of raindrops has been observed in a number of experiments (Houghton, 1985) and the results are reported in Fig. 14.2, i.e. falling velocity versus drop diameter D. From the graph, it appears that around $D = 4$ mm drops reach the critical speed of 9 m s^{-1}, which

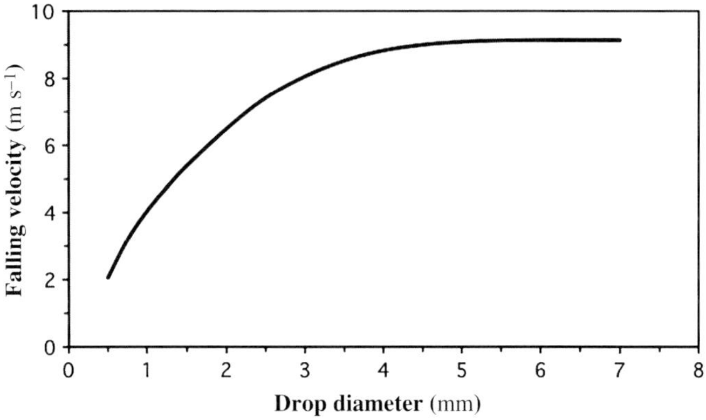

FIGURE 14.2 Terminal velocity of free falling drops.

increases turbulence and air drag, so that the terminal velocity remains unchanged at this top speed. When drops reach $D = 7$ mm, the exceedingly high forces generated by friction and the instability for turbulent wake generated behind the falling drops exceed the resistance of the surface tension and the drops will explode, i.e. self-splashing into a number of smaller droplets. Unfortunately, there are not easy methods to measure the drop diameter and only approximate evaluations can be made with a crude estimate of the drop diameter and other variables.

However, even in the case of a successful tentative of knowing the rainfall that crosses a vertical plane this result may hardly be applied to the actual case of a true surface, although with a very simple geometry, e.g. the facade of a building. A sophisticated simulation is needed. In the actual case of a building, the structural elements perturb the wind field and, consequently, the horizontal drag changes as well as the resulting path of raindrops. For instance, the upper part of buildings is hit by raindrops that have been dragged till the last few seconds by a wind stream that is passed nearly undisturbed above the obstacle; the lower part is hit by raindrops that in the last part of their path have been dragged away by a wind stream turning around the building and becoming parallel to its surface. Under such circumstances, droplets are deviated laterally and dispersed and those that reach the surface for their inertia are only a small fraction of the incoming rain. This is why the upper part of buildings is generally washed out and cleaned better than the lower part (Fig. 14.3). In the case of a monument also the situation is worse, because the complex three-dimensional shape of the monument and the local eddy perturbations make extremely difficult to evaluate impacting rain and run-off.

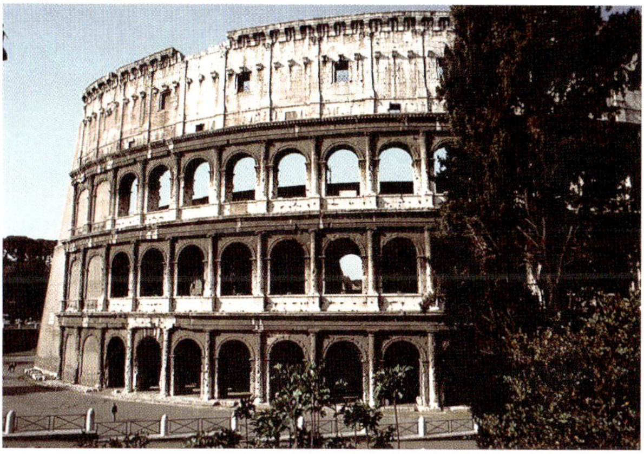

FIGURE 14.3 The efficiency of wind-borne droplets in washing the upper part of monuments is evident at the Coliseum, Rome. The upper part is washed out by rainfall, whereas in the intermediate and lower parts, dust and soot accumulate.

14.3. WET AND DRY DEPOSITION SAMPLERS

It is evident that monuments react with the chemical substances that deposit on them and not with the gases and particles that are simply suspended in the atmosphere. Most gases and aerosols may remain in the atmosphere because they have a very low deposition rate, so that a measurement in the air is useful for studies on health (in fact lungs pump and filter air with suspended pollution) but not for monument decay. For this reason, the so-called *wet and dry deposition sampler* has been invented to measure, separately, the acid rain and the airborne particles that deposit on a horizontal surface during drytime. Some slightly different devices have been invented (e.g. Georgii and Pankrath, 1982; Munn and Rodhe, 1985) but the most widespread types consist of two equal buckets, made of glass, plastic (e.g. polyvinyl chloride or polyethylene), stainless steel or other nonreactive materials, one to collect dry deposition and one for the wet one, with only one lid, which is alternatively displaced over the dry or the wet bucket, driven by a rainfall sensor. The rainfall sensor is composed of a resistance sensor with several parallel unshielded metal resistors that are transformed in a short circuit by rainwater; another common sensor is a capacitive one, where absorbed water changes the dielectric capacity in a capacitor. Another system is to interrupt an infrared beam emitted by a diode, but several other possibilities exist. After a week, or another chosen time interval, it is possible to pick up the two buckets and analyse separately the content of the dry bucket, i.e. dust and particles, and the acid rain collected in the wet one. This device is obviously located in an open area, far from obstacles or pollution sources.

The collection vessels are equipped with an external metal ring with needles, or crossed wires, in order to avoid landing and staying of birds on the funnel edge thus contaminating the sample with their droppings.

This crude device has been largely used in the past two decades but presents the following problems. Although the lid is over the wet bucket during sunshine, the bucket becomes hot and some water evaporates, changing the pH of the wet sample. In addition, condensation nuclei, dust (especially Saharan dust in Europe) or other solid particles that fall with the rainfall may eventually dissolve and buffer the solution changing the pH and the ionic composition. Real-time measurements automatically carried out during the precipitation event are preferable (Camuffo et al., 1984, 1988; Camuffo, 1990). Also biological life, especially algae, may develop in collected rainwater perturbing the chemical equilibrium.

The wet/dry sensor that controls the lid movement is not very fast and is provided with a selected lag to confirm the signal and avoid false alarms. As a consequence, the dry bucket receives the earliest raindrops that often are the most polluted ones. A problem is that the dry/wet sensor may also respond to dew, frost and fog, inappropriately opening the cover of the wet bucket and collecting there, for several hours especially during the night-time, the dry deposition.

It is not clear what the dry bucket collects. Substantially, it gathers at the bottom all the coarse particles that deposit via gravitational settling and in the outer and inner surface, it attracts the charged particles that are driven by the electric field generated by the bucket when it is hit by solar radiation or is rubbed by the wind friction. In sunny days, the bucket is always heated by solar radiation; it becomes hot and thermophoresis tends to weakly counteract the other deposition mechanisms. For this reason, the bucket interacts with the suspended particles with a variable selective action on the deposition mechanisms.

In any case, the dry deposit that is collected with this method is in some way representative of the coarse and giant particles, i.e. the so-called *dustfall*, which accumulates on the upward horizontal surface of monument and is mainly composed of soot, dust, pollens and fibres. This is very different from the measurement of the suspended particulate matter performed with high-volume samplers that are composed of filter, motor blower and flow meter. Such measurement practically consist in sampling all the suspended particles that are larger than the average aerodynamic size of the filter pores, independent of whether these particles will later deposit or not.

A simpler, earlier version of the wet and dry deposition sampler is the so-called *bulk precipitation sampler* composed of only one funnel and precipitation collector that remains uncovered for the whole sampling period, collecting whatever is depositing, either in the dry or the wet phase. This instrument is intended to monitor the deposit that will form on a monument, or a plant, irrespective of the distinction between the dry or the wet phase. However, the approximation is crude and the representativeness uncertain. In addition, being without an automatic cover, the evaporation loss may be very important in hot, windy climates or dry periods. Like the dry collector of the wet and dry sampler, the bulk sampler may be contaminated by local dust, which may significantly alter the pH. Although this method had some success in the past, the data are uncertain and for this reason difficult to interpret. Sound information on this subject is difficult to obtain and further research is needed to devise reliable instruments for automatic monitoring of pollutants that will deposit onto selected surfaces.

References

Weather Precipitation measurements

Burt, S., 2012. The Weather Observer's Handbook. Cambridge University Press, Cambridge, 456 pp.

Houghton, D.D., 1985. Handbook of Applied Meteorology. Wiley, New York, 1461 pp.

Landsberg, H.E., 1981. The Urban Climate. Academic Press, New York, 275 pp.

Lanza, L., Leroy, M., Alexandropulos, C., Stagi, L., Waubon, W., 2006. WMO Laboratory Intercomparison of Rainfall Intensity Gauges. Instruments and Observing Methods. Report No. 84. WMO/TD No 1304, Geneva.

Michaelides, S. (Ed.), 2008. Precipitation: Advances in Measurement, Estimation, and Prediction. Springer Verlag, Berlin, 540 pp.

Srivastava, G.P., 2008. Surface Meteorological Instruments and Measurement Practices. Atlantic, Nice Printing Press, New Delhi, 435 pp.

UK Meteorological Office, 1981. Handbook of Meteorological Instruments. In: Measurement of Precipitation and Evaporation, vol. 5. Her Majesty's Stationary Office, London.

Vuerich, E., Monesi, C., Lanza, L., Stagi, L., Lanzinger, E., 2009. WMO Field Intercomparison of Rainfall Intensity Gauges, Vigna di Valle, Italy, October 2007–April 2009. WMO-IOM Report 99, Geneva.

WMO, 1966. International Meteorological Vocabulary. WMO Technical Publication No 182 TP.91, World Meteorological Organization, Geneva.

WMO, 1984. Compendium of Lecture Notes for Training Class IV Meteorological Personnel. In: Meteorology, vol. 2. WMO Technical Publication No 266, World Meteorological Organization, Geneva, 455 pp.

WMO, 1986. Compendium of Lecture Notes on Meteorological Instruments for Training Class III and Class IV Meteorological Personnel, vol. 1. WMO Technical Publication No 622, Geneva.

WMO, 2008a. Guide to Meteorological Instruments and Methods of Observation. WMO Technical Publication No 8, World Meteorological Organisation, Geneva.

WMO, 2008b. Guide to Hydrological Practices. In: Hydrology – from Measurement to Hydrological Information, sixth ed., vol. 1. WMO Technical Publication No 168, World Meteorological Organization, Geneva.

Precipitation on monuments, wet and dry deposition samplers

Georgii, H.W., Pankrath, J., 1982. Deposition of Atmospheric Pollutants. Reidel, Dordrecht, 217 pp.

Camuffo, D., 1990. Acidic precipitation research in Italy. In: Bresser, A.H.M., Salomons, W. (Eds.), Acidic Precipitation, Advances in Environmental Science, vol. 5. Springer Verlag, New York, pp. 229–265.

Camuffo, D., Del Monte, M., Ongaro, A., 1984. The pH of atmospheric precipitation at Venice, related to both the dynamics of precipitation events and weathering of monuments. Science Total Environment 40, 125–140.

Camuffo, D., Bernardi, A., Zannetti, M., 1988. Analysis of the real-time measurement of the pH of rainfall at Padova, Italy: seasonal variation and meteorological aspects. Science Total Environment 67, 187–200.

Leysen, L., Roekens, E., Van Grieken, R., 1989. Air-pollution-induced chemical decay of a sandy limestone Cathedral in Belgium. Science Total Environment 74, 263–287.

Munn, R.E., Rodhe, H., 1985. Compendium of meteorology, vol. II, part 6. In: Air Chemistry and Air Pollution Meteorology. World Meteorological Organisation, WMO No.364, Geneva, p. 209.

Appendix 1: List of Fundamental Constants Met in This Book

Latin Letters

$a = 7.5$, $b = 237.3\,°C$: Magnus–Tetens coefficients for water vapour in equilibrium with the liquid phase

$a = 9.5$, $b = 265.5\,°C$: Magnus–Tetens coefficients for water vapour in equilibrium with the solid phase

$a = 7.65$, $b = 243.12\,°C$: Magnus–Tetens coefficients revised by Sonntag for water vapour in equilibrium with the liquid phase

$A = 6.2 \times 10^{-4}\,K^{-1}$ and $A = 6.667 \times 10^{-4}\,K^{-1}$: psychrometer coefficients

$c_v = 0.331$ cal $g^{-1}\,K^{-1}$: specific heat of water vapour at constant volume

$c_p = 0.441$ cal $g^{-1}\,K^{-1}$: specific heat of water vapour at constant pressure

$c_{pd} = 0.240$ cal $g^{-1}\,K^{-1} = 1.003$ J $g^{-1}\,K^{-1}$: isobaric specific heat of dry air

$\mathscr{D}_{aw} = 0.239\,cm^2\,s^{-1}$ at $T = 8\,°C$: coefficient of mutual diffusion between dry air and water vapour

$e = 2.71828182845904\ldots$: *Neper* number; $1/e = 0.36787945\ldots$; $1-(1/e) = 0.6321206\ldots$

$e_{sat}(0) = 6.11$ hPa: saturation pressure of water vapour at 0 °C

$g = 980.665\,cm\,s^{-2}$: Earth's acceleration of gravity (standard gravity)

$h = 6.62606957 \times 10^{-27}$ erg s: Planck constant

$k = 1.3806488 \times 10^{-16}$ erg K^{-1}: Boltzmann constant

$k_T = 6.26 \times 10^{-5}$ cal $cm^{-1}\,s^{-1}\,K^{-1}$ (or 26.2 mW $m^{-1}\,K^{-1}$): thermal conductivity of air at 300 K

$K_f = 1.86$ K mol^{-1}: freezing point depression constant for water

$L_v = 538.7$ cal g^{-1}: latent heat of vaporization of water vapour

$M_a = 28.966$ g mol^{-1}: mass of dry air

$M_{CO_2} = 44.01$ g mol^{-1}: mass of CO_2

$M_H = 1.00794$ g mol^{-1} mass of H

$M_{N_2} = 28.02$ g mol^{-1}: mass of N_2

$M_O = 15.9994$ g mol^{-1}: mass of O

$M_{O_2} = 31.9988$ g mol^{-1}: mass of O_2

$M_v = 18.016$ g mol^{-1}: mass of water vapour

$N = 6.023 \times 10^{23}$: Avogadro number

$p = 1013$ hPa: average atmospheric pressure at 0 °C and sea level

$Ra \sim 10^9$: critical Rayleigh number

$Rf \approx 0.2$: critical Richardson number

$\Re = 8.3169 \times 10^7$ erg mol^{-1} K^{-1} = 1.986 cal mol^{-1} K^{-1}: universal gas constant

$\Re_a = 0.2870 \times 10^7$ erg g^{-1} K^{-1} = 0.06857 cal g^{-1} K^{-1}: constant of dry air

$\Re_v = 0.4615 \times 10^7$ erg g^{-1} K^{-1} = 0.1102 cal g^{-1} K^{-1}: constant of water vapour

$v_e = 1.12 \times 10^4$ m s^{-1}: escape velocity from the Earth

$v_S = 344$ m s^{-1}: velocity of sound

Greek Letters

$\alpha = 1/273.15 = 0.003661$

$\beta_1 = 1.81 \times 10^{-4}$ °C^{-1}: cubic coefficient of thermal expansion of mercury

$\beta_2 = 2.53 \times 10^{-5}$ °C^{-1}: cubic coefficient of thermal expansion of thermometric glass

$\Gamma = dT/dz = 1$°C/100 m: dry adiabatic gradient

$\delta_\odot(0) = -23° \, 27'$: solar declination at winter solstice

$\eta = 1307$ μPa s at 10 °C and 1002 μPa s at 20 °C: dynamic viscosity of water

$\lambda = 0.065$ μm: mean free path of air molecules at standard temperature and pressure

$\mu \approx 1.7 \times 10^{-4}$ P (i.e. g cm^{-1} s^{-1}) at $T = 0$ °C: coefficient of molecular viscosity of air, also called dynamic viscosity

$\nu = 0.15$ cm^2 s^{-1}: kinematic viscosity of air

$\rho_a = 1.205 \times 10^{-3}$ g cm^{-3}: density of air at standard temperature and pressure

$\rho_p = 1$ g cm^{-3}: density of water droplets

$\rho_q = 2.65$ g cm^{-3}: density of quartz granules

$\sigma = 0.286 \times 10^{-10}$ cal cm^{-2} min^{-1} K^{-4} = 5.67 erg cm^{-2} s^{-1} K^{-4}: Stefan constant

$\sigma = 75.6$ erg cm^{-2} at $T = 273$ K and $\sigma = 72.2$ erg cm^{-2} at $T = 293$ K: surface tension of water

$\sigma_{sl} = 17.2$ erg cm^{-2}: surface tension of water in equilibrium between the solid and liquid phases.

τ: typically of the order 1–10 dyne cm^{-2} surface stress of the PBL

Appendix 2: Summary of Key Equations to Calculate Humidity Variables

It may be useful to find summarized the key equations that enable the calculation of relative humidity and other related variables from three basic parameters, including atmospheric pressure p. In the case of approximate calculations and in the absence of barometric readings, p may be substituted with the standard sea-level average $p = 1013$ hPa or with the approximation $p = 1000$ hPa to make easier calculations. The following formulae, already reported in another publication (Camuffo, 2004), are recommended by the EU standard EN16242: 2012 '*Conservation of cultural property – Procedures and instruments for measuring humidity in the air and moisture exchanges between air and cultural property.*'

A1. Available instruments: psychrometer, barometer

Input parameters: air temperature t (°C), wet bulb air temperature t_w (°C), barometric pressure p (hPa)

Water vapour pressure e (A.1)

$$e = 6.112 \times \left(10^{\frac{7.65 t_w}{243.12 + t_w}} - 1.068 \times 10^{-4} p(t - t_w)\right) (\mathrm{hPa})$$

Humidity mixing ratio MR (A.2)

$$MR = 3801.5 \times \frac{10^{\frac{7.65 t_w}{243.12 + t_w}} - 1.068 \times 10^{-4} p(t - t_w)}{p - 6.112 \times \left(10^{\frac{7.65 t_w}{243.12 + t_w}} - 1.068 \times 10^{-4} p(t - t_w)\right)} \left(\mathrm{g\,kg^{-1}}\right)$$

Absolute humidity AH (A.3)

$$AH = 1344.6 \times \frac{10^{\frac{7.65 t_w}{243.12 + t_w}} - 1.068 \times 10^{-4} p(t - t_w)}{273.15 + t} \left(\mathrm{g\,m^{-3}}\right)$$

Relative humidity RH (A.4)

$$RH = 100 \times \frac{10^{\frac{7.65 t_w}{243.12 + t_w}} - 1.068 \times 10^{-4} p(t - t_w)}{10^{\frac{7.65 t}{243.12 + t}}} (\%)$$

Dew point temperature DP (A.5)

$$DP = \frac{243.12 \times \ln\left(10^{\frac{7.65 t_w}{243.12 + t_w}} - 1.068 \times 10^{-4} p(t - t_w)\right)}{17.62 - \ln\left(10^{\frac{7.65 t_w}{243.12 + t_w}} - 1.068 \times 10^{-4} p(t - t_w)\right)} (^\circ\mathrm{C})$$

A2. Available instruments: thermometer, *RH* hygrometer, barometer

Input parameters: air temperature t (°C), relative humidity RH (%), barometric pressure p (hPa)

Water vapour pressure e (A.6)

$$e = 0.06112 \times 10^{\frac{7.65 t}{243.12 + t}} \times RH \ (\mathrm{hPa})$$

Humidity mixing ratio MR (A.7)

$$MR = 38.015 \times \frac{10^{\frac{7.65 t}{243.12 + t}} \times RH}{p - \left(0.06112 \times 10^{\frac{7.65 t}{243.12 + t}} \times RH\right)} \left(\mathrm{g\,kg^{-1}}\right)$$

Absolute humidity AH (A.8)

$$AH = 13.44 \times \frac{10^{\frac{7.65 t}{243.12 + t}}}{273.15 + t} \times RH \left(\mathrm{g\,m^{-3}}\right)$$

Dew point temperature DP (A.9)

$$DP = \frac{243.12 \times \ln\left(10^{\frac{7.65t}{243.12+t}} \times \frac{RH}{100}\right)}{17.62 - \ln\left(10^{\frac{7.65t}{243.12+t}} \times \frac{RH}{100}\right)} \ (^{\circ}\mathrm{C})$$

A3. Available instruments: dew point hygrometer, thermometer, barometer

Input parameters: air temperature t (°C), dew point temperature t_d (°C), barometric pressure p (hPa)

Water vapour pressure e (A.10)

$$e = 6.112 \times 10^{\frac{7.65t_d}{243.12+t_d}} \ (\mathrm{hPa})$$

Humidity mixing ratio MR (A.11)

$$MR = 3801.5 \times \frac{10^{\frac{7.65t_d}{243.12+t_d}}}{p - 6.112 \times 10^{\frac{7.65t_d}{243.12+t_d}}} \ \left(\mathrm{g\,kg^{-1}}\right)$$

Absolute humidity AH (A.12)

$$AH = 1344.6 \times \frac{10^{\frac{7.65t_d}{243.12+t_d}}}{273.15 + t} \ \left(\mathrm{g\,m^{-3}}\right)$$

Relative humidity RH (A.13)

$$RH = 100 \times 10^{\frac{7.65t_d}{243.12+t_d} - \frac{7.65t}{243.12+t}} \ (\%)$$

Reference

Camuffo, D., 2004. Thermodynamics for cultural heritage. In: Martini, M., Milazzo, M., Piacentini, M. (Eds.), Physics Methods in Archaeometry, International School of Physics 'Enrico Fermi', Varenna. IOS Press, Amsterdam, pp. 37–98.

Appendix 3: Essential Glossary

Accuracy: the extent to which the measurement approaches the true measurand value, often expressed as the ratio of the error to the full-scale output, indicated as $\pm$percent of the range.

Ambient conditions: the microclimate and other environmental conditions (e.g. rainfall, dew, freezing/thawing cycles, deliquescence crystallization cycles, ventilation and turbulence, UV, visible and IR radiation) of the medium surrounding the sensor.

Bimodal distribution: a distribution with two distinct modes (see *mode*).

Damping: when an instrument is subjected to a forcing step function, the output may reach asymptotically the final value (first-order measuring instruments) or oscillate around it with overshoots and undershoot (second-order measuring instruments). Damping is the attenuation of the signal that smoothes the response of second-order instruments. Selected damping attenuation to a signal may be obtained with an electronic filter or mechanically with a friction. Mechanical damping is especially important for wind vanes.

Distance constant: the length of the path travelled by the wind in the time interval required for an output of a wind sensor to reach $1 - e^{-1} = 63.2\%$ of a step change of wind speed.

Drift: gradual deviation from the original characteristics. The deviation increases with time and severity of ambient conditions (e.g. temperature, pressure and humidity).

Hysteresis: the maximum difference in output, when the same measurand value of the parameter is reached first with increasing values and then with decreasing ones. It is generally expressed as percent of the full-scale output.

Linearity: the closeness of a calibration curve or an instrument response to a straight line.

Mean: the arithmetic average $<x>$ of a set of N readings of the variable x:

$$<x> = \frac{1}{N}\sum x_i$$

Median: the middle value in a set of values.

Measurand: the item that should be measured.

Mode: the value (or the reading) that occurs most frequently.

Percentile: a value in the range of set of data that separates the range into two groups, so that a given percentage of data lies below this value.

Precision: the closeness of the output to the true measurand value.

Range: the interval of values that are intended to be measured, that are potentially measurable, that have been measured, specified by their upper and lower limits.

Reliability: the probability that a transducer will satisfactorily perform its function. It can be measured by the variance of repeated measurements of the same value of the measurand parameter, taken under the same conditions.

Repeatability: the ability to reproduce the same output when measuring the same value of the measurand variable, taken under the same conditions. It can be expressed as the maximum difference of the output readings or in $\pm$ percent of the range.

Representativeness: also when the measure is not affected by instrumental errors, the output does not always characterize the actual value of the measurand parameter in the site and the time instant under investigation. This happens typically when the value of the parameter is not homogeneously distributed or the measurement is taken at a point where the measurand has a singular value. Measurements are done to define the parametric values of a surface or a site and observations are limited to one or few points that should characterize a wider space. A measurement that characterizes the value of the variable in its context, as expected, is called representative.

Resolution: the ratio of the output step change to the full-scale output, generally expressed in percent. It represents the smallest detectable change of the measurand parameter. If the output varies with continuity, the resolution is infinite.

Response time: the length of time required for the output to rise to a percentage of its final value, when the measurand parameter has undergone a step change. When the percentage is 63.2%, it is called *time constant*. About 95% of the step change is reached after three time constants. Of course, the response time (and the time constant) is independent of the span of the output change.

Sensor: the functional part of the instrument, which couples it to the measurand variable (i.e. the input) generating a mechanical or electrical signal that is the output of the sensor.

Sensitivity: the ratio of the change in the transducer output to a change in the value of the measurand parameter.

Specificity to the response: the capability of an instrument to respond only to the changes of the variable that it is designed to measure and be insensitive to all the other ones.

Threshold: the minimum value of the measurand variable that can be detected by the transducer. If the variable does not start from zero, the *threshold* is the smallest change of the variable that results in a measurable change in the output. However, in this case it is better to speak in terms of *resolution*.

Time constant: the time required to detect and indicate $1 - e^{-1} = 63.2\%$ of a step change. See also *response time*.

Relevant Objects, Museums, Monuments etc Exemplified in Figures

D

F

G

H

K

I

L

T

U

V

W

Index

Note: Page numbers with "*f*" denote figures; "*t*" tables.

R

S

X

Y

Z